Current Research on Hyperspectral and Multispectral Imaging and Their Applications in Precision Agriculture II

Current Research on Hyperspectral and Multispectral Imaging and Their Applications in Precision Agriculture II

Guest Editors

Jinling Zhao
Chuanjian Wang

Basel • Beijing • Wuhan • Barcelona • Belgrade • Novi Sad • Cluj • Manchester

Guest Editors

Jinling Zhao
National Engineering
Research Center for
Agro-Ecological Big Data
Analysis & Application
Anhui University
Hefei
China

Chuanjian Wang
School of Internet
Anhui University
Hefei
China

Editorial Office
MDPI AG
Grosspeteranlage 5
4052 Basel, Switzerland

This is a reprint of the Special Issue, published open access by the journal *Agronomy* (ISSN 2073-4395), freely accessible at: https://www.mdpi.com/journal/agronomy/special_issues/9F30XL4Y6P.

For citation purposes, cite each article independently as indicated on the article page online and as indicated below:

Lastname, A.A.; Lastname, B.B. Article Title. *Journal Name* **Year**, *Volume Number*, Page Range.

ISBN 978-3-7258-3017-6 (Hbk)
ISBN 978-3-7258-3018-3 (PDF)
https://doi.org/10.3390/books978-3-7258-3018-3

Contents

Article

Estimation of Relative Chlorophyll Content in Spring Wheat Based on Multi-Temporal UAV Remote Sensing

Qiang Wu [1], Yongping Zhang [1,*], Zhiwei Zhao [1,*], Min Xie [1] and Dingyi Hou [2]

1 College of Agronomy, Inner Mongolia Agricultural University, Huhhot 010019, China
2 Chifeng Forest and Grassland Protection and Development Center, Chifeng 024005, China
* Correspondence: imauzyp@163.com (Y.Z.); imauzzw@163.com (Z.Z.)

Abstract: Relative chlorophyll content (SPAD) is an important index for characterizing the nitrogen nutrient status of plants. Continuous, rapid, nondestructive, and accurate estimation of SPAD values in wheat after heading stage can positively impact subsequent nitrogen fertilization management strategies, which regulate grain filling and yield quality formation. In this study, the estimation of SPAD of leaf relative chlorophyll content in spring wheat was conducted at the experimental base in Wuyuan County, Inner Mongolia in 2021. Multispectral images of different nitrogen application levels at 7, 14, 21, and 28 days after the wheat heading stage were acquired by DJI P4M UAV. A total of 26 multispectral vegetation indices were constructed, and the measured SPAD values of wheat on the ground were obtained simultaneously using a handheld chlorophyll meter. Four machine learning algorithms, including deep neural networks (DNN), partial least squares (PLS), random forest (RF), and Adaptive Boosting (Ada) were used to construct SPAD value estimation models at different time from heading growth stages. The model's progress was evaluated by the coefficient of determination (R^2), root mean square error (RMSE), and mean absolute error (MAPE). The results showed that the optimal SPAD value estimation models for different periods of independent reproductive growth stages of wheat were different, with PLS as the optimal estimation model at 7 and 14 days after heading, RF as the optimal estimation model at 21 days after heading, and Ada as the optimal estimation model at 28 d after heading. The highest accuracy was achieved using the PLS model for estimating SPAD values at 14 d after heading (training set $R^2 = 0.767$, RMSE = 3.205, MAPE = 0.060, and $R^2 = 0.878$, RMSE = 2.405, MAPE = 0.045 for the test set). The combined analysis concluded that selecting multiple vegetation indices as input variables of the model at 14 d after heading stage and using the PLS model can significantly improve the accuracy of SPAD value estimation, provides a new technical support for rapid and accurate monitoring of SPAD values in spring wheat.

Keywords: wheat; machine learning; SPAD; vegetation indices

Citation: Wu, Q.; Zhang, Y.; Zhao, Z.; Xie, M.; Hou, D. Estimation of Relative Chlorophyll Content in Spring Wheat Based on Multi-Temporal UAV Remote Sensing. *Agronomy* **2023**, *13*, 211. https://doi.org/10.3390/agronomy13010211

Academic Editors: Jinling Zhao and Chuanjian Wang

Received: 14 December 2022
Revised: 6 January 2023
Accepted: 9 January 2023
Published: 10 January 2023

1. Introduction

Spring wheat is a major crop in northern China, and its growth and yield are critical for ensuring food security in the region. Chlorophyll, a pigment essential for photosynthesis in plants, has a strong influence on the nitrogen nutrition status, photosynthetic capacity, and yield of crops, and is a key parameter reflecting crop growth. Accurate and rapid estimation of chlorophyll levels can effectively assess the growth environment, water, and fertilizer management of crops, informing subsequent field management decisions and yield prediction [1–3]. Chemical methods are the traditional approach for measuring chlorophyll content, but they are laborious, destructive, and slow [4]. In addition, chlorophyll extracted from plant leaves is susceptible to decomposition by light, resulting in inaccurate measurement. The manual handheld chlorophyll meter, while faster than chemical methods, can only provide information on the chlorophyll content of a single leaf and does not account for the vertical heterogeneity within the canopy. Therefore, accurate measurement over a large area in time and space is not achievable with manual handheld chlorophyll meters.

Unmanned aerial vehicles (UAVs) have emerged as a promising remote sensing platform for obtaining physiological and biochemical traits of crops due to their mobility, flexibility, wide coverage, and high spatial and temporal resolution [5,6]. Spectral images acquired by UAVs in combination with algorithmic models have been shown to effectively monitor chlorophyll content [7–9]. While hyperspectral cameras have high inversion capabilities, they also have disadvantages, such as high cost, poor convenience, and complex operation processes. In contrast, multispectral images are less expensive and easier to control in flight and have equivalent inversion capabilities to hyperspectral images. Therefore, the use of UAV multispectral images for the acquisition and accurate inversion of crop growth parameters has great theoretical and practical value. Several studies have been conducted using UAV multispectral remote sensing technology to monitor the physical and chemical parameters of crops. Zhou et al. [10] developed a SPAD value inversion model for winter wheat using stepwise regression, principal component regression, and ridge regression. Niu et al. [11] employed two visible vegetation indices and four multispectral vegetation indices, along with stepwise regression and random forest regression methods, to estimate the SPAD values of winter wheat. Mao et al. [12] utilized two multispectral sensors with varying spectral response functions (Multiple Camera Array MCA and Sequoia) to obtain multispectral images of maize flowering under different nitrogen application levels and developed a more accurate estimation model by calculating vegetation indices and regressing them on the ground SPAD values.

The growth of wheat can be divided into three stages: the foundation stage, which begins at seed emergence and ends at stem elongation; the construction stage, which starts at the first node detachable from flowering and is a critical stage for yield; and the production stage, which begins at flowering and ends at ripening. Within the construction stage, there is also a period of nutritional and reproductive growth in parallel. After wheat heading, nutritional growth largely ceases, and the plant enters the independent reproductive growth stage. In previous research, the use of UAV multispectral imaging to estimate chlorophyll content has mostly focused on the flowering or early filling stages [13–15], with fewer studies examining the multi-temporal variation of wheat SPAD within the independent reproductive growth stage after heading. However, as the plant progresses through different growth stages, the optimal estimation model may change due to changes in the modeling data set. Therefore, this study employed separate modeling for different growth periods of wheat in order to achieve higher estimation accuracy.

The Hetao irrigation area, located in the northwestern part of China, is a region characterized by aridity and semi-aridity, and has limited water resources. In recent years, various water-saving irrigation patterns have been introduced in the region to replace conventional irrigation methods. These changes in irrigation patterns may impact the growth and development of wheat, which can be reflected in changes in canopy reflectance and SPAD values. Previous research on SPAD estimation of spring wheat in the Hetao irrigation area has not considered the effects of different irrigation modes on SPAD values. This study aims to develop an estimation model for SPAD values in spring wheat under both conventional and water-saving irrigation modes in the Hetao irrigation area, in order to improve the general applicability of the model for large-scale satellite remote sensing applications in the region.

2. Materials and Methods

2.1. Study Site and Experimental Design

In this study, two irrigation modes (conventional irrigation and water-saving irrigation) and six nitrogen fertilizer application rates were studied in a field trial. Multispectral images were collected using a UAV equipped with multispectral sensors at four time points (7, 14, 21, and 28 days after wheat heading) and were combined with ground measurements. Four machine learning regression models (DNN, PLS, RF, and Ada) were used to determine the optimal period and model for estimating SPAD values after the heading stage of spring wheat in the Hetao irrigation area of Inner Mongolia. The results of this study provide theoretical support for remote sensing monitoring of SPAD values in this region.

This study was conducted at Wuyuan Agricultural Technology Extension Center (107°35′ N, 40°30′50″ E, elevation 1028 m a.s.l.), located in Bayannur City, Inner Mongolia, China, during 2021 (location is shown in Figure 1). The region has a temperate continental monsoon climate. The soil type at the experimental site was loam, with baseline fertility level of organic matter 17.65 g/kg, alkaline nitrogen 57.45 mg/kg, available phosphorus 26.83 mg/kg, available potassium 152.42 mg/kg, and pH = 7.32. The spring wheat cultivar "Yongliang 4" was selected for the study. The experiment used a split-plot design, with irrigation as the main plot and nitrogen (N) application as the subplot. There were two irrigation modes: conventional irrigation (four times at tillering stage, jointing stage, flowering stage, and early grain filling stage) and water-saving irrigation (two times at jointing stage and flowering stage), each with a volume of 900 m^3/ha using flood irrigation. The N application subplot had six levels: CK (no fertilizer), N0 (0 kg/ha), N1 (75 kg/ha), N2 (150 kg/ha), N3 (225 kg/ha), N4 (300 kg/ha). The experiment had a total of 12 treatments with three replications, resulting in 36 experimental plots of 42 m^2 each. Phosphorus fertilizer was applied as a base fertilizer at sowing, and the sowing rate was set at 375 kg/ha. Rainfall and temperature data are shown in Figure 2.

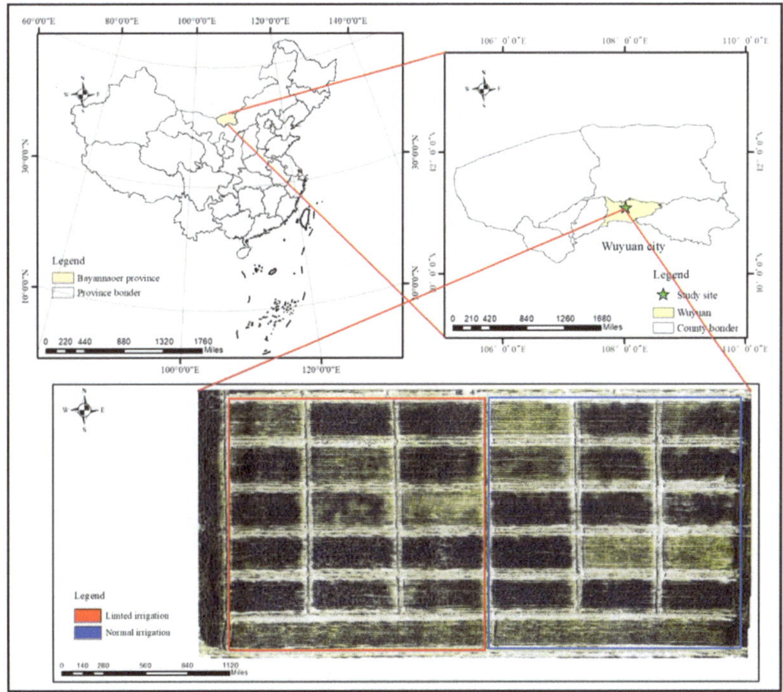

Figure 1. Geographical location of the experimental site. Red frame is 2W treatment; Blue frame is 4W treatment.

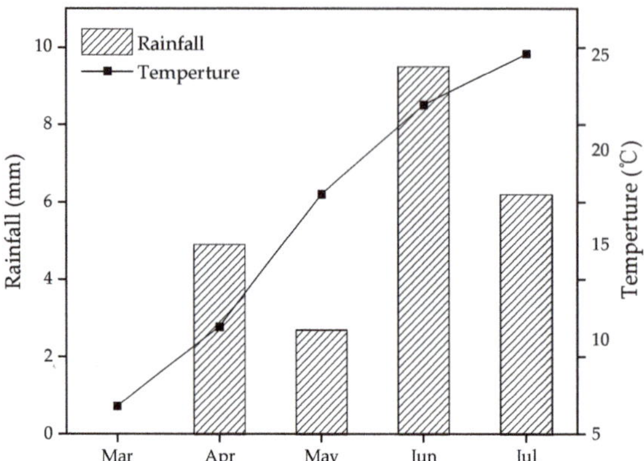

Figure 2. Rainfall and temperature of experimental site. Mar: March, Apr: April, Jun: June, Jul: July.

2.2. UAV Multispectral Data Acquisition and Processing

Multispectral image data were collected at 7, 14, 21, and 28 days after heading stage of the wheat plant using a DJI Phantom 4 multispectral drone (Da-Jiang Innovations, Shenzhen, China).The drone (P4M, Figure 3) integrates five multispectral sensors (blue B: 450 ± 16 nm, green G: 560 ± 16 nm, red R: 650 ± 16 nm, red edge RE: 730 ± 16 nm 16 nm, near infrared NIR: 840 ± 26 nm) and one RGB visible light sensor. To avoid hotspot phenomenon in the images, the images were acquired between 9:00 and 11:00 a.m. on clear and windless days, with the takeoff location fixed and kept consistent each time. Before takeoff, the UAV was manually placed directly above the three reflectivity gray plates of 20%, 40%, and 60%, and reflectivity plate photos were taken. The flight path was automatically planned by DJI GS Pro after calculating the current solar azimuth, with a flight altitude of 30 m, a heading overlap of 85%, and a collateral overlap of 80%. The D-RTK 2 high-precision GNSS mobile station was used to assist the positioning of the UAV and improve the positioning accuracy of the UAV itself. After the flight, DJI Terra was used to perform radiometric correction of the images acquired during the mission, followed by image stitching to obtain a single-band reflectivity orthophoto.

Figure 3. A photo for Phantom 4 multispectral.

2.3. Construction and Selection of Spectral Indices

The reflectance of each treatment plot was extracted by the zonal statistics function of ENVI, and the following vegetation indices (VIs, Vegetable Indices) were calculated (Table 1).

Table 1. Vegetation indices and calculation method.

Index Name	Calculation Formula	References
Leaf chlorophyll index	$LCI = (R_{nir} - R_{rededge})/(R_{nir} + R_{red})$	[16]
Difference vegetation index	$DVI = R_{nir} - R_{red}$	[16]
Enhanced Vegetation Index	$EVI = 2.5 \times (R_{nir} - R_{red})/(R_{nir} + 6 \times R_{red} - 7.5 \times R_{blue} + 1)$	[16]
Green Normalized Difference Vegetation	$GNDVI = (R_{nir} - R_{green})/(R_{nir} + R_{green})$	[17]
Ratio Between NIR and Green Bands	$VI_{(nir/green)} = R_{nir}/R_{green}$	[18]
Ratio Between NIR and Red Bands	$VI_{(nir/red)} = R_{nir}/R_{red}$	[19]
Ratio Between NIR and Red Edge Bands	$VI_{(nir/rededge)} = R_{nir}/R_{rededge}$	[20]
Napierian Logarithm of The Red Edge	$ln_{RE} = 100 \times (ln_{nir} - ln_{red})$	[21]
Modified Soil-Adjusted Vegetation Index 1	$MSAVI1 = (1 + L)\left(\frac{R_{nir} - R_{red}}{R_{nir} + R_{red} + L}\right)(L = 0.1)$	[22]
Modified Soil-Adjusted Vegetation Index 2	$MSAVI2 = R_{nir} + 0.5 - \sqrt{(2 \times R_{nir} + 1)^2 - 8 \times (R_{nir} - R_{red})}/2$	[22]
Optimized Soil-Adjusted Vegetation Index	$OSAVI = (1 + 0.16) \times \frac{(R_{nir} - R_{red})}{(R_{nir} + R_{red} + 0.16)}$	[23]
Modified Triangular Vegetation Index 2	$MTVI2 = \frac{1.5 \times [1.2 \times (R_{nir} - R_{green}) - 2.5 \times (R_{red} - R_{green})]}{\sqrt{(2 \times R_{nir} + 1)^2 - (6 \times R_{nir} - 5 \times \sqrt{R_{red}}) - 0.5}}$	[24]
Normalized Difference Red Edge Index	$NDRE = \frac{(R_{nir} - R_{rededge})}{(R_{nir} + R_{rededge})}$	[25]
Normalized Difference Vegetation Index	$NDVI = \frac{(R_{nir} - R_{red})}{(R_{nir} + R_{red})}$	[26]
Modified Simple Radio	$MSR = (R_{nir} - R_{red} - 1)/(\sqrt{R_{nir} + R_{red}} + 1)$	[27]
Soil-Adjusted Vegetation Index	$SAVI = \frac{(R_{nir} - R_{red})}{(R_{nir} + R_{red} + 0.5)} \times (1 + 0.5)$	[28]
Simplified Canopy Chlorophyll Content Index	$SCCCI = \frac{NDRE}{NDVI}$	[29]
Modified Chlorophyll Absorption Reflectance Index	$MCARI = \left(R_{rededge} - R_{red} - 0.2 \times (R_{rededge} - R_{green})\right) \times \left(\frac{R_{rededge}}{R_{red}}\right)$	[30]
Modified Chlorophyll Absorption Reflectance Index 2	$MCARI2 = 1.5 \times \frac{(2.5 \times (R_{nir} - R_{rededge}) - 1.3 \times (R_{nir} - R_g))}{(2 \times (R_{nir} + 1)^2 - (6 \times R_{nir} - 5 \times (R_{red})^2) - 0.5)}$	[31]
Transformed Chlorophyll Absorption Reflectance Index	$TCARI = 3 \times \left((R_{rededge} - R_{red}) - 0.2 \times (R_{rededge} - R_{green}) \times \left(\frac{R_{rededge}}{R_{red}}\right)\right)$	[32]
Normalized Difference Index	$NDI = \frac{(R_{nir} - R_{rededge})}{(R_{nir} + R_{red})}$	[33]
Red-Edge Chlorophyll Index 1	$Cl1 = \frac{R_{nir}}{R_{rededge}} - 1$	[34]
Red-Edge Chlorophyll Index 2	$Cl2 = \frac{R_{rededge}}{R_{green}} - 1$	[35]
Structure-Insensitive Pigment Index	$SIPI = \frac{(R_{nir} - R_{blue})}{(R_{nir} + R_{red})}$	[36]
TCARI/OSAVI	$\frac{TCARI}{OSAVI}$	[31]
MCARI/OSAVI	$\frac{MCARI}{OSAVI}$	[31]

2.4. Ground Data Acquisition and Processing

During images collection by the UAV, five wheat plants were selected in each plot according to the "five-point sampling method", and the SPAD values of leaf tip, leaf middle, and leaf base of the flag leaf of the wheat plant were measured using a SPAD 502Plus chlorophyll meter (Konica Minolta, Tokyo, Japan). The average value was taken as the SPAD value of the plant.

2.5. Construction of Regression Model

In this study, four regression models were implemented in Python for the estimation of SPAD values as follows.

Deep neural networks (DNN) is a neural network containing multiple hidden layers (at least 3 layers), which has more hidden layers and a stronger fitting ability compared to traditional neural networks.

Partial least squares (PLS) draws on the advantages of statistical methods, such as correlation analysis, principal component analysis, and multiple linear regression, and is widely used in hyperspectral inversion estimation with high correlation between independent variables because of its strong ability to remove autocorrelation between features.

Random forest regression (RF) is a machine learning algorithm that uses multiple decision trees to train and predict samples and has strong anti-interference ability. It also has the advantages of fast training speed and no processing of input data.

Adaptive Boosting (Ada) is one of the representative algorithms of Boosting in integrated learning, which mainly changes to obtain different test samples by controlling the weights of sample distribution.

2.6. Segmentation of Dataset and Accuracy Evaluation

The samples of each period were randomly divided into training set and test set at the ratio of 7:3, and K-fold cross validation (K = 5) was used to optimize the model. Five-fold cross-validation involves dividing the original training set into 5 groups, using each subset of data as a validation set in turn, and using the remaining 4 subsets of data as the training set. The results from K groups are then summed and averaged to reduce the error of the training set and improve the generalization ability of the model by avoiding the inclusion of test data during the training process.

The accuracy of the model is evaluated by three metrics: the coefficient of determination (R^2), the root means square error (RMSE), and the mean absolute prediction error (MAPE). R^2 is used to indicate the degree of fit between the estimated and measured values, with a value closer to 1 indicating a higher accuracy of the model fit. RMSE reflects the deviation of the estimated value from the measured value, with a smaller value indicating a higher accuracy of the model fit. MAPE is the average of absolute errors, which more accurately reflects the actual errors in the prediction value.

$$R^2 = 1 - \frac{\sum_{i=1}^{n}(y_i - \hat{y}_i)^2}{\sum_{i=1}^{n}(y_i - \overline{y}_i)^2} \tag{1}$$

$$RMSE = \sqrt{\frac{\sum_{i=1}^{n}(y_i - \hat{y}_i)^2}{n}} \tag{2}$$

$$MAPE = \frac{100\%}{n} \sum_{i=1}^{n} \left| \frac{\hat{y}_i - y_i}{y_i} \right| \tag{3}$$

y_i is the observed value, $\overline{y}_i$ is the mean of the observed values, $\hat{y}_i$ is the model predicted value, and n is the number of samples.

3. Results

3.1. Basic Statistical Information of Measured SPAD Values

The SPAD values of different treatments are shown in Table 2. As the reproductive process advances, different N treatments under the 2W treatment showed a trend of increasing then decreasing, the maximum SPAD value appeared 21 days after heading, and the maximum SPAD value of different N treatments under the 4W treatment appeared 7 days after heading. With the increase in nitrogen application, the SPAD values after different heading periods showed a trend of increasing and then decreasing.

Table 2. Basic statistics of the field measurements at different stages.

Irrigation	N Treatment	7 d	14 d	21 d	28 d
	N0	41.99 bcd	43.42 bcd	43.11 d	36.33 bc
	N5	46.90 abc	47.75 abc	47.05 bcd	42.99 ab
2W	N10	48.42 abc	50.23 ab	51.89 ab	43.82 ab
	N15	50.10 ab	51.05 a	55.50 a	45.29 a
	N20	48.58 abc	46.05 abcd	51.53 abc	46.51 a
	CK	39.95 cd	36.26 de	33.46 e	14.73 e
	N0	41.70 bcd	35.75 ef	28.47 e	10.47 e
	N5	46.61 abc	42.75 bcde	45.09 cd	24.35 d
4W	N10	47.97 abc	41.80 cde	45.81 bcd	29.48 cd
	N15	51.15 a	46.77 abcd	50.62 abc	34.55 c
	N20	41.65 bcd	45.75 abcd	46.13 bcd	26.15 d
	CK	36.29 cd	29.63 f	31.37 e	12.87 e

Alphabets within columns followed by the same letter are statistically insignificant at the 0.05 level.

3.2. Correlation Analysis of SPAD Values and Vegetation Indices

The correlation coefficients between SPAD values and each vegetation index at different periods after heading are shown in Table 3. The highest correlation coefficient was observed at 7 days after heading, followed by 21 days. Except for 7 days after heading, the correlation coefficient for the rest of the period showed 2W < 4W. Under the 2W treatment, the vegetation indices with the highest correlation coefficients at 7, 14, 21, and 28 days after tapping were MCARI2, MSAVI2, SCCCI, and MCARI. The highest correlation coefficients under 4W were MCARI2, MSAVI1, CL1, and DVI.

Table 3. Correlation coefficients between spectral vegetation indices and SPAD.

Indices	7 d		14 d		21 d		28 d	
	2W	4W	2W	4W	2W	4W	2W	4W
DVI	0.912	0.867	0.650	0.867	0.701	0.830	0.708	0.802
EVI	0.908	0.866	0.695	0.871	0.708	0.856	0.727	0.798
NDVI	0.842	0.815	0.790	0.871	0.703	0.867	0.740	0.790
GNDVI	0.882	0.825	0.776	0.870	0.741	0.903	0.741	0.779
NDRE	0.912	0.846	0.739	0.863	0.764	0.925	0.749	0.774
LCI	0.912	0.846	0.739	0.863	0.764	0.925	0.749	0.774
OSAVI	0.879	0.847	0.745	0.873	0.714	0.876	0.734	0.799
VI(NIR/G)	0.893	0.867	0.671	0.830	0.765	0.919	0.742	0.756
VI(NIR/R)	0.842	0.863	0.637	0.819	0.761	0.918	0.726	0.754
VI(NIR/RE)	0.917	0.863	0.698	0.849	0.765	0.926	0.748	0.766
lnRE	0.856	0.844	0.735	0.853	0.747	0.916	0.742	0.782
MSAVI1	0.870	0.840	0.759	0.873	0.713	0.875	0.736	0.797
MSAVI2	0.729	0.706	0.815	0.855	0.589	0.796	0.734	0.719
MTVI2	0.895	0.868	0.677	0.863	0.711	0.859	0.722	0.801
MSR	0.915	0.859	0.640	0.866	0.692	0.801	0.679	0.793
SAVI	0.898	0.860	0.706	0.872	0.712	0.867	0.727	0.802
SCCCI	0.924	0.840	0.723	0.864	0.772	0.925	0.231	−0.652
MCARI	−0.824	−0.782	−0.785	−0.854	−0.739	−0.881	−0.783	0.463
MCARI2	0.929	0.870	0.663	0.853	0.764	0.910	0.746	0.750
TCARI	−0.817	−0.850	−0.638	−0.830	−0.766	−0.921	−0.562	0.682
NDI	0.902	0.840	0.755	0.866	0.758	0.920	0.749	0.779
CL1	0.917	0.863	0.698	0.849	0.765	0.926	0.748	0.766
CL2	0.869	0.850	0.710	0.840	0.759	0.910	0.743	0.763
SIPI	0.834	0.809	0.779	0.865	0.713	0.867	0.735	0.788
TCARI/OSAVI	−0.826	−0.841	−0.689	−0.846	−0.753	−0.919	−0.766	−0.618
MCARI/OSAVI	−0.826	−0.786	−0.797	−0.870	−0.695	−0.847	−0.757	−0.742

3.3. Model Development and Evaluation

3.3.1. Estimation of SPAD Values after Heading 7 d

Table 4 and Figure 4 shows the evaluation of the accuracy of the SPAD estimation models for 7 d after wheat heading stage. the R^2 of the training set and the test set of the four models are above 0.70, with the accuracy of the training set higher accuracy than the test set. Using the accuracy of the test set as the evaluation criterion of the models, the accuracy of the models was found to be in the following order: PLS > RF > Ada > DNN. The R^2, RMSE, and MAPE values of the test set of the PLS model were 0.762, 3.048, and 0.052, respectively, which were 7.32%, 2.28%, and 7.17% higher than the R^2 of the RF, Ada, and DNN models, respectively. RMSE decreased by 9.25%, 3.33%, and 9.20%, and MAPE decreased by 21.25%, 0%, and 7.69%. Based on these results, it can be concluded that the PLS model had the highest accuracy and stability in estimating the SPAD values of wheat 7 d after heading stage.

Table 4. Accuracy assessment of different estimation models at 7 d after heading.

Model	Training Set			Test Set		
	R^2	RMSE	MAPE	R^2	RMSE	MAPE
DNN	0.754	2.784	0.048	0.710	3.359	0.063
PLS	0.786	2.595	0.043	0.762	3.048	0.052
RF	0.957	1.169	0.021	0.745	3.153	0.052
Ada	0.968	0.829	0.014	0.711	3.357	0.056

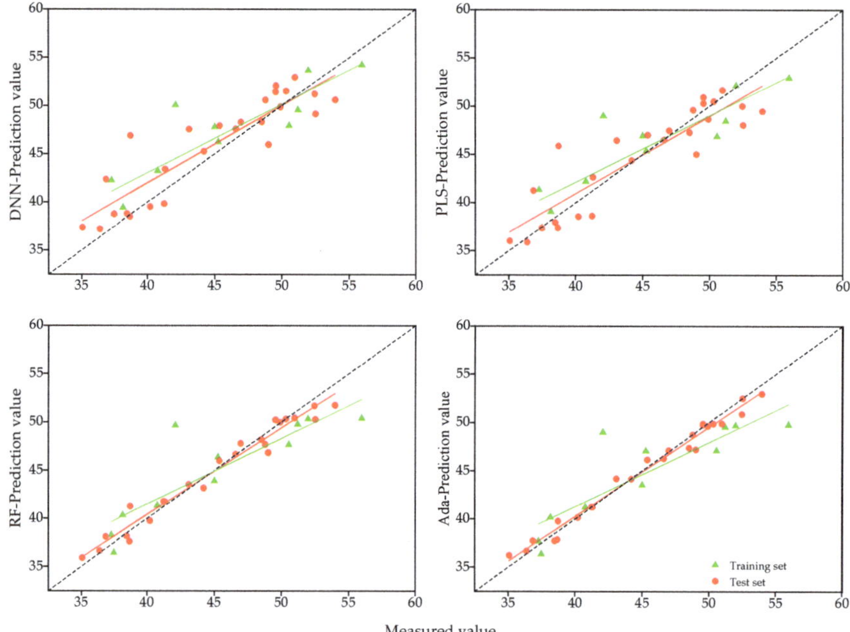

Figure 4. Calibration and validation results of different estimation models at 7 d after heading.

3.3.2. Estimation of SPAD Values after Heading 14 d

The accuracy of the SPAD estimation models for 14 d after wheat heading stage was evaluated and presented in Table 5 and Figure 5. The R^2 values for the training and test sets of the four models are above 0.75, and the accuracy of the training set of the DNN and PLS models was lower than that of the test set, while the accuracy of the training set of the RF and Ada models was higher than that of the test set. When using the accuracy of the test set as the evaluation criterion for the models, the accuracies of different models were found to be PLS > Ada > RF > DNN in descending order. The R^2, RMSE, and MAPE values for the test set of PLS model were 0.878, 2.405, and 0.045, which were 12.13%, 10.72%, and 7.33% higher than the R^2 values of the RF, Ada, and DNN models, respectively, and the RMSE is reduced by 25.05%, 23.41%, and 18.28%, and the MAPE decreased by 44.44%, 53.33%, and 35.56%. The comprehensive analysis concluded that the PLS model had the highest accuracy and stability in estimating the SPAD values of wheat 14 d after heading stage.

Table 5. Accuracy assessment of different estimation models at 14 d after heading.

Model	Training Set			Test Set		
	R^2	RMSE	MAPE	R^2	RMSE	MAPE
DNN	0.716	3.538	0.067	0.783	3.209	0.065
PLS	0.767	3.205	0.060	0.878	2.405	0.045
RF	0.924	1.835	0.036	0.793	3.140	0.069
Ada	0.934	1.711	0.029	0.818	2.943	0.061

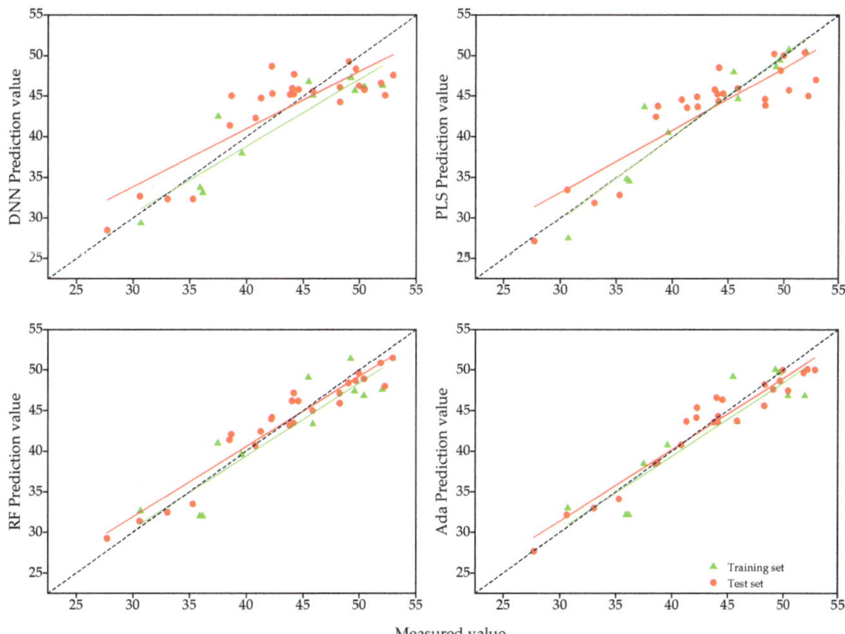

Figure 5. Calibration and validation results of different estimation models at 14 d after heading.

3.3.3. Estimation of SPAD Values after Heading 21 d

The accuracy of SPAD estimation models for 21 d after wheat heading stage was shown in Table 6 and Figure 6. The R^2 of training set and test set of all four models were above 0.65, with the training set accuracy of the DNN and PLS models higher than the test set, while the training set accuracy of the RF and Ada models were lower than the test set. The accuracy of different models, from largest to smallest, was DNN > RF > Ada > PLS. The R^2 value of the DNN test set model was 0.737, the RMSE was 4.806, and the MAPE was 0.086, which were 12.18%, 1.80%, and 5.74% higher than the R^2 values of the PLS, RF, and Ada models, respectively. The RMSE was 12.41%, 2.42%, −11.02%, and the MAPE decreased by 28.33%, 18.10%, and −4.88%. Based on this analysis, it can be concluded that the DNN model was the most accurate and stable in estimating the SPAD values of wheat 21 d after heading stage.

Table 6. Accuracy assessment of different estimation models at 21 d after heading.

Model	Training Set			Test Set		
	R^2	RMSE	MAPE	R^2	RMSE	MAPE
DNN	0.881	2.922	0.044	0.737	4.806	0.086
PLS	0.777	4.009	0.080	0.657	5.487	0.120
RF	0.678	4.815	0.097	0.724	4.925	0.105
Ada	0.684	5.028	0.092	0.697	4.329	0.082

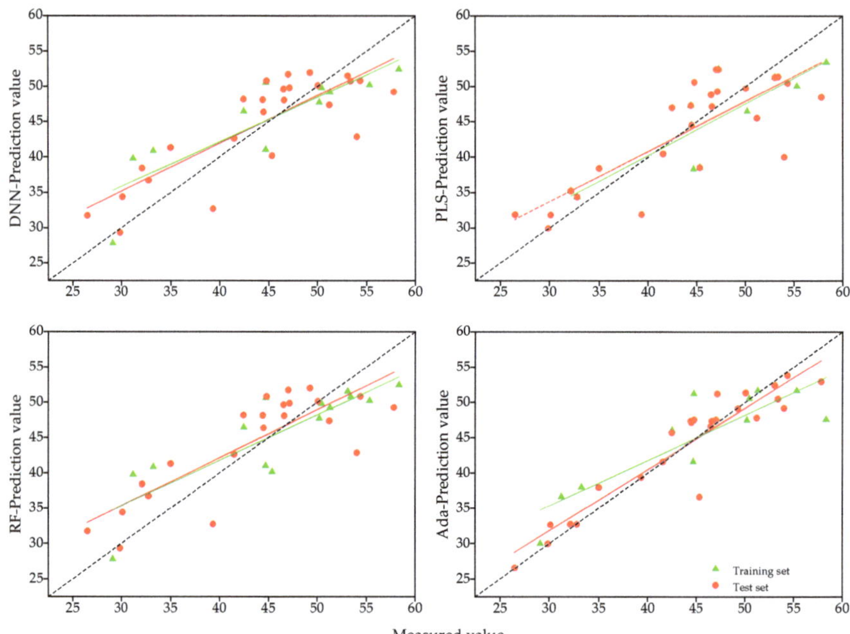

Figure 6. Calibration and validation results of different estimation models at 21 d after heading.

3.3.4. Estimation of SPAD Values after Heading 28 d

The accuracy of the SPAD estimation model for 28 d after the wheat heading stage was evaluated and presented in Table 7 and Figure 7. The R^2 values for the training and test set of all four models are above 0.65, and the accuracy of the training set is higher than the test set for all models except the DNN model. When using the test set accuracy as the evaluation criterion, the Ada model was found to be the most accurate, followed by RF, PLS, and DNN in descending order. The R^2 value for the Ada model was 0.815, the RMSE was 5.904, and the MAPE was 0.237, which were 20.38%, 14.63%, and 14.47% higher, respectively, than those of the DNN, PLS, and RF models. The RMSE was reduced by 62.77%, 60.67%, and 60.59%, and the MAPE was reduced by 22.80%, 20.74%, and 31.30% for the Ada model compared to the other models. Overall, the Ada model was found to have the highest accuracy and stability in estimating the SPAD values of wheat 28 d after heading stage.

Table 7. Accuracy assessment of different estimation models at 28 d after heading.

Model	Training Set			Test Set		
	R^2	RMSE	MAPE	R^2	RMSE	MAPE
DNN	0.691	7.054	0.262	0.677	7.801	0.307
PLS	0.713	6.803	0.299	0.711	7.383	0.299
RF	0.926	3.442	0.115	0.712	7.368	0.345
Ada	0.971	2.168	0.067	0.815	5.904	0.237

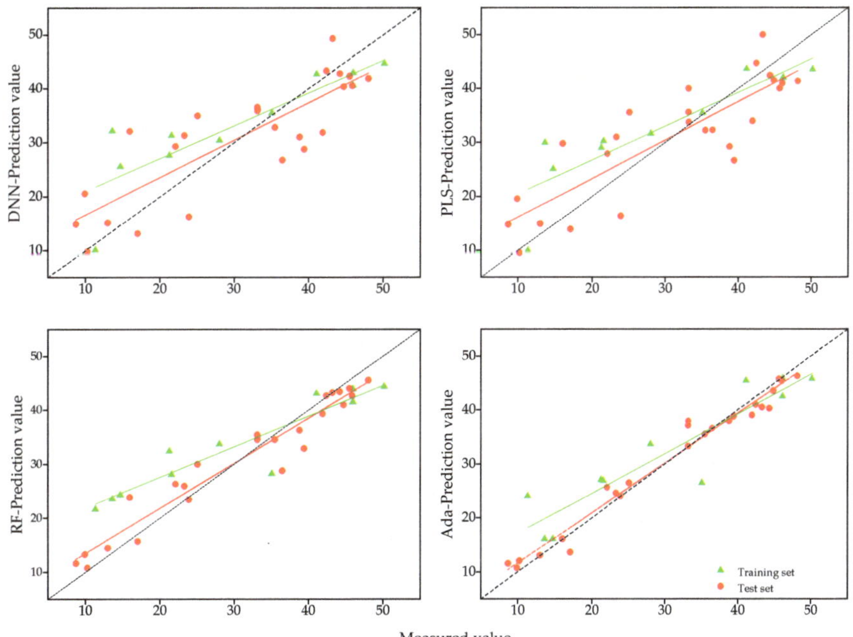

Figure 7. Calibration and validation results of different estimation models at 28 d after heading.

3.4. Comparison of Accuracy of Four Estimation Models at Different Growth Stages

As the growth process progressed, the accuracy of the four models demonstrated an overall trend of increasing then decreasing and then increasing again. The optimal model for estimating SPAD values varied depending on the growth stage (Figure 8). The model with the highest R^2 value at 7 and 14 days after heading was PLS, the model with the highest R^2 value at 21 d after heading was RF, and the model with the highest R^2 value at 28 d after heading was Ada. The model with the lowest RMSE value at 7 and 14 days after heading was PLS, and the model with the lowest RMSE value at 21 and 28 days after heading was Ada. The model with the lowest MAPE value at 7 days after heading was PLS and RF, the lowest at 14 days after heading was PLS, and the lowest at 21 and 28 days after heading was Ada.

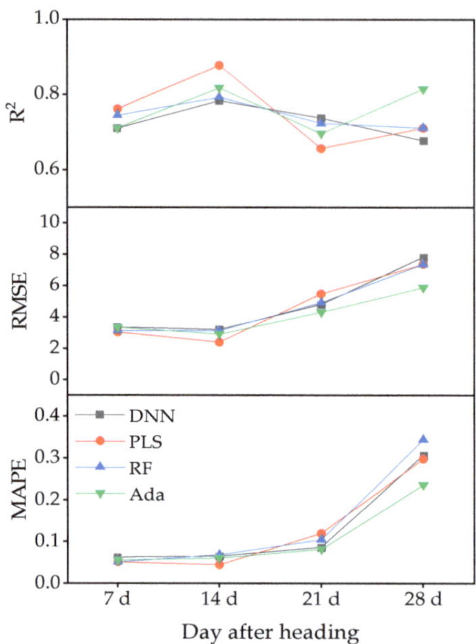

Figure 8. Comparison of estimation accuracy of four models test set at different growth stages.

4. Discussion

Leaf SPAD value is an important indicator for characterizing the nitrogen nutrition status of plants [37], and several studies have been conducted to estimate the SPAD value of crop plants using the unmanned aerial vehicle (UAV) remote sensing technology platform in order to diagnose crop nitrogen nutrition and provide a reference basis for subsequent field nitrogen fertilization management [38,39]. Fast access to the growth conditions of crops in farmland is an important aspect of smart agriculture for fast sensing and intelligent decision making. Although there are few studies that focus on the independent reproductive growth stage (from heading to maturity) of wheat, monitoring leaf SPAD values during this stage and adopting appropriate water and fertilizer management can coordinate the development of population and individual main stem and tillers, and have significant regulatory effects on late grain filling, final yield, and grain quality formation. In this study, multispectral images were acquired every 7 days after the wheat heading stage using an unmanned aerial vehicle, and vegetation indices were extracted to construct separate SPAD estimation models for wheat.

Vegetation indices have been found to be correlated with agronomic traits making them an important alternative to traditional agronomic parameters [40]. This study demonstrated that the measured SPAD values of wheat at various growth stages had strong correlations and linear sensitivity with most vegetation indices. However, this study also found that the correlation between the corresponding SPAD values and the vegetation indices gradually decreased as the reproductive process progressed. This may be due to the fact that during the nutritional growth stage of wheat, most of the absorbed and accumulated nitrogen is stored in nutritional organs, such as leaves [41], leading to a higher spectral reflectance sensitivity of the canopy. During the independent reproductive growth stage, nitrogen is gradually transferred from nutritional organs to the spike, the leaves senesce and degrade [42], resulting in a decrease in spectral reflectance sensitivity.

The spectral characteristics of the vegetation canopy are influenced by numerous physical and biochemical variables, which exhibit different behaviors at different growth stages [43,44]. As a result, the accuracy of the estimation models constructed based on the extraction of vegetation indices from spectral reflectance varies across fertility periods.

Previous research has shown that the accuracy of estimation models tends to decrease at the heading, flowering, and filling stages of winter wheat [45]. Another study found that the accuracy of three regression models for estimating SPAD values in winter wheat increased, then decreased, and then increased again during the four growth periods from nodulation to flowering [10]. In the present study, the accuracy of the model constructed at the four growth stages showed relatively consistent performance, except for the SPAD value estimation model at 21 d of heading stage. The width of a violin plot represents the probability density of the data. As shown in Figure 9c, the distribution of spectral indices at 21 d after heading displayed an obvious concentration and an oversaturation phenomenon, leading to a decreased sensitivity of vegetation indices and subsequently increased error and decreased model accuracy in the estimation process. In contrast, after 28 d of heading stage, plants treated with low levels of nitrogen exhibited senescence, chlorophyll decomposition, and decreased chlorophyll content, while those treated with high levels of nitrogen maintained high chlorophyll content, which to some extent weakened the oversaturation phenomenon and thus improved model accuracy.

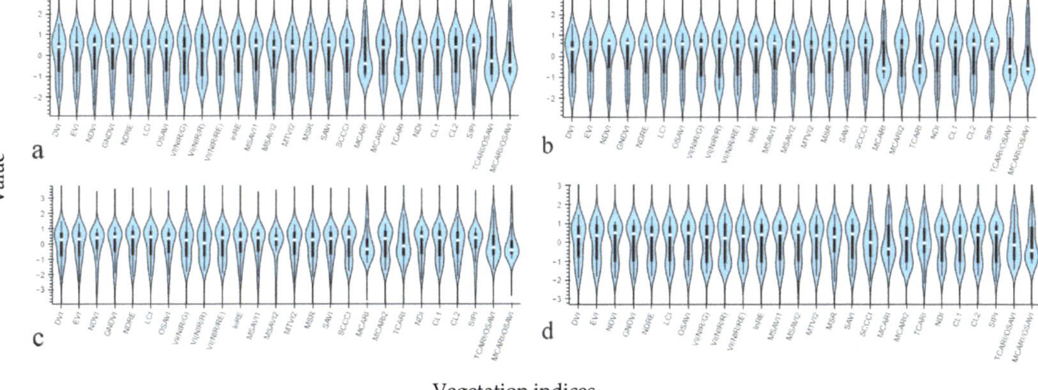

Figure 9. Violin plots of the spectral indices. (**a–d**) are violin plots of vegetation indices at 7, 14, 21, and 28 days after heading, respectively.

Compared with using a single vegetation index as the model input variable, the estimation method based on multiple vegetation indices can improve the model accuracy to a certain extent. In a multi-year study by Kooistra et al. [46] to estimate chlorophyll content of potato leaves, the regression model obtained with a single vegetation index as an input variable had a maximum R^2 of 0.641. Multiple vegetation indices were used in this study to jointly participate in the estimation, and the model with the lowest precision among the four periods, $R^2 = 0.657$, was also above this precision. Yuan et al. [47] found in their study on estimating chlorophyll content of plant seedlings that most vegetation indices, although significantly correlated with SPAD values, had low R^2 for regression models with a single vegetation index as the explanatory variable, and the accuracy of the regression models was significantly improved when multiple vegetation indices were used as explanatory variables. In this study, all four regression models based on 26 vegetation indices demonstrated good performance, with test set R^2 ranging from 0.657 to 0.878, RMSE ranging from 2.934 to 7.801, and MAPE ranging from 0.045 to 0.345. This suggests that the inclusion of multiple vegetation indices in the regression models allowed for the incorporation of more valid spectral information, resulting in improved estimation accuracy.

The accuracy of different models for estimating SPAD values of spring wheat varied significantly across different data sets. An appropriate estimation model can effectively capture the relationship between vegetation indices and SPAD values and maintain stable model structure and parameters. Previous studies have shown the effectiveness of machine learning algorithms in plant nutrition diagnosis [48,49], and it is important to identify the best model for estimating SPAD at different stages of fertility in spring wheat after heading stage. In this study, the optimal model was PLS at 7 and 14 days after heading, RF at 21 d after heading, and Ada at 28 d after heading. The R^2 of the training set of the Ada and RF models at 7, 14, and 28 days was significantly higher than that of the test set, and showed a significant overfitting phenomenon, probably due to the high autocorrelation among the 26 vegetation indices of the input, which was influenced by the multicollinearity of the input feature variables. The regression modeling method of PLSR converts a set of highly correlated independent variables into a set of mutually independent, non-linearly related principal component variables by extracting principal components in the process of establishing regressions, which can effectively capture most of the information of the original data and eliminate the covariance among vegetation indices, resulting in the best estimation accuracy of all four models being obtained at 7 d and 14 d after sampling. In the case of deep learning models, a large amount of diverse data is typically required for model training in order to understand the relationship between data and estimates, and the number of samples in this study was not sufficient to support a deep learning network with multiple hidden layers, leading to a low level of estimation accuracy for the DNN model. However, the estimation accuracy (R^2) for all four periods was found to be greater than 0.677, indicating the strong potential for the DNN model to be used for estimation. This finding is in line with the results of Liu et al. [50].

5. Conclusions

The optimal SPAD estimation models were different for the four periods after the spring wheat heading stage, with PLS as the optimal model at 7 and 14 d after heading stage, RF at 21 d after heading stage, and Ada at 28 d after heading stage, where the highest accuracy was achieved by using the PLS model to estimate SPAD values at 14 d after heading stage (training set R^2 = 0.767, RMSE = 3.205, MAPE = 0.060, and test set R^2 = 0.878, RMSE = 2.405, MAPE = 0.045).

Further studies could include validation of the model's performance on additional datasets, assessment of its practical usefulness and potential adoption by farmers through field trials, integration with precision agriculture tools and technologies, such as drones or sensors, and the optimization of fertilization and irrigation practices to improve wheat yield and quality.

Author Contributions: Conceptualization, Q.W. and Y.Z.; methodology, Z.Z.; software, M.X.; validation, Z.Z. and Y.Z.; formal analysis, Z.Z.; investigation, D.H.; resources, M.X.; data curation, Z.Z.; writing—original draft preparation, Q.W.; writing—review and editing, Y.Z.; visualization, Y.Z.; supervision, Y.Z.; project administration, Y.Z.; funding acquisition, Y.Z. All authors have read and agreed to the published version of the manuscript.

Funding: This research was funded by Inner Mongolia "science and technology" action focus on special "Research and Application of Key Technologies for Production and Processing of Durum Wheat and Products in Hetao irrigation area" (NMKJXM202111-3) and Inner Mongolia Natural Science Foundation of China "Research on nitrogen nutrition diagnosis of spring wheat in Hetao irrigation area based on UAV mapping technology" (2021MS03089).

Data Availability Statement: Not applicable.

Acknowledgments: We truly appreciate the assistance from the Wuyuan Agricultural Technology Extension Center at Bayannur, Inner Mongolia, China.

Conflicts of Interest: The authors declare no conflict of interest.

References

1. Jiang, H.L.; Yang, H.; Chen, X.P.; Wang, S.D.; Li, X.K.; Liu, K.; Cen, Y. Research on Accuracy and Stability of Inversing Vegetation Chlorophyll Content by Spectral Index Method. *Spectrosc. Spectr. Anal.* **2015**, *35*, 975–981.
2. Yuan, Y.C.; Zhou, Y.; Song, Y.F.; Xu, Z.; Wang, K.J. Estimation Method of Wheat Canopy Chlorophyll Based on Information Entropy Feature Selection. *Trans. Chin. Soc. Agric. Mach.* **2022**, *53*, 186–195.
3. Liu, T.; Zhang, H.; Wang, Z.Y.; He, C.; Zhang, Q.G.; Jiao, Y.Z. Estimation of the leaf area index and chlorophyll content of wheat using UAV multi-spectrum images. *Trans. Chin. Soc. Agric. Eng.* **2021**, *37*, 65–72.
4. Deng, S.Q.; Zhao, Y.; Bai, X.Y.; Li, X.; Sun, Z.D.; Liang, J.; Sun, Z.H.; Cheng, S. Inversion of chlorophyll and leaf area index for winter wheat based on UAV image segmentation. *Trans. Chin. Soc. Agric. Eng.* **2022**, *38*, 136–145.
5. Yi, H.; Li, F.; Yang, H.B.; Li, Y. Estimation of canopy chlorophyll in potato based on UAV hyperspectral images. *J. Plant Nutr. Fertil.* **2021**, *27*, 2184–2195.
6. Ma, L.F.; Hu, N.Y.; Li, W.; Qin, W.L.; Huang, S.B.; Wang, Z.M.; Li, F.; Yu, K. Using multispectral drone data to monitor maize's response to various irrigation modes. *J. Plant Nutr. Fertil.* **2022**, *28*, 743–753.
7. Li, Z.F.; Su, J.X.; Fei, C.; Li, Y.Y.; Liu, N.N.; Fan, H.; Chen, B. Estimation of chlorophyll content in sugar beet under drip irrigation based on hyperspectral data. *J. Agric. Resour. Environ.* **2020**, *37*, 761–769.
8. Zhou, M.G.; Shao, G.M.; Zhang, L.Y.; Liu, Z.K.; Han, W.T. Multi-spectral Inversion of SPAD Value of Winter Wheat Based on Unmanned Aerial Vehicle Remote Sensing. *Water Sav. Irrig.* **2019**, *9*, 40–45.
9. Wei, Q.; Zhang, B.Z.; Wei, Z.; Han, X.; Duan, C.F. Estimation of Canopy Chlorophyll Content in Winter Wheat by UAV Multispectral Remote Sensing. *J. Triticeae Crops* **2020**, *40*, 365–372.
10. Zhou, M.G.; Shao, G.M.; Zhang, L.Y.; Yao, X.M.; Han, W.T. Inversion of SPAD value of winter wheat by multispectral remote sensing of unmanned aerial vehicles. *Trans. Chin. Soc. Agric. Eng.* **2020**, *36*, 125–133.
11. Niu, Q.L.; Feng, H.K.; Zhou, X.G.; Zhu, J.Q.; Yong, B.B.; Li, H.Z. Combining UAV Visible Light and Multispectral Vegetation Indices for Estimating SPAD Value of Winter Wheat. *Trans. Chin. Soc. Agric. Mach.* **2021**, *52*, 183–194.
12. Mao, Z.H.; Deng, L.; Sun, J.; Zhang, A.W.; Chen, X.Y.; Zhao, Y. Research on the Application of UAV Multispectral Remote Sensing in the Maize Chlorophyll Prediction. *Spectrosc. Spectr. Anal.* **2018**, *38*, 2923–2931.
13. Han, W.T.; Peng, X.S.; Zhang, L.Y.; Niu, Y.X. Summer Maize Yield Estimation Based on Vegetation Index Derived from Multi-temporal UAV Remote Sensing. *Trans. Chin. Soc. Agric. Mach.* **2020**, *51*, 148–155.
14. Chen, Q.; Xu, H.G.; Cao, Y.B.; Duan, F.Y.; Chen, Z. Grain Yield Prediction of Winter Wheat Using Multi-temporal UAV Based on Multispectral Vegetation Index. *Trans. Chin. Soc. Agric. Mach.* **2021**, *52*, 160–167.
15. Zhou, X.; Zheng, H.B.; Xu, X.Q.; He, J.Y.; Ge, X.K.; Yao, X.; Cheng, T.; Zhu, Y.; Cao, W.X.; Tian, Y.C. Predicting grain yield in rice using multi-temporal vegetation indices from UAV-based multispectral and digital imagery. *ISPRS J. Photogramm. Remote Sens.* **2017**, *130*, 246–255. [CrossRef]
16. Huete, A.; Justice, C.; Liu, H. Development of vegetation and soil indices for MODIS-EOS. *Remote Sens. Environ.* **1994**, *49*, 224–234. [CrossRef]
17. Gitelson, A.A.; Merzlyak, M.N. Remote sensing of chlorophyll concentration in higher plant leaves. *Adv. Space Res.* **1998**, *22*, 689–692. [CrossRef]
18. Gamon, J.A.; Surfus, J.S. Assessing leaf pigment content and activity with a reflectometer. *New Phytol.* **1999**, *143*, 105–117. [CrossRef]
19. Birth, G.S.; McVey, G.R. Measuring the color of growing turf with a reflectance spectrophotometer 1. *Agron. J.* **1968**, *60*, 640–643. [CrossRef]
20. Jasper, J.; Reusch, S.; Link, A. Active sensing of the N status of wheat using optimized wavelength combination: Impact of seed rate, variety and growth stage. *Precis. Agric.* **2009**, *9*, 23–30.
21. Filella, I.; Penuelas, J. The red edge position and shape as indicators of plant chlorophyll content, biomass and hydric status. *Int. J. Remote Sens.* **1994**, *15*, 1459–1470. [CrossRef]
22. Qi, J.; Chehbouni, A.; Huete, A.R.; Kerr, Y.H.; Sorooshian, S. A modified soil adjusted vegetation index. *Remote Sens. Environ.* **1994**, *48*, 119–126. [CrossRef]
23. Rondeaux, G.; Steven, M.; Baret, F. Optimization of soil-adjusted vegetation indices. *Remote Sens. Environ.* **1996**, *55*, 95–107. [CrossRef]
24. Xing, N.C.; Huang, W.J.; Xie, Q.Y.; Shi, Y.; Ye, H.C.; Dong, Y.Y.; Wu, M.Q.; Sun, G.; Jiao, Q.J. A Transformed Triangular Vegetation Index for Estimating Winter Wheat Leaf Area Index. *Remote Sens.* **2019**, *12*, 16. [CrossRef]
25. Sims, D.A.; Gamon, J.A. Relationships between leaf pigment content and spectral reflectance across a wide range of species, leaf structures and developmental stages. *Remote Sens. Environ.* **2002**, *81*, 337–354. [CrossRef]
26. Chen, J.; Jönsson, P.; Tamura, M.; Gu, Z.H.; Matsushita, B.; Eklundh, L. A simple method for reconstructing a high-quality NDVI time-series data set based on the Savitzky–Golay filter. *Remote Sens. Environ.* **2004**, *91*, 332–344. [CrossRef]
27. Chen, J.M. Evaluation of vegetation indices and a modified simple ratio for boreal applications. *Can. J. Remote Sens.* **1996**, *22*, 229–242. [CrossRef]
28. Huete, A.R. A soil-adjusted vegetation index (SAVI). *Remote Sens. Environ.* **1988**, *25*, 295–309. [CrossRef]
29. Fitzgerald, G.; Rodriguez, D.; O'Leary, G. Measuring and predicting canopy nitrogen nutrition in wheat using a spectral index—The canopy chlorophyll content index (CCCI). *Field Crops Res.* **2010**, *116*, 318–324. [CrossRef]

30. Kimura, R.; Okada, S.; Miura, H.; Kamichika, M. Relationships among the leaf area index, moisture availability, and spectral reflectance in an upland rice field. *Agric. Water Manag.* **2004**, *69*, 83–100. [CrossRef]
31. Wu, C.Y.; Niu, Z.; Tang, Q.; Huang, W.J. Estimating chlorophyll content from hyperspectral vegetation indices: Modeling and validation. *Agric. For. Meteorol.* **2008**, *148*, 1230–1241. [CrossRef]
32. Devadas, R.; Lamb, D.W.; Simpfendorfer, S.; Backhouse, D. Evaluating ten spectral vegetation indices for identifying rust infection in individual wheat leaves. *Precis. Agric.* **2009**, *10*, 459–470. [CrossRef]
33. Zha, Y.; Gao, J.; Ni, S. Use of normalized difference built-up index in automatically mapping urban areas from TM imagery. *Int. J. Remote Sens.* **2003**, *24*, 583–594. [CrossRef]
34. Ju, C.H.; Tian, Y.C.; Yao, X.; Cao, W.X.; Zhu, Y.; Hannaway, D. Estimating leaf chlorophyll content using red edge parameters. *Pedosphere* **2010**, *20*, 633–644. [CrossRef]
35. Clevers, J.G.P.W.; Gitelson, A.A. Remote estimation of crop and grass chlorophyll and nitrogen content using red-edge bands on Sentinel-2 and-3. *Int. J. Appl. Earth Obs. Geoinf.* **2013**, *23*, 344–351. [CrossRef]
36. Verrelst, J.; Schaepman, M.E.; Koetz, B.; Kneubühler, M. Angular sensitivity analysis of vegetation indices derived from CHRIS/PROBA data. *Remote Sens. Environ.* **2008**, *112*, 2341–2353. [CrossRef]
37. Wang, W.; Cheng, Y.K.; Ren, Y.; Zhang, Z.H.; Geng, H.G. Prediction of Chlorophyll Content in Multi-Temporal Winter Wheat Based on Multispectral and Machine Learning. *Front. Plant Sci.* **2022**, *13*, 896408. [CrossRef]
38. Fu, Z.P.; Yu, S.S.; Zhang, J.Y.; Xi, H.; Gao, Y.; Lu, R.H.; Zheng, H.B.; Zhu, Y.; Cao, W.X.; Liu, X.J. Combining UAV multispectral imagery and ecological factors to estimate leaf nitrogen and grain protein content of wheat. *Eur. J. Agron.* **2022**, *132*, 126405. [CrossRef]
39. Revill, A.; Florence, A.; MacArthur, A.; Hoad, S.P.; Rees, R.M.; Williams, M. The value of Sentinel-2 spectral bands for the assessment of winter wheat growth and development. *Remote Sens.* **2019**, *11*, 2050. [CrossRef]
40. Basso, B.; Cammarano, D.; De, V.P. Remotely sensed vegetation indices: Theory and applications for crop management. *Riv. Ital. Agrometeorol.* **2004**, *1*, 36–53.
41. El-Hendawy, S.E.; Hassan, W.M.; Al-Suhaibani, N.A.; Schmidhalter, U. Spectral assessment of drought tolerance indices and grain yield in advanced spring wheat lines grown under full and limited water irrigation. *Agric. Water Manag.* **2017**, *182*, 1–12. [CrossRef]
42. Bowman, B.C.; Chen, J.; Zhang, J.; Wheeler, J.; Wang, Y.; Zhao, W.; Nayak, S.; Heslot, N.; Bockelman, H.; Bonman, J.M. Evaluating grain yield in spring wheat with canopy spectral reflectance. *Crop Sci.* **2015**, *55*, 1881–1890. [CrossRef]
43. Asner, G.P. Biophysical and biochemical sources of variability in canopy reflectance. *Remote Sens. Environ.* **1998**, *64*, 234–253. [CrossRef]
44. Ustin, S.L. Remote sensing of canopy chemistry. *Proc. Natl. Acad. Sci. USA* **2013**, *110*, 804–805. [CrossRef] [PubMed]
45. Zheng, H.B.; Li, W.; Jiang, J.L.; Liu, Y.; Cheng, T.; Tian, Y.C.; Zhu, Y.; Cao, W.X.; Zhang, Y.; Yao, X. A comparative assessment of different modeling algorithms for estimating leaf nitrogen content in winter wheat using multispectral images from an unmanned aerial vehicle. *Remote Sens.* **2018**, *10*, 2026. [CrossRef]
46. Kooistra, L.; Clevers, J.G.P.W. Estimating potato leaf chlorophyll content using ratio vegetation indices. *Remote Sens. Lett.* **2016**, *7*, 611–620. [CrossRef]
47. Yuan, Y.; Wang, X.F.; Shi, M.M.; Wang, P. Performance comparison of RGB and multispectral vegetation indices based on machine learning for estimating *Hopea hainanensis* SPAD values under different shade conditions. *Front. Plant Sci.* **2022**, *13*, 2615. [CrossRef]
48. Peng, Y.; Fan, M.; Song, J.Y.; Cui, T.T.; Li, H. Assessment of plant species diversity based on hyperspectral indices at a fine scale. *Sci. Rep.* **2018**, *8*, 4776. [CrossRef]
49. Peng, M.M.; Han, W.T.; Li, C.Q.; Huang, S.J. Improving the Spatial and Temporal Estimation of Maize Daytime Net Ecosystem Carbon Exchange Variation Based on Unmanned Aerial Vehicle Multispectral Remote Sensing. *IEEE J. Sel. Top. Appl. Earth Obs. Remote Sens.* **2021**, *14*, 10560–10570. [CrossRef]
50. Liu, S.B.; Jin, X.L.; Nie, C.W.; Wang, S.Y.; Yu, X.; Cheng, M.H.; Shao, M.C.; Wang, Z.X.; Tuohuti, N.; Bai, Y.; et al. Estimating leaf area index using unmanned aerial vehicle data: Shallow vs. deep machine learning algorithms. *Plant Physiol.* **2021**, *187*, 1551–1576. [CrossRef]

Article

CA-BIT: A Change Detection Method of Land Use in Natural Reserves

Bin Jia [1,2], **Zhiyou Cheng** [1,3], **Chuanjian Wang** [1,3,*], **Jinling Zhao** [1,3] and **Ning An** [1,2]

[1] National Engineering Research Center for Analysis and Application of Agro-Ecological Big Data, Anhui University, Hefei 230601, China
[2] School of Electronic Information Engineering, Anhui University, Hefei 230601, China
[3] School of Internet, Anhui University, Hefei 230039, China
* Correspondence: wcj_si@ahu.edu.cn

Abstract: Natural reserves play a leading role in safeguarding national ecological security. Remote sensing change detection (CD) technology can identify the dynamic changes of land use and warn of ecological risks in natural reserves in a timely manner, which can provide technical support for the management of natural reserves. We propose a CD method (CA-BIT) based on the improved bitemporal image transformer (BIT) model to realize the change detection of remote sensing data of Anhui Natural Reserves in 2018 and 2021. Resnet34-CA is constructed through the combination of Resnet34 and a coordinate attention mechanism to effectively extract high-level semantic features. The BIT module is also used to efficiently enhance the original semantic features. Compared with the overall accuracy of the existing deep learning-based CD methods, that of CA-BIT is 98.34% on the natural protected area CD datasets and 99.05% on LEVIR_CD. Our method can effectively satisfy the need of CD of different land categories such as construction land, farmland, and forest land.

Keywords: natural reserves; remote sensing; change detection; deep learning; residual attention network; bitemporal image transformer

Citation: Jia, B.; Cheng, Z.; Wang, C.; Zhao, J.; An, N. CA-BIT: A Change Detection Method of Land Use in Natural Reserves. *Agronomy* **2023**, *13*, 635. https://doi.org/10.3390/agronomy13030635

Academic Editor: Francisco Manzano Agugliaro

Received: 12 January 2023
Revised: 16 February 2023
Accepted: 18 February 2023
Published: 23 February 2023

1. Introduction

Natural reserves are the core carrier of ecological construction and occupy the primary position in safeguarding national ecological security. Anhui Province has implemented the Guidance on Establishing a Natural Reserves System with National Parks as the Mainstay [1] to build a natural reserves system with a reasonable layout, complete types, and perfect functions [2]. The rapid development of industrialization and urbanization has made the contradiction between ecological protection and economic development increasingly prominent. The fragmentation of environmental patches caused by human activities has greatly threatened biodiversity, which seriously affects the management, protection effect, and healthy development of natural reserves in Anhui Province. A large number of natural reserves with large area, wide distribution, complex geographical environment, complicated construction projects, and few supervision staff are present in Anhui Province. Timely detection and supervising of various illegal human activities in the natural reserves are difficult by traditional ground investigation means.

Remote sensing image change detection (CD) identifies changes in the Earth's surface by analyzing satellite images acquired at different times over the same geographical area [3]. Remote sensing image CD technology has always been widely utilized to record and monitor changes and maintain the sustainable development of the earth environment. At present, remote sensing image CD is widely used in many fields, such as urbanization detection [4], environmental monitoring [5–7], disaster assessment [8], and other fields.

The increasing popularity of high-resolution remote sensing images has expanded the potential applications of CD in high-resolution bitemporal images. At present, a deep convolutional neural network (DCNN) is successfully applied to high-resolution remote sensing image analysis and CD tasks due to its strong advanced feature extraction capability [9]. The high-level semantic features of each temporal image are extracted on a CNN-based structure [10–12], and the final change map is generated by clustering or a threshold-based classification method. Obtaining contextual content over space and time is critical to identifying associated changes in multi-time high-resolution images. Thus, the latest CD models have been focused on increasing the receptive field, which is defined as the size of the area in the input where the feature is generated. Therefore, a CD model with stacked convolution layers [13,14] and the application of extended convolution and attention mechanism [15] is proposed. Currently, squeeze and excitation (SE) [16], the bottleneck attention module (BAM) [17], and convolutional block attention module (CBAM) [18] are mainly used on the mobile network design. However, SE only considers internal access information and ignores the importance of the spatial structure of the target in vision. BAM and CBAM attempt to introduce positional information through global pooling on the channels. However, this way can only capture local information rather than obtaining long-range dependent information. By contrast, the coordinate attention mechanism (CA) mechanism captures not only cross-channel information but also orientation perception and position-sensitive information [19]. The attention-based approach is effective in global information modeling but has difficulty relating remote spatiotemporal details.

The recent success of transformers (i.e., nonlocal self-attention) in natural language processing (NLP) has led researchers to apply transformers to various computer vision tasks. Chen Hao et al. [20] used the bitemporal image transformer (BIT) module, which can model the context information in token-based space–time, to enhance the original features. However, they failed to efficiently extract high-level semantic features. Bandara et al. [21] applied a layered transformer encoder (TE) with a lightweight *MLP* decoder to the CD task. The multi-level differential features can be effectively combined, but the spatiotemporal details cannot be efficiently linked.

In summary, we propose a CA-BIT model that combines the residual attention network (ResNet34-CA) and BIT for land use CD in natural reserves. Unlike BIT, our CA-BIT adds the CA to ResNet34 to improve image feature extraction and obtain better change recognition results. The model can provide automatic CD technical support for the daily supervision and environmental supervision of natural reserves. It is also important regarding the timely warning of ecological risks of natural reserves in Anhui Province.

The remainder of the paper is organized as follows. In Section 2, the architectural details of the proposed network are introduced. The study area, data, and experimental environment configuration are presented in Section 3. In Section 4, the proposed method is compared with other different deep learning models, and the CD results are analyzed. Some conclusions are presented in Section 5.

2. Network Model

2.1. CA-BIT Model Overview

The overall framework of the CA-BIT model is shown in Figure 1. First, the images T_1 and T_2 are input into the residual attention network (ResNet34-CA), and the feature map $X^i \in R^{H \times W \times C} (i \in 1, 2)$ is obtained for each image, where H, W, and C denote the height, width, and channel size of the feature map, respectively. Next, the resulting feature map X^i is fed into the BIT module to generate enhanced features X_T^i. Then, fusing the feature map X^i yields the feature map X_{new}^i. The resulting feature map X_{new}^i is fed to the prediction part to produce pixel-level predictions.

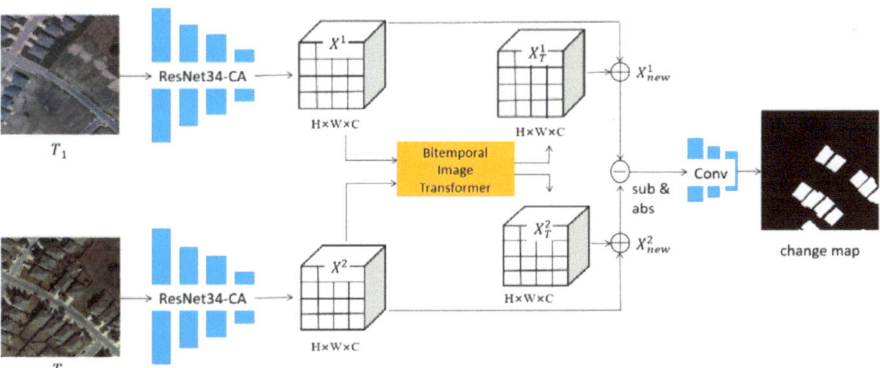

Figure 1. Framework of the CA-BIT model. We first use the image features extracted by the ResNet34-CA backbone. Then, we enhance the image features by the BIT module and fuse the features. Finally, our prediction part produces pixel-level predictions by feeding the computed the feature difference images into a shallow CNN.

2.2. Residual Attention Network

The backbone part of this network is to extract the bitemporal image feature maps through a ResNet34 network [22] combined with a CA residual module (Figure 2). We use CA to build the CA residual module, which adds CA after two layers of 3×3 convolution in the residual block structure. Then, we apply it to the ResNet34 network and construct it as ResNet34-CA to better extract the high-level semantic feature maps.

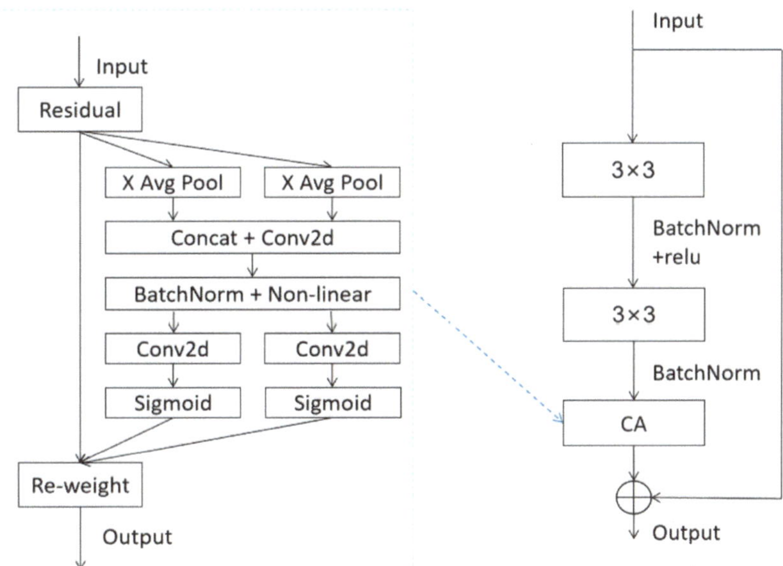

Figure 2. CA residual module. The CA is inserted into the residual block. Note: "X Avg Pool" and "Y Avg Pool" refer to 1D horizontal global pooling and 1D vertical global pooling, respectively.

2.3. Bitemporal Image Transformer

We refer to the BIT model [20], and feature fusion is added to the original BIT module. The overall block diagram is shown in Figure 3. The high-level concept of change objects of interest can be expressed in terms of several visual words, that is, semantic tokens. For this purpose, we represent bitemporal images as several tokens and use a transformer encoder

(TE) [23] to model the context in a compact token-based space–time. We then refine the original features through a transformer decoder (TD) and fuse the original features through a jump connection, where the learned context-rich markers are fed back into the pixel space.

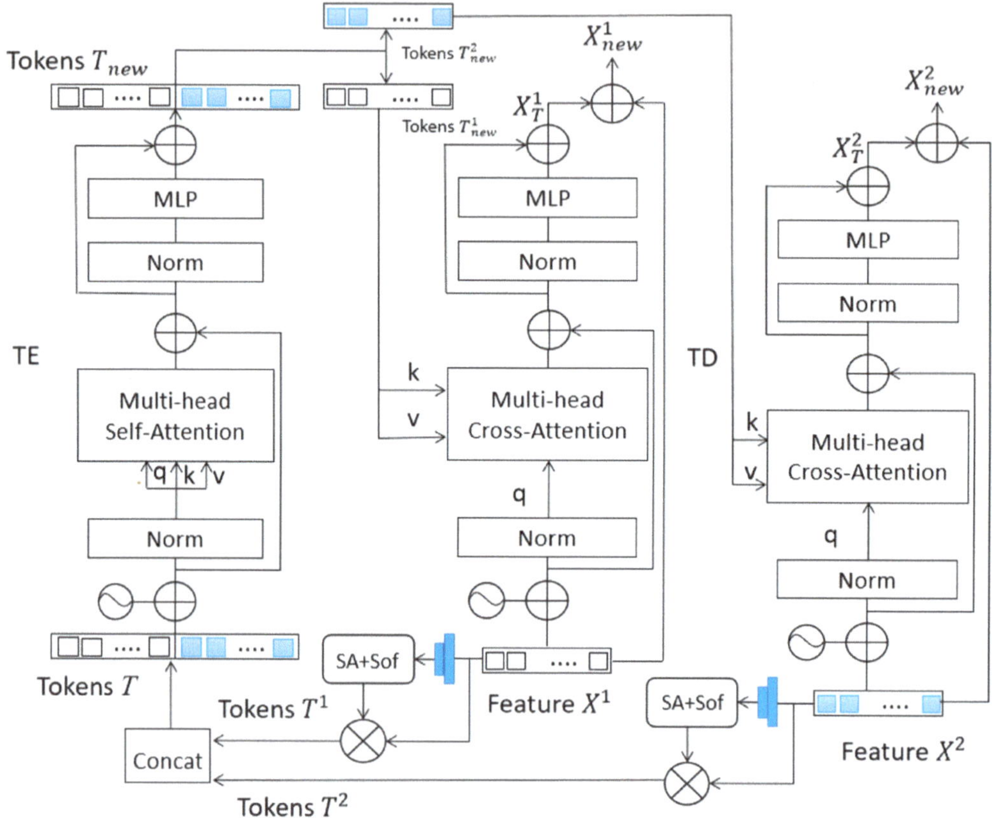

Figure 3. Bitemporal image transformer module and feature fusion.

The BIT learns a set of spatial attention (SA) maps by concentrating the feature mapping space into a set of features to obtain compact semantic tokens. Let $X^1, X^2 \in R^{HW \times C}$ be the bitemporal feature graph of the input. Let $T^1, T^2 \in R^{L \times C}$ be two groups of tokens, where L ($L << HW$) is the size of the set of vocabularies for the token.

For each pixel on the feature map X^1, X^2, we utilize a point-wise convolution to obtain L semantic groups, and each group represents one semantic concept. Then, the Softmax (Sof) function is used on the HW dimension of each semantic group to calculate the SA map. The weighted average sum of the pixels in X^1, X^2 is calculated using the attention mapping to obtain a compact set of vocabularies of size L, that is, semantic tokens $T^1, T^2 \in R^{L \times C}$. The formula is as follows:

$$T^1 = (A^1)^T X^1 = (\sigma(\phi(X^1; W)))^T X^1 \tag{1}$$

$$T^2 = (A^2)^T X^2 = (\sigma(\phi(X^2; W)))^T X^2 \tag{2}$$

where $\phi(\cdot)$ indicates the point-wise convolution with the learnable kernel $W \in R^{C \times L}$, and $\sigma(\cdot)$ is the Sof function that normalizes each semantic group to obtain the attention map $A^1, A^2 \in R^{HW \times L}$. T^1, T^2 is calculated from the multiplication of A^1, A^2 and X^1, X^2.

2.3.1. Transformer Encoder

After concatenating the two semantic token sets T^1, T^2 into one token set $T \in R^{2L \times C}$, we then model the context between these tokens with the TE part. The TE part can utilize the global semantic relationships in token-based space–time to generate context-rich token representations for each time. As shown on the left side of Figure 3, the token T is fed into the TE part to obtain a new token set $T_{new} \in R^{2L \times C}$.

The TE part consists of multi-head self-attention (MSA) and multi-layer perceptron (*MLP*) blocks. Different from the original transformer that uses the post-norm residual unit, we use the validated ViT [24] pre-norm residual unit, that is, the normalized layer is added before the MSA/*MLP*.

In each layer l, the self-attentive input is a triple (query Q, key K, and value V) that is calculated from the input $T^{(l-1)} \in R^{2L \times C}$ as follows:

$$Q = T^{(l-1)} W^q \tag{3}$$

$$K = T^{(l-1)} W^k \tag{4}$$

$$V = T^{(l-1)} W^v \tag{5}$$

where $W_q^{l\ 1}, W_k^{l-1}, W_v^{l\ 1} \in R^{C \times d}$ are the three learnable parameters of the linear projection layers, and d is the channel dimension of the three layers. An attention head is expressed as

$$Att(Q, K, V) = \sigma(\frac{QK^T}{\sqrt{d}})V \tag{6}$$

where $\sigma(\cdot)$ represents the Sof function running on the channel dimension.

The MSA block of the TE part executes multiple stand-alone attention heads in parallel, and it connects the output and then projects it to obtain the final value. The formula is

$$\begin{aligned} MSA&\left(T^{(l-1)}\right) \\ &= Concat(head_1, \cdots, head_h)W^O \\ where\ head_j &= Att\left(T^{(l-1)}W_j^q, T^{(l-1)}W_j^k, T^{(l-1)}W_j^v\right) \end{aligned} \tag{7}$$

where *head* is the number of attention heads, and $W_j^q, W_j^k, W_j^v \in R^{C \times d}, W^O \in R^{hd \times C}$ represent the linear projection matrix.

The *MLP* block of the TE part is composed of two linear transformation layers, with an activation of Gaussian error linear unit [25] in the middle. The dimension of input and output is C, and the inner layer is $2C$. The formula is

$$MLP\left(T^{(l-1)}\right) = GELU\left(T^{(l-1)}W_1\right)W_2 \tag{8}$$

where $W_1 \in R^{C \times 2C}, W_2 \in R^{2C \times C}$ represent the linear projection matrix.

2.3.2. Transformer Decoder

A new token set $T_{new} \in R^{2L \times C}$ is obtained from the TE part, split into two sets of context-rich tokens $T_{new}^1, T_{new}^2 \in R^{L \times C}$, and input into the TD part composed of multi-head cross-attention (MA) and *MLP*. The changes in interest are well revealed by the fact that these context-rich markers contain compact high-level semantic information. Then, we need to project the representation of the concept back into pixel space to obtain pixel-level features. Given a series of features $X^i (i = 1,2)$, the TD part utilizes the relationship between each pixel and the set of tokens T_{new}^i to obtain the refined features X_T^i.

In MA, the keys and values are from the token set T_{new}^i, and the queries are from the image feature X^i. In each layer l, the MA is formally defined as

$$MA\left(X^{i,(l-1)}, T_{new}^i\right) = Concat(head_1, \cdots, head_h)W^O,$$
$$where\ head_j = Att\left(X^{i,(l-1)}W_j^q, T_{new}^i W_j^k, T_{new}^i W_j^v\right) \tag{9}$$

where $W_j^q, W_j^k, W_j^v \in R^{C \times d}, W^O \in R^{hd \times C}$ represent the linear projection matrix, and head is the number of attention heads.

The feature map X^i ($i = 1,2$) and refined features X_T^i obtained from the TD part are fused to obtain the feature map X_{new}^i.

$$X_{new}^i = FeatureX^i + X_T^i \tag{10}$$

2.4. Prediction Part

The CA-BIT model extracts the resulting high-level semantic feature X_{new}^i, which uses a very shallow FCN for change recognition. The predictor head generates the predicted change probability map $P \in R^{H_0 \times W_0 \times 2}$ (H_0 and W_0 are the height and width of the original image, respectively) according to the feature map X_{new}^1, X_{new}^2, which is given by

$$P = \sigma(g(up(D))) = \sigma\left(g\left(up\left(\left|X_{new}^1 - X_{new}^2\right|\right)\right)\right) \tag{11}$$

where $D \in R^{H \times W \times C}$ is the element absolute subtraction value of two feature maps. Upsampling is conducted for $up : R^{H \times W \times C} \rightarrow R^{H_0 \times W_0 \times C}$, and the classifier is changed to $g : R^{H_0 \times W_0 \times C} \rightarrow R^{H_0 \times W_0 \times 2}$. $\sigma(\cdot)$ represents the Sof function.

2.5. Loss Function

In the training phase, the network parameters are optimized by minimizing the cross-entropy loss. The loss function is formally defined as

$$L = \frac{1}{H_0 \times W_0} \sum_{h=1, w=1}^{H,W} l(P_{hw}, Y_{hw}) \tag{12}$$

where $l(P_{hw}, y) = -\log(P_{hwy})$ is the cross-entropy loss, and P_{hw} is the label of the pixel at the position (h,w).

3. Data and Experiments

3.1. Study Area and Data Sources

3.1.1. Overview of the Study Area

Anhui is a provincial administrative region of the People's Republic of China, which is located in the Yangtze River Delta region of East China ($114°54'$—$119°37'$, $29°41'$—$34°38'$), with a total area of 140,100 km^2. The province has established more than 300 protected nature areas at various levels, including nature reserves, geological parks, scenic spots, forest parks, and wetland parks. These areas play an important role in the conservation of biodiversity and protection of natural heritage, which improve ecological environment quality and maintain ecological security.

3.1.2. Experimental Data

The experimental data come from the satellite remote sensing detection project of human activities in natural reserves in Anhui Province, including the high-resolution remote sensing images and the human activity vector database of natural reserves in Anhui Province. The data comprise the high-resolution remote sensing image of Beijing No. 2 acquired in 2018 with a spatial resolution of 1 m and the high-resolution remote sensing image of Gaojing No. 1 acquired in 2021 with a spatial resolution of 0.5 m. Owing to the huge amount of data, the study sample area is reasonably selected according to the distribution of the change type, and the data contain 685,703,168 pixels. The map of Wanfoshan–Longhekou Reservoir (Wanfo Lake) in Shucheng County, Lu'an City, Anhui Province is synthesized using high-resolution data and made by ArcGIS10.4 software. It is shown in Figure 4, and the red box is the sample area (Figure 5).

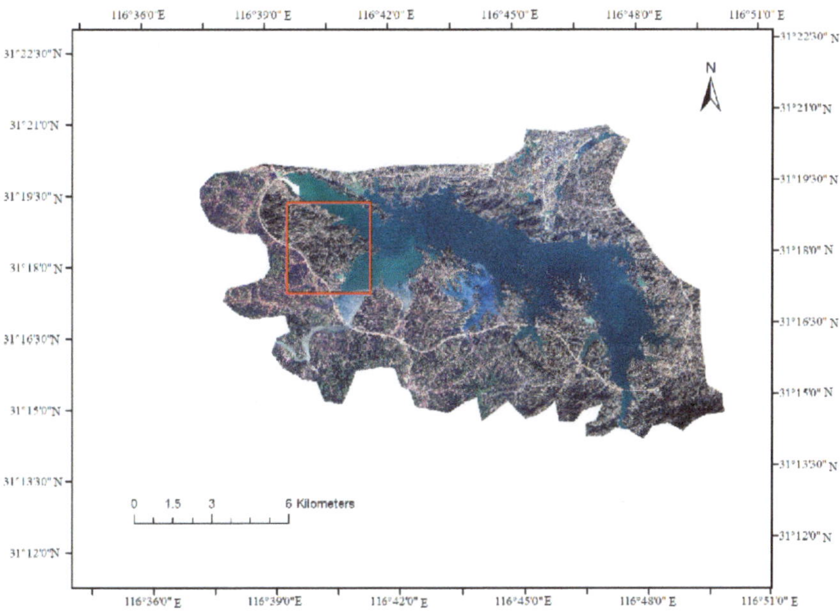

Figure 4. Wanfoshan–Longhekou Reservoir (Wanfo Lake) Scenic Area in Shucheng County, Lu'an City, Anhui Province (Beijing No. 2).

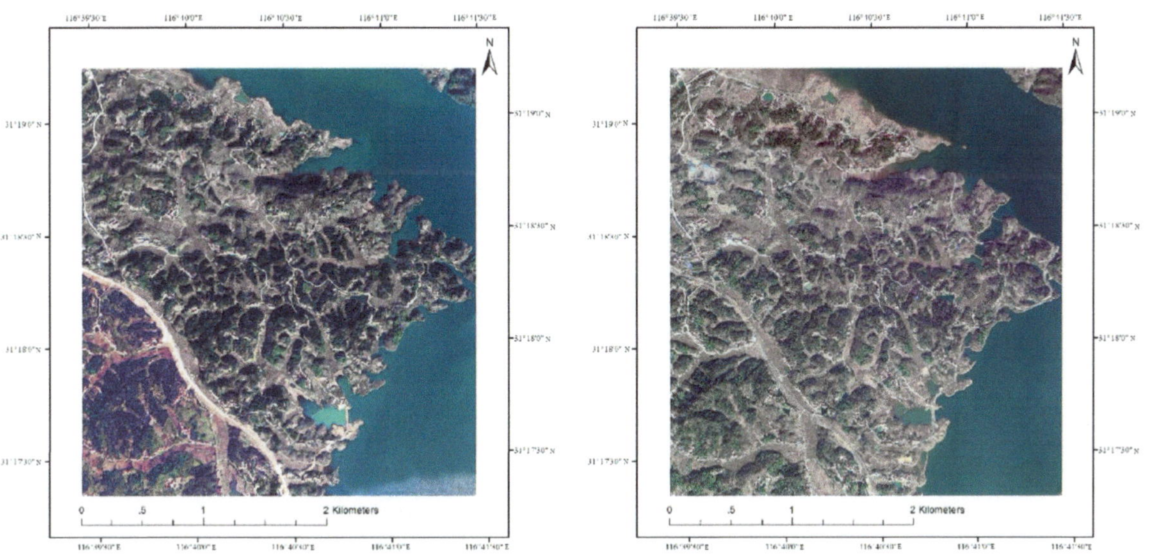

Figure 5. Data of sample area.

The reference basis of land use types in natural reserves in Anhui Province is the Technical Specification for Remote Sensing Monitoring of Human Activities in natural reserves (HJ1156-2021). Considering the actual needs of the Anhui natural satellite remote sensing detection project, 10 different classes are distinguished (mineral resources development, industrial development, energy development, tourism development, transport development, aquaculture development, agricultural development, settlements and other activities, forest management, and engineering construction). Then, by adjusting the change map spot of natural reserves, two types of label data for the change/no change in natural reserves are made, in which the numbers of changed and unchanged pixels are 26,316,288 and 659,386,879, respectively. The remote sensing images and label data are trimmed to a size of 256 pixels × 256 pixels with no overlap, and they are randomly divided into the training set, test set, and validation set in a ratio of 7:2:1, with 7322, 2092, and 1046 pairs of data blocks, respectively.

We also conduct experiments on another public CD dataset. The LEarning, VIsion and Remote sensing CD (LEVIR_CD) is a large-scale public building CD dataset. It contains 637 ultra-high-resolution (VHR, 0.5 m/pixel) Google Earth image patches on a size of 1024 pixels × 1024 pixels. Owing to GPU memory capacity limitations, we cut the image into small blocks of a size of 256 pixels × 256 pixels with no overlap and randomly divide the dataset (train, test, and validation). Therefore, we obtain 7120, 2048, and 1024 pairs of blocks for train, test, and validation, respectively.

3.2. Experimental Environment Configuration and Evaluation Indicators

3.2.1. Experimental Environment Configuration

Our proposed model uses PyTorch as a deep learning framework and uses JetBrains PyCharm 2020 as the development platform; the development language is Python3.8, and all models are trained and tested on computers configured as Intel Core (TM) i7-10700K CPU and NVIDIA GeForce GTX 3080 Ti graphics cards. The same experimental parameters are utilized in the experiments, including momentum (0.99), weight decay (0.0005), batch size (8), running epochs (200), and initial learning (0.01). The gradient descent optimization method used for the optimization model is stochastic gradient descent [26]. The data enhancement strategy includes random flipping, rotating, and Gaussian blur while loading the image pairs. For simplicity, the best model validated after each training phase is used to evaluate the test set.

3.2.2. Evaluating Indicator

We use the accuracy of the evaluation index precision, recall, harmonized mean of precision and recall (*F1*), intersection ratio (*IoU*), and overall accuracy (*OA*) to evaluate the accuracy for quantifying the effect of the evaluation model. The calculation formula of the evaluation index is

$$precision = \frac{TP}{TP + FP} \tag{13}$$

$$recall = \frac{TP}{TP + FN} \tag{14}$$

$$F1 = \frac{2}{recall^{-1} + precision^{-1}} \tag{15}$$

$$IoU = \frac{TP}{TP + FN + FP} \tag{16}$$

$$OA = \frac{TP + TN}{TP + TN + FN + FP} \tag{17}$$

where *TP*, *TN*, *FP*, and *FN* express the number of true positive, true negative, false-positive, and false-negative, respectively.

4. Experimental Results and Analysis

Six models, namely, full convolutional Siamese difference (FC-Siam-Di), full convolutional Siamese connection (FC-Siam-Conc), dual-task constrained deep Siamese convolution network (DTCDSCN), ResNet34_CA network, BIT, and transformer-based Siam network (ChangeFormer), are selected for comparison to verify the confidence of our CD model. We use our natural reserve CD dataset and the LEVIR_CD to verify the above-mentioned CD network.

The results of the accuracy evaluation of the various methods are shown in Table 1. The sample area is analyzed qualitatively to show the results of the CD task more intuitively. Figure 6 illustrates the CD results of the sample areas of the natural reserves. Figure 6a shows the results of visual interpretation of reference changes based on high-resolution remote sensing imagery. Figure 7 illustrates the CD results of the public building CD dataset LEVIR_CD.

Table 1. Evaluation of CD accuracy under different methods.

Method	Natural Reserve CD					LEVIR_CD				
	Precision	Recall	F1	IoU	OA	Precision	Recall	F1	IoU	OA
FC-Siam-Conc	37.25	64.76	47.3	30.98	96.02	91.99	76.77	83.69	71.96	98.49
FC-Siam-Di	39.18	50.8	44.24	28.41	96.47	89.53	83.31	86.31	75.92	98.67
ResNet34-CA	43.55	44.68	44.11	28.29	96.87	86.13	80.63	83.29	71.36	98.35
BIT	69.37	41.55	51.97	35.11	97.20	89.24	89.37	89.31	80.68	98.92
DTCDSCN	52.30	73.3	61.04	43.93	97.42	88.53	86.83	87.67	78.05	98.77
ChangeFormer	68.55	53.89	60.34	43.21	98.04	92.05	88.80	90.40	82.48	99.04
CA-BIT	74.61	60.32	66.71	50.05	98.34	92.30	88.72	90.48	82.61	99.05

Note: All values are reported as a percentage (%). Black in bold indicates the best, and blue in bold is the second.

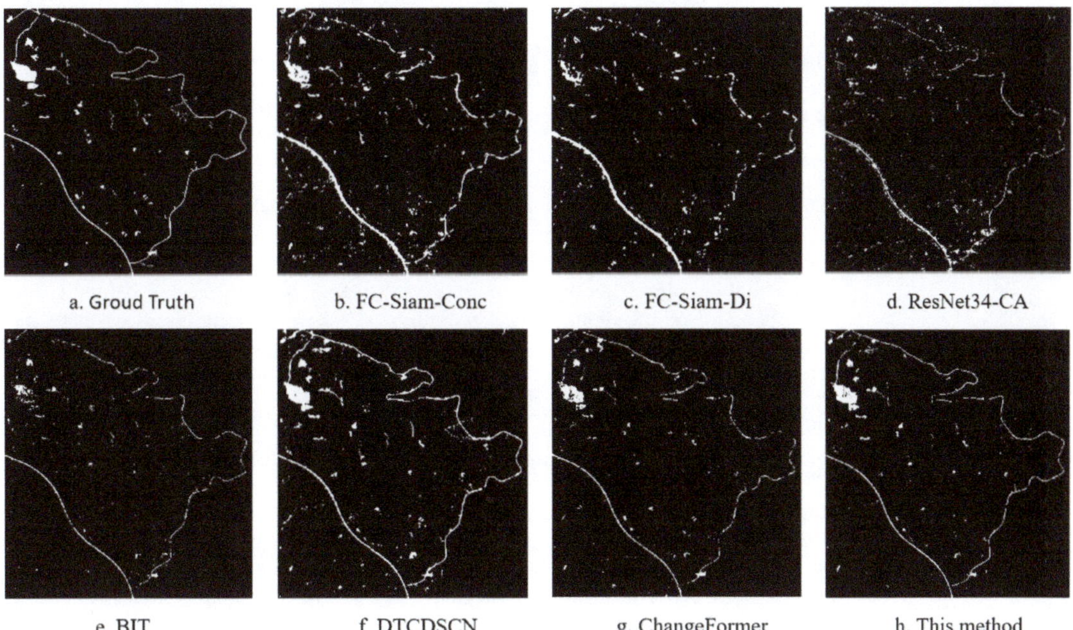

a. Groud Truth   b. FC-Siam-Conc   c. FC-Siam-Di   d. ResNet34-CA

e. BIT   f. DTCDSCN   g. ChangeFormer   h. This method

Figure 6. CD results under different methods (sample area).

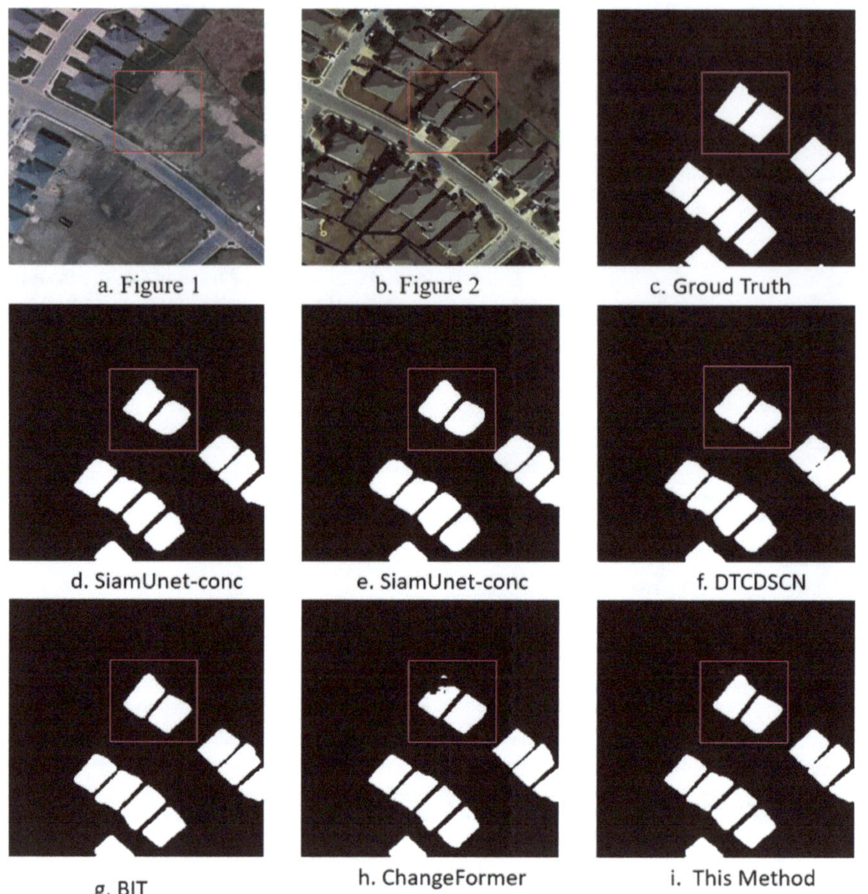

Figure 7. CD results under different methods (LEVIR_CD).

As observed from the comparison effect of Figure 6 and Table 1 in the natural reserves CD dataset, the early full convolution of Siamese (difference and connection) network changes has the lowest detection accuracy. The results of the change identification can only roughly extract the approximate extent of the change, but it is difficult to obtain detailed results of the change of real geographical entities. The ResNet34-CA network with attention mechanism added to the residual network increases the receiving domain but fails to effectively connect the dual time characteristics, which results in low accuracy, and large blocks of change areas are under-judged in change recognition. The BIT network adds a multi-layer transformer after the CNN, and it models the context information in token-based space–time to enhance the original semantic features. However, it ignores the effective extraction of high-level semantic features of the image. Obtaining good boundaries is difficult due to the dramatically changing areas (Figure 6e); the $F1$ is 0.5197 and the IoU is 0.3511. DTCDSCN uses a multi-scale feature connectivity approach to increase channel attention and SA in the deep Siamese FCN, which results in more distinguished features and higher CD accuracy. There is a clear edge profile and also some misjudgements in the change recognition. The transformer-based Siamese network ChangeFormer uses a hierarchical TE and a simple MLP decoder to effectively utilize the dual-temporal image multi-level differential features to achieve high overall accuracy, delicately extract areas of change, and obtain good boundaries. The CD accuracy of the CA-BIT algorithm is the best overall with an *OA* of 0.9834, an *F1* of 0.6671, and an *IoU* of 0.5005, and obtains detailed, realistic results of changes in geographical entities.

Table 1 shows that the present model still achieves better CD accuracy in the LEVIR_CD public datasets than the latest ChangeFormer networks. As shown in Figure 7, compared with the detailed performance in the red box, the present model can better express the small ground material changes, which further proves its effectiveness and feasibility.

In conclusion, the CA-BIT method can be used in nature reserve CD datasets. This model can effectively exert the advantages of high-resolution data to enrich spatial information. It can also obtain good CD results by extracting the semantic features and enhancing bitemporal features. The proposed model has achieved a relatively good CD effect compared with other methods. However, it also needs to be improved in the actual application of human activity detection in natural reserves.

5. Conclusions

Natural reserves have many change categories, and the number of change samples is much smaller than unchanged samples. To address these problems, we propose a remote sensing image CD model CA-BIT that combines ResNet34-CA and BIT. The CA-BIT model not only effectively extracts the global semantic features but also models the context information in token-based space–time to enhance the original semantic features. As a result, it reveals the changes in interest in the presence of dual-temporal images. In the natural conservation area CD datasets, the CA-BIT model works better than other recent deep learning-based models. The CA-BIT model still has better applicability and robustness in the public CD dataset LEVIR_CD.

Author Contributions: Conceptualization, B.J., Z.C. and C.W.; methodology, B.J., Z.C. and C.W.; formal analysis, C.W. and J.Z.; data curation, C.W.; writing—original draft preparation, B.J.; writing—review and editing, Z.C., J.Z. and N.A.; visualization, N.A.; supervision, Z.C. and C.W.; funding acquisition, C.W. and N.A. All authors have read and agreed to the published version of the manuscript.

Funding: This research was funded by the Natural Science Foundation of China (31971789), the Excellent Scientific Research and Innovation Team (2022AH010005), and the National Key Research and Development Project (2017YFB050420).

Institutional Review Board Statement: Not applicable.

Informed Consent Statement: Not applicable.

Data Availability Statement: https://justchenhao.github.io/LEVIR/ (accessed on 10 January 2023).

Acknowledgments: We thank all editors and reviewers for their valuable comments and suggestions, which improved this manuscript.

Conflicts of Interest: The authors declare no conflict of interest.

References

1. Authorless, Guiding Opinions on the Establishment of a Nature Reserve System with National Parks as the Mainstay, issued by the State Office of the Central Government, Green China. 2019; pp. 26–32. Available online: http://www.gov.cn/zhengce/2019-06/26/content_5403497.htm (accessed on 10 January 2023).
2. Yang, Z.; Guo, S.; Lin, S.; Xu, Q. Analysis of the quantity type and spatial overlap of natural reserves in Guangdong Province. *For. Environ. Sci.* **2021**, *37*, 54–60. [CrossRef]
3. Singh, A. Review Article Digital change detection techniques using remotely-sensed data. *Int. J. Remote Sens.* **1989**, *10*, 989–1003. [CrossRef]
4. Huang, X.; Zhang, L.; Zhu, T. Building Change Detection from Multitemporal High-Resolution Remotely Sensed Images Based on a Morphological Building Index. *IEEE J. Sel. Top. Appl. Earth Obs. Remote Sens.* **2013**, *7*, 105–115. [CrossRef]
5. Chen, C.-F.; Son, N.-T.; Chang, N.-B.; Chen, C.-R.; Chang, L.-Y.; Valdez, M.; Centeno, G.; Thompson, C.A.; Aceituno, J.L. Multi-Decadal Mangrove Forest Change Detection and Prediction in Honduras, Central America, with Landsat Imagery and a Markov Chain Model. *Remote Sens.* **2013**, *5*, 6408–6426. [CrossRef]
6. Fang, J.; Zhigang, Y.; Shengcai, G.; Qihu, X.; Yingqin, L.; Yali, S. Remote sensing monitoring of human activities in Guangdong Province in national natural reserves based on high-resolution images. *Guangdong For. Sci. Technol.* **2022**, *2*, 038.

7.   Zhai, P.; Li, S.; Hu, Y. Object-oriented land cover change detection of collaborative optical and radar remote sensing data. *J. Agric. Engineer.* **2021**, *37*, 216–224.

8.   Brunner, D.; Lemoine, G.; Bruzzone, L. Earthquake Damage Assessment of Buildings Using VHR Optical and SAR Imagery. *IEEE Trans. Geosci. Remote Sens.* **2010**, *48*, 2403–2420. [CrossRef]

9.   Shi, W.; Zhang, M.; Zhang, R.; Chen, S.; Zhan, Z. Change Detection Based on Artificial Intelligence: State-of-the-Art and Challenges. *Remote Sens.* **2020**, *12*, 1688. [CrossRef]

10.   Daudt, R.C.; Le Saux, B.; Boulch, A. Fully Convolutional Siamese Networks for Change Detection. In Proceedings of the 2018 25th IEEE International Conference on Image Processing (ICIP), Athens, Greece, 7–10 October 2018; pp. 4063–4067. [CrossRef]

11.   El Amin, A.M.; Liu, Q.; Wang, Y. Zoom out CNNs features for optical remote sensing change detection. In Proceedings of the 2017 2nd International Conference on Image, Vision and Computing (ICIVC), Chengdu, China, 2–4 June 2017; pp. 812–817. [CrossRef]

12.   Lei, T.; Zhang, Y.; Lv, Z.; Li, S.; Liu, S.; Nandi, A.K. Landslide Inventory Mapping From Bitemporal Images Using Deep Convolutional Neural Networks. *IEEE Geosci. Remote Sens. Lett.* **2019**, *16*, 982–986. [CrossRef]

13.   Chen, J.; Yuan, Z.; Peng, J.; Chen, L.; Huang, H.; Zhu, J.; Liu, Y.; Li, H. DASNet: Dual Attentive Fully Convolutional Siamese Networks for Change Detection in High-Resolution Satellite Images. *IEEE J. Sel. Top. Appl. Earth Obs. Remote Sens.* **2021**, *14*, 1194–1206. [CrossRef]

14.   Chen, H.; Shi, Z. A Spatial-Temporal Attention-Based Method and a New Dataset for Remote Sensing Image Change Detection. *Remote Sens.* **2020**, *12*, 1662. [CrossRef]

15.   Liu, Y.; Pang, C.; Zhan, Z.; Zhang, X.; Yang, X. Building Change Detection for Remote Sensing Images Using a Dual-Task Constrained Deep Siamese Convolutional Network Model. *IEEE Geosci. Remote Sens. Lett.* **2020**, *18*, 811–815. [CrossRef]

16.   Hu, J.; Shen, L.; Albanie, S.; Sun, G. Squeeze-and-Excitation Networks. *IEEE Trans. Pattern Anal. Mach. Intell.* **2019**, *42*, 2011–2023. [CrossRef]

17.   Park, J.; Woo, S.; Lee, J.-Y.; Kweon, I.S. BAM: Bottleneck Attention Module. *arXiv* **2018**, arXiv:1807.06514.

18.   Woo, S.; Park, J.; Lee, J.-Y.; Kweon, I.S. CBAM: Convolutional block attention module. In Proceedings of the European Conference on Computer Vision, Munich, Germany, 8–14 September 2018. [CrossRef]

19.   Hou, Q.; Zhou, D.; Feng, J. Coordinate Attention for Efficient Mobile Network Design. *arXiv* **2021**, arXiv:2103.02907.

20.   Chen, H.; Qi, Z.; Shi, Z. Remote Sensing Image Change Detection with Transformers. *IEEE Trans. Geosci. Remote Sens.* **2021**, *60*, 1–14. [CrossRef]

21.   Bandara, W.G.C.; Patel, V.M. A Transformer-Based Siamese Network for Change Detection. *arXiv* **2022**, arXiv:2201.01293.

22.   He, K.; Zhang, X.; Ren, S.; Sun, J. Deep residual learning for image recognition. In Proceedings of the IEEE Computer Society Conference on Computer Vision and Pattern Recognition (CVPR), Las Vegas, NV, USA, 27–30 June 2016; pp. 770–778. [CrossRef]

23.   Vaswani, A.; Shazeer, N.; Parmar, N.; Jones, L.; Gomez, A.; Kaiser, L.; Polosukhin, I. Attention Is All You Need. *arXiv* **2017**, arXiv:1706.03762.

24.   Dosovitskiy, A.; Beyer, L.; Kolesnikov, A.; Weissenborn, D.; Zhai, X.; Unterthiner, T.; Dehghani, M.; Minderer, M.; Heigold, G.; Gelly, S.; et al. An Image is Worth 16×16 Words: Transformers for Image Recognition at Scale. *arXiv* **2020**, arXiv:2010.11929.

25. Hendrycks, D.; Gimpel, K. Gaussian Error Linear Units (GELUs). *arXiv* 2016. [CrossRef]
26. Robbins, H.; Monro, S. A Stochastic Approximation Method. *Ann. Math. Stat.* **1951**, *22*, 400–407. [CrossRef]

Article

Identification of Soybean Planting Areas Combining Fused Gaofen-1 Image Data and U-Net Model

Sijia Zhang [1], Xuyang Ban [2], Tian Xiao [2], Linsheng Huang [2], Jinling Zhao [2,*], Wenjiang Huang [3] and Dong Liang [2,*]

[1] School of Internet, Anhui University, Hefei 230039, China
[2] National Engineering Research Center for Analysis and Application of Agro-Ecological Big Data, Anhui University, Hefei 230601, China
[3] Key Laboratory of Digital Earth Science, Aerospace Information Research Institute, Chinese Academy of Sciences, Beijing 100094, China
* Correspondence: zhaojl@ahu.edu.cn (J.Z.); dliang@ahu.edu.cn (D.L.)

Abstract: It is of great significance to accurately identify soybean planting areas for ensuring agricultural and industrial production. High-resolution satellite remotely sensed imagery has greatly facilitated the effective extraction of soybean planting areas but novel methods are required to further improve the identification accuracy. Two typical planting areas of Linhu Town and Baili Town in Northern Anhui Province, China, were selected to explore the accurate extraction method. The 10 m multispectral and 2 m panchromatic Gaofen-1 (GF-1) image data were first fused to produce training, test, and validation data sets after the min–max standardization and data augmentation. The deep learning U-Net model was then adopted to perform the accurate extraction of soybean planting areas. Two vital influencing factors on the accuracies of the U-Net model, including cropping size and training epoch, were compared and discussed. Specifically, three cropping sizes of 128×128, 256×256, and 512×512 px, and 20, 40, 60, 80, and 100 training epochs were compared to optimally determine the values of the two parameters. To verify the extraction effect of the U-Net model, comparison experiments were also conducted based on the SegNet and DeepLabv3+. The results show that U-Net achieves the highest *Accuracy* of 92.31% with a Mean Intersection over Union (*mIoU*) of 81.35%, which is higher than SegNet with an improvement of nearly 4% in *Accuracy* and 10% on *mIoU*. In addition, the *mIoU* has been also improved by 8.89% compared with DeepLabv3+. This study provides an effective and easily operated approach to accurately derive soybean planting areas from satellite images.

Keywords: remote sensing; data augmentation; parameter optimization; planting area extraction; deep learning

Citation: Zhang, S.; Ban, X.; Xiao, T.; Huang, L.; Zhao, J.; Huang, W.; Liang, D. Identification of Soybean Planting Areas Combining Fused Gaofen-1 Image Data and U-Net Model. *Agronomy* **2023**, *13*, 863. https://doi.org/10.3390/agronomy13030863

Academic Editor: Camilla Dibari

Received: 17 January 2023
Revised: 7 March 2023
Accepted: 13 March 2023
Published: 15 March 2023

1. Introduction

Soybean (*Glycine max* (L.) Merr.) is one of the most important oil-bearing crops around the world. It is also one of China's major food crops, which can be used to provide valuable oil and protein constituents for both humans and livestock. The largest production areas are in China's three northeastern provinces [1]. To make a decision on the cultivation and trade of soybean, it is highly important to figure out the planting areas and spatial distribution. Traditional methods mainly rely on manual measurement and statistical sampling to achieve statistical data, which are time-consuming, susceptible to subjective judgment, labor-intensive, etc. The advancement of earth-observing techniques has greatly improved the monitoring and extraction of crop planting and growth information, especially at a large spatial scale. Remote sensing (RS) technology can provide spatial, spectral, and temporal information of soybean, with macroscopic and dynamic characteristics. When RS technology is applied to the monitoring of soybean, the specific properties and features can be derived from various sensors.

With the development of satellite RS technology, RS images have gradually become a main data source for extracting crop planting information. RS technology has been widely used with soybean crops including estimating the planting areas [2], yield estimation [3], growth monitoring [4], detection and classification of diseases and insect pests [5], etc. It is obvious that a precise understanding of soybean planting areas and their geographical distribution is the prerequisite for various applications. In most previous studies, single-source RS images have been adopted to extract soybean planting areas, mainly including the MODerate Resolution Imaging Spectroradiometer (MODIS) series of satellites of Gaofen, Landsat, and Sentinel. Chang et al. [6] applied a 500-meter time-sequential composite MODIS to estimate corn and soybean areas for the dominant production areas of the USA by taking advantage of low spatial and high temporal resolution MODIS data. Huang et al. [7] identified the corn and soybean cropping areas using the random forest (RF) classifier and multi-temporal 16-meter-resolution GF-1 Wide Field of View (WFV) imagery. Zhong et al. [8] developed an innovative phenology-based classification method to map corn and soybean via over 100 Landsat TM and ETM+ images. Multiple sets of input variables and RF classifiers were jointly used to achieve accuracies higher than 88%. Zhu et al. [9] integrated multi-temporal Sentinel-1/2 microwave and optical multispectral data to map the spatial distribution of soybean through a stepwise hierarchical extraction strategy. The RF proved to be superior to a Back-Propagation Neural Network (BPNN) and Support Vector Machine (SVM). It can be found that multitemporal features of satellite imagery are mainly used to identify soybean information. The overall accuracies are generally lower than 90%.

In recent years, the rapid development of Unmanned Aerial Vehicles (UAV) has provided higher-resolution remote sensing imagery for soybean monitoring. Ranđelović et al. [10] adopted the vegetation indices (VIs) derived from three-channel UAV images of Red, Green, and Blue (RGB) bands to predict soybean plant density. In addition to commonly used machine learning algorithms, some Convolutional Neural Networks (CNN)-based methods have been also used in accurately extracting soybean planting areas. CNN consist of three layers, a convolutional layer, a max pooling layer, and a fully connected layer, which can greatly improve the classification accuracy of single or multiple objects by learning deep features. Habibi et al. [11] used the You Only Look Once version 3 (YOLOv3) object detection algorithm to accurately measure actual soybean plant density, showing higher accuracy than the partial least squares and RF methods. Yang et al. [12] collected RGB and multispectral images using a quad-rotor UAV and employed the U-Net model to improve the soybean recognition accuracy. The results show that the accuracy of the U-Net was the best when compared with DeepLabv3+, RF, and SVM. In comparison, with centimeter-level spatial resolution UAV imagery, the spatial resolution has been improved from a hundred-meter to sub-meter resolution for spaceborne satellites. The improvement in spatial resolution has facilitated the use of deep learning algorithms to improve the extraction accuracies of crop planting areas [13].

There are several influencing factors on the extraction of soybean planting areas, such as RS images, classifiers, planting structure, and the area of the study space, etc. Soybean and corn are the two crops that are generally misclassified during their growing seasons. Both crops show similar spectral and textural features at the initial growth stages but the indicative features can be explored in their middle and late growth seasons [14]. It is highly necessary to find the discriminative features between soybean and corn using remote-sensing technology. The selection of imaging time for RS images is essential for accurately identifying soybean planting areas. In addition to the RS images, classifiers are also factors affecting accuracy. For example, the Simple Non-Iterative Clustering segmentation method and the Continuous Naive Bayes classifier were used to map the soybeans and corn in Paraná state, Brazil, with a minimum global accuracy of 90% [15]. Xu et al. [16] developed a deep learning approach, named DeepCropMapping (DCM), to dynamically map corn and soybean. The DCM model significantly outperformed the Transformer, RF, and Multilayer Perceptron (MLP) methods. To improve the monitoring and classification performance, multitemporal and multispectral RS data are generally input into the classifiers [17]. It is inevitable that high computing power is required to generate a large number of training samples for obtaining reliable and accurate performance. To increase work efficiency in practice, it is significant to map the soybean in a relatively short time. In this study, a 2-meter-resolution fused GF-1 image with the appropriate imaging time was used to identify soybean plating areas via the CNN-based U-Net model. The model is composed of a contracting path and an expansive path, in which the U-shaped architecture and skip connections are the outstanding features [18]. It is simple, efficient, easy to understand, and customizable.

Our highlights for this study are: (1) Fusing the 8-meter multispectral and 2-meter panchromatic GF-1 satellite images to assist in the production of high-quality training samples. (2) Using the intelligible and certified U-Net model to identify soybean planting areas. (3) Comparing and discussing the two crucial parameters, image cropping size and training epoch, which greatly affect the accuracy of the U-Net model to find out the optimal values. The main objective of this study was to optimally determine a U-Net model for accurately extracting soybean planting areas, with the best cropping sizes and training epochs based on high-resolution GF-1 fused imagery. An additional objective was to validate the accuracy of the model by comparing the SegNet and DeepLabv3+. The two networks are mainstream CNN architectures for image segmentation, which have been usually adopted as comparison models.

2. Materials and Methods

2.1. Study Area

The study area was located in Guoyang County, spanning from 33°27′ to 33°47′ N and 115°53′ to 116°33′ E, Bozhou City, Northern Anhui Province, China (Figure 1). The soybean planting area reached 71,086.67 ha in 2022. Linhu Town and Biaoli Town were selected as the study areas, which are important soybean planting areas in Guoyang County. The two towns have flat terrain and four distinct seasons, with an annual rainfall of about 800 mm. A warm–temperate semi-humid monsoon climate and sufficient sunlight are beneficial to the growth of crops such as soybean, wheat, maize, etc.

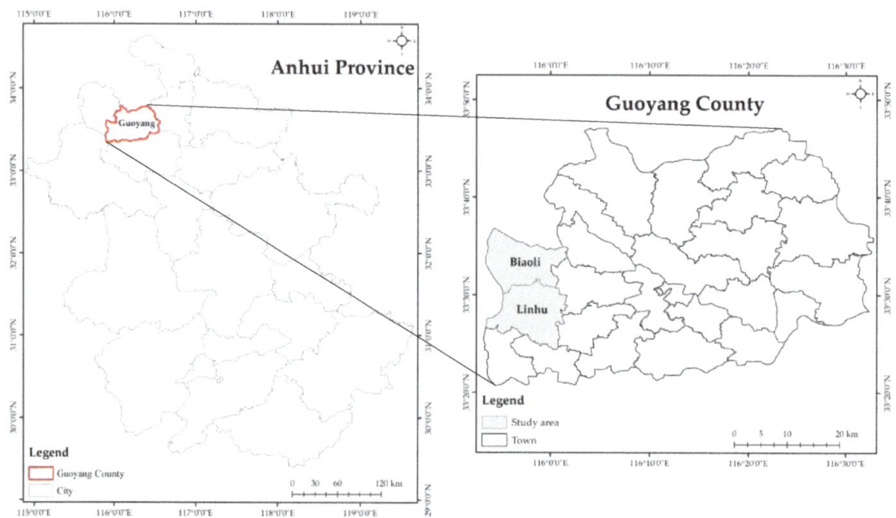

Figure 1. Geographic locations of Biaoli Town and Linhu Town, Guoyang County, Bozhou City, Anhui Province, China.

2.2. Growth Stages of Soybean

Growth stages of soybeans are divided into vegetative growth stages and reproductive growth stages. There are distinctive spectral and spatial characteristics for soybeans at different stages. In addition, other crops (e.g., corn, sorghum, cotton) may also show similar characteristics to soybeans, which causes trouble in accurately identifying soybean planting areas. It is highly important to figure out the phenological stages for selecting appropriate remote-sensing images [4]. When applying remote-sensing technology to the classification of soybeans, it is vital for finding out the optimal imagery. In this study area, soybeans are generally sown in mid-to-late June and harvested in late September or early October (Table 1).

Table 1. Primary soybean growth stages in the study area.

<table>
<tr><td>Month</td><td colspan="2">June</td><td colspan="3">July</td><td colspan="3">August</td><td colspan="3">September</td><td colspan="2">October</td></tr>
<tr><td rowspan="3">Phenological stage</td><td>Mid-</td><td>Late</td><td>Early</td><td>Mid-</td><td>Late</td><td>Early</td><td>Mid-</td><td>Late</td><td>Early</td><td>Mid-</td><td>Late</td><td>Early</td><td>Mid-</td></tr>
<tr><td>Sowing</td><td></td><td></td><td>Third node</td><td></td><td></td><td>Blooming</td><td></td><td></td><td></td><td></td><td></td><td></td></tr>
<tr><td></td><td>Seedling emergence</td><td></td><td></td><td colspan="2">Side branch</td><td></td><td></td><td></td><td>Podding</td><td></td><td>Maturity</td><td></td></tr>
</table>

2.3. Data Sources and Preprocessing

The GF-1 optical satellite was launched on 26 April 2013, which is the first satellite of the China High-resolution Earth Observation System (CHEOS). The satellite has broken through key technologies of high spatial resolution, combination of multispectral and wide coverage, etc., which was widely used in various fields (e.g., modern agriculture, disaster prevention and reduction, resource and environment monitoring) [19–21]. According to Table 1 and the optimal temporal selection for identifying soybeans [15], GF-1 satellite images (Table 2) with 2-meter multispectral and 8-meter panchromatic spectral bands were acquired on 18 August 2019 at blooming and podding stages of soybean. The in situ experiments were also simultaneously carried out to collect the ground truth data. More specifically, the samples were randomly selected in a large soybean field and positioned using a sub-meter GeoXH2008 handheld GPS receiver (Trimble, Westminster, CO, USA). The positioned samples were then overlayed on the GF-1 imagery to select training datasets. The radiometric correction and orthorectification were first carried out, and then the NNDiffuse Pan Sharpening algorithm, proposed by the Rochester Institute of Technology (RIT), was used to generate a 2-meter resolution fused image in ENVI (The Environment for Visualizing Images, Exelis Visual Information Solutions, Inc., Broomfield, CO, USA) 5.3 software.

Table 2. Technical parameters of payloads for GF-1 satellite.

Band Number	Band Name	Spectral Range (μm)	Spatial Resolution (m)	Revisit Cycle	Swath (km)
P	Panchromatic	0.45–0.90	2		
B1	Blue	0.45–0.52			60
B2	Green	0.52–0.59	8	4 days	(Two
B3	Red	0.63–0.69			cameras)
B4	NIR	0.77–0.89			

2.4. Dataset Production

(1) Image cropping

For a satellite remote sensing image, the size and data volume are much larger than a picture photographed by a handheld camera. When a complete image is directly input into a classifier, especially for a CNN-based method, the computing capacity is generally incapable of supporting the complex and interactive algorithms for a computer. In addition, the target objects of soybeans are randomly distributed in a remote-sensing image. Consequently, the image must be cropped into small tiles to accelerate the training and classification. As a kind of data preprocessing, cropping can produce new data by cropping the central pixels of an image. Considering the town-based images in this study, three cropping sizes of 128×128, 256×256, and 512×512 px images were obtained and compared (Figure 2). It can be found that the 256×256 px images were the most suitable size as the training, validation, and test data sets.

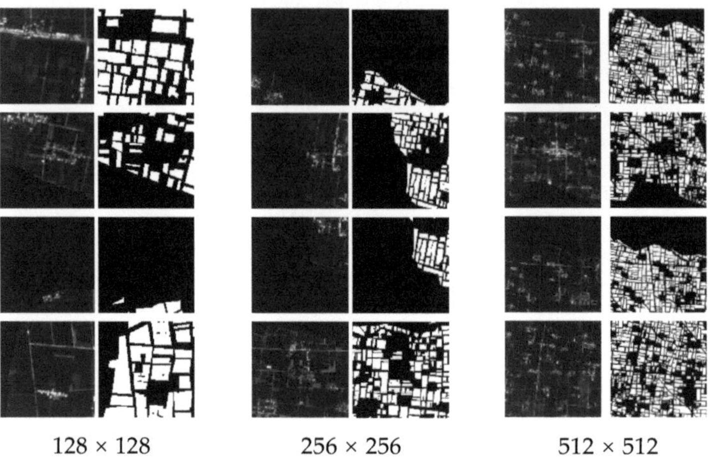

Figure 2. Comparison of three image cropping sizes of 128×128, 256×256, and 512×512 px.

(2) Training and test datasets

Two folders were created to deposit training and test datasets, which were named train and test. The original GF-1 images and datasets used in the experiment were also deposited in both folders. For the training folder, the cropped images, two folders of original images named "src" and labeled samples named "label" were placed in the folder. For the test folder, the folder named "prediction" was also created. When producing the datasets, the cropped images were in one-to-one correspondence with corresponding labels. Each image dataset was named using the natural numbers with the format *.png. Similarly, the labeled samples were also named in accordance with the same rules (Figure 3).

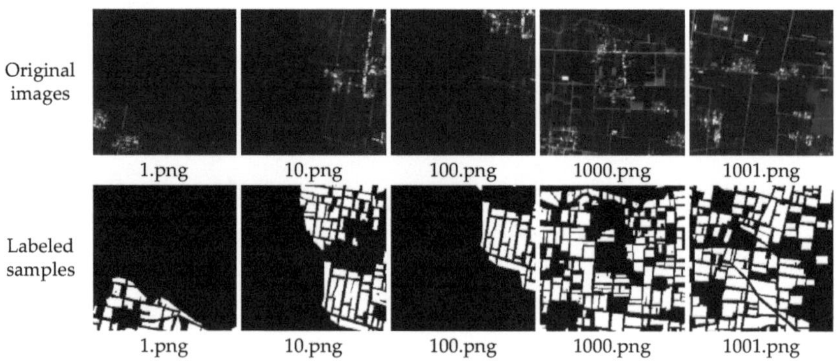

Figure 3. Original images and corresponding labeled samples.

(3) Data normalization

Normalization is an important procedure for optimizing neural networks, which can effectively reduce the influence among different datasets and improve the interactive speed of training models [22]. In this study, the min–max normalization was adopted to perform the data normalization, which can perform the linear transformations of original data [23]. The pixel values of an image in the range (0, 255) were normalized to (0, 1). The specific formula is shown in Equation (1).

$$x_{new} = \frac{x - x_{min}}{x_{max} - x_{min}}$$ (1)

where x_{new} is the normalized value; x is the old value, x_{max} is the maximum value of original pixel; and x_{min} is the minimum value of original pixel.

(4)   Data augmentation

Insufficient training datasets will cause several severe problems such as unbalanced samples, over-fitting, or poor generalization ability in neural networks [24]. It is highly necessary to carry out the data augmentation because we manually produce the training samples. In addition, they are random and irregular for the spatial distribution of soybean planting areas. To fully train the U-Net model, various samples are required to be derived from data augmentation. Specifically, the scale transformation, horizontal and vertical flip, rotation, etc., were used to obtain more training datasets.

2.5. U-Net Model

The U-Net model was first proposed by Ronneberger et al. based on the fully convolutional networks (FCN) [25]. More specifically, convolutional layers and pooling layers are used to extract features and transposed convolutional layers are adopted to revert image sizes. The model was originally applied to the sematic segmentation of medical images. Afterward, it was widely used in crop detection and classification [26–28], due to the advantages of the encoder-decoder structure and skipping networks. Previous studies have shown that U-Net model has strong feature extraction abilities and good segmentation effects even if the sample size is small. As shown in Figure 4, the structure is composed of contracting path (encoder) and expansive path (decoder). In the down-sampling process, every two convolutional layers form a convolution block and there are in total five convolution blocks. In each up-sampling process, the convolution feature maps to the two convolution layers are reduced, whose number is from the encoding path. In the process of feature extraction, the size of a remote sensing image will be reduced when passing through a pooling layer every time. For the U-Net network structure, feature extraction and up-sampling are contacted to form a U-shaped structure.

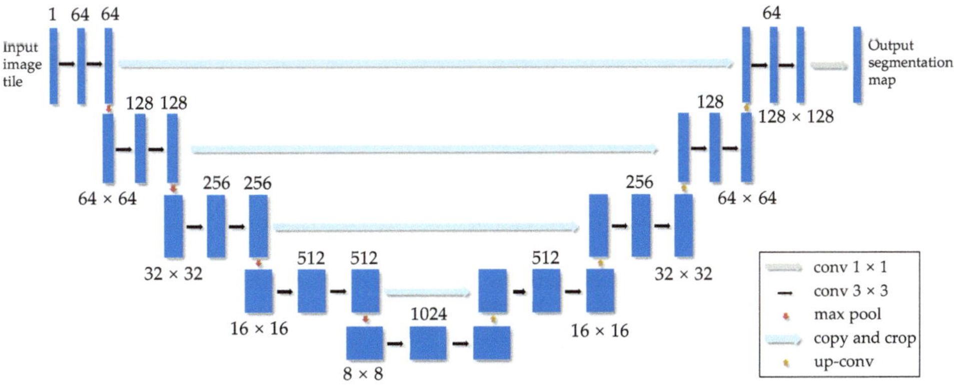

Figure 4. Structure diagram of the U-Net model used in this study.

2.6. Evaluation Metrics

The confusion matrix-based *Accuracy*, *Recall*, *Precision*, F1-score (*F1*), Intersection over Union (*IoU*) and Mean Intersection over Union (*MIoU*) were selected as the evaluation metrics to assess accuracy (Table 3) [29,30]. *Accuracy* is generally used to evaluate the overall accuracy of the model. Recall is used to evaluate the classification effect of test set. *F1* is used to comprehensively evaluate the classification performance. *IoU* is used to calculate the coincidence degree between target and prediction areas. *MIoU* is the evaluation index to assess the prediction results of a network. The closer its value approaches 1, the better the segmentation effect of the network is.

Table 3. Evaluation metrics for assessing accuracy.

Metric	Formula	Variable Explanation
Accuracy	$Accuracy = \frac{TP+TN}{TP+TN+FP+FN}$	*TP* (true positives) are the number of pixels correctly classified as soybean planting areas; *TN* (true negatives) represent the number of background pixels that are predicted as background pixels; *FP* (false positives) represent the number of pixels that are background pixels but misclassified as soybean planting areas; and *FN* (false negatives) are the number of pixels that are soybean planting areas but misclassified as background pixels.
Recall	$Recall = \frac{TP}{TP+FN}$	
Precision	$Precision = \frac{TP}{TP+FP}$	
F1-score	$F1 = 2 * \frac{Precision*Recall}{Precision+Recall}$	
IoU	$IoU = \frac{P_{ii}}{\sum_{j=0}^{k} p_{ij} + \sum_{j=0}^{k} p_{ji} - P_{ii}}$	k is the number of categories; P_{ii} is the number of correctly identified pixels of category i; p_{ij} is the number of pixels that are category i but predicted as the category j; and p_{ji} is the number of pixels that are category j but predicted as the category i.
MIoU	$MIoU = \frac{1}{k} \sum_{i=0}^{k} \frac{P_{ii}}{\sum_{j=0}^{k} p_{ij} + \sum_{j=0}^{k} p_{ji} - P_{ii}}$	

3. Results and Discussion

3.1. Model Training

The curves of loss and accuracy based on the U-Net model are shown in Figure 5. As seen in Figure 5a, the loss value gradually decreases with continuous training. When the training epochs are 20 and 40, there is a little fluctuation but it maintains about 0.0004 when the number of training epochs reaches 50. As shown in Figure 5b, the training accuracy steadily increases with the progression of training. When the epoch reaches 40, the *Accuracy* exceeds 99% and reaches 99.51% as the epoch increases to 60. The *Accuracy* finally reaches 99.69%, indicating that the model has high classification accuracy and achieves a good training performance. Sixty epochs can be a good choice, after comprehensively considering the training accuracy and speed. In addition, during the training process, it is also important to adjust the model parameters. The learning rate and batch size are the two crucial parameters [31]. The learning rate greatly influences the minimum final convergence of the model. In this study, we used Adam to adjust it from the default value and its value was finally set to 0.0001. Batch size affects the model's performance, large values lead to a poor generalization ability but small values affect the model's convergence. It was finally set to four through a debugging process. The epoch of 100 was used to train the model.

We also mapped the training results for epochs 1, 10, 20, 30, 40, 50, 60, 70, 80, 90, and 100. It is found that the prediction result is extremely poor for one epoch. Some samples have no corners and many areas are not predicted and some pixels are not soybean areas, showing that the model is not well-trained at the beginning. With the increase in epochs, the prediction result gradually shows a better performance. When it reaches 60, the classification has already shown a good result, in which the training result reaches a high coincidence with the labeled samples.

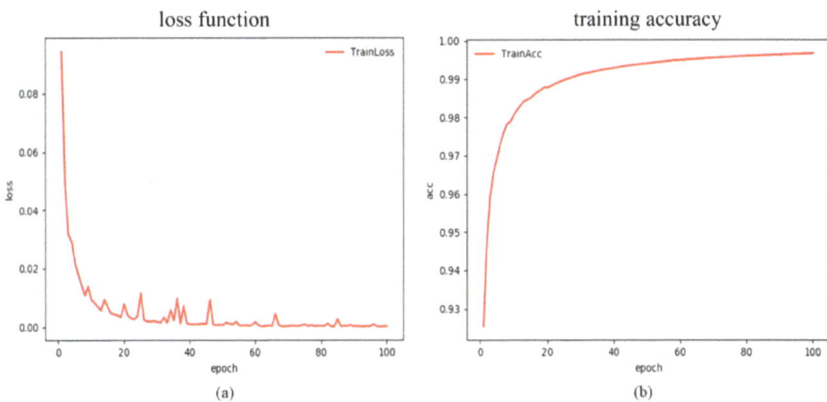

Figure 5. Curves of (**a**) loss and (**b**) accuracy.

3.2. Influence of Cropping Size on Prediction Accuracy

In order to explore the optimal cropping size, three sizes of 128 × 128, 256 × 256, and 512 × 512 px were adopted to comparatively analyze its influence on accuracy. A total of 143 test images for the 128 × 128 px, 42 for the 256 × 256 px, and 12 for the 512 × 512 px were cropped. Figures 6–8 compare the original images, labeled samples, and predicted images for the three cropping sizes, respectively. In order to assure comparability, the network parameters of the U-Net model were the same in the comparison experiments, namely, the batch size was set to four and the epochs were set to 100. As shown in Table 4, all the evaluation metrics are the highest for the cropping size of 256 × 256 px. There are great differences in accuracy among the three cropping sizes, indicating that cropping sizes have a nonnegligible influence on the U-Net model [32].

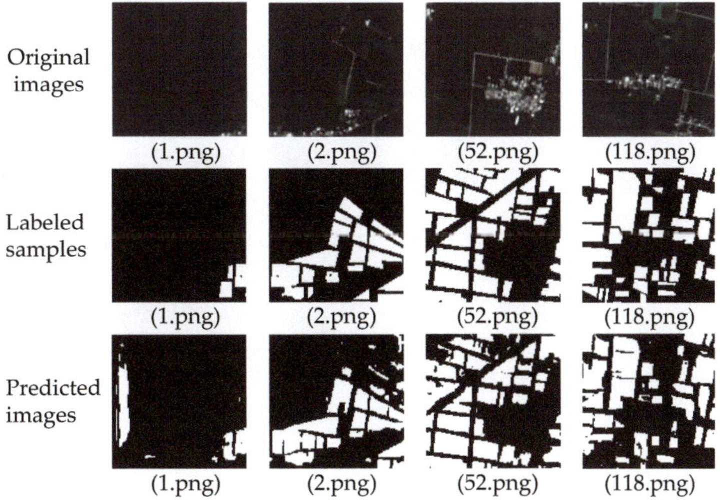

Figure 6. The test results under the 128 × 128 px cropping size.

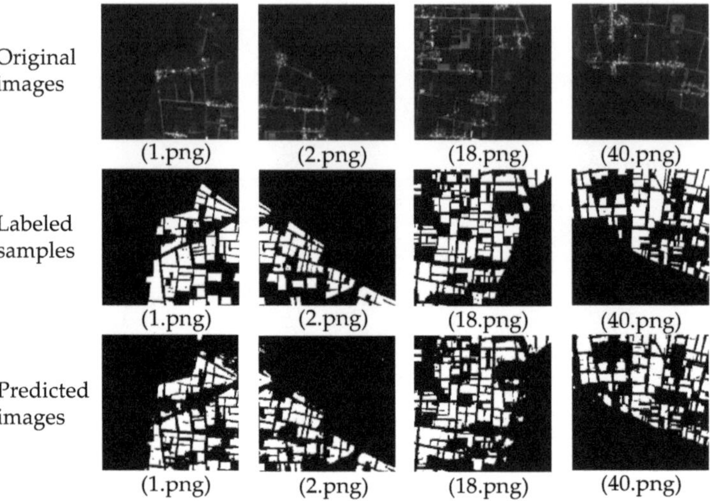

Figure 7. The test results under the 256 × 256 px cropping size.

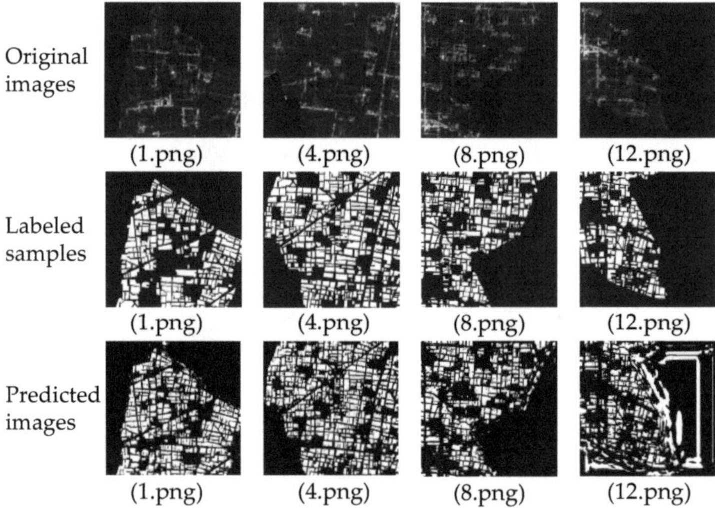

Figure 8. The test results under the 512 × 512 px cropping size.

Table 4. Comparisons of accuracy under different cropping sizes.

Cropping Size	Accuracy (%)	Recall (%)	Precision (%)	F1 (%)	IoU (%)	MIoU (%)
128 × 128	88.75	75.85	80.73	78.21	64.22	75.06
256 × 256	92.31	85.43	82.52	83.95	72.34	81.35
512 × 512	77.46	53.13	71.82	61.08	43.96	58.29

Figure 6 shows some test results under the 128 × 128 px cropping size. The overall prediction effect is fairly good, however, the edges and corners in some images are not accurately predicted. For the predicted image of 1.png, a large area is classified as soybean planting areas but they are not really soybeans according to the labeled image. It can be found that more pixels are misclassified as soybean planting areas. For the predicted image of 2.png, some small areas are identified as soybean planting areas, with an *Accuracy* of 88.75% and *MIoU* of 75.06%. As a whole, the integrity and completeness are worse for the four images.

In comparison with Figure 6, the *Accuracy* reaches 92.31% under the 256 × 256 px cropping size (Figure 7). The *MIoU* reaches 81.35% and the training accuracy of the model reaches 99.69%. It is obvious that there are only subtle differences between predicted images and labeled samples. The primary reason is that there are textural differences in some corners, which causes the model not to achieve the ideal effect. More than 80% of *MIoU* indicates that the overall performance achieves the desired results.

When the cropping size reaches 512 × 512 px, there are large misclassification areas (Figure 8), especially for the 12.png. More pixels are incorrectly predicted, resulting in an *Accuracy* of less than 80%. The *MIoU* is only 58.29%, showing that the training effect is not satisfactory.

3.3. Influence of Training Epochs on Prediction Accuracies

When training the datasets, the predicted map of each epoch was saved. It can be found that with the increase in training epochs, the training effect becomes better. An optimal model can be achieved when the accuracy reached its peak. As shown in Figure 9, the test results are optimal when the training epoch is 60, and the predicted soybean planting areas are closest to the labeled samples. When the training epoch is less than 60, the test results show more or less misclassified pixels to a varying degree, especially for the 24.png test image. A few background areas are misinterpreted as soybean planting areas. This phenomenon shows that insufficient training will lead to underfitting. When the training epoch reaches 80 and 100, the misclassification phenomena are slightly reduced compared with the epoch of 60, however, more predicted pixels are produced. Combined with Figure 5, when the training epoch reaches 60, the accuracy and loss reach the most stable state. Excessive training does not increase the training effect and the epoch of 60 was selected to train and test the model.

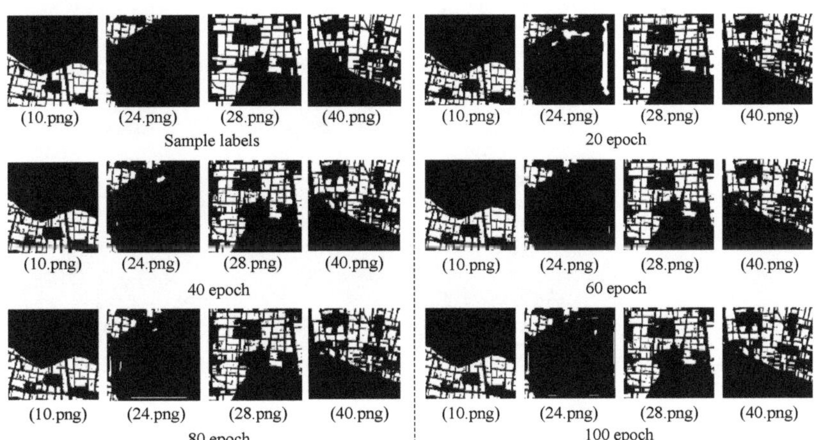

Figure 9. Visual comparison of extracted soybeans at different training epochs.

Figure 10 shows that the *Accuracy*, *Recall*, *F1*, *IoU*, and *MIoU* show a first increase and then a decrease in trend with the increase in training epochs. When the epochs reach 60, the prediction accuracies are the highest with an *MIoU* of 81.36%, showing that the prediction achieves good performance.

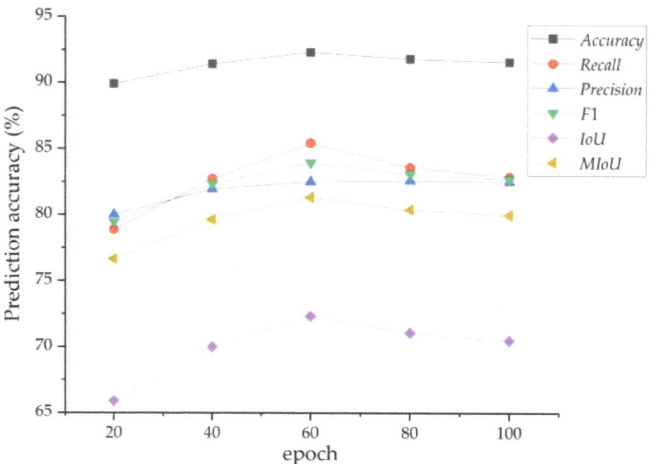

Figure 10. Comparison of prediction accuracies at different training epochs.

3.4. Comparison of Prediction Accuracies among Different Models

SegNet [33] and DeepLabv3+ [34] deep learning network models were selected as the comparative experiments to verify the U-Net model. The training epoch was set to 100, the batch size was set to 4, and the cropping size was 256 × 256 px for the three models. Four typical images of soybean planting areas were selected to compare the three models, and their original images and corresponding labels are shown in Figure 11.

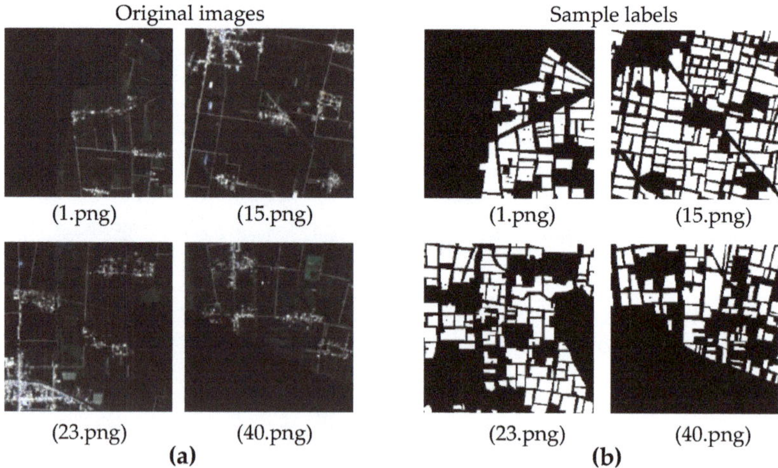

Figure 11. (a) Original sample images and (b) labeled samples.

As shown in Figure 12, the overall spatial distributions of soybean planting areas are similar for the U-Net, SegNet, and DeepLabv3+ models but there are actually significant differences between them. It is obvious that the most similar test results to the labeled samples are from the U-Net model. There are also some additional pixels that are misclassified soybean planting areas for U-Net and SegNet models but the extraction effect of U-Net is better. More soybean planting areas are not fully identified for the SegNet model. The relatively regular shaped soybean fields cannot be embodied. When analyzing the predicted results from DeepLabv3+ model, it can be found that the boundaries of most extracted areas are smoothed and do not show the angular shapes of soybean planting areas. In general, the *Accuracy* of the U-Net model reaches 92% and is improved by 4% than the SegNet model. The *MIoU* reaches 81.35%, 71.31%, and 72.46%, respectively, for the U-Net, SegNet, and DeepLabv3+, which is, respectively, improved by 10% and 9% for U-Net compared with SegNet and DeepLabv3+.

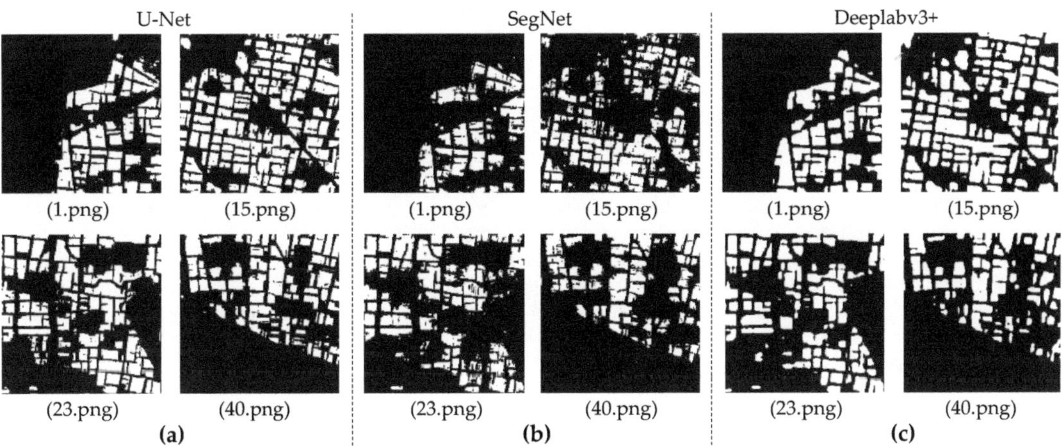

Figure 12. Comparison of identified soybean planting areas using the three models: (**a**) U-Net, (**b**) SegNet, (**c**) DeepLabv3+.

4. Conclusions

It is necessary to detect images with good resolution in an advanced phase of the vegetative cycle. The 2-meter-resolution fused GF-1 imagery and blooming and podding stages are selected to accurately identify soybean planting areas. In addition, considering the parameter optimization of the U-Net model, different cropping sizes, training epochs, and comparative models are adopted to explore the influence of the U-Net model on the identification accuracy. The best U-Net model is obtained by comparing three cropping sizes of 128 × 128, 256 × 256, and 512 × 512 px and 20, 40, 60, 80, and 100 training epochs. The comparative analysis shows that the extraction effect is optimal when the cropping size is 256 × 256 px, with an *Accuracy* of 92.31% and *MIoU* of 81.35%. Afterward, the training epoch was set to 100 and every 20 epochs were extracted as comparisons. When the number of training epochs reaches 60, the prediction result is optimal. After comparing the SegNet and DeepLabv3+ models, the extraction effect of the U-Net model proved to be the best. The cropping images were used to compare the extraction results rather than a complete administrative division. When the separate identified images were merged into an administrative division, there were some problems to be solved, such as seams and boundaries among different soybean fields. This study can provide a methodological reference for extracting planting areas of soybean or other crops. In a future study, sub-meter remote sensing imagery can be used to further validate our model. In addition, improved U-Net models by adding an attention mechanism, residual module, multi-scale features, etc., can be adopted to further improve the identification accuracy of soybean planting areas.

Author Contributions: J.Z. and D.L. conceived and designed the experiments; S.Z. and T.X. performed the experiments; S.Z., X.B. and L.H. analyzed the data; J.Z., W.H. and D.L. wrote and proofread the paper. All authors have read and agreed to the published version of the manuscript.

Funding: This research was funded by the National Key Research and Development Program of China (2019YFE0115200), the Natural Science Foundation of Anhui Province (2008085MF184), Science and Technology Major Project of Anhui Province (202003a06020016), and the Excellent Scientific Research and Innovation Team (2022AH010005).

Data Availability Statement: Not applicable.

Conflicts of Interest: The authors declare no conflict of interest.

References

1. Liu, X.; Jin, J.; Wang, G.; Herbert, S.J. Soybean yield physiology and development of high-yielding practices in Northeast China. *Field Crop. Res.* **2008**, *105*, 157–171. [CrossRef]
2. da Silva Junior, C.A.; Leonel-Junior, A.H.S.; Rossi, F.S.; Correia Filho, W.L.F.; de Barros Santiago, D.; de Oliveira-Júnior, J.F.; Teodoro, P.E.; Lima, M.; Capristo-Silva, G.F. Mapping soybean planting area in midwest Brazil with remotely sensed images and phenology-based algorithm using the Google Earth Engine platform. *Comput. Electron. Agr.* **2020**, *169*, 105194. [CrossRef]
3. Monteiro, L.A.; Ramos, R.M.; Battisti, R.; Soares, J.R.; Oliveira, J.C.; Figueiredo, G.K.; Lamparelli, R.A.C.; Nendel, C.; Lana, M.A. Potential use of data-driven models to estimate and predict soybean yields at national scale in Brazil. *Int. J. Plant Prod.* **2022**, *16*, 691–703. [CrossRef]
4. Diao, C. Remote sensing phenological monitoring framework to characterize corn and soybean physiological growing stages. *Remote Sens. Environ.* **2020**, *248*, 111960. [CrossRef]
5. Santos, L.B.; Bastos, L.M.; de Oliveira, M.F.; Soares, P.L.M.; Ciampitti, I.A.; da Silva, R.P. Identifying nematode damage on soybean through remote sensing and machine learning techniques. *Agronomy* **2022**, *12*, 2404. [CrossRef]
6. Chang, J.; Hansen, M.C.; Pittman, K.; Carroll, M.; DiMiceli, C. Corn and soybean mapping in the United States using MODIS time-series data sets. *Agron. J.* **2007**, *99*, 1654–1664. [CrossRef]
7. Huang, J.; Hou, Y.; Su, W.; Liu, J.; Zhu, D. Mapping corn and soybean cropped area with GF-1 WFV data. *Trans. Chin. Soc. Agric. Eng.* **2017**, *33*, 164–170.
8. Zhong, L.; Gong, P.; Biging, G.S. Efficient corn and soybean mapping with temporal extendability: A multi-year experiment using Landsat imagery. *Remote Sens. Environ.* **2014**, *140*, 1–13. [CrossRef]
9. Zhu, M.; She, B.; Huang, L.; Zhang, D.; Xu, H.; Yang, X. Identification of soybean based on Sentinel-1/2 SAR and MSI imagery under a complex planting structure. *Ecol. Inform.* **2022**, *72*, 101825. [CrossRef]
10. Ranđelović, P.; Đorđević, V.; Milić, S.; Balešević-Tubić, S.; Petrović, K.; Miladinović, J.; Đukić, V. Prediction of soybean plant density using a machine learning model and vegetation indices extracted from RGB images taken with a UAV. *Agronomy* **2020**, *10*, 1108. [CrossRef]
11. Habibi, L.N.; Watanabe, T.; Matsui, T.; Tanaka, T.S. Machine learning techniques to predict soybean plant density using UAV and satellite-based remote sensing. *Remote Sens.* **2021**, *13*, 2548. [CrossRef]
12. Yang, Q.; She, B.; Huang, L.; Yang, Y.; Zhang, G.; Zhang, M.; Hong, Q.; Zhang, D. Extraction of soybean planting area based on feature fusion technology of multi-source low altitude unmanned aerial vehicle images. *Ecol. Inform.* **2022**, *70*, 101715. [CrossRef]
13. Zhao, J.; Wang, J.; Qian, H.; Zhan, Y.; Lei, Y. Extraction of winter-wheat planting areas using a combination of U-Net and CBAM. *Agronomy* **2022**, *12*, 2965. [CrossRef]
14. Shen, Y.; Li, Q.; Du, X.; Wang, H.; Zhang, Y. Indicative features for identifying corn and soybean using remote sensing imagery at middle and later growth season. *Natl. Remote Sens. Bull.* **2022**, *26*, 1410–1422.
15. Paludo, A.; Becker, W.R.; Richetti, J.; Silva, L.C.D.A.; Johann, J.A. Mapping summer soybean and corn with remote sensing on Google Earth Engine cloud computing in Parana state–Brazil. *Int. J. Digital Earth* **2020**, *13*, 1624–1636. [CrossRef]
16. Xu, J.; Zhu, Y.; Zhong, R.; Lin, Z.; Xu, J.; Jiang, H.; Huang, J.; Li, H.; Lin, T. DeepCropMapping: A multi-temporal deep learning approach with improved spatial generalizability for dynamic corn and soybean mapping. *Remote Sens. Environ.* **2020**, *247*, 111946. [CrossRef]
17. Seo, B.; Lee, J.; Lee, K.D.; Hong, S.; Kang, S. Improving remotely-sensed crop monitoring by NDVI-based crop phenology estimators for corn and soybeans in Iowa and Illinois, USA. *Field Crop. Res.* **2019**, *238*, 113–128. [CrossRef]
18. Solórzano, J.V.; Mas, J.F.; Gao, Y.; Gallardo-Cruz, J.A. Land use land cover classification with U-net: Advantages of combining sentinel-1 and sentinel-2 imagery. *Remote Sens.* **2021**, *13*, 3600. [CrossRef]
19. Yao, Y.; Liang, S.; Fisher, J.B.; Zhang, Y.; Cheng, J.; Chen, J.; Jia, K.; Zhang, X.; Bei, X.; Shang, K.; et al. A novel NIR–red spectral domain evapotranspiration model from the Chinese GF-1 satellite: Application to the Huailai agricultural region of China. *IEEE Trans. Geosci. Remote Sens.* **2020**, *59*, 4105–4119. [CrossRef]
20. Sun, W.; Tian, Y.; Mu, X.; Zhai, J.; Gao, P.; Zhao, G. Loess landslide inventory map based on GF-1 satellite imagery. *Remote Sens.* **2017**, *9*, 314. [CrossRef]
21. Li, J.; Chen, X.; Tian, L.; Huang, J.; Feng, L. Improved capabilities of the Chinese high-resolution remote sensing satellite GF-1 for monitoring suspended particulate matter (SPM) in inland waters: Radiometric and spatial considerations. *ISPRS J. Photogramm. Remote Sens.* **2015**, *106*, 145–156. [CrossRef]
22. Sola, J.; Sevilla, J. Importance of input data normalization for the application of neural networks to complex industrial problems. *IEEE Trans. Nucl. Sci.* **1997**, *44*, 1464–1468. [CrossRef]
23. Saranya, C.; Manikandan, G. A study on normalization techniques for privacy preserving data mining. *Int. J. Eng. Technol.* **2013**, *5*, 2701–2704.
24. Wambugu, N.; Chen, Y.; Xiao, Z.; Tan, K.; Wei, M.; Liu, X.; Li, J. Hyperspectral image classification on insufficient-sample and feature learning using deep neural networks: A review. *Int. J. Appl. Earth Observ. Geoinform.* **2021**, *105*, 102603. [CrossRef]

25. Ronneberger, O.; Fischer, P.; Brox, T. U-net: Convolutional networks for biomedical image segmentation. In *International Conference on Medical Image Computing and Computer-Assisted Intervention*; Springer: Cham, Switzerland, 2015; pp. 234–241.
26. Freudenberg, M.; Nölke, N.; Agostini, A.; Urban, K.; Wörgötter, F.; Kleinn, C. Large scale palm tree detection in high resolution satellite images using U-Net. *Remote Sens.* **2019**, *11*, 312. [CrossRef]
27. Liu, G.; Bai, L.; Zhao, M.; Zang, H.; Zheng, G. Segmentation of wheat farmland with improved U-Net on drone images. *J. Appl. Remote Sens.* **2022**, *16*, 034511. [CrossRef]
28. Zhang, S.; Zhang, C. Modified U-Net for plant diseased leaf image segmentation. *Comput. Electron. Agric.* **2023**, *204*, 107511. [CrossRef]
29. Liu, X.; Liu, X.; Wang, Z.; Huang, G.; Shu, R. Classification of laser footprint based on random forest in mountainous area using GLAS full-waveform features. *IEEE J. Sel. Topics Appl. Earth Observ. Remote Sens.* **2022**, *15*, 2284–2297. [CrossRef]
30. Behera, S.K.; Rath, A.K.; Sethy, P.K. Fruits yield estimation using Faster R-CNN with MIoU. *Multimed. Tools Appl.* **2021**, *80*, 19043–19056. [CrossRef]
31. Lee, S.; He, C.; Avestimehr, S. Achieving small-batch accuracy with large-batch scalability via Hessian-aware learning rate adjustment. *Neural Netw.* **2023**, *158*, 1–14. [CrossRef]
32. Dong, X.; Lei, Y.; Wang, T.; Thomas, M.; Tang, L.; Curran, W.J.; Liu, T.; Yang, X. Automatic multiorgan segmentation in thorax CT images using U-net-GAN. *Med. Phys.* **2019**, *46*, 2157–2168. [CrossRef]
33. Badrinarayanan, V.; Kendall, A.; Cipolla, R. Segnet: A deep convolutional encoder-decoder architecture for image segmentation. *IEEE Trans. Pattern Anal. Mach. Intell.* **2017**, *39*, 2481–2495. [CrossRef]
34. Chen, L.C.; Zhu, Y.; Papandreou, G.; Schroff, F.; Adam, H. Encoder-decoder with atrous separable convolution for semantic image segmentation. In Proceedings of the European Conference on Computer Vision (ECCV), Munich, Germany, 8–14 September 2018; pp. 801–818.

MDPI

Article

Estimation of Fv/Fm in Spring Wheat Using UAV-Based Multispectral and RGB Imagery with Multiple Machine Learning Methods

Qiang Wu [1], Yongping Zhang [1,*], Min Xie [1,*], Zhiwei Zhao [1], Lei Yang [2], Jie Liu [3] and Dingyi Hou [4]

[1] College of Agronomy, Inner Mongolia Agricultural University, Huhhot 010019, China
[2] Bayannaoer Academy of Agricultural and Animal Sciences, Bayannaoer 015000, China
[3] School of Biological Science and Technology, Baotou Teachers' College, Baotou 014031, China
[4] Chifeng Forest and Grassland Protection and Development Center, Chifeng 024005, China
* Correspondence: imauzyp@imau.edu.cn (Y.Z.); xiemin@imau.edu.cn (M.X.)

Abstract: The maximum quantum efficiency of photosystem II (Fv/Fm) is a widely used indicator of photosynthetic health in plants. Remote sensing of Fv/Fm using MS (multispectral) and RGB imagery has the potential to enable high-throughput screening of plant health in agricultural and ecological applications. This study aimed to estimate Fv/Fm in spring wheat at an experimental base in Hanghou County, Inner Mongolia, from 2020 to 2021. RGB and MS images were obtained at the wheat flowering stage using a Da-Jiang Phantom 4 multispectral drone. A total of 51 vegetation indices were constructed, and the measured Fv/Fm of wheat on the ground was obtained simultaneously using a Handy PEA plant efficiency analyzer. The performance of 26 machine learning algorithms for estimating Fv/Fm using RGB and multispectral imagery was compared. The findings revealed that a majority of the multispectral vegetation indices and approximately half of the RGB vegetation indices demonstrated a strong correlation with Fv/Fm, as evidenced by an absolute correlation coefficient greater than 0.75. The Gradient Boosting Regressor (GBR) was the optimal estimation model for RGB, with the important features being RGBVI and ExR. The Huber model was the optimal estimation model for MS, with the important feature being MSAVI2. The Automatic Relevance Determination (ARD) was the optimal estimation model for the combination (RGB + MS), with the important features being SIPI, ExR, and VEG. The highest accuracy was achieved using the ARD model for estimating Fv/Fm with RGB + MS vegetation indices on the test sets (Test set MAE = 0.019, MSE = 0.001, RMSE = 0.024, R^2 = 0.925, RMSLE = 0.014, MAPE = 0.026). The combined analysis suggests that extracting vegetation indices (SIPI, ExR, and VEG) from RGB and MS remote images by UAV as input variables of the model and using the ARD model can significantly improve the accuracy of Fv/Fm estimation at flowering stage. This approach provides new technical support for rapid and accurate monitoring of Fv/Fm in spring wheat in the Hetao Irrigation District.

Keywords: Fv/Fm; UAV; remote sensing; wheat; machine learning

Citation: Wu, Q.; Zhang, Y.; Xie, M.; Zhao, Z.; Yang, L.; Liu, J.; Hou, D. Estimation of Fv/Fm in Spring Wheat Using UAV-Based Multispectral and RGB Imagery with Multiple Machine Learning Methods. *Agronomy* **2023**, *13*, 1003. https://doi.org/10.3390/agronomy13041003

Academic Editors: Jinling Zhao and Chuanjian Wang

Received: 20 February 2023
Revised: 25 March 2023
Accepted: 28 March 2023
Published: 29 March 2023

1. Introduction

Spring wheat is an important crop in northern China, with a wide planting range and high yields. The Hetao Irrigation District is one of the major production areas for spring wheat in China, and it plays a significant role in ensuring food security [1]. Chlorophyll fluorescence has proven to be a useful indicator of photosynthetic system health and is widely employed in assessing photosynthesis [2]. Fv/Fm, a commonly used chlorophyll fluorescence parameter, represents the ratio of variable fluorescence (Fv) to maximum fluorescence (Fm) of chlorophyll. Fv represents the fluorescence emitted by open PSII reaction centers, whereas Fm represents the maximum fluorescence emitted by fully open PSII reaction centers under saturating light conditions. Fv/Fm reflects the efficiency of energy transfer within the PSII antenna and the proportion of open PSII reaction centers [3,4]. The

Fv/Fm parameter is critical in understanding the physiological status of plants and is a measure of a plant's capacity to convert light energy into chemical energy, which can offer insights into the health and productivity of crops [5]. In the case of spring wheat, Fv/Fm can be utilized to monitor crop growth and identify potential stress factors, such as water and nutrient deficiencies, that may adversely impact crop yield [6]. Traditionally, Fv/Fm has been estimated using portable instruments such as pulse amplitude modulated (PAM) fluorometers, which necessitate darkening the leaf with a leaf clamp for 15–20 min before measurement. PAM measurements can be laborious and time-consuming, particularly when large areas require monitoring. To overcome these limitations, researchers have explored the use of remote sensing technology.

Remote sensing is a powerful tool for monitoring plant growth. Remote sensing approaches retrieve chlorophyll fluorescence, which is excited by the absorption of sunlight, using spectral reflectance. Since the fluorescence emission spectrum is superimposed on leaf or canopy reflectance that can be obtained by handheld, ground-mounted, aerial, or space-borne sensors, remote sensing technique opens a new way for upscaling chlorophyll fluorescence from leaf to landscape levels. Fv/Fm has been estimated in many studies. Zhao et al. [7] collected spectral data and Fv/Fm values from potato leaves using a hyperspectral imaging system and a closed chlorophyll fluorescence imaging system, decomposed the spectral data by continuous wavelet transform (CWT), and developed an estimation model using partial least squares. Yi et al. [8] used hyperspectral and PAM fluorescence data along with correlation and regression analyses to develop Fv/Fm estimation models for aspen and cherry leaves. Jia et al. [9] calculated vegetation index using hyperspectral data to estimate Fv/Fm for wheat through linear regression. With these Fv/Fm estimation studies, the time-consuming issue of traditional fluorescence determination methods were addressed, but ground-based hyperspectral data collection was not only expensive but also incapable of estimating Fv/Fm at high spatial and temporal resolution. Unmanned aerial vehicle (UAVs) equipped with RGB sensors and multispectral sensors may solve the problem at low cost [10]. Most research on estimating Fv/Fm has primarily relied on ground-based hyperspectral measurements, with few studies employing unmanned aerial vehicles equipped with multispectral and visible sensors. This study fills a critical gap in the literature by demonstrating the feasibility of using UAV-based remote sensing to estimate Fv/Fm, offering new opportunities for efficient and high-throughput monitoring of plant health and productivity.

Machine learning techniques have revolutionized the field of data analysis by identifying complex patterns and trends that are often challenging to detect using traditional methods. In recent years, the application of machine learning methods to analyze data acquired by UAVs has gained significant traction [11–15]. However, the majority of previous studies have focused on using a limited number of machine learning methods (one to four) to estimate the desired parameters, with few investigations comparing the performance of more than twenty different techniques. Given the subtle variation of Fv/Fm and the limited spectral resolution of multispectral and RGB sensors compared to hyperspectral, it is crucial to employ multiple machine learning methods to achieve higher accuracy. This approach allows for the exploitation of the complementarity of various algorithms and enables robust and comprehensive estimation of Fv/Fm from remote sensing data. In this study, the goal is to estimate Fv/Fm in spring wheat using UAV-acquired RGB and MS remote sensing data by multiple machine learning methods to improve accuracy, which is important for rapid and accurate detection of wheat stress and timely adjustment of field management measures.

2. Materials and Methods

2.1. Study Site and Experimental Design

During the wheat flowering stage, both RGB and MS remote sensing images were obtained, resulting in the calculation of 51 vegetation indices (comprising 25 RGB and 26 multispectral). Following this, critical spectral features were extracted, while multicollinearity was eliminated, and feature selection was conducted to estimate the Fv/Fm values. An array of 26 machine learning techniques were utilized, with their respective performances assessed based on accuracy, stability, and interpretability. In conclusion, a high-precision UAV remote sensing monitoring model for the Fv/Fm of spring wheat in the Hetao Irrigation District was developed, thus providing a robust scientific foundation and theoretical underpinning for local agricultural advancement.

The study was conducted from 2020 to 2021 at the experimental field of the Bayannur Institute of Agriculture and Animal Husbandry Science, located in the Inner Mongolia Autonomous Region at 40°04′ N, 10°03′ E and an altitude of 1038 m above sea level (Figure 1). The soil type at the experimental site was loam, and the baseline fertility information is presented in Table 1. A split-plot design was utilized, with nitrogen (N) fertilizer application methods serving as the main plot and cultivar as the subplot. The main plot, which included five levels (CK, N1, N2, N3, and N4), featured various N application methods. N1 (0.8/0.2), N2 (0.7/0.3), N3 (0.5/0.5), and N4 (0.3/0.7) had the same N application rate of 180 kg/ha but differed in seeding fertilizer rates and follow-up fertilizer rates, while CK had no fertilizer applied. The subplot comprised three cultivars of spring wheat: "Baimai 13", "Nongmai 730", and "Nongmai 482". The experiment included a total of 15 treatments with three replications, resulting in 45 experimental plots, each measuring 12 m^2. The plots were arranged in randomized groups. The sowing rate was set at 300 kg/ha. Phosphorus fertilizer was applied as a basal fertilizer during sowing, and no potassium fertilizer was applied during the entire reproductive phase. Three flood irrigations were performed at the tillering, heading, and grain filling stages, each with a volume of 900 m^3/ha.

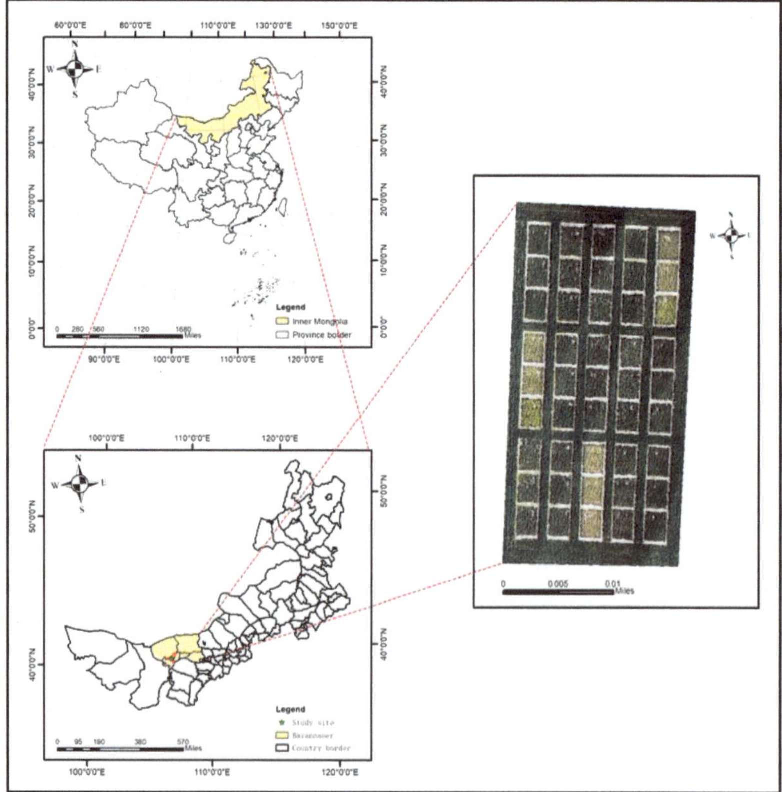

Figure 1. Geographical location of the experimental site.

Table 1. Soil base fertility information of the experimental site.

Year	Organic Matter (g/kg)	Alkaline-N (mg/kg)	Available-P (mg/kg)	Available-K (mg/kg)	PH
2020	14.31	59.63	21.32	117.71	7.62
2021	13.94	56.27	20.83	110.47	7.59

2.2. UAV Multispectral Data Acquisition and Processing

Remote images were obtained during the flowering stage of the wheat plant (12 June 2020; 15 June 2021) using a DJI Phantom 4 multispectral drone (Da-Jiang Innovations, Shenzhen, China). The drone (P4M, Figure 2) features 6 CMOS, including 1 color sensor (ISO: 200–800) for visible (RGB) imaging and 5 monochrome sensors (Table 2) for multispectral (MS) imaging. The images were acquired on clear and windless days, with a fixed and consistent takeoff location. The D-RTK 2 (Da-Jiang Innovations, Shenzhen, China) high-precision GNSS mobile station was utilized to assist in the positioning of the UAV and enhance its positioning accuracy. Prior to takeoff, the UAV was manually placed directly above three reflectivity gray plates of 20%, 40%, and 60%, and reflectivity plate photos were taken. The flight path was automatically planned by DJI GS Pro (Da-Jiang Innovations, Shenzhen, China) after calculating the current solar azimuth, with a flight altitude of 30 m, the ground sampling distance was 1.59 cm/pixel, a heading overlap of 85%, and a collateral overlap of 80%. Following the flight, DJI Terra (Da-Jiang Innovations, Shenzhen, China) was used to perform radiometric correction for multispectral images, followed by image stitching to generate a single-band reflectivity orthophoto. RGB images were stitched to produce color ortho images without a radiation correction step.

Figure 2. A photo for Phantom 4 multispectral.

Table 2. Spectral parameters of multispectral sensors.

Band	Center Wavelength/nm	Bandwidth/nm
Blue (B)	450	16
Green (G)	560	16
Red (R)	650	16
Red Edge (RE)	730	16
Infrared (NIR)	840	26

2.3. Construction and Selection of Spectral Indices

The digital number values for each RGB band and the reflectance of each MS band in each treatment plot were extracted using the zonal statistics function of ENVI. Subsequently, two types of vegetation indices (VIs) were computed, as presented in Tables 3 and 4.

Table 3. Vegetation indices based on RGB digital number value and calculation method.

Index Name	Calculation Formula	References
Red (R), Green (G), Blue (B)	Raw digital number value of each band	/
Normalized Red	$r = R/(R + G + B)$	/
Normalized Green	$g = G/(R + G + B)$	/
Normalized Blue	$b = B/(R + G + B)$	/
Green Red Ratio Index	$GRRI = G/R$	/
Green Blue Ratio Index	$GBRI = G/B$	/
Red Blue Ratio Index	$RBRI = R/B$	/
Excess Red Vegetation Index	$ExR = 1.4 \times r - g$	[16]
Excess Green Vegetation Index	$ExG = 2 \times g - r - b$	[16]
Excess Blue Vegetation Index	$ExB = 1.4 \times b - g$	[16]
Excess Green Minus Excess Red Index	$ExGR = ExG - ExR$	[16]
Woebbecke Index	$WI = (G - B)/(G + R)$	[17]
Normalized Difference Index	$NDI = (r - g)/(r + g + 0.01)$	[17]
Color Intensity	$INT = (R + B + G)/3$	[18]
Green Leaf Index 1	$GLI1 = (2 \times G - R - B)/(2 \times G + R + B)$	[19]
Green Leaf Index 2	$GLI2 = (2 \times G - R + B)/(2 \times G + R + B)$	[19]
Vegetative Index	$VEG = G/\left(R^{(2/3)} \times b^{(1/3)}\right)$	[20]
Combination	$COM = 0.25 \times ExG + 0.3 \times ExGR + 0.33 \times CIVE + 0.12 \times VEG$	[21]
Color Index of Vegetation	$CIVE = 0.441 \times r - 0.811 \times g + 0.3856 \times b + 18.79$	[22]
Normalized Green–Red Vegetation Index	$NGRVI = (G - R)/(G + R)$	[23]
Kawashima Index	$IKAW = (R - B)/(R + B)$	[24]
Visible-band difference vegetation Index	$VDVI = (2 \times g - r - b)/(2 \times g + r + b)$	[25]
Visible Atmospherically Resistance Index	$VARI = (g - r)/(g + r - b)$	[26]
Principal Component Analysis Index	$IPCA = 0.994 \times |R - B| + 0.961 \times |G - B| + 0.914 \times |G - R|$	[27]
Modified Green Red Vegetation Index	$MGRVI = (G^2 - R^2)/(G^2 + R^2)$	[28]
Red Green Blue Vegetation Index	$RGBVI = (G^2 - B \times R)/(G^2 + B \times R)$	[28]

Table 4. Vegetation indices based on MS sing-band reflectance and calculation method.

Index Name	Calculation Formula	References
Difference vegetation index	$DVI = R_{nir} - R_{red}$	[29]
Enhanced Vegetation Index	$EVI = 2.5 \times (R_{nir} - R_{red})/(R_{nir} + 6 \times R_{red} - 7.5 \times R_{blue} + 1)$	[29]
Leaf chlorophyll index	$LCI = (R_{nir} - R_{rededge})/(R_{nir} + R_{red})$	[29]
Green Normalized Difference Vegetation	$GNDVI = (R_{nir} - R_{green})/(R_{nir} + R_{green})$	[30]
Ratio Between NIR and Green Bands	$VI_{(nir/green)} = R_{nir}/R_{green}$	[31]
Ratio Between NIR and Red Bands	$VI_{(nir/red)} = R_{nir}/R_{red}$	[32]
Ratio Between NIR and Red Edge Bands	$VI_{(nir/rededge)} = R_{nir}/R_{rededge}$	[33]
Napierian Logarithm of The Red Edge	$ln_{RE} = 100 \times (ln_{nir} - ln_{red})$	[34]
Modified Soil-Adjusted Vegetation Index 1	$MSAVI1 = (1 + L)\left(\frac{R_{nir} - R_{red}}{R_{nir} + R_{red} + L}\right)(L = 0.1)$	[35]
Modified Soil-Adjusted Vegetation Index 2	$MSAVI2 = R_{nir} + 0.5 - \sqrt{(2 \times R_{nir} + 1)^2 - 8 \times (R_{nir} - R_{red})}/2$	[35]
Optimized Soil-Adjusted Vegetation Index	$OSAVI = (1 + 0.16) \times \frac{(R_{nir} - R_{red})}{(R_{nir} + R_{red} + 0.16)}$	[36]
Modified Triangular Vegetation Index 2	$MTVI2 = \frac{1.5 \times [1.2 \times (R_{nir} - R_{green}) - 2.5 \times (R_{red} - R_{green})]}{\sqrt{(2 \times R_{nir} + 1)^2 - (6 \times R_{nir} - 5 \times \sqrt{R_{red}}) - 0.5}}$	[37]
Normalized Difference Red Edge Index	$NDRE = \frac{(R_{nir} - R_{rededge})}{(R_{nir} + R_{rededge})}$	[38]
Normalized Difference Vegetation Index	$NDVI = \frac{(R_{nir} - R_{red})}{(R_{nir} + R_{red})}$	[39]

Table 4. *Cont.*

Index Name	Calculation Formula	References
Modified Simple Radio	$MSR = (R_{nir} - R_{red} - 1)/\left(\sqrt{R_{nir} + R_{red}} + 1\right)$	[40]
Soil-Adjusted Vegetation Index	$SAVI = \frac{(R_{nir} - R_{red})}{(R_{nir} + R_{red} + 0.5)} \times (1 + 0.5)$	[41]
Simplified Canopy Chlorophyll Content Index	$SCCCI = \frac{NDRE}{NDVI}$	[42]
Modified Chlorophyll Absorption Reflectance Index	$MCARI = \left(R_{rededge} - R_{red} - 0.2 \times \left(R_{rededge} - R_{green}\right)\right) \times \left(\frac{R_{rededge}}{R_{red}}\right)$	[43]
Structure-Insensitive Pigment Index	$SIPI = \frac{(R_{nir} - R_{blue})}{(R_{nir} + R_{red})}$	[44]
Transformed Chlorophyll Absorption Reflectance Index	$TCARI = 3 \times \left(\left(R_{rededge} - R_{red}\right) - 0.2 \times \left(R_{rededge} - R_{green}\right) \times \left(\frac{R_{rededge}}{R_{red}}\right)\right)$	[45]
Normalized Difference Index	$NDI = \frac{(R_{nir} - R_{rededge})}{(R_{nir} + R_{red})}$	[46]
Red-Edge Chlorophyll Index 1	$CI1 = \frac{R_{nir}}{R_{rededge}} - 1$	[47]
Red-Edge Chlorophyll Index 2	$CI2 = \frac{R_{rededge}}{R_{green}} - 1$	[48]
Modified Chlorophyll Absorption Reflectance Index 2	$MCARI2 = 1.5 \times \frac{\left(2.5 \times \left(R_{nir} - R_{rededge}\right) - 1.3 \times \left(R_{nir} - R_{green}\right)\right)}{\left(2 \times (R_{nir} + 1)^2 - \left(6 \times R_{nir} - 5 \times (R_{red})^2\right) - 0.5\right)}$	[49]
TCARI/OSAVI	$\frac{TCARI}{OSAVI}$	[49]
MCARI/OSAVI	$\frac{MCARI}{OSAVI}$	[49]

2.4. Fluorescence Data Acquisition and Processing

The collection of fluorescence data was synchronized with the UAV flight, and the fluorescence information of wheat canopy leaves was obtained using the Handy PEA plant efficiency analyzer (Hansatech Instruments Ltd., Norfolk, UK). In each plot, 20 leaves were randomly collected, and the average value was adopted as a representative value of the plant. Prior to collection, the target leaves were subjected to a dark treatment for 20 min using the leaf clips that were provided with the instrument.

2.5. Construction of Regression Model

In this study, 26 machine learning regression models (listed in Table 5) were developed using PyCaret to estimate Fv/Fm. PyCaret is a user-friendly, open source, low-code machine learning library in Python that enables users to easily prepare data, train and evaluate machine learning models, and deploy models to production. PyCaret offers various features for data preparation, feature engineering, model training and evaluation, model interpretation, and model deployment. Additionally, it includes built-in visualizations and interactive plots that help users to interpret model results.

Table 5. Machine learning models.

Model	Abbreviation	References
AdaBoost Regressor	ADA	[50]
Automatic Relevance Determination	ARD	[51]
Bayesian Ridge	BR	[52]
CatBoost Regressor	CatBoost	[53]
Decision Tree Regressor	DT	[54]
Dummy Regressor	Dummy	[55]
Elastic Net	EN	[56]
Extra Trees Regressor	ET	[57]
Extreme Gradient Boosting	EGB	[58]
Gradient Boosting Regressor	GBR	[59]
Huber Regressor	Huber	[60]
K Neighbors Regressor	KNN	[61]
Kernel Ridge	KR	[62]
Lasso Least Angle Regression	LLAR	[63]
Lasso Regression	Lasso	[64]

Table 5. *Cont.*

Model	Abbreviation	References
Least Angle Regression	LAR	[63]
Light Gradient Boosting Machine	LGBM	[65]
Linear Regression	LR	[66]
Multilayer Perceptron Regressor	MLP	[67]
Orthogonal Matching Pursuit	OMP	[68]
Passive Aggressive Regressor	PAR	[69]
Random Forest Regressor	RF	[70]
Random Sample Consensus	RANSC	[71]
Ridge Regression	Ridge	[72]
Support Vector Machine Regression	SVM	[73]
TheilSen Regressor	TR	[74]

To prepare the data, a normalization technique was applied to transform the data into a fixed range between 0 and 1, thereby ensuring that all features were on the same scale. Multicollinearity, which refers to high correlation between multiple features, was addressed by removing highly correlated features to ensure data stability.

Feature selection was performed to select key features and reduce noise, thereby enhancing the accuracy and efficiency of the algorithm. Once all the models were built, the model with the highest accuracy was selected based on the accuracy ranking, and hyperparameter optimization was conducted to further improve model accuracy. This study used the feature selection scheme of the embedding method, implemented by calling the SelectFromModel API in sklearn, relying on the algorithmic model of random forests.

2.6. Segmentation of Dataset and Accuracy Evaluation

The 90 samples were randomly partitioned into a training set and a test set in a 0.7/0.3 ratio; a K-fold cross-validation (K = 5) was employed to train and optimize the model. Seven indicators were utilized to assess the accuracy of the model in the test set:

MAE (Mean Absolute Error) is a measure of the average magnitude of the errors in a set of predictions, without considering their direction. It measures the average absolute difference between the actual and predicted values. A smaller MAE indicates a more accurate prediction;

MSE (Mean Squared Error) is a measure of the average magnitude of the errors in a set of predictions, considering both the magnitude and direction of the errors. It measures the average of the squared differences between the actual and predicted values. A smaller MSE indicates a more accurate prediction;

R^2 (Coefficient of Determination) is a statistical measure that represents the proportion of the variance in the dependent variable that is predictable from the independent variables. In regression analysis, R^2 is used to evaluate the goodness of fit of the model. Generally, it ranges from 0 to 1, with a higher value indicating a better fit. An R^2 of 1 indicates that the model perfectly predicts the target variable, while an R^2 of 0 indicates that the model does not explain any variance in the target variable;

RMSE (Root Mean Squared Error) is the square root of MSE. It is a measure of the average magnitude of the errors in a set of predictions, considering both the magnitude and direction of the errors. A smaller RMSE indicates a more accurate prediction;

RMSLE (Root Mean Squared Logarithmic Error) is similar to RMSE, but instead of taking the difference between the actual and predicted values, it takes the logarithmic difference. It is used in cases where the target variable has a skewed distribution;

MAPE (Mean Absolute Percentage Error) is a measure of the average error as a percentage of the actual values. It measures the average percentage difference between the actual and predicted values. A lower MAPE indicates a more accurate prediction;

TT (Total Time) is defined as the amount of time spent on building the machine learning model. The value of TT represents the computational cost of constructing the model, including the time spent on training and validating the model. A smaller value of TT indicates that the model has a lower computational cost and can be built more efficiently. This can be beneficial in scenarios where the model needs to be constructed in a timely manner, or when computational resources are limited. A lower TT value also indicates that the model may be more scalable and can be trained on larger datasets without excessive computational cost.

$$\text{MAE} = \frac{1}{n} \sum_{i=1}^{n} |y_i - \hat{y}| \tag{1}$$

$$\text{MSE} = \frac{\sum_{i=1}^{n} (y_i - \hat{y}_i)^2}{n} \tag{2}$$

$$\text{RMSE} = \sqrt{\frac{\sum_{i=1}^{n} (y_i - \hat{y}_i)^2}{n}} \tag{3}$$

$$\text{R}^2 = 1 - \frac{\sum_{i=1}^{n} (y_i - \hat{y}_i)^2}{\sum_{i=1}^{n} \left(y_i - \bar{y}_i\right)^2} \tag{4}$$

$$\text{RMSLE} = \sqrt{\frac{1}{n} \sum_{i=1}^{n} \left(log(y_i + 1) - log(\hat{y}_i + 1)\right)^2} \tag{5}$$

$$\text{MAPE} = \frac{100\%}{n} \sum_{i=1}^{n} \left| \frac{\hat{y}_i - y_i}{y_i} \right| \tag{6}$$

where n is the number of samples, y_i is the observed value, $\bar{y}_i$ is the mean of the observed value, and $\hat{y}_i$ is the predicted value.

3. Results

3.1. Basic Statistical Information of the Fv/Fm Dataset

As shown in Table 6, the basic statistics of the measured Fv/Fm, indicating a range of values with a minimum of 0.550, a maximum of 0.848, a mean of 0.773, a standard deviation of 0.081, and a coefficient of variation (CV%) of 10.4. These statistics suggest that the Fv/Fm measurements demonstrate moderate variability within the total dataset. Both the training and test sets were derived from the total dataset and displayed similar ranges of Fv/Fm values. Specifically, the training set showed a minimum of 0.551, a maximum of 0.846, a mean of 0.775, a standard deviation of 0.077, and a CV% of 9.900, while the test set demonstrated a minimum of 0.550, a maximum of 0.848, a mean of 0.768, a standard deviation of 0.090, and a CV% of 11.700. The distributions of Fv/Fm values in the training and test sets were comparable, with a slightly higher mean and lower CV for the training set than the test set. These findings indicate that the training and test sets are representative of the total dataset and can be utilized for model development and validation.

Table 6. Basic statistics of the measured Fv/Fm.

Dataset	Minimum	Maximum	Mean	STDEV	CV (%)
Total Dataset	0.550	0.848	0.773	0.081	10.4
Training set	0.551	0.846	0.775	0.077	9.900
Test set	0.550	0.848	0.768	0.090	11.700

3.2. Correlation Analysis of Fv/Fm with Multispectral and RGB Vegetation Indices

The correlation coefficients between Fv/Fm values and MS, RGB vegetation indices are shown in Table 7. The highest correlation coefficients of multispectral vegetation indices were NDVI, followed by SIPI, EVI, and MSAVI2. The highest correlation coefficients of RGB vegetation indices were RGBVI, followed by VDVI, GLI, and CIVE. Most of the multispectral vegetation indices and half of RGB vegetation indices showed good correlation with Fv/Fm. Overall, the correlation between the multispectral vegetation indices and Fv/Fm was higher than that of the RGB vegetation indices.

Table 7. Correlation coefficients between MS, RGB vegetation indices, and Fv/Fm.

Multispectral Vegetation Indices	Correlation Coefficient	RGB Vegetation Indices	Correlation Coefficient
DVI	0.857	b	−0.679
EVI	0.869	g	0.827
NDVI	0.899	r	0.283
GNDVI	0.888	GRRI	0.309
NDRE	0.850	GBRI	0.737
LCI	0.797	RBRI	0.502
OSAVI	0.892	INT	−0.816
VI(NIR/G)	0.784	GRVI	0.319
VI(NIR/R)	0.724	NDI	−0.324
VI(NIR/RE)	0.807	WI	0.769
lnRE	0.86	IKAW	0.501
MSAVI1	0.895	GLI	0.832
MSAVI2	0.896	GLI2	−0.126
MTVI2	0.863	VARI	−0.55
MSR	0.849	ExR	−0.435
SAVI	0.88	ExG	0.827
SCCCI	0.756	ExB	−0.743
MCARI	−0.793	ExGR	0.829
MCARI2	0.754	VEG	0.767
TCARI	−0.71	IPCA	−0.821
NDI	0.865	CIVE	−0.831
CL1	0.807	COM	0.827
CL2	0.835	RGBVI	0.843
SIPI	0.898	MGRVI	0.324
TCARI/OSAVI	−0.784	VDVI	0.832
MCARI/OSAVI	−0.893		

3.3. Important Features Selected after Data Pre-Processing

The process of feature selection was conducted to identify the most crucial variables that significantly influence the estimation of the chlorophyll fluorescence parameter Fv/Fm in spring wheat using UAV remote sensing. Table 8 illustrates the significant features that were selected after the application of data pre-processing techniques, including the visible light vegetation indices (RGB), multispectral vegetation indices (MS), and a combination of both (RGB + MS). The pre-processing methodology involved eliminating feature collinearity and implementing tree-based feature selection methods. The table highlights that RGBVI and ExR were the vital features for the RGB dataset, MSAVI2 was the only critical feature for the MS dataset, and SIPI, ExR, and VEG were the essential features for the RGB + MS dataset.

Table 8. Important features selected after data pre-processing.

VIs Type	Important Features
RGB	RGBVI, ExR
MS	MSAVI2
RGB + MS	SIPI, ExR, VEG

3.4. Model Based on RGB VIs Development and Evaluation

Table 9 presented the results of the performance evaluation of 26 machine learning models for estimating vegetation indices using RGB vegetation indices. The models were ranked according to their R^2 accuracy scores, with the top performing model listed first.

Evaluation metrics such as mean absolute error (MAE), mean squared error (MSE), root mean squared error (RMSE), relative mean squared logarithmic error (RMLSE), mean absolute percentage error (MAPE), and computation time (TT) were used to assess model performance. The gradient boosting regression (GBR) model achieved the highest accuracy with an R^2 score of 0.800, followed closely by the random forest (RF) model with an R^2 score of 0.795. Most of the other models performed relatively poorly, with R^2 scores ranging from 0.789 to −116.050. Notably, the worst performing models, including lasso, elastic net, least angle regression (LLAR), dummy, support vector machine (SVM), and kernel ridge regression (KR), had negative R^2 scores. In addition to the R^2 scores, the evaluation metrics indicated that the best performing models also had the lowest MAE, MSE, RMSE, RMLSE, and MAPE scores, demonstrating the models' ability to make accurate predictions with low error rates. However, there was considerable variation in computation time among the models, with some taking significantly longer than others. In conclusion, the results suggest that GBR models are the most accurate and efficient for estimating Fv/Fm using RGB vegetation indices.

Table 9. Accuracy assessment of different estimation models based on RGB vegetation indices.

Model	MAE	MSE	RMSE	R^2	RMLSE	MAPE	TT (s)
GBR	0.023	0.001	0.033	0.800	0.019	0.032	0.010
RF	0.024	0.001	0.033	0.795	0.019	0.032	0.058
XGB	0.026	0.001	0.034	0.789	0.020	0.036	0.106
Catboost	0.024	0.001	0.034	0.785	0.020	0.033	0.192
ET	0.024	0.001	0.034	0.782	0.020	0.033	0.048
KNN	0.026	0.001	0.035	0.771	0.020	0.035	0.008
ADA	0.028	0.001	0.035	0.767	0.021	0.038	0.018
Huber	0.030	0.002	0.039	0.711	0.022	0.040	0.006
Ridge	0.031	0.002	0.039	0.707	0.023	0.041	0.006
LR	0.030	0.002	0.039	0.707	0.023	0.040	0.006
LAR	0.030	0.002	0.039	0.707	0.023	0.040	0.006
BR	0.030	0.002	0.039	0.707	0.023	0.041	0.006
OMP	0.031	0.002	0.039	0.707	0.022	0.041	0.004
ARD	0.031	0.002	0.039	0.707	0.022	0.041	0.006
DT	0.031	0.002	0.040	0.688	0.023	0.042	0.004
TR	0.036	0.003	0.049	0.572	0.029	0.051	0.178
PAR	0.042	0.003	0.050	0.528	0.029	0.056	0.006
LGBM	0.049	0.004	0.063	0.283	0.036	0.067	0.018
RANSC	0.043	0.005	0.065	0.195	0.037	0.062	0.008
MLP	0.055	0.005	0.067	0.158	0.038	0.072	0.014
lasso	0.058	0.006	0.076	−0.059	0.045	0.082	0.006
EN	0.058	0.006	0.076	−0.059	0.045	0.082	0.006
LLAR	0.058	0.006	0.076	−0.059	0.045	0.082	0.008
Dummy	0.058	0.006	0.076	−0.059	0.045	0.082	0.004
SVM	0.072	0.006	0.077	−0.104	0.044	0.093	0.006
KR	0.792	0.632	0.795	−116.050	0.523	1.030	0.006

3.5. Model Based on MS VIs Development and Evaluation

Table 10 presented an accuracy assessment of 26 machine learning models for estimating vegetation indices using multi-spectral data. The top seven models with the highest R^2 scores, ranging from 0.860 to 0.849, were Huber, LR, Ridge, LAR, OMP, BR, ARD, and TR. These models had the lowest MAE, MSE, RMSE, and RMLSE values among all the models, indicating that they produced the most accurate estimates of vegetation indices. The computation time for these models ranged from 0.004 to 0.316 s, with the LR model having the longest computation time. The next group of models, with R^2 scores ranging from 0.794 to 0.684, included KNN, RF, CatBoost, ADA, GBR, ET, and PAR. These models had higher MAE, MSE, RMSE, and RMLSE values than the top models, indicating that they produced less accurate estimates of vegetation indices. The computation time for these models ranged from 0.006 to 0.190 s. The last group of models, with R^2 scores ranging from 0.668 to -0.567, included XGB, DT, RANSC, LGBM, SVM, Lasso, EN, LLAR, Dummy, MLP, and KR. These models had the lowest R^2 scores and the highest MAE, MSE, RMSE, and RMLSE values, indicating that they produced the least accurate estimates of Fv/Fm. The computation time for these models ranged from 0.004 to 0.298 s, with the KR model having the longest computation time. Overall, the Huber model was the most accurate for estimating Fv/Fm using multispectral vegetation indices.

Table 10. Accuracy assessment of different estimation models based on MS vegetation indices.

Model	MAE	MSE	RMSE	R^2	RMLSE	MAPE	TT (s)
Huber	0.021	0.001	0.027	0.860	0.015	0.028	0.006
LR	0.022	0.001	0.027	0.860	0.015	0.029	0.316
Ridge	0.022	0.001	0.027	0.860	0.015	0.029	0.264
LAR	0.022	0.001	0.027	0.860	0.015	0.029	0.004
OMP	0.022	0.001	0.027	0.860	0.015	0.029	0.004
BR	0.022	0.001	0.027	0.860	0.015	0.029	0.004
ARD	0.022	0.001	0.027	0.860	0.015	0.029	0.006
TR	0.022	0.001	0.028	0.849	0.016	0.029	0.020
KNN	0.026	0.001	0.033	0.794	0.019	0.035	0.006
RF	0.032	0.001	0.038	0.737	0.022	0.043	0.054
Catboost	0.033	0.002	0.039	0.723	0.022	0.045	0.190
ADA	0.031	0.002	0.039	0.716	0.023	0.042	0.012
GBR	0.034	0.002	0.040	0.704	0.023	0.046	0.010
ET	0.034	0.002	0.041	0.695	0.023	0.046	0.044
PAR	0.035	0.002	0.041	0.684	0.023	0.046	0.006
XGB	0.036	0.002	0.042	0.668	0.024	0.049	0.132
DT	0.037	0.002	0.043	0.652	0.025	0.050	0.006
RANSC	0.035	0.003	0.047	0.546	0.027	0.049	0.006
LGMB	0.045	0.004	0.062	0.299	0.036	0.064	0.016
SVM	0.064	0.005	0.069	0.097	0.039	0.082	0.006
Lasso	0.058	0.006	0.076	-0.059	0.045	0.082	0.298
EN	0.058	0.006	0.076	-0.059	0.045	0.082	0.006
LLAR	0.058	0.006	0.076	-0.059	0.045	0.082	0.004
Dummy	0.058	0.006	0.076	-0.059	0.045	0.082	0.004
MLP	0.076	0.008	0.090	-0.567	0.052	0.101	0.016
KR	0.784	0.619	0.787	-113.552	0.528	1.024	0.006

3.6. Model Based on RGB and MS VIs Development and Evaluation

Table 11 presented an accuracy assessment of 26 machine learning models for the estimation of vegetation indices (VIs) using both RGB and multispectral data. The ARD and OMP models achieved the highest R^2 accuracy scores of 0.868, followed closely by the Ridge, LR, LAR, BR, and Huber models, all with R^2 values of 0.858. The Tr model obtained an R^2 of 0.849, while RANSC, KNN, RF, ET, ADA, Catboost, and XGB models had R^2 values ranging from 0.830 to 0.723. The GBR and DT models had R^2 values of 0.721 and 0.690, respectively. The PAR model had an R^2 of 0.593, indicating lower accuracy than the previous models. On the other hand, the LGMB, Lasso, EN, LLAR, and Dummy models had negative R^2 values, indicating poor accuracy. Moreover, the SVM and MLP models had low R^2 values of -0.073 and -0.292, respectively. The KR model had the worst performance with high MAE, MSE, RMSE, and RMLSE values and a very low R^2 value of -118.159. In summary, the ARD and OMP models demonstrated the same highest accuracy for estimating Fv/Fm using both MS and RGB data. However, the computation time of OMP was found to be higher than that of ARD. Therefore, the optimal model is ARD.

Table 11. Accuracy assessment of different estimation models based on RGB and MS indices.

Model	MAE	MSE	RMSE	R^2	RMLSE	MAPE	TT (s)
ARD	0.021	0.001	0.026	0.868	0.015	0.028	0.006
OMP	0.021	0.001	0.026	0.868	0.015	0.028	0.007
Ridge	0.022	0.001	0.027	0.858	0.015	0.029	0.006
LR	0.022	0.001	0.027	0.858	0.015	0.029	0.514
LAR	0.022	0.001	0.027	0.858	0.015	0.029	0.006
BR	0.022	0.001	0.027	0.858	0.015	0.029	0.006
Huber	0.022	0.001	0.027	0.857	0.016	0.029	0.006
TR	0.022	0.001	0.028	0.849	0.016	0.030	0.178
RANSC	0.024	0.001	0.030	0.830	0.017	0.032	0.010
KNN	0.023	0.001	0.030	0.826	0.017	0.031	0.006
RF	0.025	0.001	0.033	0.800	0.019	0.033	0.052
ET	0.025	0.001	0.034	0.785	0.020	0.033	0.050
ADA	0.027	0.001	0.036	0.756	0.021	0.037	0.020
Catboost	0.026	0.001	0.036	0.753	0.021	0.036	0.214
XGB	0.027	0.002	0.038	0.723	0.022	0.037	0.092
GBR	0.028	0.002	0.039	0.721	0.022	0.037	0.012
DT	0.032	0.002	0.041	0.690	0.024	0.043	0.006
PAR	0.039	0.002	0.047	0.593	0.027	0.051	0.006
LGBM	0.043	0.003	0.055	0.429	0.032	0.060	0.018
Lasso	0.058	0.006	0.076	−0.059	0.045	0.082	0.006
EN	0.058	0.006	0.076	−0.059	0.045	0.082	0.008
LLAR	0.058	0.006	0.076	−0.059	0.045	0.082	0.006
Dummy	0.058	0.006	0.076	−0.059	0.045	0.082	0.004
SVM	0.071	0.006	0.076	−0.073	0.043	0.092	0.006
MLP	0.063	0.008	0.082	−0.292	0.047	0.085	0.020
KR	0.799	0.646	0.803	−118.159	0.512	1.042	0.006

3.7. Hyperparameter Optimization of the Highest Accuracy Model in Three Modes

Table 12 and Figure 3 show the hyperparameter-optimized performance of the most accurate machine learning models in three different image types (RGB, MS and MS + RGB). The results show that the performance of these models varies according to the different image types and data sets. Notably, the ARD model achieves the highest accuracy on the test set of MS + RGB VIs mode (test set MAE = 0.019, MSE = 0.001, RMSE = 0.024, R^2 = 0.925, RMSLE = 0.014 and MAPE = 0.026).

Table 12. Hyperparameter optimization of the highest accuracy model in three image types.

Image Type	Model	Dataset	MAE	MSE	RMSE	R^2	RMLSE	MAPE
RGB	GBR	Training set	0.014	0.000	0.021	0.925	0.012	0.019
		Test set	0.027	0.001	0.036	0.838	0.020	0.037
MS	Huber	Training set	0.021	0.001	0.027	0.878	0.015	0.028
		Test set	0.018	0.001	0.025	0.920	0.014	0.024
MS + RGB	ARD	Training set	0.021	0.001	0.026	0.882	0.015	0.028
		Test set	0.019	0.001	0.024	0.925	0.014	0.026

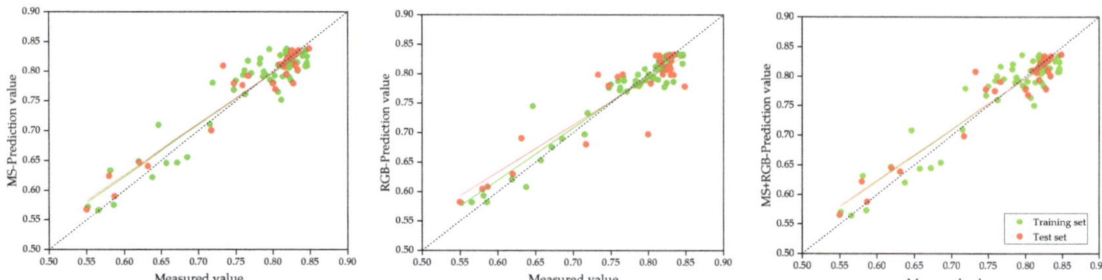

Figure 3. Calibration and validation results of the highest accuracy model in MS, RGB, RGB + MS data after hyperparameter optimization.

4. Discussion

Chlorophyll fluorescence is a potent indicator of photosynthesis, where Fv/Fm serves as a measure of the maximum photochemical efficiency of photosystem II in chloroplasts and can be utilized as an indicator of plant health [75]. For many plant species, the optimal Fv/Fm range is approximately between 0.79 and 0.84, with lower values indicating higher plant stress [76]. The wheat flowering period is a critical stage of growth and development, as it signifies the transition of wheat plants from vegetative to reproductive growth stages, where the focus of growth shifts from leaves and roots to flowers and seeds [77]. During this period, wheat plants require significant amounts of nutrients and energy to support the normal development and maturation of flowers and seeds [78]. Therefore, providing sufficient water and nutrients is crucial for wheat plants during the flowering period. Stress during this phase can significantly affect the yield and quality of wheat [79]. Monitoring Fv/Fm during the flowering period is crucial in detecting stress and implementing timely agricultural management practices to optimize wheat production and achieve high yields.

In this study, both multispectral vegetation indices and RGB vegetation indices were found to be effective in estimating Fv/Fm values. The R^2 of test set was 0.920 for multispectral vegetation indices and 0.838 for RGB vegetation indices, indicating that the former had a higher accuracy in estimating Fv/Fm values. Multispectral images acquired by UAVs can provide information about the spectral reflectance of vegetation, which changes simultaneously in the canopy when stress occurs [80], and this can be used to estimate Fv/Fm. The key idea of this approach is that Fv/Fm is related to the fluorescence yield of photosystem II, which in turn is related to the chlorophyll content of leaves. The chlorophyll content can be estimated from the reflectance in the red and near-infrared (NIR) bands [81]. The important spectral index MSAVI2 extracted from the multispectral estimation in this study was calculated precisely from the red and NIR bands. In addition to multispectral images, RGB images can also be used to estimate Fv/Fm values by color information. In general, the color of plants changes when they are under stress, and although some color changes are difficult for the human eye to observe, color values can be quantified by computer technology [82]. However, due to the lack of vegetation-sensitive red edge and infrared bands, the RGB images do not provide enough spectral information, so the estimated Fv/Fm is not as accurate as the MS images.

This study shows that the machine learning models have high accuracy and stability and can effectively use RGB and MS data to estimate Fv/Fm, which is consistent with the findings of other studies [83] using machine learning for remote sensing estimation. The effects of different machine learning models and different types of vegetation indices on the accuracy of Fv/Fm estimation are obvious. Firstly, different types of vegetation indices can provide different information in estimating Fv/Fm, and different machine learning models have different adaptation and fitting ability to different datasets [84]. The MS + RGB model exhibits superior accuracy compared to both the MS and RGB models in estimating Fv/Fm. This suggests that the integration of RGB and MS data provides benefits in enhancing the accuracy of the estimation. Additionally, the findings suggest that utilizing multiple data sources can enhance the accuracy of the model as compared to relying solely on single-source data. These results are consistent with previous studies that have reported improved accuracy in multi-source data estimation [85,86]. However, the improvement in accuracy when combining (MS + RGB) is only marginally better than using MS alone in this study. The ARD model cannot be quantified for the percentage contribution of RGB and MS, so the random-forest-model-based feature importance assessment was implemented, the result was shown in Figure 4. The importance of multispectral vegetation index in the model was higher than that of RGB vegetation indices, and the RGB contributed less valid information.

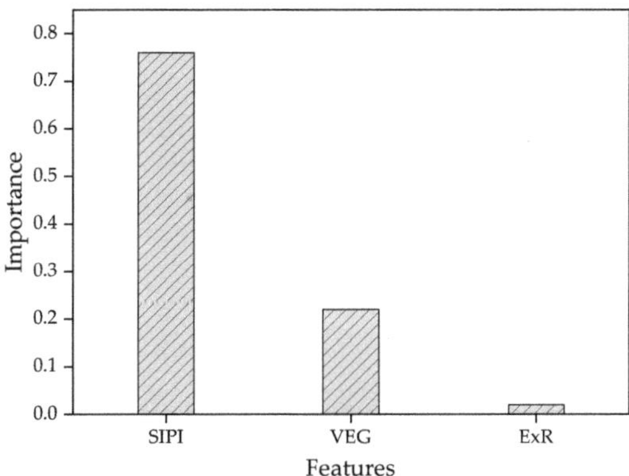

Figure 4. Feature importance assessment based on random forest model.

Zhao et al. [7] developed two estimation models for Fv/Fm of potato leaves with test set R^2 of 0.807 and 0.822 and RMSE of 0.018 and 0.017, respectively. Yi et al. [8] developed a generalized estimation model of Fv/Fm for poplar leaves and cherry leaves with $R^2 = 0.88$. In a study [9] estimating Fv/Fm in winter wheat, training set $R^2 = 0.50$, RMSE = 0.012, test set $R^2 = 0.55$, RMSE = 0.014. In these previous estimation studies of Fv/Fm based on hyperspectral, the highest R^2 was of 0.88, and the highest accuracy combined model without hyperparameter optimization in this study had R^2 of 0.868, which shows that the estimation accuracy of multispectral and RGB is not as good as hyperspectral. However, Yang et al. [87] showed that hyperparameter optimization can be helpful in improving the estimation accuracy of the model. In this study, the test set R^2 of both the MS estimation and the estimation of the MS and RGB combination reaches 0.92 after hyperparameter optimization, indicating that the accuracy gap caused by the sensors can be narrowed or even surpassed by algorithm optimization. Negative values of R^2 were observed in all three estimation models, which are usually uncommon, indicating that the prediction results using the model are worse than the estimation results using the mean, because the mean reflects the central tendency of the data, while the prediction results of the model deviate more from the true value than the mean [88]. Notably, the performance of linear regression (LR), ridge regression (Ridge), least angle regression (LAR), and orthogonal matching pursuit (OMP) models was almost identical. Moreover, the reason for this was that only MSAVI2 was screened in the multispectral vegetation indices and LR, Ridge, LAR, and OMP are all linear models, indicating that different linear models may not have a significant effect on the accuracy of estimation when a single vegetation index was used as a characteristic variable.

The present study has important practical implications for agriculture and environmental monitoring. By demonstrating the feasibility of using UAV-based remote sensing to estimate Fv/Fm in spring wheat, our research offers new opportunities for efficient and high-throughput monitoring of plant health and productivity. This can help to improve crop management strategies, enhance yield and quality of agricultural products, and reduce the environmental impact of farming practices. In addition, the integration of machine learning methods with multispectral and RGB imagery can further enhance the accuracy and reliability of Fv/Fm estimation, enabling more precise and targeted interventions in crop management.

5. Conclusions

Integration of multiple machine learning methods with multispectral and RGB imagery acquired from UAV-based remote sensing can improve the accuracy and reliability of Fv/Fm estimation in spring wheat during the flowering stage. The important features and the optimal Fv/Fm estimation models for different types remote sensing images were different: with gradient boosting regressor (GBR) as the optimal estimation model for RGB, the important features were RGBVI and ExR; with Huber as the optimal estimation model for MS, the important feature was MSAVI2; and automatic relevance determination (ARD) as the optimal estimation model for combination (MS + RGB), the important features were SIPI, ExR, VEG. The highest accuracy was achieved using the ARD model for estimating Fv/Fm with RGB + MS vegetation indices (Test set MAE = 0.019, MSE = 0.001, RMSE = 0.024, R^2 = 0.925, RMSLE = 0.014, MAPE = 0.026). Based on the results of this study, there is great potential for the use of remote sensing and machine learning for efficient and sustainable plant health monitoring and management.

In conclusion, while the present study provides a valuable contribution to the use of remote sensing and machine learning for estimating Fv/Fm in wheat during the flowering stage, there are several limitations. Firstly, the study was conducted at a single ecological site over a two-year period, which may limit the generalizability of the findings to other regions and ecosystems. Future studies conducted at multiple sites with varying environmental conditions would provide a more comprehensive understanding of the applicability of the proposed methods. Secondly, the study focused solely on the flowering stage of wheat growth, which is only one phase of the crop's development. It is possible that the performance of the proposed remote sensing and machine learning methods may differ at other stages of wheat growth, such as germination, tillering, or grain filling. Further research is needed to examine the feasibility and accuracy of the methods across multiple growth stages. Thirdly, while the proposed methods show promise for estimating Fv/Fm in wheat, it should be noted that these methods may not be directly applicable to other plant species or crops. The optimal settings and parameters for remote sensing and machine learning may vary depending on the physiological and structural characteristics of the target plants. Future studies should explore the generalizability and adaptability of the methods to other crops and plant species.

Author Contributions: Conceptualization, Q.W. and Y.Z.; methodology, Q.W.; software, M.X.; validation, Z.Z. and J.L.; formal analysis, Z.Z.; investigation, D.H.; resources, L.Y.; data curation, D.H. and J.L; writing—original draft preparation, Q.W.; writing—review and editing, Y.Z.; visualization, M.X.; supervision, Y.Z.; project administration, Y.Z.; funding acquisition, Y.Z. and M.X. All authors have read and agreed to the published version of the manuscript.

Funding: This research was funded by Inner Mongolia "science and technology" action focus on special "Yellow River Basin durum wheat industrialization capacity enhancement" (NMKJXM202201-4); Inner Mongolia "science and technology" action focus on special "Research and Application of Key Technologies for Production and Processing of Durum Wheat and Products in Hetao irrigation area" (NMKJXM202111-3) and Inner Mongolia Natural Science Foundation of China "Research on nitrogen nutrition diagnosis of spring wheat in Hetao irrigation area based on UAV mapping technology" (2021MS03089).

Data Availability Statement: Not applicable.

Acknowledgments: We sincerely appreciate the assistance of the Wheat Research Institute of the Inner Mongolia Bayannur Academy of Agriculture and Animal Science.

Conflicts of Interest: The authors declare no conflict of interest.

References

1. Zhao, Z.; Li, M.; Wu, Q.; Zhang, Y. Effects of Different Soil Moisture-Holding Strategies on Growth Characteristics, Yield and Quality of Winter-Seeded Spring Wheat. *Agronomy* **2022**, *12*, 2746. [CrossRef]
2. Baker, N.R. Chlorophyll Fluorescence: A Probe of Photosynthesis In Vivo, Annu. *Rev. Plant Biol.* **2008**, *59*, 89–113. [CrossRef]
3. Maxwell, K.; Johnson, G.N. Chlorophyll fluorescence–A practical guide. *J. Exp. Bot.* **2000**, *51*, 659–668. [CrossRef]
4. Baker, N.R.; Rosenquist, E. Applications of chlorophyll fluorescence can improve crop production strategies: An examination of future possibilities. *J. Exp. Bot.* **2004**, *55*, 1607–1621. [CrossRef]
5. Genty, B.; Briantais, J.M.; Baker, N.R. The relationship between the quantum yield of photosynthetic electron transport and quenching of chlorophyll fluorescence. *Biochim. Biophys. Acta (BBA)-Gen. Subj.* **1989**, *990*, 87–92. [CrossRef]
6. Sharma, D.K.; Andersen, S.B.; Ottosen, C.O.; Rosenqvist, E. Wheat cultivars selected for high Fv/Fm under heat stress maintain high photosynthesis, total chlorophyll, stomatal conductance, transpiration and dry matter. *Physiol. Plant.* **2015**, *153*, 284–298. [CrossRef]
7. Zhao, R.; An, L.; Song, D.; Li, M.; Qiao, L.; Liu, N.; Sun, H. Detection of chlorophyll fluorescence parameters of potato leaves based on continuous wavelet transform and spectral analysis. *Spectrochim. Acta A* **2021**, *259*, 119768. [CrossRef]
8. Yi, P.; Aoli, Z.; Tinge, Z.; Sheng, H.F.; Yan, G.; Yan, Q.T.; Ying, Z.; Kan, L. Using remotely sensed spectral reflectance to indicate leaf photosynthetic efficiency derived from active fluorescence measurements. *J. Appl. Rem. Sens.* **2017**, *11*, 026034. [CrossRef]
9. Jia, M.; Li, D.; Colombo, R.; Wanlg, Y.; Wang, X.; Cheng, T.; Zhu, Y.; Yao, X.; Xu, C.; Ouer, G.; et al. Quantifying Chlorophyll Fluorescence Parameters from Hyperspectral Reflectance at the Leaf Scale under Various Nitrogen Treatment Regimes in Winter Wheat. *Remote Sens.* **2019**, *11*, 2838. [CrossRef]
10. Liu, J.; Zhu, Y.; Tao, X.; Chen, X.; Li, X. Rapid prediction of winter wheat yield and nitrogen use efficiency using consumer-grade unmanned aerial vehicles multispectral imagery. *Front. Plant Sci.* **2022**, *13*, 1032170. [CrossRef] [PubMed]
11. Torres-Tello, J.W.; Ko, S. A novel approach to identify the spectral bands that predict moisture content in canola and wheat. *Biosyst. Eng.* **2021**, *210*, 91–103. [CrossRef]
12. Wu, Q.; Zhang, Y.; Zhao, Z.; Xie, M.; Hou, D. Estimation of Relative Chlorophyll Content in Spring Wheat Based on Multi-Temporal UAV Remote Sensing. *Agronomy* **2023**, *13*, 211. [CrossRef]
13. Wilke, N.; Siegmann, B.; Postma, J.A.; Muller, O.; Krieger, V.; Pude, R.; Rascher, U. Assessment of plant density for barley and wheat using UAV multispectral imagery for high-throughput field phenotyping. *Comput. Electron. Agric.* **2021**, *189*, 106380. [CrossRef]
14. Du, X.; Wan, L.; Cen, H.; Chen, S.; Zhu, J.; Wang, H.; He, Y. Multi-temporal monitoring of leaf area index of rice under different nitrogen treatments using UAV images. *Int. J. Precis. Agric. Aviat.* **2020**, *1*, 11–18. [CrossRef]
15. Maimaitijiang, M.; Sagan, V.; Sidike, P.; Daloye, A.M.; Erkbol, H.; Fritschi, F.B. Crop monitoring using satellite/UAV data fusion and machine learning. *Remote Sens.* **2020**, *12*, 1357. [CrossRef]
16. Meyer, G.E.; Neto, J.C. Verification of color vegetation indices for automated crop imaging applications. *Comput. Electron. Agric.* **2008**, *63*, 282–293. [CrossRef]
17. Woebbecke, D.M.; Meyer, G.E.; Von Bargen, K.; Mortensen, D.A. Color indices for weed identification under various soil, residue, and lighting conditions. *Trans. ASAE* **1995**, *38*, 259–269. [CrossRef]
18. Vincent, L. Morphological grayscale reconstruction in image analysis: Applications and efficient algorithms. *IEEE Trans. Image Process* **1993**, *2*, 176–201. [CrossRef]
19. Sellers, P.J.; Hall, F.G.; Nielsen, G.A. Vegetation/atmosphere transfer models. *Adv. Space Res.* **1987**, *7*, 149–159. [CrossRef]
20. Hague, T.; Tillett, N.D.; Wheeler, H. Automated crop and weed monitoring in widely spaced cereals. *Precis. Agric.* **2006**, *7*, 21–32. [CrossRef]
21. Guijarro, M.; Pajares, G.; Riomoros, I.; Herrera, P.J.; Burgos-Artizzu, X.P.; Ribeiro, A. Automatic segmentation of relevant textures in agricultural images. *Comput. Electron. Agric.* **2011**, *75*, 75–83. [CrossRef]
22. Daughtry, C.S.T.; Walthall, C.L.; Kim, M.S.; De Colstoun, E.B.; McMurtrey, J.E. Estimating corn leaf chlorophyll concentration from leaf and canopy reflectance. *Remote Sens. Environ.* **2000**, *74*, 229–239. [CrossRef]
23. Tucker, C.J. Red and photographic infrared linear combinations for monitoring vegetation. *Remote Sens. Environ.* **1979**, *8*, 127–150. [CrossRef]
24. Kawashima, S. A new vegetation index for the monitoring of vegetation phenology and thermal stress. *Int. J. Remote Sens.* **2002**, *23*, 2003–2017. [CrossRef]
25. Gitelson, A.A.; Kaufman, Y.J.; Stark, R.; Rundquist, D. Novel algorithms for remote estimation of vegetation fraction. *Remote Sens. Environ.* **2002**, *80*, 76–87. [CrossRef]
26. Wang, X.; Wang, M.; Wang, S.; Wu, Y. Extraction of vegetation information from visible unmanned aerial vehicle images. *Trans. Chin. Soc. Agric. Eng.* **2015**, *31*, 152–159. [CrossRef]
27. Saberioon, M.M.; Amin, M.S.M.; Anuar, A.R.; Gholizadeh, A. Assessment of rice leaf chlorophyll content using visible bands at different growth stages at both the leaf and canopy scale. *Int. J. Appl. Earth Obs. Geoinf.* **2014**, *32*, 35–45. [CrossRef]
28. Bendig, J.; Yu, K.; Aasen, H.; Bolten, A.; Bennertz, S.; Broscheit, J.; Gnyp, M.L.; Bareth, G. Combining UAV-based plant height from crop surface models, visible, and near infrared vegetation indices for biomass monitoring in barley. *Int. J. Appl. Earth Obs. Geoinf.* **2015**, *39*, 79–87. [CrossRef]

29. Huete, A.; Justice, C.; Liu, H. Development of vegetation and soil indices for MODIS-EOS. *Remote Sens. Environ.* **1994**, *49*, 224–234. [CrossRef]

30. Gitelson, A.A.; Merzlyak, M.N. Remote sensing of chlorophyll concentration in higher plant leaves. *Adv. Space Res.* **1998**, *22*, 689–692. [CrossRef]

31. Gamon, J.A.; Surfus, J.S. Assessing leaf pigment content and activity with a reflectometer. *New Phytol.* **1999**, *143*, 105–117. [CrossRef]

32. Birth, G.S.; McVey, G.R. Measuring the color of growing turf with a reflectance spectrophotometer. *Agron. J.* **1968**, *60*, 640–643. [CrossRef]

33. Jasper, J.; Reusch, S.; Link, A. Active sensing of the N status of wheat using optimized wavelength combination: Impact of seed rate, variety and growth stage. *Precis. Agric.* **2009**, *9*, 23–30. [CrossRef]

34. Filella, I.; Penuelas, J. The red edge position and shape as indicators of plant chlorophyll content, biomass and hydric status. *Int. J. Remote Sens.* **1994**, *15*, 1459–1470. [CrossRef]

35. Qi, J.; Chehbouni, A.; Huete, A.R.; Kerr, Y.H.; Sorooshian, S. A modified soil adjusted vegetation index. *Remote Sens. Environ.* **1994**, *48*, 119–126. [CrossRef]

36. Rondeaux, G.; Steven, M.; Baret, F. Optimization of soil-adjusted vegetation indices. *Remote Sens. Environ.* **1996**, *55*, 95–107. [CrossRef]

37. Xing, N.C.; Huang, W.J.; Xie, Q.Y.; Shi, Y.; Ye, H.C.; Dong, Y.Y.; Wu, M.Q.; Sun, G.; Jiao, Q.J. A Transformed Triangular Vegetation Index for Estimating Winter Wheat Leaf Area Index. *Remote Sens.* **2019**, *12*, 16. [CrossRef]

38. Sims, D.A.; Gamon, J.A. Relationships between leaf pigment content and spectral reflectance across a wide range of species, leaf structures and developmental stages. *Remote Sens. Environ.* **2002**, *81*, 337–354. [CrossRef]

39. Rouse, J.W.; Haas, R.H.; Schell, J.A.; Deering, D.W. Monitoring vegetation systems in the Great Plains with ERTS. *NASA Spec. Publ.* **1973**, *351*, 309–317.

40. Chen, J.M. Evaluation of vegetation indices and a modified simple ratio for boreal applications. *Can. J. Remote Sens.* **1996**, *22*, 229–242. [CrossRef]

41. Huete, A.R. A soil-adjusted vegetation index (SAVI). *Remote Sens. Environ.* **1988**, *25*, 295–309. [CrossRef]

42. Fitzgerald, G.; Rodriguez, D.; O'Leary, G. Measuring and predicting canopy nitrogen nutrition in wheat using a spectral index—The canopy chlorophyll content index (CCCI). *Field Crops Res.* **2010**, *116*, 318–324. [CrossRef]

43. Kimura, R.; Okada, S.; Miura, H.; Kamichika, M. Relationships among the leaf area index, moisture availability, and spectral reflectance in an upland rice field. *Agric. Water Manag.* **2004**, *69*, 83–100. [CrossRef]

44. Verrelst, J.; Schaepman, M.E.; Koetz, B.; Kneubühler, M. Angular sensitivity analysis of vegetation indices derived from CHRIS/PROBA data. *Remote Sens. Environ.* **2008**, *112*, 2341–2353. [CrossRef]

45. Devadas, R.; Lamb, D.W.; Simpfendorfer, S.; Backhouse, D. Evaluating ten spectral vegetation indices for identifying rust infection in individual wheat leaves. *Precis. Agric.* **2009**, *10*, 459–470. [CrossRef]

46. Zha, Y.; Gao, J.; Ni, S. Use of normalized difference built-up index in automatically mapping urban areas from TM imagery. *Int. J. Remote Sens.* **2003**, *24*, 583–594. [CrossRef]

47. Ju, C.H.; Tian, Y.C.; Yao, X.; Cao, W.X.; Zhu, Y.; Hannaway, D. Estimating leaf chlorophyll content using red edge parameters. *Pedosphere* **2010**, *20*, 633–644. [CrossRef]

48. Clevers, J.G.P.W.; Gitelson, A.A. Remote estimation of crop and grass chlorophyll and nitrogen content using red-edge bands on Sentinel-2 and -3. *Int. J. Appl. Earth Obs. Geoinf.* **2013**, *23*, 344–351. [CrossRef]

49. Wu, C.Y.; Niu, Z.; Tang, Q.; Huang, W.J. Estimating chlorophyll content from hyperspectral vegetation indices: Modeling and validation. *Agric. For. Meteorol.* **2008**, *148*, 1230–1241. [CrossRef]

50. Drucker, H. Improving Regressors Using Boosting Techniques. In Proceedings of the Icml; Citeseer: Princeton, NJ, USA, 1997; Volume 97, pp. 107–115.

51. Mørup, M.; Hansen, L.K. Automatic relevance determination for multi-way models. *J. Chemom. A J. Chemom. Soc.* **2009**, *23*, 352–363. [CrossRef]

52. MacKay, D.J.C. Bayesian interpolation. *Neural Comput.* **1992**, *4*, 415–447. [CrossRef]

53. Prokhorenkova, L.; Gusev, G.; Vorobev, A.; Dorogush, A.V.; Gulin, A. CatBoost: Unbiased Boosting with Categorical Features. *Adv. Neural Inf. Process. Syst.* **2018**, *31*, 6638–6648.

54. Myles, A.J.; Feudale, R.N.; Liu, Y.; Woody, N.A.; Brown, S.D. An Introduction to Decision Tree Modeling. *J. Chemom. A J. Chemom. Soc.* **2004**, *18*, 275–285. [CrossRef]

55. Wilde, J. Identification of multiple equation probit models with endogenous dummy regressors. *Econ. Lett.* **2000**, *69*, 309–312. [CrossRef]

56. Zou, H.; Hastie, T. Regularization and variable selection via the elastic net. *J. R. Stat. Soc. Ser. B Stat. Methodol.* **2005**, *67*, 301–320. [CrossRef]

57. Geurts, P.; Ernst, D.; Wehenkel, L. Extremely randomized trees. *Mach. Learn.* **2006**, *63*, 3–42. [CrossRef]

58. Chen, T.; He, T.; Benesty, M.; Khotilovich, V.; Tang, Y.; Cho, H.; Chen, K.; Mitchell, R.; Cano, I.; Zhou, T.; et al. Xgboost: Extreme Gradient Boosting. *R Package Version 0.4-2.* **2015**, *1*, 1–4.

59. Friedman, J.H. Greedy function approximation: A gradient boosting machine. *Ann. Stat.* **2001**, *29*, 1189–1232. [CrossRef]

60. Huber, P.J. Robust estimation of a location parameter. *Ann. Math. Stat.* **1964**, *35*, 73–101. [CrossRef]

61. Zhang, M.-L.; Zhou, Z.-H. ML-KNN: A Lazy Learning Approach to Multi-Label Learning. *Pattern Recognit.* **2007**, *40*, 2038–2048. [CrossRef]
62. Durrant, S.D.; Bissell, J. Performance of kernel-based regression methods in the presence of outliers. *J. Process Control* **2010**, *20*, 959–967. [CrossRef]
63. Efron, B.; Hastie, T.; Johnstone, I.; Tibshirani, R. Least angle regression. *Ann. Stat.* **2004**, *32*, 407–499. [CrossRef]
64. Tibshirani, R. Regression shrinkage and selection via the lasso. *J. R. Stat. Soc. Ser. B Stat. Methodol.* **1996**, *58*, 267–288. [CrossRef]
65. GuolinKe, Q.M.; Finley, T.; Wang, T.; Chen, W.; Ma, W.; Ye, Q.; Liu, T.-Y. Lightgbm: A Highly Efficient Gradient Boosting Decision Tree. *Adv. Neural Inf. Process. Syst.* **2017**, *30*, 52.
66. Su, X.; Yan, X.; Tsai, C.-L. Linear Regression. *Wiley Interdiscip. Rev. Comput. Stat.* **2012**, *4*, 275–294. [CrossRef]
67. Rumelhart, D.E.; Hinton, G.E.; Williams, R.J. Learning representations by back-propagating errors. *Nature* **1986**, *323*, 533–536. [CrossRef]
68. Tropp, J.A.; Gilbert, A.C. Signal Recovery from Random Measurements via Orthogonal Matching Pursuit. *IEEE Trans. Inf. Theory* **2007**, *53*, 4655–4666. [CrossRef]
69. Crammer, K.; Dekel, O.; Keshet, J.; Shalev-Shwartz, S.; Singer, Y. Online passive-aggressive algorithms. *J. Mach. Learn. Res.* **2006**, *7*, 551–585.
70. Breiman, L. Random forests. *Mach. Learn.* **2001**, *45*, 5–32. [CrossRef]
71. Fischler, M.A.; Bolles, R.C. Random sample consensus: A paradigm for model fitting with applications to image analysis and automated cartography. *Commun. ACM* **1981**, *24*, 381–395. [CrossRef]
72. Hoerl, A.E.; Kennard, R.W. Ridge regression: Biased estimation for nonorthogonal problems. *Technometrics* **1970**, *12*, 55–67. [CrossRef]
73. Drucker, H.; Burges, C.J.; Kaufman, L.; Smola, A.; Vapnik, V. Support Vector Regression Machines. *Adv. Neural Inf. Process. Syst.* **1996**, *9*, 155–161.
74. Theil, H. A Rank-Invariant Method of Linear and Polynomial Regression Analysis. *Indag. Math.* **1950**, *12*, 173.
75. Bolhar-Nordenkampf, H.R.; Long, S.P.; Baker, N.R.; Oquist, G.; Schreiber, U.L.E.G.; Lechner, E.G. Chlorophyll fluorescence as a probe of the photosynthetic competence of leaves in the field: A review of current instrumentation. *Funct. Ecol.* **1989**, *3*, 497–514. [CrossRef]
76. Kalaji, H.M.; Schansker, G.; Brestic, M.; Bussotti, F.; Calatayud, A.; Ferroni, L. Frequently asked questions about chlorophyll fluorescence, the sequel. *Photosynth. Res.* **2017**, *132*, 13–66. [CrossRef] [PubMed]
77. Acevedo, E.; Silva, P.; Silva, H. Wheat Growth and Physiology. *Bread Wheat Improv. Prod.* **2002**, *30*, 39–70.
78. Shewry, P.R. Wheat. *J. Exp. Bot.* **2009**, *60*, 1537–1553. [CrossRef]
79. Tilling, A.K.; O'Leary, G.J.; Ferwerda, J.G.; Jones, S.D.; Fitzgerald, G.J.; Rodriguez, D.; Belford, R. Remote Sensing of Nitrogen and Water Stress in Wheat. *Field Crops Res.* **2007**, *104*, 77–85. [CrossRef]
80. Guo, Y.; Chen, S.; Li, X.; Cunha, M.; Jayavelu, S.; Cammarano, D.; Fu, Y. Machine learning-based approaches for predicting SPAD values of maize using multi-spectral images. *Remote Sens.* **2022**, *14*, 1337. [CrossRef]
81. Zarco-Tejada, P.J.; Berni, J.A.; Suárez, L.; Sepulcre-Cantó, G.; Morales, F.; Miller, J.R. Imaging chlorophyll fluorescence with an airborne narrow-band multispectral camera for vegetation stress detection. *Remote Sens. Environ.* **2009**, *113*, 1262–1275. [CrossRef]
82. Liu, Y.; Hatou, K.; Aihara, T.; Kurose, S.; Omasa, K. A robust vegetation index based on different UAV RGB images to estimate SPAD values of naked barley leaves. *Remote Sens.* **2021**, *13*, 686. [CrossRef]
83. Ahmad, S.; Kalra, A.; Stephen, H. Estimating soil moisture using remote sensing data: A machine learning approach. *Adv. Water Resour.* **2010**, *33*, 69–80. [CrossRef]
84. Korotcov, A.; Tkachenko, V.; Russo, D.P.; Ekins, S. Comparison of deep learning with multiple machine learning methods and metrics using diverse drug discovery data sets. *Mol. Pharmaceut.* **2017**, *14*, 4462–4475. [CrossRef] [PubMed]
85. Yuan, Y.; Wang, X.; Shi, M.; Wang, P. Performance comparison of RGB and multispectral vegetation indices based on machine learning for estimating Hopea hainanensis SPAD values under different shade conditions. *Front. Plant Sci.* **2022**, *13*, 928953. [CrossRef] [PubMed]
86. Zhang, J. Multi-source remote sensing data fusion: Status and trends. *Int. J. Image Data Fus.* **2010**, *1*, 5–24. [CrossRef]
87. Yang, L.; Shami, A. On hyperparameter optimization of machine learning algorithms: Theory and practice. *Neurocomputing* **2020**, *415*, 295–316. [CrossRef]
88. Mittlböck, M. Calculating adjusted R^2 measures for Poisson regression models. *Comput. Meth. Prog. Biomed.* **2002**, *68*, 205–214. [CrossRef] [PubMed]

Article

Synchronous Retrieval of LAI and Cab from UAV Remote Sensing: Development of Optimal Estimation Inversion Framework

Fengxun Zheng [1], Xiaofei Wang [1], Jiangtao Ji [1], Hao Ma [1,*], Hongwei Cui [1], Yi Shi [1] and Shaoshuai Zhao [2]

[1] College of Agricultural Equipment Engineering, Henan University of Science and Technology, Luoyang 471023, China
[2] Henan Modern Agricultural Big Data Industry Technology Research Institute Co., Ltd., Zhengzhou 450046, China
*   Correspondence: mahao@haust.edu.cn

Abstract: UAV (unmanned aerial vehicle) remote sensing provides the feasibility of high-throughput phenotype nondestructive acquisition at the field scale. However, accurate remote sensing of crop physicochemical parameters from UAV optical measurements still needs to be further studied. For this purpose, we put forward a crop phenotype inversion framework based on the optimal estimation (OE) theory in this paper, originating from UAV low-altitude hyperspectral/multispectral data. The newly developed unified linearized vector radiative transfer model (UNL-VRTM), combined with the classical PROSAIL model, is used as the forward model, and the forward model was verified by the wheat canopy reflectance data, collected using the FieldSpec Handheld in Qi County, Henan Province. To test the self-consistency of the OE-based framework, we conducted forward simulations for the UAV multispectral sensors (DJI P4 Multispectral) with different observation geometries and aerosol loadings, and a total of 801 sets of validation data were obtained. In addition, parameter sensitivity analysis and information content analysis were performed to determine the contribution of crop parameters to the UAV measurements. Results showed that: (1) the forward model has a strong coupling between vegetation canopy and atmosphere environment, and the modeling process is reasonable. (2) The OE-based inversion framework can make full use of the available radiometric spectral information and had good convergence and self-consistency. (3) The UAV multispectral observations can support the synchronous retrieval of LAI (leaf area index) and Cab (chlorophyll a and b content) based on the proposed algorithm. The proposed inversion framework is expected to be a new way for phenotypic parameter extraction of crops in field environments and had some potential and feasibility for UAV remote sensing.

Keywords: crop population phenotype; optimal estimation inversion; unmanned aerial vehicle (UAV); hyperspectral; multispectral

Citation: Zheng, F.; Wang, X.; Ji, J.; Ma, H.; Cui, H.; Shi, Y.; Zhao, S. Synchronous Retrieval of LAI and Cab from UAV Remote Sensing: Development of Optimal Estimation Inversion Framework. *Agronomy* **2023**, *13*, 1119. https://doi.org/10.3390/agronomy13041119

Academic Editors: Jinling Zhao and Chuanjian Wang

Received: 27 March 2023
Revised: 9 April 2023
Accepted: 12 April 2023
Published: 14 April 2023

1. Introduction

Unmanned aerial vehicle (UAV) remote sensing has great potential in collecting crop information [1,2]. Crop phenotype data can be retrieved from the hyperspectral or multispectral sensors onboard UAVs. The phenotypic data can be further used for crop breeding (seed screening, crop–environment interaction, etc.) [3–5] and field management (growth monitoring, yield estimation, pest and disease prediction, etc.) [6–8]. However, it is still a great challenge to retrieve the crop physiological parameters from the optical reflectance data [9].

For the inversion of crop parameters, currently, there are two main approaches: the empirical model method and the physical model method [10,11].

The empirical method makes use of the crop-sensitive bands to construct spectral indices [12–14]. Then, a regression model between phenotype parameters and spectral indices can be established using a large number of measurements. Thus, the inversion accuracy of the empirical method mainly depends on the spectral indices and the number of samples [15–17]. Since the spectral indices are usually constructed from measurements at several wavelengths, these methods can hardly make use of all the observation information. Moreover, the application of the empirical model approach is limited [18–20].

The physical model method builds a forward model from the solar radiation to the crop canopy, and then to the sensor [21,22]. Then, crop parameters can be obtained by finding the minimum of the cost function. The cost function is usually a multivariable nonlinear function and mathematical tool, which is constructed by model simulation results and actual measurements, as the Least Squares method can be used to solve the cost function [23,24]. However, the iterative process requires repeated forward model calculation, which is time-consuming, and prior constraints are also required to avoid falling into local optimality [25,26]. Look-up Tables (LUTs) are widely used to replace the real-time calculation of the forward model, which can greatly reduce the computation load [27,28]. However, LUTs are not suitable for multiparameter inversion because the size of the LUT grows exponentially with the number of parameters.

At present, the physical models are rare for the UAV remote sensing of crop parameters. The PROSAIL model, widely used to simulate canopy reflectance from crop physicochemical and structural parameters, does not consider solar radiation and atmospheric environment yet [29,30]. The typical atmospheric radiative transfer models, such as 6SV (Second Simulation of a Satellite Signal in the Solar Spectrum—vector) [31], MODTRAN (moderate-resolution atmospheric transmission) [32], RT3 (the polarized radiative transfer) [33], and UNL-VRTM [34], can be used for UAV remote sensing. However, the inputs of these models do not include the crop phenotype parameters. The 6SV provides a Python interface for PROSAIL but does not have open-source code [35].

On the one hand, most of the physical models do not realize the radiation transmission from the atmosphere to the vegetation canopy [16], and atmospheric correction is a separate process that needs to be performed. On the other hand, previous studies on crop parameter inversion have focused on the establishment of spectral index and LUTs [11,27], and there is limited information on inversion methods to extract crop multiparameters synchronously. Therefore, this study considers the atmospheric composition during the establishment of the forward model and evaluates the contribution of crop parameters to the forward model. In addition, we propose an optimal inversion framework for the synchronous retrieval of crop phenotypic parameters to solve the multiparameter inversion problem. The main objectives include the following: (1) to construct a forward model by coupling the PROSAIL model and UNL-VRTM model; (2) to analyze the sensitivity and information content of the model to crop parameters and to determine retrieved parameters; (3) to use optimal estimation theory and other techniques to find the minimum of the cost function; and (4) to establish and validate the optimal inversion framework for the synchronous retrieval of crop phenotypic parameters.

In Section 1, we introduced two main approaches currently for the inversion of crop parameters and the research purpose of this paper. The components of our forward coupling model are described in Section 2, and we present the method for the synchronous retrieval of multiparameters in Section 3. In Section 4, we performed the parameter sensitivity analysis and information content analysis to determine the contribution of parameters to the model. In Section 5, we verified the accuracy of the forward model and tested the consistency of the inversion framework used to retrieve LAI and Cab. The discussion and conclusion are in Sections 6 and 7, respectively.

2. Modeling of UAV Observations

For UAV remote sensing, the observation contains the contribution of atmospheric scattering, absorption, and canopy reflection. The measured surface reflectance can be described by the following equation [36,37]:

$$R_\lambda(\mu_s, \mu_v, \phi) = T_\lambda^\downarrow R_{crop}(\mu_s, \mu_v, \phi) T_\lambda^\uparrow \tag{1}$$

where μ_s and μ_v are the cosine of the solar zenith angle (SZA) and cosine of the viewing zenith angle (VZA), respectively; ϕ refers to the relative azimuth angle; λ denotes the wavelength; T represents the atmospheric transfer term, which describes the contribution of atmospheric gases and aerosols. The superscripts $\downarrow$ and $\uparrow$ represent the direction of downwelling and upwelling, respectively. For low-altitude observation, the effect of the atmosphere on upward transmission can be ignored. R_{crop} is the canopy reflectance of the crop, which is determined by the growth stage and status of the crop.

2.1. Atmospheric Radiative Transfer

To calculate the downward radiation term $T_\lambda^\downarrow$, the UNL-VRTM model is adopted, which is an open-source atmospheric radiative transfer model [38,39]. The inputs of UNL-VRTM include the atmospheric parameters (pressure, altitude, temperature, etc.), aerosol parameters (the aerosol optical depth, particle size distribution, and complex refractive indices), and observation geometries (solar zenith angle, viewing zenith angle, and relative azimuth angle). The outputs of UNL-VRTM include not only the Stokes vector but also their sensitivities (Jacobians) with respect to aerosol and surface model parameters. However, the UNL-VRTM's built-in surface model is the kernel-driven BRDF (Bi-directional Reflectance Distribution Function) model, and none of the BRDF kernels contains the phenotypic parameters of the crop. Therefore, the UNL-VRTM model cannot be used directly as a forward model for crop parameter inversion. The main input parameters of the UNL-VRTM model are given in Table 1.

Table 1. Main input parameters of the UNL-VRTM model.

Parameter Types	Parameter Symbols	Parameter Description	Unit
Geometry	SZA	Solar zenith angle	Degrees (°)
	VZA	Viewing zenith angle	Degrees (°)
	SAA	Solar azimuthal angle	Degrees (°)
	VAA	Viewing azimuthal angle	Degrees (°)
Atmosphere	Atmospheric type	The meteorological and air density profile	–
	Pressure	Surface pressure	hPa
	Altitude	Surface altitude	m
Aerosol	AOD	Aerosol optical depth	–
	Ri	Complex refractive index of aerosol	–
	Profile	The vertical profile of aerosol	–
	PSD	Aerosol particle size distribution	–
Surface	Lambertian	Lambertian surface reflectance	–
	BRDF	Surface bidirectional reflectance	–
Spectra	Wavelength	Central wavelength of spectral channel	nm
	FWHM	Full width at half maximum of spectral channel	nm

2.2. Crop Canopy Reflectance

The classic PROSAIL model is adopted to describe the contribution of crop canopy R_{crop}. The PROSAIL model, a fusion of PROSPECT (leaf reflectance and transmittance) and SAIL (plant canopy reflectance), is widely used for quantitative inversion of vegetation parameters [40,41]. PROSAIL provides the calculation of the spectral and directional reflectance of the plant canopy [42]. However, the application of PROSAIL for crop parameter inversion requires reflectance after atmospheric correction. Traditionally, atmospheric correction is a separate process before inversion, executed by radiative transfer models, such as 6S and RT3. The inputs of the PROSAIL model and the range of parameters are given in Table 2.

Table 2. Main input parameters of the PROSAIL model.

Model	Parameter Symbols	Parameter Description	Common Value	Search Range	Unit
PROSPECT	N	Leaf structure parameter	1.3	1.2~2.8	–
	Cab	Chlorophyll a and b content	50	20~70	$\mu g \cdot cm^{-2}$
	Car	Carotenoids content	8	6~12	$\mu g \cdot cm^{-2}$
	Cw	Equivalent water thickness	0.004	0.004~0.05	cm
	Cm	Leaf mass per unit leaf area	0.012	0.003~0.027	$g \cdot cm^{-2}$
SAIL	LAI	Leaf area index	1.4	1~7	–
	ALA	Average leaf angle	15	0~90	Degrees (°)
	Hspot	Hot spot	0.01	0.01~1.0	–
	Psoil	Soil coefficient	0.1	0.1~1.0	–

2.3. The Coupling Model for Crop Parameter Inversion

2.3.1. Coupling of Models

For modeling the UAV observations, we replace the built-in surface BRDF (Bidirectional Reflectance Distribution Function) model in UNL-VRTM with the PROSAIL model. Then, the process of atmospheric correction and the process from the biochemical content to the spectral reflectance are integrated into a unified model. The coupling model can be specifically described by the following equation:

$$M_\lambda(\mu_s, \mu_v, \phi) = U(\lambda, \mu_s, \mu_v, \phi, x_{air}, P(x_{crop})) \tag{2}$$

where M denotes the coupling model, U represents the UNL-VRTM model, and P refers to the PROSAIL model. The x_{air} and x_{crop} represent the atmospheric and crop parameters, respectively.

The diagram of the coupling model is shown in Figure 1. The inputs of the coupling model include the crop physiological parameters, canopy structure parameters, and atmospheric parameters. The output of the coupling model is the intensity of radiation.

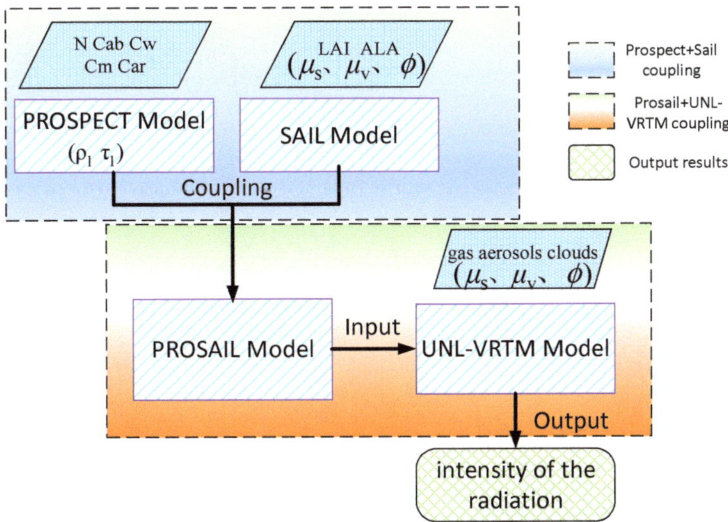

Figure 1. Diagram of the coupling model (where μ_s and μ_v are cosine of solar zenith angle and cosine of view zenith angle, respectively, and ϕ is relative azimuth angle.)

The downwelling radiation after gas absorption, Rayleigh scattering, and aerosol scattering, simulated by UNL-VRTM, is shown in Figure 2, in which the assumption of

Lambertian surface is adopted, and the surface reflectance is set to 0.2. Moreover, urban aerosol types are selected and different aerosol loadings, of which the AOD (Aerosol optical depth) changes from 0.2 to 1.0, are considered.

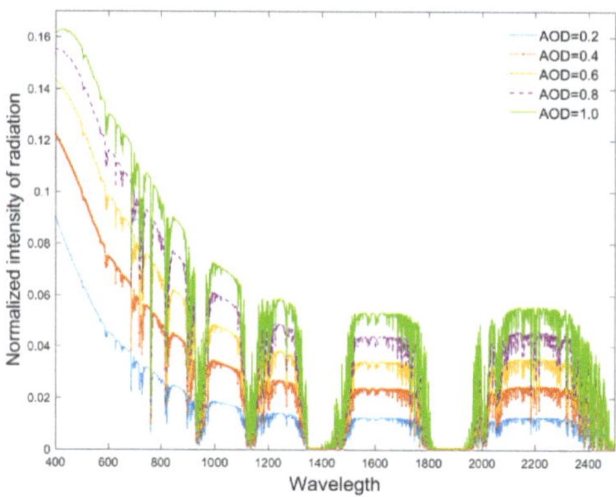

Figure 2. Forward simulation of different atmospheric aerosols by UNL-VRTM.

It can be seen from Figure 2 that the radiation gradually increased with the increase in AOD on the whole, mainly caused by aerosol scattering. However, the radiation is nearly equal regardless of the AOD in the near-infrared bands (such as 1320–1410 nm and 1800–2200 nm). For the intensity of radiation, it is found that the contribution of AOD has about a 3.0–20% difference with the increase in AOD, indicating that the value of AOD is an important input variable of the UNL-VRTM model.

Based on the values listed in Table 2, the spectral reflectance between 400 nm and 2500 nm, simulated by the PROSAIL model and the coupling model, is shown in Figure 3.

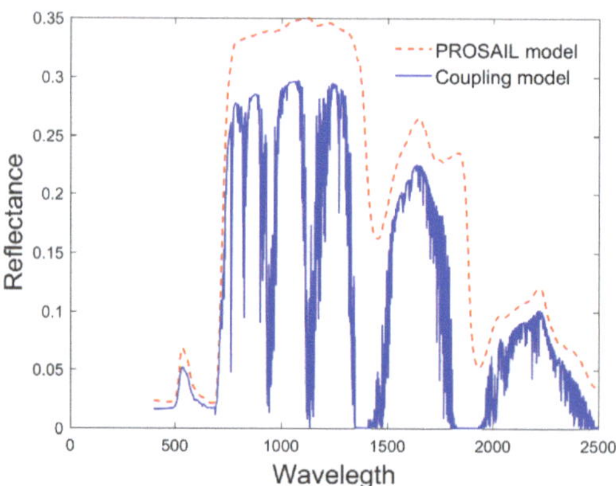

Figure 3. The spectral reflectance simulated by PROSAIL and the coupling model.

From Figure 3, the reflectance of the coupling model is smaller than the vegetation canopy reflectance simulated by the PROSAIL model at all spectral wavelengths. The decrease in reflectance at wavelengths larger than 750 nm is mainly due to aerosol absorption, and in the blue spectrum (400–450 nm) it is because of the contribution of Rayleigh scattering. Taking into account solar radiation, the spectral reflectance near 1400 and 1900 nm was equal to zero because incident and reflected energy are almost reduced by water vapor and carbon dioxide, which indicated that the forward model has a strong coupling between vegetation canopy and atmosphere environment.

2.3.2. Calculation of Jacobians

The outputs of UNL-VRTM provide the Jacobians of the measurements with respect to the surface reflectance. However, in the inversion framework, the Jacobians of the measurements with respect to the crop phenotype parameter (x_{crop}) are required. The Jacobian matrix determines the direction of convergence and the step size when solving the cost function [43]. According to the chain rule of derivative, the sensitivity transfer process of the observation vector Y to the crop parameter x_{crop} is expressed by the following equation [44–46]:

$$J = \frac{\partial Y}{\partial x_{crop}} = \frac{\partial U}{\partial P}\frac{\partial P}{\partial x_{crop}} \tag{3}$$

where $\frac{\partial U}{\partial P}$ is the Jacobian of the observation with respect to the canopy reflectance, and $\frac{\partial P}{\partial x_{crop}}$ is the Jacobian of the canopy reflectance with respect to the crop parameters, which can be calculated by the finite difference method.

3. Method for the Synchronous Retrieval of LAI and Cab

Inversion is usually defined as determining the state vector by observation vector. A complex UAV observation system can be represented by the simple mathematical model as follows:

$$Y = M(x_a, x_b) + \epsilon \tag{4}$$

where Y denotes the UAV observation vector and M denotes the forward model. ϵ is an error term including uncertainties in measurements and forward model. x_a is the state vector that contains crop parameters to be retrieved, and the choice of state vector depends on which parameter contains more information in the observation data. x_b represents other parameters.

The reverse solution of Equation (4) is the inversion process of crop parameters. Assuming that the observed vector can constrain the solution of the above equation, then theoretically x_a can be obtained when Y is given. For the UAV observation vector, due to insufficient effective observation information, it is often difficult to constrain the reverse solution process of the above equation. The state vector x_a is usually not unique, and we can obtain a statistical estimate value of x_a because of the error term ϵ.

Based on the optimal estimation theory, the inversion of crop parameters can be regarded as finding the minimum of the cost function. In most cases, it is an ill-posed problem to find the accurate value of the phenotypic parameters [47]. The a priori knowledge of crop parameters should be introduced into the inversion as a constraint term. Then, the cost function $O(x)$ is composed of two terms: the measurement term and a priori constraint term. M is always a nonlinear function for UAV remote sensing, thus solving the optimal solution of Equation (4) needs multiple iterations. The corresponding cost function in the p'th iteration of the state vector is expressed as follows [48,49].

$$O\left(x_a^p\right) = \frac{1}{2}[Y - M^p]^T S_y^{-1}[Y - M^p] + \frac{1}{2}\gamma\left(x_a^p - x_{pri}\right)^T S_{pri}^{-1}\left(x_a^p - x_{pri}\right) \tag{5}$$

where M denotes the simulation results of the forward model at the p'th iteration. S_y is the error covariance matrix, which indicates the measurement uncertainty. Lagrange multiplier γ is a regularization parameter defined following the work of Xu et al. [50]. x_a^p represents the state vector of the p'th iteration. x_{pri} and S_{pri} describe the a priori estimate and the error covariance matrix of the state vector, respectively.

Generally, finding the minimum value of $O(x)$ is a nonlinear problem and needs the gradient vector of $O(x)$, which can be further expressed as ∇

$$\nabla O\left(x_a^p\right) = J^T S_y^{-1}[Y - M^p] + \gamma S_{pri}^{-1}\left(x_a^p - x_{pri}\right) \tag{6}$$

where ∇ refers to the gradient operator, and J denotes the Jacobian matrix.

Based on multiple iterations, the state vector at the p + 1'th iteration can be described as:

$$x_a^{p+1} = x_a^p - \alpha_p H_p \nabla O\left(x_a^p\right)^T \tag{7}$$

where H is the inverse matrix of the Hessian matrix constructed with successive gradient vectors in the quasi-Newton method, and α_p is the step factor of the iteration.

The flow chart of the inversion framework is shown in Figure 4. The inputs of the inversion framework include a priori knowledge (prior estimates, upper and lower bounds), the cost function, and its gradient. Firstly, the observation vector Y consists of the hyperspectral or multispectral radiometric measurements carried by the UAV platform, and the observation error covariance matrix S_y was constructed by a given observation error. At the same time, the parameters to be retrieved and a priori estimates x_{pri} should be determined. The error covariance matrix S_{pri} is constructed according to the uncertainties of a priori estimation. After that, the cost function can be established according to Equation (5). To find the minimum value of $O(x)$, the quasi-Newton method implemented by the L-BFGS-B code was introduced. The L-BFGS-B algorithm is a highly effective tool in bounded minimization problems [51,52]. Finally, the optimal estimation of the state vector is considered to be found, if the following Equation (8) is satisfied. Otherwise, the state vector continues to be updated until the termination condition is satisfied. In other words, the crop parameters are retrieved.

$$\left\|x_a^{p+1} - x_a^p\right\|_2 \leq \text{Threshold} \tag{8}$$

where $\| \ \|_2$ denotes the L2 paradigm, and the superscript p refers to the p'th iteration.

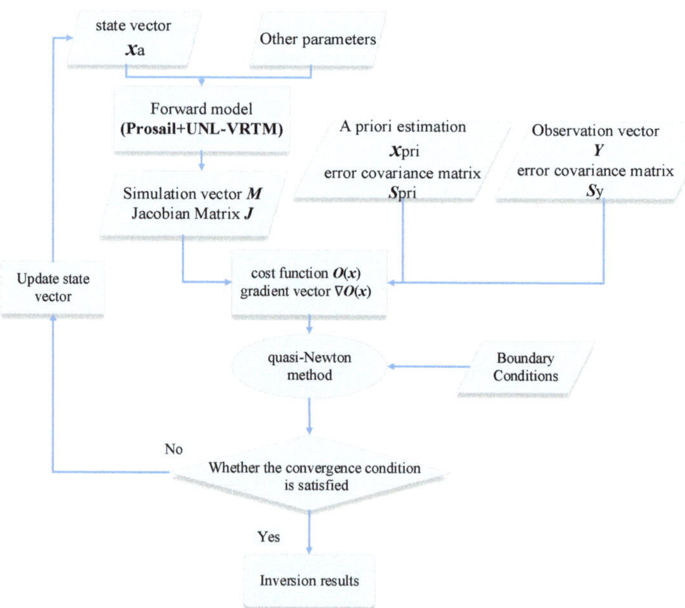

Figure 4. The optimization inversion flow chart based on iteration.

4. Model Analysis for UAV Multispectral Measurements

4.1. The Sensitivity of the Model to Crop Parameters

The multispectral sensor is widely used in the agricultural field because of its price advantage. In this paper, the band settings of UAV (P4 Multispectral, SZ DJI Technology Co., Ltd., Shenzhen, China) were adopted, as a case study, for sensitivity analysis of the model to crop parameters. The UAV is equipped with a multispectral camera with the central wavelength of 450, 560, 650, 730, and 840 nm. The band settings of UAV multispectral sensors are given in Table 3.

Table 3. The multispectral bands of UAV.

Band	Center Wavelength/nm	Bandwidth
Band1-Blue	450	16
Band2-Green	560	16
Band3-Red	650	16
Band4-RedEdge	730	16
Band5-NIR	840	26

Obviously, it is difficult to obtain all the crop parameters from the limited channels of the multispectral sensor. Thus, we must figure out exactly which parameter can be retrieved. To solve this problem, firstly, we calculate the influence of each parameter change on the model output reflectance. The simulation results are shown in Figure 5, in which we assumed that the input parameters were independent in the coupling model. For each of the parameters, the reflectance is calculated by fixing other parameter values listed in Tables 1 and 2.

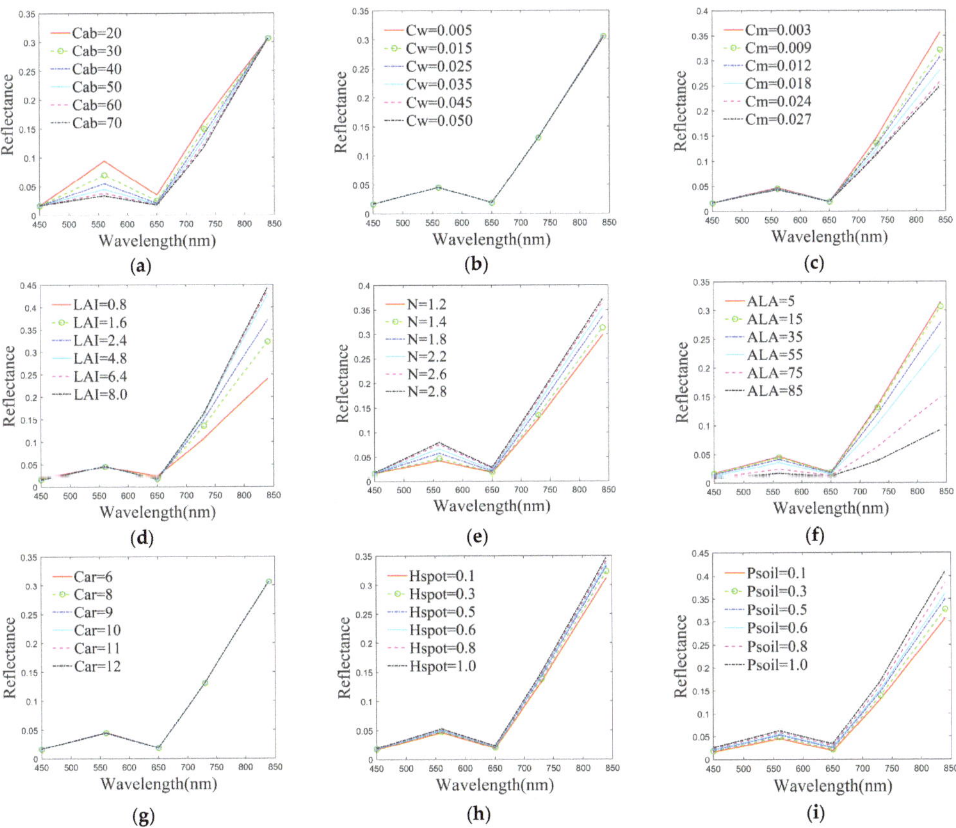

Figure 5. Spectral response under different input parameter values. (**a–d**) are the different reflectance changes of physicochemical parameters Cab, Cw, Cm and LAI, respectively. (**e,f**) are the difference reflectance changes of structure parameter N and geometry parameter ALA, respectively. (**g–i**) are the difference reflectance changes of parameters Car, Hspot and Psoil, respectively.

From Figure 5, we can see that the Cab yields a strong contribution to green band, and a weak influence to red-edge band, and the reflectance gradually decreases with the increase in Cab. It is due to the weak absorption of chlorophyll on the green band. The Cw and Cm reflect the accumulation of water and dry matter during wheat growth, respectively. However, the simulation shows that the sensor is insensitive to Cw in all the channels, and the same is true for the Car parameter on the visible band. The changes of Cm and LAI are well reflected in the reflectance of NIR band, while their contributions are opposite. Parameters related to the structure or geometry, such as the (ALA), leaf structural parameter (N), and Hspot, affect the measurement of all bands. Usually, these parameters are considered as nonstate quantities, using an empirical value with a given error during the inversion. The Psoil represents the proportion of bare soil and vegetation cover; therefore, the increase in Psoil means the enhancement of contribution from the bare soil surface. From Figure 5i, the reflectance of each band increases with the increase in Psoil, since the reflectance of bare soil has little difference at each wavelength. These results were consistent with the previous research [53–55].

4.2. Parameters Information Content Analysis

Furthermore, the information of each parameter contained in the multispectral observation is quantified using the degrees of freedom of the signal (DFS) [56]. The DFS is defined as the partial derivative of the posterior estimate values with respect to state parameters and can be calculated by the following equation:

$$\frac{\partial \hat{x}}{\partial x} = \left(J^{\mathrm{T}} S_{\mathrm{Obs}}^{-1} J + S_{\mathrm{pri}}^{-1} \right)^{-1} J^{\mathrm{T}} S_{\mathrm{Obs}}^{-1} J \tag{9}$$

Ideally, the state vector can be completely determined by measurements, that is, all the parameters can be obtained for observation data. In this case, the amount of information for each parameter is equal to 1 and the sum of the information of all parameters is equal to the number of parameters. However, subject to observation and systematic errors, the DFS is always less than 1 for each parameter. Therefore, the signal degrees of freedom can characterize the ability of observations and models to invert parameters. The closer the parameter's DFS is to 1, the more adequate information the observation contains about that parameter.

To study the contribution of the phenotypic parameters to observation, Figure 6a shows the variation of all parameters information obtained by different wavelengths' observation; in other words, the number of parameters can be retrieved from single wavelength observation. In order to describe the observation capabilities of the P4 multispectral UAV, we carried the forward simulation of multispectral bands. Finally, the DFS of each parameter is calculated according to the information content analysis method, as shown in Figure 6b.

From Figure 6a, the total DFS of different wavelengths between 400 and 1200 nm is less than 1, which indicated that the single-band observation has insufficient retrieval ability for multiple parameters. It is because the single band cannot obtain sufficient information. From Figure 6b, (1) the DFS of LAI, Cab, and N were 0.83, 0.71, and 0.58, respectively, significantly higher than Cm (DFS = 0.2). It indicated that the observation contains more effective information on these parameters and can further be better extracted. (2) The DFS of ALA, Car, and Psoil were lower than 0.2, indicating that these parameters are difficult to be retrieved by multispectral observation and can be used as a priori value in the inversion framework.

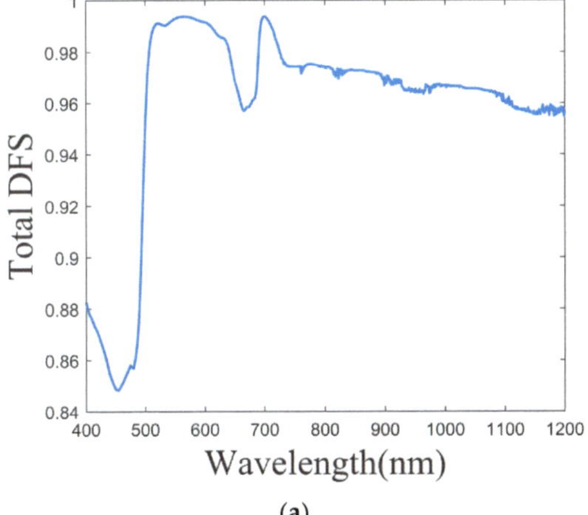

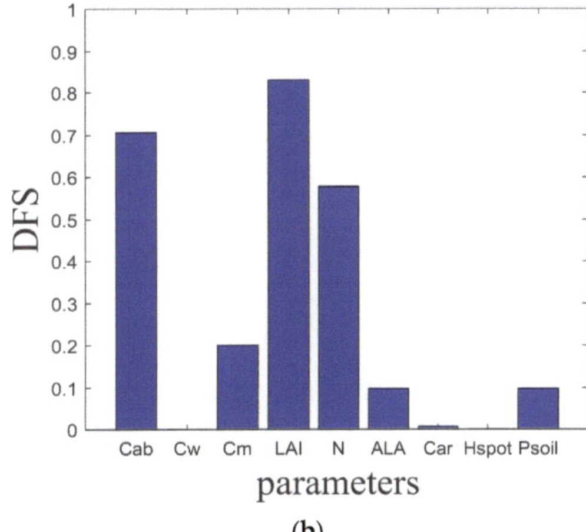

(a)

(b)

Figure 6. The DFS of crop parameters: (**a**) the DFS of different wavelengths for all parameters and (**b**) the DFS of five bands for each parameter.

4.3. Setting of State Vectors and Boundary Conditions

For crop phenotype parameters inversion, we assumed that the observation error does not vary with wavelength and observation geometry in this paper. The state and nonstate vectors of the crop phenotype parameters are composed of the following equations:

$$\begin{cases} x_a = [\text{Cab}, \text{LAI}]^T \\ x_b = [\text{N}, \text{Car}, \text{Cw}, \text{Cm}, \text{ALA}, \text{hspot}, \text{psoil}]^T \end{cases} \tag{10}$$

The boundary range of the state vector determines the interval range for finding the optimal solution. The prior estimation is based on the historical data statistics of the same period of crop parameters. The values of the nonstate vector come from the empirical values of the fertility period and sensitivity analysis results, which provide some auxiliary constraints for the optimal solution. In this paper, the boundary conditions for the LAI were $(0.5, 7)$ and for Cab were $(20, 70)$.

5. Results

5.1. Validation of the Forward Model

The study area is a national high-standard farmland of 5000 hm^2 in Qixian County, Henan Province (114.17° E, 35.6° N), where winter wheat was cultivated annually from October to June, approximately. The region has a warm temperate humid monsoon climate with warm and rainy summer and cold and dry winter. The measured data were derived from the FieldSpec Handheld (a handheld geophysical spectrometer, Analytica Spectra Devices, Inc., Boulder, CO, USA). The spectral range of the measuring instrument is 325~1075 nm, and the spectral resolution was 3 nm at wavelengths 325~700 nm. We collected the test data on 4 March 2021 at about 12:00 when the solar light intensity was stable and the weather was clear and cloudless. The measurement results can represent the true reflectance of the vegetation canopy because the instrument is closer to the target and less influenced by aerosols and water vapor. Figure 7 compares the results between the simulated data at a solar zenith angle of 5° and the measured data at a sampling test point.

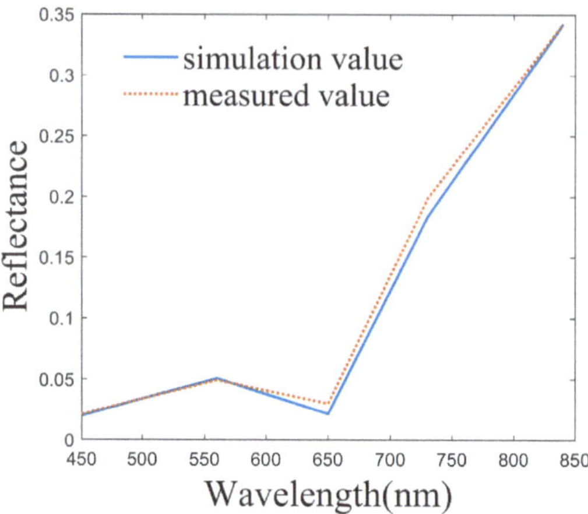

Figure 7. Comparison of simulated values and ground truth values.

It can be seen from Figure 7 that the simulated values of band 3 and band 4 were lower than the ground-measured values. However, the error between the measured and simulated reflectance at the sampling point was within 2%. The results indicated that the simulation of crop reflectance can be achieved by the coupled model from solar radiation to vegetation, further reflecting that the forward modeling process is reasonable.

5.2. Retrieval Demonstration and Self-Consistency Tests

Self-consistency tests following the inversion framework are conducted. Firstly, forward simulation datasets are obtained by the coupling model according to the bands of P4 multispectral UAV. The input parameters varied by 10% themselves. The observation geometries are adopted for the yearly solar altitude variation at mid-latitudes in the northern hemisphere [57], in which the solar zenith angle varies from $10°$ to $60°$, view zenith angle and relative azimuth are both $0°$. Secondly, to simulate the measurement error, we also added the 5% Gaussian noise to the simulation results. Finally, we obtained a total of 671 sets of validation data.

Figure 8a,b show the iterative process of apparent reflectance and the convergence of cost function for Cab = 63, LAI = 1.3, and the solar zenith angle of $15°$ in the optimized inversion. The red solid line represents the measured results from the forward model simulation.

As shown in Figure 8a, the model simulated value gradually approaches the observations with the increase in iteration times, and the best fit has been achieved through the OE algorithm at the 14th iteration. The fitting residual error is used to describe the difference between the observation and simulated values. In Figure 8a, the residual error of the last iteration is only 0.023%, and the average residual error of all simulated data is 0.3% after statistics. From Figure 8b, the cost function changes rapidly from the second to the fifth iteration and gradually approached a minimum value after the eighth iteration. The minimum value is determined by the observation error and the error in the radiation transfer and coupling process between the vegetation canopy and the atmosphere. After statistics, the average iteration times of test datasets were about 14 when the cost function reaches the minimum value. The details of the inversion process indicated that the optimal estimation inversion algorithm can realize the dynamic adjustment of inversion parameters and had good convergence in obtaining the retrieval results.

A comparison is summarized in Figure 9 between the inversion values reconstructed from the optimized inversion framework and the true value used in the dataset. The coefficient of determination (R^2) and root-mean-square error (RMSE) are used to verify the consistency of the optimized algorithm. A larger R^2 and a smaller RMSE indicate the higher accuracy of the forward model and the better consistency of the inversion framework.

The retrieved Cab and LAI synchronously under different observation geometries were compared to the true value as shown in Figure 9a,b.

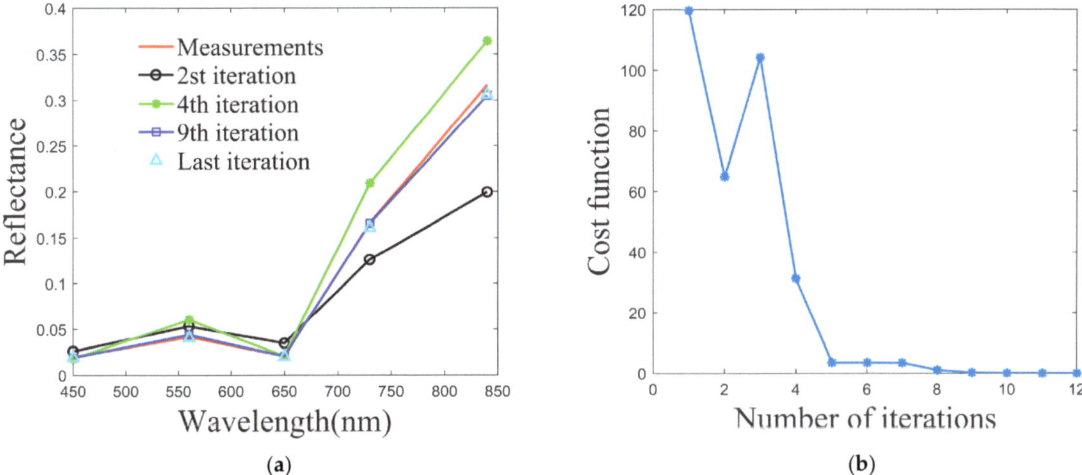

(a)

(b)

Figure 8. Retrieval process demonstration: (**a**) the illustration of the iterative process for reflectance based on the coupling model and (**b**) the description of the convergence process for the cost function.

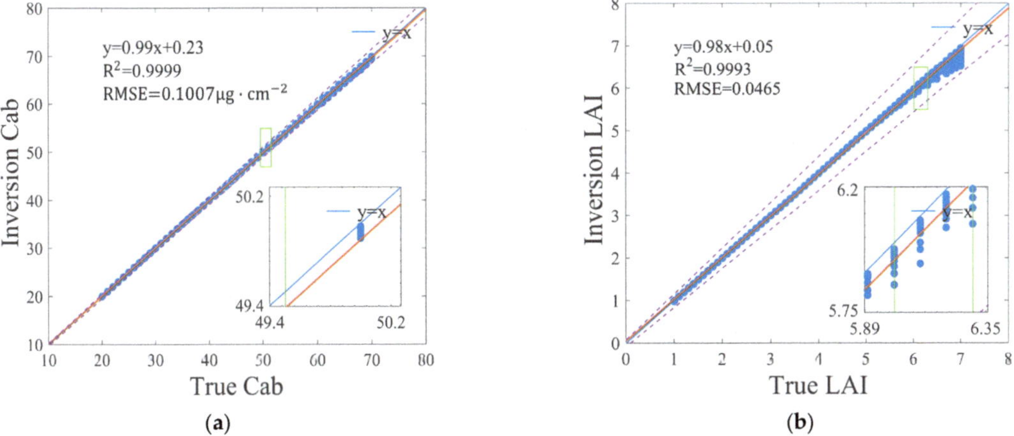

(a)

(b)

Figure 9. Validation of the Cab and LAI with the OE-iteration method: (**a**) validation of the Cab with the OE-iteration method against the true value under different observation geometries and (**b**) validation of the LAI with the OE-iteration method against the true value under different observation geometries. The blue solid line, dashed line, and orange solid line are the 1:1 line, error line, and fit line, respectively. The error lines of (**a**,**b**) are $E = x \pm 0.02x \pm 0.08$ and $E = x \pm 0.08x \pm 0.08$. The statistics of the linear fitting result, including slope, intercept, R^2, and RMSE, are listed in the upper left corner of each scatter plot.

As seen in Figure 9a, the R^2 value for Cab was 0.9999, while the RMSE was 0.1007, which indicated that the Cab had good retrieval results. Similarly, the verification results of LAI were consistent with Cab. Meanwhile, it is worth noting that the slopes of the fit lines were all close to 1. With the variation of solar zenith angle, the error lines of Cab and LAI were within $E = x \pm 0.02x \pm 0.08$ and $E = x \pm 0.08x \pm 0.08$, respectively. The error line is defined as a correlation range between the true value and the inversion value. The results showed the final model retrieval values were in good agreement with the real values.

Moreover, for the validation of the influence of different aerosol loads on the inversion results, another case was also performed. The viewing geometries, including solar zenith angle, view zenith angle, and relative azimuth angle, were $65°$, $0°$, and $120°$, respectively. The state parameters varied by 50% themselves, with different aerosol AOD from 0.1 to 1.0, a total of 130 datasets are obtained. Figure 10a,b show the scatterplots between the retrieval results and the true value under different AODs.

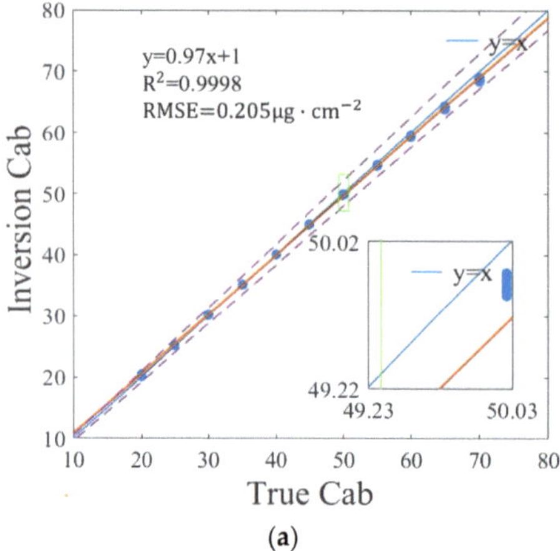
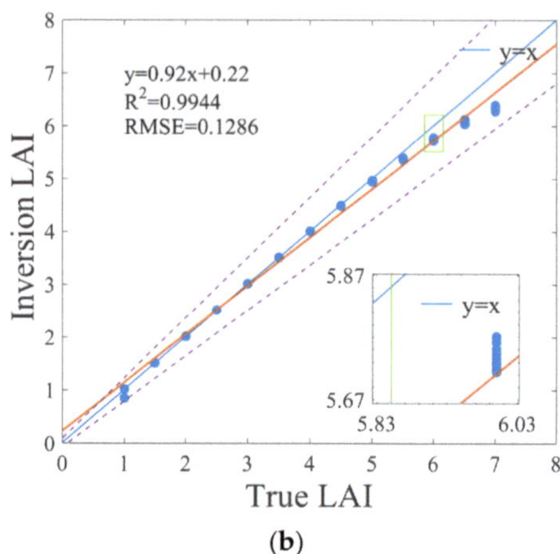

Figure 10. Validation of the Cab and LAI with the OE-iteration method: (**a**) validation of the Cab with the OE-iteration method against the true value under different AODs and (**b**) validation of the LAI with the OE-iteration method against the true value under different AODs. The blue solid line, dashed line, and orange solid line are the 1:1 line, error line, and fit line, respectively. The error lines of (**a**,**b**) are $E = x \pm 0.04x \pm 0.08$.and $E = x \pm 0.14x \pm 0.08$. The statistics of the linear fitting result, including slope, intercept, R^2, and RMSE, are listed in the upper left corner of each scatter plot.

As seen in Figure 10a, the R^2 value and RMSE for Cab were 0.9998 and 0.205, which indicated the uncertainty of the retrieved Cab increased weakly with the varying aerosol loading compared with Figure 9a. It can be seen from Figure 10b that the retrieved LAI value is lower than the true value when the LAI is larger than 6. However, the validation results were almost similar to Figure 9a,b, with the R^2 values all larger than 0.99. The results indicated that the developed inversion framework showed good performance regardless of the varied aerosol-loading situation.

6. Discussion

The PROSAIL model is sensitive to crop parameters that have been well documented in previous studies [21,40], and UAV-observed reflectance data can in turn promote LAI and Cab retrieval. Nevertheless, solar radiation and atmospheric environment information from UAV observation in physical models cannot be considered in general. Therefore, the forward model by replacing the built-in surface BRDF model in UNL-VRTM with the PROSAIL model was established in this study, which was calculated based on the optimal estimation inversion framework to obtain crop parameters retrieved. The advantage of the forward model is to achieve atmospheric correction during the radiation transfer process and save the costs in human resources, materials, and time. As a result, compared with the measured data collected by the FieldSpec Handheld, the physical transmission model from solar radiation to the crop canopy, and then to the sensor is used as the UAV observation system. Moreover, crop multiparameters retrieved from current inversion approaches cannot be estimated synchronously. Such as LUTs and machine learning algorithms are not suitable for multiparameter inversion [11,27]. Therefore, an optimal estimation inversion algorithm from UAV remote sensing data was proposed, and we can obtain a statistical estimate value of the state vector based on optimal estimation theory. The advantage of the proposed inversion algorithm is that it can extract crop LAI and Cab synchronously from UAV observation.

Based on the optimal estimation theory, we can find the best solution in the continuous state space. Theoretically, the inversion accuracy of crop parameters by the optimal estimation method is higher than LUTs [58]. However, the optimal estimation algorithm requires calculating forward transmission timely and achieves multiple iterations. Thus, it took a lot of time to complete the inversion. After statistics, the average iteration times of test datasets were about 14. We also need to improve the computational efficiency of the inversion algorithm with a larger amount of data.

From the calculation process of the cost function and its gradient vector, it can be seen that our optimal algorithm is related to a priori constraints and the observation data. The algorithm test was carried out based on simulated data in our inversion framework. Although we added noise to the simulated data, the uncertainty of the actual observation data is more complex than the simulated data in fact. The accuracy of the inversion result not only is affected by the uncertainty of observation data, but also depends on prior constraint values (including nonstate parameters value, the initial value of the state parameters, and their boundary conditions). The prior values are derived from historical data statistics of the same period of crop parameters. Reasonable prior estimates and error assumptions can constrain the inversion results to a relatively optimal range and further speed up the iterative computation. Furthermore, we need to optimize the parameter of the measured data and prior constraints and carry out the inversion verification of the actual observation data.

7. Conclusions

In this study, we used a physical model to establish a multiparameter inversion framework for crop phenotypes based on optimal estimation inversion theory. Firstly, we unified PROSAIL and UNL-VRTM models to achieve synchronous atmospheric correction. Secondly, multiple phenotypic parameters inversion framework was established based on the optimized estimation method with prior constraints. Then, the contribution of parameters to the model was obtained by parameter sensitivity analysis, and the parameters to be retrieved were determined according to the information content analysis method. Finally, the optimized estimation inversion framework self-consistency test was performed, and the model inversion values were validated against the true values under different observation geometries and aerosol loadings. The results showed that the coefficients of determination (R^2) between the inversion values and the true values were above 0.99, indicating that UAV multispectral observations can support the inversion of LAI and Cab. It also indicated that the inversion framework can make full use of the available radiation spectral information and has good stability. The findings from this study suggest that our proposed forward model has a strong coupling between vegetation reflectance and atmosphere. Moreover, there were good convergence and consistency for the inversion framework in obtaining inversion LAI and Cab. The optimal inversion framework for the synchronous retrieval of crop phenotypic parameters is expected to be performed on actual observation data in the future and can be applied to the monitoring of crop parameters in the field environment.

Author Contributions: Conceptualization, F.Z., J.J. and H.M.; methodology, F.Z. and X.W.; validation, F.Z. and X.W.; formal analysis, F.Z. and J.J.; investigation, F.Z. and X.W.; resources, F.Z., X.W. and S.Z.; data curation, F.Z. and X.W.; writing—original draft preparation, F.Z.; writing—review and editing, F.Z., X.W. and H.C.; supervision, Y.S.; project administration, H.C., Y.S. and J.J.; funding acquisition, J.J. and H.M. All authors have read and agreed to the published version of the manuscript.

Funding: This research was funded by the Key specialized research and development breakthrough in Henan province (No. 222102110105; No. 222102110215); Major science and technology project of Henan Province (No. 221100110800); the Key Scientific Research Projects of Colleges and Universities in Henan Province (No. 21A416003).

Data Availability Statement: Not applicable.

Acknowledgments: We are very grateful to Jun Wang and Xiaoguang Xu for the UNL-VRTM radiative transfer code. The data and Fortran codes used for the forward simulations are available on the UNL-VRTM website (http://www.unl-vrtm.org, accessed on 23 March 2021). We acknowledge the L-BFGS-B software package developed by Ciyou Zhu, Jorge Nocedal, and Richard Byrd.

Conflicts of Interest: The authors declare no conflict of interest.

References

1. Yang, G.J.; Liu, J.G.; Zhao, C.J.; Li, Z.H.; Huang, Y.B.; Yu, H.Y.; Xu, B.; Yang, X.D.; Zhu, D.M.; Zhang, X.Y.; et al. Unmanned aerial vehicle remote sensing for field-based crop phenotyping: Current status and perspectives. *Front. Plant Sci.* **2017**, *8*, 1111. [CrossRef] [PubMed]
2. Zhang, H.D.; Wang, L.Q.; Tian, T.; Yin, J.H. A review of unmanned aerial vehicle low-altitude remote sensing (uav-lars) use in agricultural monitoring in China. *Remote Sens.* **2021**, *13*, 1221. [CrossRef]
3. Tattaris, M.; Reynolds, M.P.; Chapman, S.C. A direct comparison of remote sensing approaches for high-throughput phenotyping in plant breeding. *Front. Plant Sci.* **2016**, *7*, 1131. [CrossRef]
4. Haghighattalab, A.; Perez, L.G.; Mondal, S.; Singh, D.; Schinstock, D.; Rutkoski, J.; Ortiz-Monasterio, I.; Singh, R.P.; Goodin, D.; Poland, J. Application of unmanned aerial systems for high throughput phenotyping of large wheat breeding nurseries. *Plant Methods* **2016**, *12*, 35. [CrossRef] [PubMed]
5. Gracia-Romero, A.; Kefauver, S.C.; Fernandez-Gallego, J.A.; Vergara-Diaz, O.; Nieto-Taladriz, M.T.; Araus, J.L. UAV and ground image-based phenotyping: A proof of concept with durum wheat. *Remote Sens.* **2019**, *11*, 1244. [CrossRef]
6. Tao, H.L.; Feng, H.K.; Xu, L.J.; Miao, M.K.; Long, H.L.; Yue, J.B.; Li, Z.H.; Yang, G.J.; Yang, X.D.; Fan, L.L. Estimation of crop growth parameters using UAV-based hyperspectral remote sensing data. *Sensors* **2020**, *20*, 1296. [CrossRef]
7. Feng, A.J.; Zhou, J.F.; Vories, E.D.; Sudduth, K.A.; Zhang, M.N. Yield estimation in cotton using UAV-based multi-sensor imagery. *Biosyst. Eng.* **2020**, *193*, 101–114. [CrossRef]
8. Ma, C.Y.; Liu, M.X.; Ding, F.; Li, C.C.; Cui, Y.Q.; Chen, W.N.; Wang, Y.L. Wheat growth monitoring and yield estimation based on remote sensing data assimilation into the SAFY crop growth model. *Sci. Rep.* **2022**, *12*, 5473. [CrossRef]
9. Velusamy, P.; Rajendran, S.; Mahendran, R.K.; Naseer, S.; Shafiq, M.; Choi, J.G. Unmanned aerial vehicles (UAV) in precision agriculture: Applications and challenges. *Energies* **2022**, *15*, 217. [CrossRef]
10. Liu, K.; Zhou, Q.B.; Wu, W.B.; Tian, X.; Tang, H.J. Estimating the crop leaf area index using hyperspectral remote sensing. *J. Integr. Agric.* **2016**, *15*, 475–491. [CrossRef]
11. Darvishzadeh, R.; Atzberger, C.; Skidmore, A.; Schlerf, M. Mapping grassland leaf area index with airborne hyperspectral imagery: A comparison study of statistical approaches and inversion of radiative transfer models. *ISPRS J. Photogramm. Remote Sens.* **2011**, *66*, 894–906. [CrossRef]
12. Liang, L.; Huang, T.; Di, L.P.; Geng, D.; Di, G.; Yan, J.; Wang, S.G.; Wang, L.J.; Li, L.; Chen, B.Q.; et al. Influence of different bandwidths on LAI estimation using vegetation indices. *IEEE J. Sel. Top. Appl. Earth Obs. Remote Sens.* **2020**, *13*, 1494–1502. [CrossRef]
13. Viña, A.; Gitelson, A.A.; Nguy-Robertson, A.L.; Peng, Y. Comparison of different vegetation indices for the remote assessment of green leaf area index of crops. *Remote Sens. Environ.* **2011**, *115*, 3468–3478. [CrossRef]
14. Yu, K.; Lenz-Wiedemann, V.; Chen, X.P.; Bareth, G. Estimating leaf chlorophyll of barley at different growth stages using spectral indices to reduce soil background and canopy structure effects. *ISPRS J. Photogramm. Remote Sens.* **2014**, *97*, 58–77. [CrossRef]
15. Lin, H.; Liang, L.; Zhang, L.P.; Du, P.J. Wheat leaf area index inversion with hyperspectral remote sensing based on support vector regression algorithm. *Trans. Chin. Soc. Agric. Eng.* **2013**, *29*, 139–146.
16. Liang, L.; Di, L.P.; Zhang, L.P.; Deng, M.X.; Qin, Z.H.; Zhao, S.H.; Lin, H. Estimation of crop LAI using hyperspectral vegetation indices and a hybrid inversion method. *Remote Sens. Environ.* **2015**, *165*, 123–134. [CrossRef]
17. Verrelst, J.; Muñoz, J.; Alonso, L.; Delegido, J.; Rivera, J.P.; Camps-Valls, G.; Moreno, J. Machine learning regression algorithms for biophysical parameter retrieval: Opportunities for sentinel-2 and -3. *Remote Sens. Environ.* **2012**, *118*, 127–139. [CrossRef]
18. Lu, X.P.; Wang, X.X.; Zhang, X.J.; Wang, J.; Yang, Z.N. Winter wheat leaf area index inversion by the genetic algorithms neural network model based on SAR data. *Int. J. Digit. Earth* **2022**, *15*, 362–380. [CrossRef]
19. Saddik, A.; Latif, R.; Elhoseny, M.; El Ouardi, A. Real-time evaluation of different indexes in precision agriculture using a heterogeneous embedded system. *Sustain. Comput. Inform. Syst.* **2021**, *30*, 100506. [CrossRef]
20. Saddik, A.; Latif, R.; El Ouardi, A.; Alghamdi, M.I.; Elhoseny, M. Improving Sustainable Vegetation Indices Processing on Low-Cost Architectures. *Sustainability* **2022**, *14*, 2521. [CrossRef]

21. Duan, S.B.; Li, Z.L.; Wu, H.; Tang, B.H.; Ma, L.L.; Zhao, E.Y.; Li, C.R. Inversion of the PROSAIL model to estimate leaf area index of maize, potato, and sunflower fields from unmanned aerial vehicle hyperspectral data. *Int. J. Appl. Earth Obs. Geoinf.* **2014**, *26*, 12–20. [CrossRef]

22. Upreti, D.; Huang, W.J.; Kong, W.P.; Pascucci, S.; Pignatti, S.; Zhou, X.F.; Ye, H.C.; Casa, R. A comparison of hybrid machine learning algorithms for the retrieval of wheat biophysical variables from sentinel-2. *Remote Sens.* **2019**, *11*, 481. [CrossRef]

23. Rivera, J.P.; Verrelst, J.; Leonenko, G.; Moreno, J. Multiple cost functions and regularization options for improved retrieval of leaf chlorophyll content and LAI through Inversion of the PROSAIL Model. *Remote Sens.* **2013**, *5*, 3280–3304. [CrossRef]

24. He, W.; Yang, H.; Pan, J.J.; Xu, P.P. Exploring Optimal Design of Look-Up Table for PROSAIL Model Inversion with Multi-Angle MODIS Data. In Proceedings of the Land Surface Remote Sensing, Kyoto, Japan, 21 November 2012; SPIE: Bellingham, WA, USA, 2012; Volume 8524, pp. 327–339.

25. Richter, K.; Hank, T.B.; Vuolo, F.; Mauser, W.; D'Urso, G. Optimal exploitation of the sentinel-2 spectral capabilities for crop leaf area index mapping. *Remote Sens.* **2012**, *4*, 561–582. [CrossRef]

26. Sun, J.; Wang, L.; Shi, S.; Li, Z.H.; Yang, J.; Gong, W.; Wang, S.Q.; Tagesson, T. Leaf pigment retrieval using the PROSAIL model: Influence of uncertainty in prior canopy-structure information. *Crop J.* **2022**, *10*, 1251–1263. [CrossRef]

27. Miraglio, T.; Adeline, K.; Huesca, M.; Ustin, S.; Briottet, X. Joint use of PROSAIL and DART for fast LUT building: Application to gap fraction and leaf biochemistry estimations over sparse oak stands. *Remote Sens.* **2020**, *12*, 2925. [CrossRef]

28. Darvishzadeh, R.; Matkan, A.A.; Ahangar, A.D. Inversion of a radiative transfer model for estimation of rice canopy chlorophyll content using a lookup-table approach. *IEEE J. Sel. Top. Appl. Earth Obs. Remote Sens.* **2012**, *5*, 1222–1230. [CrossRef]

29. Sun, B.; Wang, C.F.; Yang, C.H.; Xu, B.D.; Zhou, G.S.; Li, X.Y.; Xie, J.; Xu, S.J.; Liu, B.; Zhang, J.; et al. Retrieval of rapeseed leaf area index using the PROSAIL model with canopy coverage derived from UAV images as a correction parameter. *Int. J. Appl. Earth Obs. Geoinf.* **2021**, *102*, 102373. [CrossRef]

30. Berger, K.; Atzberger, C.; Danner, M.; D'Urso, G.; Mauser, W.; Vuolo, F.; Hank, T. Evaluation of the PROSAIL model capabilities for future hyperspectral model environments: A review study. *Remote Sens.* **2018**, *10*, 85. [CrossRef]

31. Bassani, C.; Sterckx, S. Calibration of satellite low radiance by AERONET-OC products and 6SV model. *Remote Sens.* **2021**, *13*, 781. [CrossRef]

32. Rosas, J.; Houborg, R.; McCabe, M.F. Sensitivity of Landsat 8 surface temperature estimates to atmospheric profile data: A study using MODTRAN in dryland irrigated systems. *Remote Sens.* **2017**, *9*, 988. [CrossRef]

33. Shang, H.; Chen, L.; Breon, F.M.; Letu, H.; Li, S.; Wang, Z.; Su, L. Impact of cloud horizontal inhomogeneity and directional sampling on the retrieval of cloud droplet size by the POLDER instrument. *Atmos. Meas. Tech.* **2015**, *8*, 4931–4945. [CrossRef]

34. Zheng, F.X. *Aerosol Multi-Parameter Optimal Retrieval from Multi-Angle Polarization Satellite Data*; University of Chinese Academy of Sciences: Beijing, China, 2019.

35. Gómez-Dans, J.L.; Lewis, P.E.; Disney, M. Efficient emulation of radiative transfer codes using gaussian processes and application to land surface parameter inferences. *Remote Sens.* **2016**, *8*, 119. [CrossRef]

36. Mishchenko, M.I.; Travis, L.D.; Lacis, A.A. *Scattering, Absorption, and Emission of Light by Small Particles*; Cambridge University Press: Cambridge, UK, 2002.

37. Kaufman, Y.J.; Tanré, D.; Gordon, H.R.; Nakajima, T.; Lenoble, J.; Frouin, R.; Grassl, H.; Herman, B.M.; King, M.D.; Teillet, P.M. Passive remote sensing of tropospheric aerosol and atmospheric correction for the aerosol effect. *J. Geophys. Res. Atmos.* **1997**, *102*, 16815–16830. [CrossRef]

38. Hou, W.; Li, Z.; Wang, J.; Xu, X.; Goloub, P.; Qie, L. Improving remote sensing of aerosol microphysical properties by near-infrared polarimetric measurements over vegetated land: Information content analysis. *J. Geophys. Res. Atmos.* **2018**, *123*, 2215–2243. [CrossRef]

39. Wang, J.; Xu, X.G.; Ding, S.G.; Zeng, J.; Spurr, R.; Liu, X.; Chance, K.; Mishchenko, M. A numerical testbed for remote sensing of aerosols, and its demonstration for evaluating retrieval synergy from a geostationary satellite constellation of GEO-CAPE and GOES-R. *J. Quant. Spectrosc. Radiat. Transf.* **2014**, *146*, 510–528. [CrossRef]

40. Adeluyi, O.; Harris, A.; Verrelst, J.; Foster, T.; Clay, G.D. Estimating the phenological dynamics of irrigated rice leaf area index using the combination of PROSAIL and gaussian process regression. *Int. J. Appl. Earth Obs. Geoinf.* **2021**, *102*, 102454. [CrossRef]

41. Jacquemoud, S.; Verhoef, W.; Baret, F.; Bacour, C.; Zarco-Tejada, P.J.; Asner, G.P.; François, C.; Ustin, S.L. PROSPECT+SAIL models: A review of use for vegetation characterization. *Remote Sens. Environ.* **2009**, *113*, S56–S66. [CrossRef]

42. Tripathi, R.; Sahoo, R.N.; Sehgal, V.K.; Tomar, R.K.; Chakraborty, D.; Nagarajan, S. Inversion of PROSAIL model for retrieval of plant biophysical parameters. *J. Indian Soc. Remote Sens.* **2012**, *40*, 19–28. [CrossRef]

43. Hou, W.; Wang, J.; Xu, X.; Reid, J.S.; Han, D. An algorithm for hyperspectral remote sensing of aerosols: 1. Development of theoretical framework. *J. Quant. Spectrosc. Radiat. Transf.* **2016**, *178* (Suppl. C), 400–415. [CrossRef]

44. Zheng, F.X.; Li, Z.Q.; Hou, W.Z.; Qie, L.L.; Zhang, C. Aerosol retrieval study from multiangle polarimetric satellite data based on optimal estimation method. *J. Appl. Remote Sens.* **2020**, *14*, 014516. [CrossRef]

45. Zheng, F.X.; Hou, W.Z.; Li, Z.Q. Improvement of Aerosol Fine Mode Fraction Retrieval from Skylight Measurements by Degree of Linear Polarization: Information Content Analysis. In Proceedings of the AOPC 2020: Optical Spectroscopy and Imaging; and Biomedical Optics, Beijing, China, 5 November 2020; SPIE: Bellingham, WA, USA, 2020; Volume 11566, p. 1156602.

46. Spurr, R.; Wang, J.; Zeng, J.; Mishchenko, M.I. Linearized T-matrix and Mie scattering computations. *J. Quant. Spectrosc. Radiat. Transf.* **2012**, *113*, 425–439. [CrossRef]

47. Combal, B.; Baret, F.; Weiss, M.; Trubuil, A.; Macé, D.; Pragnère, A.; Myneni, R.; Knyazikhin, Y.; Wang, L. Retrieval of canopy biophysical variables from bidirectional reflectance. *Remote Sens. Environ.* **2003**, *84*, 1–15. [CrossRef]
48. Dubovik, O.; Herman, M.; Holdak, A.; Lapyonok, T.; Tanré, D.; Deuzé, J.L.; Ducos, F.; Sinyuk, A.; Lopatin, A. Statistically optimized inversion algorithm for enhanced retrieval of aerosol properties from spectral multi-angle polarimetric satellite observations. *Atmos. Meas. Tech.* **2011**, *4*, 975–1018. [CrossRef]
49. Wu, L.; Hasekamp, O.; Van Diedenhoven, B.; Cairns, B. Aerosol retrieval from multiangle, multispectral photopolarimetric measurements: Importance of spectral range and angular resolution. *Atmos. Meas. Tech.* **2015**, *8*, 2625–2638. [CrossRef]
50. Xu, X.G.; Wang, J. Retrieval of aerosol microphysical properties from AERONET photopolarimetric measurements: 1. Information content analysis. *J. Geophys. Res. Atmos.* **2015**, *120*, 7059–7078. [CrossRef]
51. Byrd, R.H.; Lu, P.H.; Nocedal, J.; Zhu, C.Y. A limited memory algorithm for bound constrained optimization. *SIAM J. Sci. Comput.* **1995**, *16*, 1190–1208. [CrossRef]
52. Xiao, Y.H.; Zhang, H.C. Modified subspace limited memory BFGS algorithm for large-scale bound constrained optimization. *J. Comput. Appl. Math.* **2008**, *222*, 429–439. [CrossRef]
53. Su, W.; Wu, J.Y.; Wang, X.S.; Xie, Z.X.; Zhang, Y.; Tao, W.C.; Jin, T. Retrieving corn canopy leaf area index based on sentinel-2 image and PROSAIL model parameter calibration. *Spectrosc. Spectr. Anal.* **2021**, *41*, 1891–1897.
54. Jia, J.Q. Study on the Inversion Method of Leaf Area Index of Summer Maize at Different Growth Stages Based on GF-2 Satellite. Northwest University: Kirkland, WA, USA, 2018.
55. Wan, L.; Zhang, J.F.; Dong, X.Y.; Du, X.Y.; Zhu, J.P.; Sun, D.W.; Liu, Y.F.; He, Y.; Cen, H.Y. Unmanned aerial vehicle-based field phenotyping of crop biomass using growth traits retrieved from PROSAIL model. *Comput. Electron. Agric.* **2021**, *187*, 106304. [CrossRef]
56. Hou, W.Z.; Wang, J.; Xu, X.G.; Reid, J.S. An algorithm for hyperspectral remote sensing of aerosols: 2. Information content analysis for aerosol parameters and principal components of surface spectra. *J. Quant. Spectrosc. Radiat. Transf.* **2017**, *192* (Suppl. C), 14–29. [CrossRef]
57. Wang, G.A.; Mi, H.T.; Deng, T.H.; Li, Y.N.; Li, L.X. Calculation of the change range of the sun high angle and the azimuth of sunrise and sunset in one year. *Meteorol. Environ. Sci.* **2007**, *30*, 161–164.
58. Jeong, U.; Kim, J.; Ahn, C.; Torres, O.; Liu, X.; Bhartia, P.K.; Spurr, R.J.; Haffner, D.; Chance, K.; Holben, B.N. An optimal-estimation-based aerosol retrieval algorithm using OMI near-UV observations. *Atmos. Chem. Phys.* **2016**, *16*, 177–193. [CrossRef]

Article

Application of UAV RGB Images and Improved PSPNet Network to the Identification of Wheat Lodging Areas

Jinling Zhao [1], Zheng Li [2], Yu Lei [1,*] and Linsheng Huang [1]

[1] National Engineering Research Center for Analysis and Application of Agro-Ecological Big Data, Anhui University, Hefei 230601, China
[2] School of Electronic and Information Engineering, Anhui University, Hefei 230601, China
* Correspondence: leiyu@ahu.edu.cn

Abstract: As one of the main disasters that limit the formation of wheat yield and affect the quality of wheat, lodging poses a great threat to safety production. Therefore, an improved PSPNet (Pyramid Scene Parsing Network) integrating the Normalization-based Attention Module (NAM) (NAM-PSPNet) was applied to the high-definition UAV RGB images of wheat lodging areas at the grain-filling stage and maturity stage with the height of 20 m and 40 m. First, based on the PSPNet network, the lightweight neural network MobileNetV2 was used to replace ResNet as the feature extraction backbone network. The deep separable convolution was used to replace the standard convolution to reduce the amount of model parameters and calculations and then improve the extraction speed. Secondly, the pyramid pool structure of multi-dimensional feature fusion was constructed to obtain more detailed features of UAV images and improve accuracy. Then, the extracted feature map was processed by the NAM to identify the less significant features and compress the model to reduce the calculation. The U-Net, SegNet and DeepLabv3+ were selected as the comparison models. The results show that the extraction effect at the height of 20 m and the maturity stage is the best. For the NAM-PSPNet, the MPA (Mean Pixel Accuracy), MIoU (Mean Intersection over Union), Precision, Accuracy and Recall is, respectively, 89.32%, 89.32%, 94.95%, 94.30% and 95.43% which are significantly better than the comparison models. It is concluded that NAM-PSPNet has better extraction performance for wheat lodging areas which can provide the decisionmaking basis for severity estimation, yield loss assessment, agricultural operation, etc.

Keywords: unmanned aerial vehicle; wheat lodging areas; PSPNet; deep learning; normalization-based attention module

Citation: Zhao, J.; Li, Z.; Lei, Y.; Huang, L. Application of UAV RGB Images and Improved PSPNet Network to the Identification of Wheat Lodging Areas. *Agronomy* **2023**, *13*, 1309. https://doi.org/10.3390/agronomy13051309

Academic Editor: Yanbo Huang

Received: 13 March 2023
Revised: 27 April 2023
Accepted: 5 May 2023
Published: 6 May 2023

1. Introduction

Wheat is one of the most important food crops and plays a vital role in determining global food security. In 2017, the global wheat output was 753 million tons, and the planting area was 221 million hectares. With the growth of the global population, the wheat output needs to increase by more than 70% by 2050 to ensure food security [1]. As one of the three major food crops in the world, wheat is an important source of phytochemicals, such as starch, protein, vitamins, and dietary fiber [2]. Lodging is a problem in wheat production which limits grain yield for a long time [3]. The use of plant growth regulators can to some extent prevent wheat lodging but natural factors, such as sudden strong winds, frost, pests, and diseases as well as human factors, such as improper fertilization, can also cause wheat lodging which can affect wheat yield and quality. A large number of stem lodging destroyed the normal canopy structure of crops, resulting in the degradation of crop canopy, and the reduction of photosynthetic activity and mechanized harvesting efficiency led to the loss of grain yield. Therefore, accurately extracting the lodging areas after wheat lodging has important significance for wheat variety screening, yield loss assessment, and guidance for agricultural insurance.

In addition, based on the lodging areas, farmers can be guided to take timely remedial measures to minimize losses. The traditional lodging monitoring method requires investigators to use rulers, GPS, and other tools to conduct field investigations to obtain information about the location and area of the lodging crops. The efficiency is low, especially for irregular lodging areas, which cannot be accurately measured. Traditional methods of identifying the wheat lodging areas and location rely heavily on the ground manual measurement evaluation and random sampling [4], the agricultural management personnel are required to conduct measurement and sampling analysis in the field to quantify the percentage of lodging and the severity of lodging. Generally, it is necessary to carry out preliminary disaster assessment, comprehensive investigation, and review sampling assessment to complete the assessment of the degree of crop lodging damage. The larger the soil area involved, the longer the time required [5], unable to meet the actual needs of large-scale land collapse disaster assessment [6], and in the process of artificial evaluation, wheat cannot be replanted, and even in the process of field measurement, it may cause secondary damage to crops. In addition, the wheat lodging area is manually demarcated by agricultural management personnel which is highly subjective and often controversial. With the rapid development of remote sensing technology, advanced technologies, such as satellite remote sensing and unmanned aerial vehicle remote sensing, are more and more widely used in crop phenotype monitoring [7]. Remote sensing technology can quickly obtain image information and spatial information of large-scale farmland and has been widely used in crop lodging monitoring in recent years [8]. Researchers used satellite remote sensing data as data source, combined with machine learning and depth school algorithm to extract crop lodging areas. Dai et al. [9] used the multi-parameter information in Sentinel-1SAR image and field lodging samples to identify rice lodging caused by heavy rainfall and strong wind and constructed a decision tree model to extract the rice lodging area. The results showed that the overall accuracy of extracting the lodging rice area was 84.38%; Sun et al. [10] carried out experiments using Sentine-2A satellite images and proposed a remote sensing winter wheat extraction method based on medium-resolution image object-oriented and in-depth learning which proved the feasibility and effectiveness of object-oriented classification method in extracting winter wheat planting area, and the precision of this method reached 93.1%. In order to realize deep learning and large-scale monitoring of wheat lodging, Tang et al. [11] proposed a semantic segmentation network model called pyramid transposed convolution network (PTCNet). Using GF-2 (Gaofen 2) satellite data as the data source, the intersection of F1 score of PTCNet and wheat lodging extraction reached 85.31% and 74.38%, respectively, which is superior to SegNet, FPN and other networks.

In recent years, with the development of UAV (Unmanned Aerial Vehicle) remote sensing technology, it has been widely used in precision agriculture because of its convenient operation, low cost, fast speed, high spatial and temporal resolution and the ability to observe in a large area. With its advantages of flexible operation, adjustable flight height and low cost, UAVs can obtain high-frequency, multi-altitude and high-precision image data and play an important role in wheat lodging monitoring and evaluation. At present, the monitoring and extraction of wheat lodging areas are mostly based on single growth stage or single-height UAV images, and the multi-scale and multi-temporal features of multi-height and multi-growth period images are not fully integrated which limits the further improvement of extraction accuracy. The emergence of new remote sensing technology based on unmanned aerial vehicle platform has been favored by remote sensing workers, and agricultural workers are more hopeful about its application prospects of agricultural remote sensing [12]. UAV remote sensing has made some achievements in the field of intelligent agriculture. In order to improve the accuracy of lodging recognition under complex field conditions, Guan et al. [13] selected maize fields with different growth stages as the study area used UAV to obtain maize images, established the data set of lodging and non-lodging and proposed an effective and fast feature screening method (AIC method). After feature screening, BLRC (binary logical regression classification), MLC (Maximum

Likelihood Classification) and RFC (Random Forest Classification) distinguish lodging and non-lodging maize based on the selected features. The results show that the proposed screening method has a good extraction effect. Zhang et al. [14] applied nitrogen fertilizer at different levels to induce different lodging conditions in the wheat field and used UAV to obtain RGB and multispectral images of different wheat growth stages. Based on these two types of images, a new method combining migration learning and DeepLabv3+ network is proposed to extract the lodging areas of different wheat growth stages and obtain better extraction results. In order to explore the evaluation of maize lodging disaster, Zheng et al. [15] used the full convolution segmentation network of deep learning technology to extract the maize lodging area, and the score of the network test set reached more than 90%. Mardanisamani et al. [16] proposed a DCNN network for lodging classification using rape and wheat. Through migration learning, the classic deep learning network was used to pre-train the lodging model of crops. The final training results were compared with VGG, AlexNet, ResNet and other networks. The results showed that the proposed network was more suitable for lodging classification. Yang et al. [17] proposed a mobile U-Net model that combines lightweight neural network with depth separable convolution and U-Net model to overcome the problems of low wheat recognition accuracy and poor real-time performance, using UAV to obtain RGB images to build data sets. The proposed model is superior to FCN and U-Net classic network in segmentation accuracy and processing speed. Varela et al. [18] collected time series multispectral UAV imagery and compared the performance of 2D-CNN and 3D-CNN architectures to estimate lodging detection and severity in sorghum. The result shows that integration of spatial and spectral features with 3D-CNN architecture is helpful to improve assessment of lodging severity.

However, most of the current research focuses on the remote sensing image of a single crop growth period or a single height. In order to dynamically monitor the wheat lodging areas and explore the impact of the depth learning network in a multi-dimensional way, this research collects the UAV-based RGB images of wheat lodging areas of two growth periods and flight heights as the data source, and makes three improvements on the basis of the Pyramid Scene Parsing Network (PSPNet) network and NAM (Normalization-based Attention Module): (1) Fusing the lightweight neural network and introducing the residual connection; (2) Constructing a pyramid pool structure of multi-dimensional feature fusion; and (3) Introducing the NAM attention module. The extraction performance of NAM-PSPNet network for wheat lodging areas are comparatively analyzed and discussed.

2. Materials and Methods

2.1. Study Area

The research area is located in Guohe Farm, Lujiang County, Hefei City, Anhui Province. It is located in the middle of Anhui Province, between the Yangtze River and Huaihe River, and the west wing of the Yangtze River Delta, between $30°56'$–$32°33'$ N, $116°40'$–$117°58'$ E (Figure 1). It is in the subtropical warm monsoon climate zone. The annual climate characteristics are: four distinct seasons, mild climate, moderate rainfall, annual average rainfall of 992 mm, average temperature of 13–20 °C, suitable for wheat growth. Wheat is one of the main crops planted in this area. The wheat variety (Ningmai 13) was sown in October 2020 and belongs to the winter wheat. In addition to basic cultivation, it also carries out management in the process of wheat growth, such as weeding, fertilization, pest control, etc.

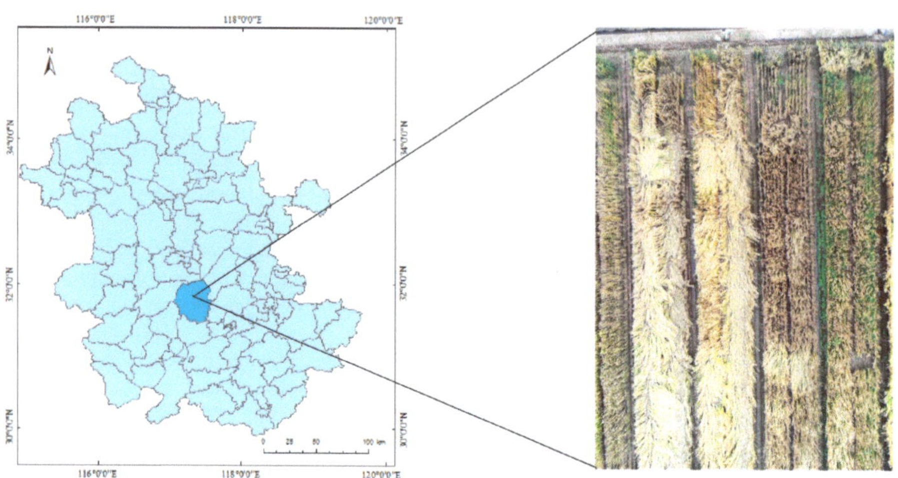

Figure 1. Geographical location of Guohe Farm, Lujiang County, Hefei City, Anhui Province, China.

2.2. Data Acquisition and Preprocessing

(1) Acquisition of UAV images. The wheat lodging was collected at the wheat grain-filling stage on 4 May 2021 and the mature stage on 24 May 2021. The flight platform is DJI Phantom 4 pro, and the image acquisition equipment is shown in Figure 2. The onboard RGB sensor, including red, green and blue bands, captures images with red, blue and green spectral channels with a resolution of 5472 × 3648 pixels. The flight altitude of UAV is set to 20 m and 40 m. Before the UAV flight, the DJI GS Pro software was used to plan the route, first select the flight area of the farm, set the flight altitude of 20 m and 40 m, the flight speed of 2 m/s and 4 m/s, the heading overlap of 85%, the side overlap of 85% and the shooting mode is set to take photos at equal time intervals. After all parameters are set, the UAV wheat data will be obtained. The day of shooting is sunny and breezy which meets the requirements of UAV flight and remote sensing data acquisition.

Figure 2. The picture of DJI Phantom 4 pro for collecting UAV images.

(2) Data annotation. The obtained UAV wheat image during the grain-filling period includes two kinds of data at the height of 20 m and 40 m, and the size of the original wheat lodging image collected is 5472 × 3648 and 8049 × 5486, respectively. Due to hardware constraints and adaptation to subsequent deep learning training, the images are uniformly cropped to the same size. The obtained UAV image has high spatial

accuracy and can clearly see the lodging information of wheat. Therefore, Expert Visual Interpretation (EVI) is selected to obtain the landmark data of real lodging areas. LabelMe [19] is selected as the annotation software to manually label the data which is an open annotation tool created by MIT Computer Science and Artificial Intelligence Laboratory (MIT CSAIL). It is a JavaScript annotation tool for online image annotation. Compared with traditional image annotation tools, its advantage is that we can use it anywhere. In addition, it can also help label images without installing or copying large data sets in the computer. According to this software, real ground lodging information can be obtained. The lodging areas were marked in LabelMe to generate a JSON file which was then converted to a PNG format image for deep learning network training preparation. As shown in Figure 3, the red areas represent the lodging areas and the black represents the non-lodging areas.

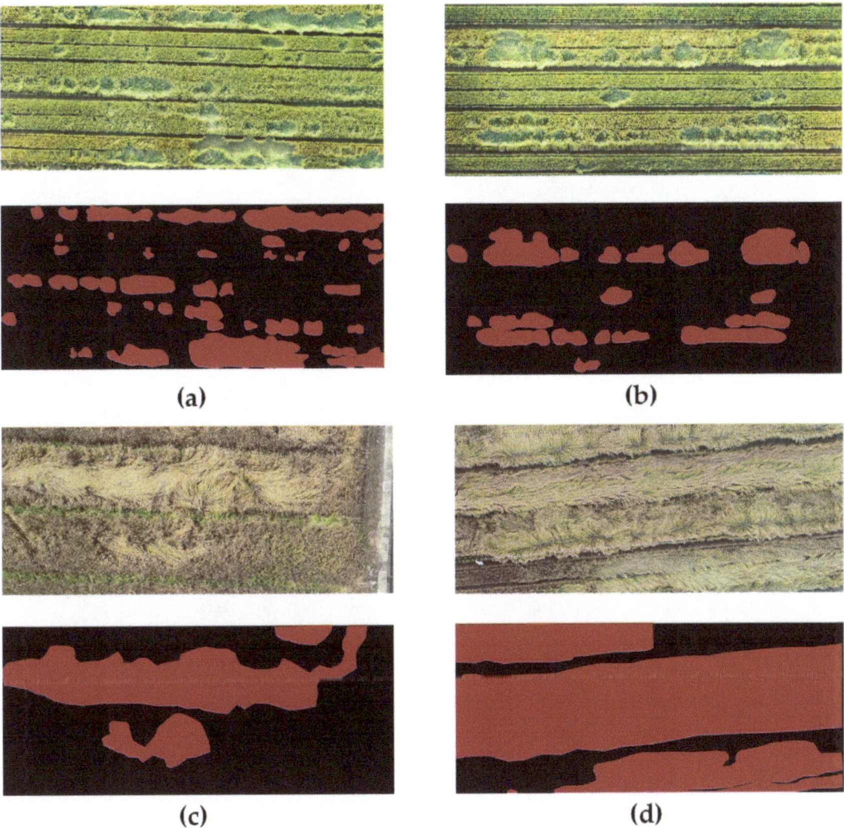

Figure 3. (**a**) Original map and label map of wheat during grain-filling period of 20 m UAV, (**b**) Original map and label map of wheat during grain-filling period of 40 m UAV, (**c**) Original map and label map of mature wheat of 20 m UAV, (**d**) Original map and label map of mature wheat of 40 m UAV.

3. Methodology

3.1. Technical Workflow

The flow chart of the research method for extraction of wheat lodging area is shown in Figure 4. First, UAV-based RGB images of wheat lodging areas at different growth stages and flight heights are collected. Secondly, the VOC format dataset was produced. Then, the UAV data are trained by using the NAM-PSPNet network and traditional split network method; Finally, the lodging areas of wheat were extracted, and the results were compared and analyzed with the U-net, SegNet and DeeplabV3+.

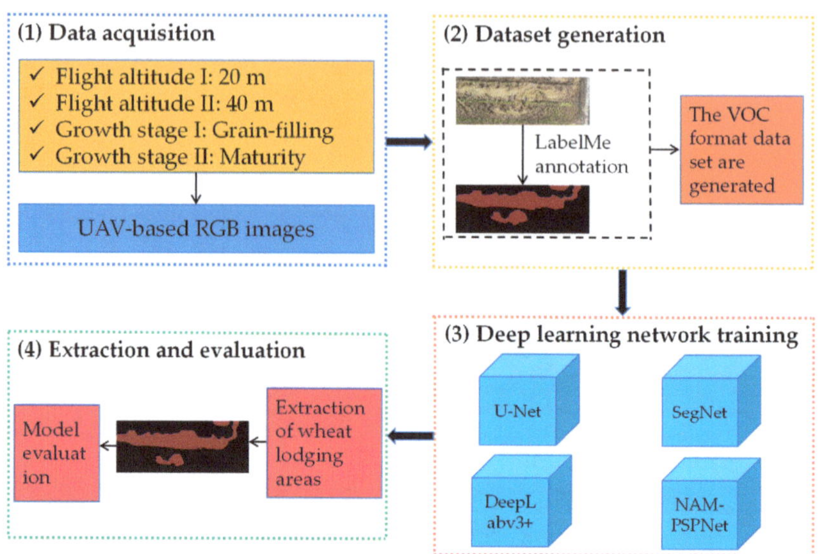

Figure 4. Technical roadmap of the study.

3.2. NAM Attention Module

NAM is based on the standardized attention module which is an efficient and lightweight attention mechanism. It adopts the same integration mode as the Convolutional Block Attention Module (CBAM) [20] and redesigns the channel and spatial attention sub-module, mainly to improve efficiency by suppressing less significant features. Recognition of less significant features is the key compressing the model, thus reducing the amount of computation. The NAM [21] module uses the weight factor to improve the attention mechanism, uses the scale factor of the batch to standardize, uses the standard deviation to express the importance of the weight and identifies the less significant features. The NAM structure diagram is shown in Figure 5.

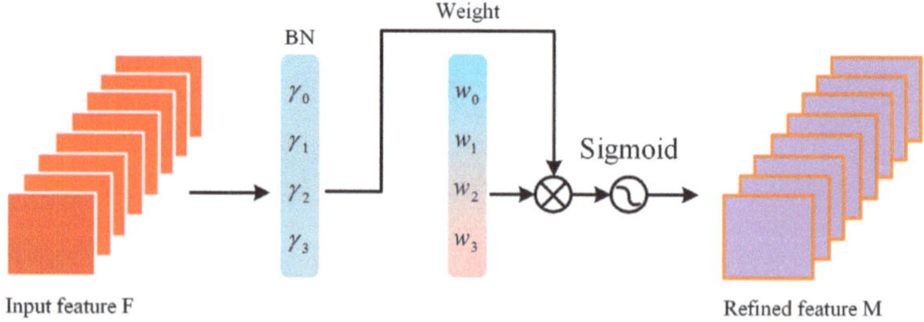

Figure 5. Structure of NAM attention module.

3.3. Structure of the NAM-PSPNet

PSPNet was proposed by Zhao et al. [22] to develop global context information through context aggregation based on different regions. In this paper, taking the PSPNet network framework as the main body, the following improvements are made as shown in Figure 6:

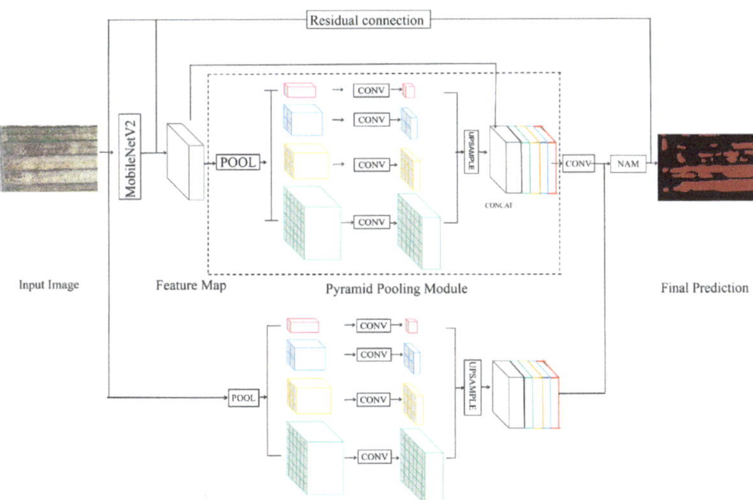

Figure 6. Structure diagram of the NAM-PSPNet network.

(1) The lightweight neural network MobileNetV2 is used to replace ResNet as the feature extraction backbone network, MobileNetV2 is used as the backbone feature extraction network and the deep separable convolution is used to replace the standard convolution. That is, the deep separable convolution with a step of one is used at the beginning and end of the feature extraction network, and the deep separable convolution with a step of 2 is used for three consecutive times in the middle layer. Reduce the amount of model parameters and calculation and improve the segmentation speed. The residual connection is introduced to extract the features that may be ignored by the original PSPNet.

(2) Construct a pyramid pool structure of multi-dimensional feature fusion. Using the method of U-Net [23] multi-dimensional feature cascade for reference, the multi-scale feature is further extracted.

(3) Add the NAM attention module to recognize the less obvious features, so as to compress the model, reduce the calculation and improve the segmentation accuracy.

4. Training of NAM-PSPNet and Evaluation Metrics

4.1. Image-Label Datasets

Establish the data set of wheat lodging for dynamic monitoring of wheat lodging area and the dataset of the impact of different scales on wheat lodging area extraction: First of all, the wheat images taken by UAV are screened, and the images with the same proportion of wheat lodging and no lodging, the images without image distortion caused by the interference of natural factors during the acquisition process, and the images with moderate exposure and good image quality are screened; Secondly, in order to more effectively increase the diversity of the data set, improve the robustness of the segmentation model and avoid the over-fitting phenomenon of the model due to insufficient data volume in the training process, the expansion of the data set for each group of image data is realized through the expansion methods of flipping, rotating, brightness and contrast enhancement; Then, the original image of the UAV and the corresponding tag image are clipped using the sliding window method, and the size of each sliding window is set to 256 × 256; that is, the size of the final training image is 256 pixels × 256 pixels; Finally, the original map and label map of the UAV at the height of 20 m and 40 m in the wheat grain-filling stage and mature stage are made into a VOC format dataset. The data is divided into training sets and test sets according to the ratio of 7:3, and the training set is divided into training and training verification parts according to the ratio of 8:2. Store the randomly selected data set with the corresponding proportion in the ImageSets folder as a txt file, store the original image data and label data in the JPEGImages and SegmentationClass folders, respectively.

4.2. Model Training

The model training is based on the Windows10 operating system. The selected programming language is Python 3.9 programming language. At the same time, the Python in-depth learning framework is adopted, and TensorFlow is used as the back end. It runs under the Intel Xeon Gold 6248R processor, 192Gb memory, NVIDIA Quadro P4000 graphics card and GPU (CUDA 10.0) environment. The NAM-PSPNet model is trained to determine the parameters in the training process and initialize the parameters of the NAM-PSPNe model. The images and labels of the training set of UAV wheat image data of different growth periods and different heights are input into the NAM-PSPNe model, and finally the wheat lodging area is predicted. The parameters of all models in the training paper are shown in Table 1.

Table 1. Network model training parameter configuration.

Parameter	Value
Batch_size	32
Learning_rate	5×10^{-3}
Input_shape	256×256
Num_classes	2
Epoch	150

The LOSS used in training consists of two parts: Cross Entropy Loss and Dice Loss. Cross Entropy Loss is the common cross entropy loss. It is used when the semantic segmentation platform uses Softmax to classify pixels. Dice loss takes the evaluation index of semantic segmentation as the loss. Dice coefficient is a set similarity measurement function which is usually used to calculate the similarity of two samples, and the value range is within the scope of [0, 1].

The calculation formula is expressed as:

$$s = \frac{2|X \cap Y|}{|X| + |Y|} \tag{1}$$

where X and Y represent the predicted result and the true result, respectively, and the Dice coefficient is the product of the predicted result and the true result plus two, divided by the predicted result plus the true result. Its value is between zero and one. The larger the predicted result is, the greater the coincidence between the predicted result and the actual result is. The bigger the Dice coefficient is, the better. If it is a loss, the smaller the better; so make Dice loss = $1 -$ Dice, you can use loss as the loss of semantic segmentation.

4.3. Evaluation Metrics

The evaluation indicators of wheat lodging area extraction used here are the MPA (Mean Pixel Accuracy), MIoU (Mean Intersection over Union), Accuracy, Precision and Recall. The extraction results of wheat lodging can be divided into TP is the lodging feature of correct detection, TN is the non-lodging feature of correct detection, FP is the non-lodging feature of false detection as lodging feature and FN is the lodging feature of wheat that is false detection as non-lodging feature.

MPA is the average value of the ratio of correctly classified pixel points and all pixel points in the two categories of lodging and non-lodging.

$$MPA = \frac{1}{k+1} \times \sum_{i=0}^{k} \frac{p_{ii}}{\sum_{j=0}^{k} p_{ij}} \tag{2}$$

where k represents the number of categories, p_{ii} represents the number of pixels that are correctly classified and p_{ij} represents the number of pixels that belong to class i but are predicted to be class j.

MIoU calculates the intersection ratio between the real lodging of the tag and the lodging predicted by the system.

$$MIoU = \frac{1}{k+1}\sum_{i=0}^{k}\frac{TP}{TP+FN+FP} \tag{3}$$

where k represents the number of categories. The MIoU reflects the coincidence degree between the predicted image and the real image. The closer the ratio is, the higher the degree of coincidence is and the higher the quality of semantic segmentation is.

Accuracy is used to evaluate the global accuracy of the model.

$$Accuracy = \frac{TP+TN}{TP+TN+FP+FN} \tag{4}$$

Precision represents the proportion of the number of pixels correctly extracted as the wheat lodging area in the image to be tested.

$$Precision = \frac{TP}{TP+FP} \tag{5}$$

Recall is the proportion of all wheat pixels in the test set that are correctly recognized as wheat lodging area pixels.

$$Recall = \frac{TP}{TP+FN} \tag{6}$$

4.4. Comparison Models

In order to dynamically monitor the wheat lodging area and analyze the extraction of wheat lodging area from different scale data, the data sets of 20 m grain-filling wheat, 20 m mature wheat, 40 m grain-filling wheat and 40 m mature wheat were input into the network training. U-Net, SegNet and DeepLabv3+ are classic models in semantic segmentation networks. These three models were selected as comparative experimental models. At the end of the model training, the optimal training weight is obtained to predict the wheat lodging area.

5. Results and Discussion

In order to dynamically monitor the lodging status of wheat, the wheat lodging area was extracted from the UAV images of the 20 m wheat grain filling stage and the mature stage. The four extraction models of U-Net, SegNet, DeepLabv3+ and NAM-PSPNet are used. The four deep-learning network models all use the training set, verification set and test set made of the same wheat field lodging data. The results obtained are shown in Table 2, showing the wheat lodging extraction results of different deep learning networks.

Table 2. Extraction effect evaluation.

Approach	Growth Period	MPA (%)	MioU (%)	Accuracy (%)	Precision (%)	Recall (%)
U-Net	Grain-filling	84.19	72.91	89.95	84.46	79.52
	Maturity	90.79	83.81	91.60	91.58	92.77
SegNet	Grain-filling	81.88	71.25	85.33	82.48	76.83
	Maturity	83.84	72.70	89.29	83.22	77.59
DeepLabv3+	Grain-filling	88.87	81.81	91.66	85.36	80.55
	Maturity	91.44	84.07	92.01	86.43	81.40
NAM-PSPNet	Grain-filling	93.73	88.96	94.12	93.78	94.82
	Maturity	94.62	89.32	94.95	94.30	95.43

It can be seen from the table that the MPA and MIoU values of U-Net reached 84.19% and 72.91%, respectively, the MPA and MIoU values of SegNet reached 81.88% and 71.25% The MPA and MIoU values of DeepLabv3+ reached 88.87% and 81.81%, and the MPA and MIoU values of NAM-PSPNet reached 93.73% and 88.96%, respectively, in the extraction of wheat lodging areas using the four deep learning models. The NAM-PSPNet network is superior to the other three networks in the extraction of lodging data during the grouting period and also shows better results in the accuracy, precision and recall evaluation indicators. The extraction effect obtained by inputting the data of wheat maturity into the network training can be obtained. Because the lodging of wheat at maturity is easier to distinguish and the lodging texture feature is easier to predict by network learning, the data extraction result of wheat maturity is better than that of the data at grain-filling stage.

The extraction results of wheat lodging data at different growth stages show that the extraction effect of wheat lodging at the mature stage is better than that at the grain-filling stage. In order to explore the extraction of multi-scale wheat lodging areas in different spatial dimensions, the data of UAV wheat heights of 20 m and 40 m are used as input. The effects of different spatial scales on the extraction of wheat lodging were discussed using the data of mature wheat at two heights. Different deep learning networks are trained, and MPA and MIoU are used as evaluation indicators. The results are shown in Table 3.

Table 3. Evaluation and comparison among the NAM-PSPNet and other methods.

Flight Altitude	Evaluation Metrics	U-Net	SegNet	DeepLabv3+	NAM-PSPNet
20 m	MPA (%)	90.79	81.88	91.44	94.62
	MIoU (%)	83.81	71.25	84.07	89.32
40 m	MPA (%)	89.65	83.49	89.81	94.35
	MIoU (%)	82.64	72.19	81.18	88.96

The four deep-learning models can be used to extract the lodging of wheat at different heights. It can be seen that the extraction effect of the lodging data of the 20 m wheat UAV is better than that of the 40 m wheat UAV. Therefore, the lodging area prediction is performed for the 20 m mature UAV data, and the lodging area prediction diagram is shown in Figure 7. It can be seen from the prediction diagram that the edges of some prediction results in the U-Net network prediction diagram are too angular; The prediction results of the SegNet network, although the overall results are in good agreement with the label chart, have some prediction errors and less prediction; The prediction results of DeepLabv3+ network also have some wrong predictions and few predictions; The prediction results of NAM-PSPNet network are closest to the sample label data. By comparison and analysis, it can be concluded that NAM-PSPNet is superior to the other three networks in the extraction of the lodging area of UAV wheat data at different heights.

The analysis above demonstrates that wheat is more susceptible to damage during the process from the grain-filling stage to the maturity stage. Extracting information on the overwhelmed areas of wheat during this stage can provide a dynamic understanding of the overall damage situation which can offer timely warnings to local farmers, reduce their losses, ensure food security and be of great significance for guiding agricultural insurance. Using UAV images of different scales to extract fallen wheat areas and integrating multi-scale and multi-temporal characteristics of different heights and fertility stages, this provides relevant experience for the development of smart agriculture.

To evaluate the effectiveness of the proposed method in this paper, we compared the wheat inversion region extraction results of the original PSPNet and NAM-PSPNet using mature wheat at 20 m and 40 m heights as the dataset for this experiment. The experimental results are presented in Table 4.

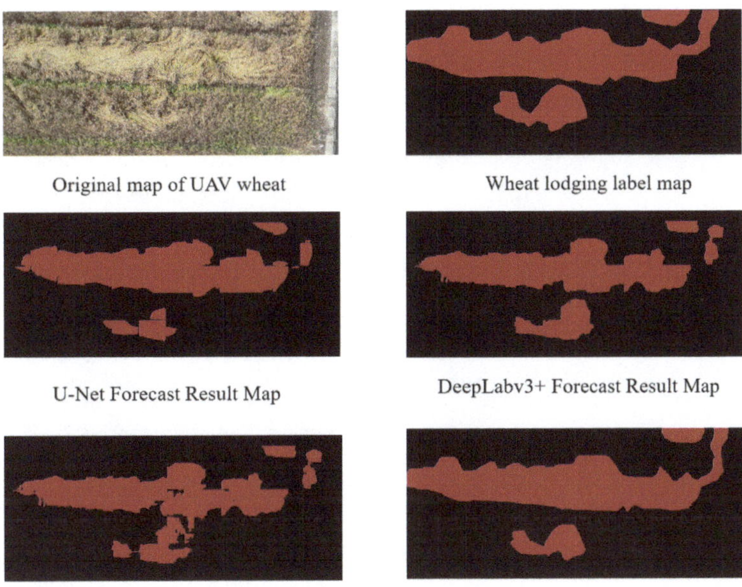

Original map of UAV wheat	Wheat lodging label map
U-Net Forecast Result Map	DeepLabv3+ Forecast Result Map
SegNet Forecast Result Map	NAM-PSPNet Forecast Result Map

Figure 7. UAV wheat original map, label map and network prediction map.

Table 4. Evaluation and comparison between PSPNet and NAM-PSPNet.

Flight Altitude	Evaluation Metrics	PSPNet	NAM-PSPNet
20 m	MPA (%)	92.15	94.62
	MIoU (%)	85.33	89.32
40 m	MPA (%)	92.07	94.35
	MIoU (%)	85.28	88.96

Based on the experimental results, it is evident that PSPNet achieved an MPA of 92.15% and an MIoU of 85.33% in the extraction of wheat inversion region at 20 m maturity which were 2.47% and 3.99% lower than those of NAM-PSPNet, respectively. Similarly, in the extraction of wheat inversion region at 40 m maturity, PSPNet achieved an MPA and MIoU of 92.07% and 85.28%, respectively, which were 2.28% and 3.68% lower than those of NAM-PSPNet. These findings demonstrate the effectiveness and superiority of NAM-PSPNet in wheat inversion region extraction. In our study, the UAV-based RGB images of wheat lodging areas at different growth stages and heights are used as the research object. Based on the NAM-PSPNet model, a better extraction effect is obtained for wheat lodging areas. Compared with the Ref. [4], which uses an improved U-Net network for rice lodging area extraction from UAV RGB and multi-spectral images, although better extraction results are obtained, multiple heights and growth stages of images are not integrated, limiting the accuracy improvement. The Ref. [14] combined transfer learning with DeepLabv3+ to extract wheat lodging, but the accuracy still needs improvement compared to our study. The proposed method in this study provides a relevant reference for the agricultural UAV remote sensing field and offers a related experience for those engaged in remote sensing image processing.

6. Conclusions

The application prospect of UAV low altitude remote sensing in intelligent agriculture is very broad. Extraction of the wheat lodging areas via UAV-based RGB images is also one of the most critical research topics in agricultural remote sensing research. With the change of wheat growth period, the extraction of wheat lodging area can show the change of wheat lodging trend to agricultural technicians in time, so that agricultural technicians can take remedial measures in time and reduce losses. In this paper, DJI Phantom 4 pro is used to obtain the RGB images of UAV wheat at the height of 20 m and 40 m at the grain-filling stage and maturity stage and to produce the UAV wheat lodging data sets at different growth stages and different heights. In our study, the NAM-PSPNet network is proposed to monitor the dynamic wheat lodging areas, and the images of wheat lodging area extracted from different scale data is also discussed. It can be seen from the comparison experiment that the extraction effect of the proposed model is better, and the evaluation index is better than other comparison algorithms. In the experimental design, the acquired data sets have errors in the process of manual annotation. Future research should explore the networks with stronger network fault tolerance and the use of semi-supervised methods to improve extraction accuracy and deal with more complex application scenarios.

Author Contributions: Y.L. and L.H. conceived and designed the experiments; J.Z. and Z.L. performed the experiments; J.Z. analyzed the data; J.Z., Z.L., Y.L. and L.H. wrote the paper. All authors have read and agreed to the published version of the manuscript.

Funding: This research was funded by the Industry–University Collaborative Education Project of Ministry of Education (220606353282640), Provincial Quality Engineering Project of Anhui Provincial Department of Education (2021jyxm0060), Natural Science Foundation of Anhui Province (2008085MF184), National Natural Science Foundation of China (31971789) and Science and Technology Major Project of Anhui Province (202003a06020016).

Data Availability Statement: Not applicable.

Conflicts of Interest: The authors declare no conflict of interest.

References

1. Zhang, H.-J.; Li, T.; Liu, H.-W.; Mai, C.-Y.; Yu, G.-J.; Li, H.-L.; Yu, L.-Q.; Meng, L.-Z.; Jian, D.-W.; Yang, L.; et al. Genetic progress in stem lodging resistance of the dominant wheat cultivars adapted to Yellow-Huai River Valleys winter wheat zone in China since 1964. *J. Integr. Agric.* **2020**, *19*, 438–448. [CrossRef]
2. Shewry, P.R.; Hey, S.J. The contribution of wheat to human diet and health. *Food Energy Secur.* **2015**, *4*, 178–202. [CrossRef] [PubMed]
3. Shah, L.; Yahya, M.; Shah, S.M.A.; Nadeem, M.; Ali, A.; Ali, A.; Wang, J.; Riaz, M.W.; Rehman, S.; Wu, W.; et al. Improving lodging resistance: Using wheat and rice as classical examples. *Int. J. Mol. Sci.* **2019**, *20*, 4211. [CrossRef] [PubMed]
4. Zhao, X.; Yuan, Y.; Song, M.; Ding, Y.; Lin, F.; Liang, D.; Zhang, D. Use of unmanned aerial vehicle imagery and deep learning unet to extract rice lodging. *Sensors* **2019**, *19*, 3859. [CrossRef] [PubMed]
5. Yang, M.D.; Boubin, J.G.; Tsai, H.P.; Tseng, H.H.; Hsu, Y.C.; Stewart, C.C. Adaptive autonomous UAV scouting for rice lodging assessment using edge computing with deep learning EDANet. *Comput. Electron. Agric.* **2020**, *179*, 105817. [CrossRef]
6. Liu, T.; Li, R.; Zhong, X.; Jiang, M.; Jin, X.; Zhou, P.; Liu, S.; Sun, C.; Guo, W. Estimates of rice lodging using indices derived from UAV visible and thermal infrared images. *Agric. For. Meteorol.* **2018**, *252*, 144–154. [CrossRef]
7. Burkart, A.; Aasen, H.; Alonso, L.; Menz, G.; Bareth, G.; Rascher, U. Angular dependency of hyperspectral measurements over wheat characterized by a novel UAV based goniometer. *Remote Sens.* **2015**, *7*, 725–746. [CrossRef]
8. Li, Z.; Chen, Z.; Ren, G.; Li, Z.; Wang, X. Estimation of maize lodging area based on Worldview-2 image. *Trans. Chin. Soc. Agric. Eng.* **2016**, *32*, 1–5.
9. Dai, X.; Chen, S.; Jia, K.; Jiang, H.; Sun, Y.; Li, D.; Zheng, Q.; Huang, J. A decision-tree approach to identifying paddy rice lodging with multiple pieces of polarization information derived from Sentinel-1. *Remote Sens.* **2022**, *15*, 240. [CrossRef]
10. Sun, Y.; Liu, P.; Zhang, Y.; Song, C.; Zhang, D.; Ma, X. Extraction of winter wheat planting area in Weifang based on Sentinel-2A remote sensing image. *J. Chin. Agric. Mech.* **2022**, *43*, 98–105.
11. Tang, Z.; Sun, Y.; Wan, G.; Zhang, K.; Shi, H.; Zhao, Y.; Chen, S.; Zhang, X. Winter wheat lodging area extraction using deep learning with GaoFen-2 satellite imagery. *Remote Sens.* **2022**, *14*, 4887. [CrossRef]
12. Gao, L.; Yang, G.; Yu, H.; Xu, B.; Zhao, X.; Dong, J.; Ma, Y. Winter Wheat Leaf Area Index Retrieval Based on UAV Hyperspectral Remote Sensing. *Trans. Chin. Soc. Agric. Eng.* **2016**, *32*, 113–120.
13. Tang, Z.; Sun, Y.; Wan, G.; Zhang, K.; Shi, H.; Zhao, Y.; Chen, S.; Zhang, X. A quantitative monitoring method for determining Maize lodging in different growth stages. *Remote Sens.* **2020**, *12*, 3149.
14. Zhang, D.; Ding, Y.; Chen, P.; Zhang, X.; Pan, Z.; Liang, D. Automatic extraction of wheat lodging area based on transfer learning method and deeplabv3+ network. *Comput. Electron. Agric.* **2020**, *179*, 105845. [CrossRef]

15. Zheng, E.G.; Tian, Y.F.; Chen, T. Region extraction of corn lodging in UAV images based on deep learning. *J. Henan Agric. Sci.* **2018**, *47*, 155–160.
16. Mardanisamani, S.; Maleki, F.; Kassani, S.H.; Rajapaksa, S.; Duddu, H.; Wang, M.; Shirtliffe, S.; Ryu, S.; Josuttes, A.; Zhang, T.; et al. Crop lodging prediction from UAV-acquired images of wheat and canola using a DCNN augmented with handcrafted texture features. In Proceedings of the IEEE/CVF Conference on Computer Vision and Pattern Recognition Workshops, Long Beach, CA, USA, 16–20 June 2019.
17. Yang, B.; Zhu, Y.; Zhou, S. Accurate wheat lodging extraction from multi-channel UAV images using a lightweight network model. *Sensors* **2021**, *21*, 6826. [CrossRef] [PubMed]
18. Varela, S.; Pederson, T.L.; Leakey, A.D.B. Implementing spatio-temporal 3D-convolution neural networks and UAV time series imagery to better predict lodging damage in sorghum. *Remote Sens.* **2022**, *14*, 733. [CrossRef]
19. Torralba, A.; Russell, B.C.; Yuen, J. LabelMe: Online image annotation and applications. *Proc. IEEE* **2010**, *98*, 1467–1484. [CrossRef]
20. Woo, S.; Park, J.; Lee, J.Y.; Kweon, I.S. Cbam: Convolutional block attention module. In Proceedings of the European Conference on Computer Vision (ECCV), Munich, Germany, 8–14 September 2018; pp. 3–19.
21. Liu, Y.; Shao, Z.; Teng, Y.; Hoffmann, N. NAM: Normalization-based attention module. *arXiv* **2021**, arXiv:2111.12419.
22. Zhao, H.; Shi, J.; Qi, X.; Wang, X.; Jia, J. Pyramid scene parsing network. In Proceedings of the IEEE Conference on Computer Vision and Pattern Recognition, Honolulu, HI, USA, 21–26 July 2017; pp. 2881–2890.
23. Ronneberger, O.; Fischer, P.; Brox, T. U-net: Convolutional networks for biomedical image segmentation. In *Medical Image Computing and Computer-Assisted Intervention–MICCAI 2015: 18th International Conference, Munich, Germany, 5–9 October 2015; Proceedings, Part III 18*; Springer International Publishing: Cham, Switzerland, 2015; pp. 234–241.

 *agronomy*

Article

Investigation of the Detectability of Corn Smut Fungus (*Ustilago maydis* DC. Corda) Infection Based on UAV Multispectral Technology

László Radócz [1], Atala Szabó [1], András Tamás [1], Árpád Illés [1,*], Csaba Bojtor [1], Péter Ragán [1], Attila Vad [2], Adrienn Széles [1], Endre Harsányi [1] and László Radócz [3]

[1] Institute of Land Use, Engineering and Precision Farming Technology, Faculty of Agricultural and Food Sciences and Environmental Management, University of Debrecen, H-4032 Debrecen, Hungary; szelesa@agr.unideb.hu (A.S.)

[2] Institutes for Agricultural Research and Educational Farm (IAREF), Farm and Regional Research Institutes of Debrecen (RID), Experimental Station of Látókép, University of Debrecen, H-4032 Debrecen, Hungary

[3] Institute of Plant Protection, Faculty of Agricultural and Food Sciences and Environmental Management, University of Debrecen, H-4032 Debrecen, Hungary

* Correspondence: illes.arpad@agr.unideb.hu

Abstract: Corn smut fungus (*Ustilago maydis* [DC.] Corda) is a globally widespread pathogen affecting both forage and sweet maize hybrids, with higher significance in sweet maize. Remote sensing technologies demonstrated favorable results for disease monitoring on the field scale. The study focused on the changes in vegetation index (VI) values influenced by the pathogen. In this study, four hybrids, two forage maize and two sweet maize hybrids were examined. Artificial infection was carried out at three different doses: a low (2500 sporidium number/mL), medium (5000 sporidium number/mL) and high dose (10,000 sporidium number/mL) with a non-infected control plot for each hybrid. The experimental plots were monitored using a multispectral UAV sensor of five monochrome channels on three different dates, i.e., 7, 14 and 21 days after infection. Five different indices (NDVI, GNDVI, ENDVI, LCI, and NDRE) were determined in Quantum GIS 3.20. The obtained results demonstrated that the infection had a significant effect on the VI values in sweet maize hybrids. A high-dose infection in the Dessert R 73 hybrid resulted in significantly lower values compared to the non-infected hybrids in three indices (NDVI, LCI and GNDVI). In the case of the NOA hybrids, GNDVI and ENDVI were able to show significant differences between the values of the infection levels.

Keywords: corn smut fungus; disease monitoring; multispectral imaging; remote sensing

Citation: Radócz, L.; Szabó, A.; Tamás, A.; Illés, Á.; Bojtor, C.; Ragán, P.; Vad, A.; Széles, A.; Harsányi, E.; Radócz, L. Investigation of the Detectability of Corn Smut Fungus (*Ustilago maydis* DC. Corda) Infection Based on UAV Multispectral Technology. *Agronomy* **2023**, *13*, 1499. https://doi.org/10.3390/agronomy13061499

Academic Editors: Jinling Zhao and Chuanjian Wang

Received: 11 May 2023
Revised: 24 May 2023
Accepted: 26 May 2023
Published: 30 May 2023

1. Introduction

Maize is one of the most important crops in Hungarian agriculture. In 2021, the maize production area in Hungary was 1,076,061 ha (hectares) [1,2], which is almost one-quarter of the Hungarian arable land. Therefore, experiments and various tests related to maize can potentially be decisive, not only from a scientific aspect but also from an economic point of view. In order to carry out a complex field examination of maize, it is essential to evaluate the optimal level of weather conditions, agrotechnics, sowing time and applied fertilizer [3,4]. The nutrient supply and, more specifically, the amount and quality of nitrogen used are of key importance during plant production [5]. Plant pests can cause significant yield losses; therefore, the planning and implementation of appropriate plant protection procedures is one of the most important factors in plant management [6]. One of the most widespread and well-known pathogens of cultivated maize is the corn smut fungus *Ustilago Maydis* [DC.] CORDA [7]. Pathogens and pests can also cause various symptoms in plants. Changes may appear in the plant tissues, such as the deformation and discoloration of the leaves, the formation of necrotic spots and the drying and bending of the plants [8].

Remote sensing can be used to monitor and detect various symptoms, especially in the case of diseases and pests causing decreased leaf chlorophyll (Chl) and carotenoid (Car) content, mainly caused by viral, bacterial and fungal infections [9].

Agricultural UAVs (Unmanned Aerial Vehicles) provide new opportunities for pathogen assessment in precision farming [10,11]. These devices can be used in the appropriate environmental conditions [12], making it possible to collect large amounts of data in a cost-effective way [13]. Remote sensing technologies can be divided into three main groups in terms of the observation and detection of plant diseases and pests [9].

The first group includes spectral systems based on VIS-SWIR visible light and near-infrared (e.g., Multispectral, Hyperspectral, RGB and NDVI-NDRE) cameras [14].

The second group involves fluorescent and thermal systems, which are based on different thermal cameras (Thermal Imaging Cameras). This method is also suitable for the presymptom detection of pathogens. For example, Gómez (2014) observed changes in rose leaves 5 days after *Peronospora sparsa* infection, before they even produced visible symptoms [15]. Furthermore, it is also possible to detect apple tree scab, *Venturia inequalis*, before its symptoms appear on the crop [16].

The third is SAR (Synthetic Aperture Radar) LiDAR Light Detection and Ranging Laser-based Remote Sensing in Hungarian. Using these methods, morphological and structural changes mainly caused by pathogenic and harmful organisms can be monitored [9]. Currently, the most common vegetation index is the NDVI (chlorophyll content, biomass, etc.) [17]. The vegetation index (NDVI, GNDVI, NDRE, etc.) values clearly show the health status of the observed plant population since the reflectance of the leaves in the visible spectrum (400–700 nm) is low due to the absorbed photosynthetic pigments (mainly chlorophylls and carotenoids) [17]. The reflectance of a plant in a state of stress increases, which can be easily detected with the appropriate tools, sensor systems and instruments. If the chlorophyll content of a leaf decreases, it can absorb less light, meaning that the reflectance increases in the visible spectra, for example on channel RED, but decreases in the near-infrared NIR [18]. Polischuk et al. (1997) performed measurements based on spectral reflectance for the early diagnosis of disease symptoms. In the case of *Nicotiana debneyi* plants, 10 days after artificial infection with tomato mosaic tobamovirus, the reflectance of the leaves decreased (chlorophyll content (Chl) reduction). However, visually detectable symptoms were observed only 2 weeks after the infection [19]. For example, Kuska (2015) examined the development of powdery mildew with a hyperspectral camera, also based on spectral reflectance. Changes were detected on the leaf surface already on the fourth day after infection before the symptoms on the leaves became visible [20]. A plant under stressful conditions (caused by different biotic or abiotic stressors) reacts with defense mechanisms, such as a decrease in the leaf area index (LAI), a change in the chlorophyll content, or a change in the surface temperature. Thus, the spectral changes that occur differ from the values of healthy, stress-free vegetation [21].

The symptoms of corn smut disease caused by the fungus *Ustilago maydis* [DC.] Corda, which is involved in this study, can affect all parts of the maize plant, including the stem, leaf, cob and crest, and it can also reduce the chlorophyll content [22].

At first, pigment destruction and a yellowish color appear, followed by tissue destruction and the appearance of necrotic spots (leaf color changes) [23,24].

It is important to mention that the pathogen has also an effect on increasing different plant hormones, such as AUXIN and CYTOKININ, [25,26], and also plant stress hormones, which are expressed due to the infection, such as APX, MDA, SOD and POD, are higher in corn smut infected (CSI) plants [27]. The spread and importance of the disease are enhanced by the global warming trend, as it prefers dry, warm conditions [28], which are increasingly typical in Hungary and Central Europe as a region [29].

This paper focuses on the changes caused by the infection of the corn smut fungi and their detectability using measurements based on the VIS-SWIR remote sensing technologies with a multispectral camera system.

2. Materials and Methods

2.1. Study Area

The study site is located in the eastern part of Hungary, in Debrecen. A total of 16 plots were created in the experimental area of the University of Debrecen (47°33′6.52″ N. 21°36′5.59″ E). Two sweet maize and two forage maize hybrids (P 9025, ARMAGNAC) and sweet maize (NOA, DESSERT R 73) were involved in the study, as shown in Figure 1. The sowing date was the same for all four hybrids—29 April 2021. The soil has a pH of 7.46 and Arany's plasticity index is 44. The total soil organic matter content was around 2.4% in the topsoil (0–30 cm). The total amount of water in the 0–100 cm depth was measured before the growing season began (8.3.2021). The average of the total water volume in the wide examined profile (0–100 cm) was 20.02 m/m%. The largest amount (23.42 m/m%) of water was located between 10 and 40 cm depth. In the deeper region (50–100 cm) of the soil, the moisture content was lower (18.6 m/m%) than the soil surface (0–50 cm). The soil samples were dried at 105 Celsius degrees for 5 days in a drying chamber. The data in the weather and soil condition database were measured and collected by the University's own weather station at the experiment site.

Figure 1. University of Debrecen, Böszörményi Street campus experimental station. Experiment area: The plots are the same size, approximately 20 m^2, with randomized placement. The size of the experiment area is 400 m^2. Specifically, 5 m long and 4 m wide. Row spacing is 75 cm, 5 rows in total, 17.5 cm spacing between plants within 1 plot, 142–143 plants sown per plot.

The artificial (*Ustilago Maydis* [DC.] Corda) infection was carried out with three different spore concentrations (spore numbers were 2500/mL, 5000/mL, 10,000/mL and +non-infected control).

The temperature was higher than the average of the last 30 years (19.72 °C). The maximum temperature was higher than 30 °C 19 times between 15.05 and 15.07. Between 06.16 and 06.30, the maximum temperature was above 30 °C every day. The average humidity was low in the middle part of the examination period. After the artificial inoculation of the plants, the relative humidity was higher (70%) than the average of the two-month period (Figures 2 and 3).

During the period of examination, the amount of precipitation was lower than the average of the last 30 years (136.1 mm). The rainfall distribution was not equal between 15.05 and 15.07. The major part of the precipitation was concentrated in the first and the last part of the examined two months. The total amount of precipitation was 89.1 mm (Figure 4).

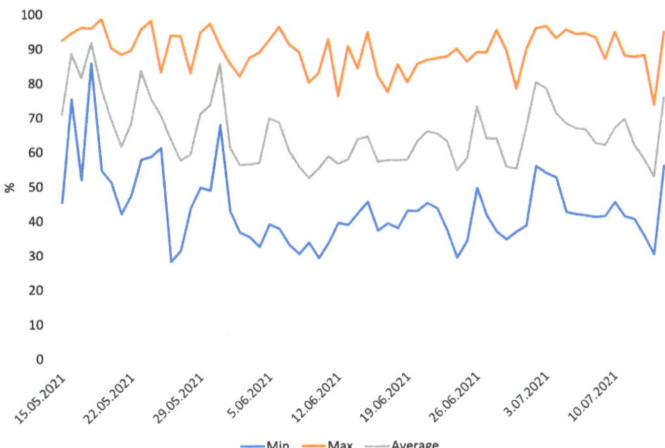

Figure 2. Air humidity in 2021 at the experiment site.

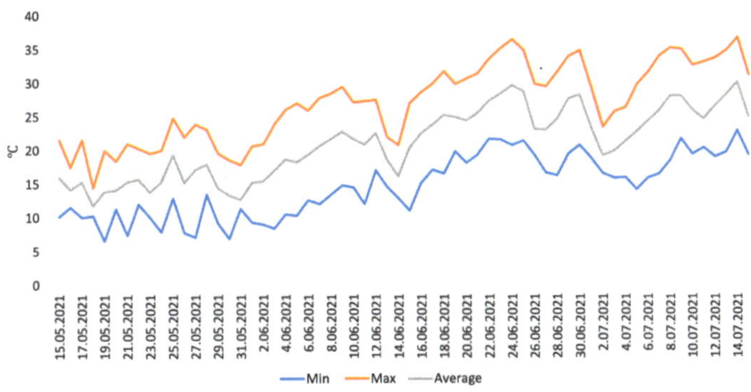

Figure 3. Air temperature in 2021 at the experiment site.

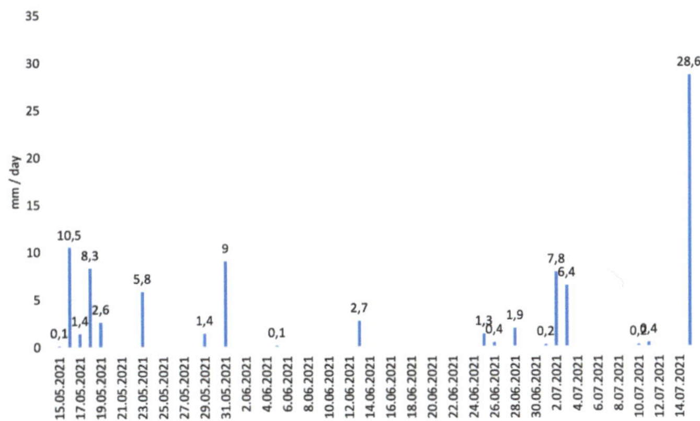

Figure 4. Rainfall in 2021 at the experiment site.

The area was kept weed-free during the tests. Therefore, the values of different weed species could not affect the values on the examined plots. The rows and row spacings were kept clear, and the use of herbicides was avoided. Only mechanical tools were used in this small area.

Three different levels of infection were carried out in the experiment. The plants were treated with the fungus in the 4–5-leaf (V5) phenophase on the 3rd of June. A total of 2 mL of spore suspension was injected into the part of the stem between the second and third node with a mass injection device [27]. The plot IDs and the doses for each plot are shown in Table 1. Cross-infection was not possible in the plots because the reproductive spores of the fungus are formed much later in the vegetative phase, and we observed the earlier stages when only chlorotic changes would appear, and the artificial infection was carried out directly with a suspension produced from compatible fungal strains.

Table 1. The Location of plot IDs and infection levels.

Hybrid	Control	Low Dose	Medium Dose	High Dose
Dessert R 73	4	13	11	6
NOA	16	10	1	7
Armagnac	3	5	14	12
P 9025	9	15	8	2

Artificial infection was performed on 3 June 2021.

- Low concentration: 2500 spores/mL
- Medium concentration: 5000 spores/mL
- High concentration: 10,000 spores/mL

2.2. Experimental Devices and Image Acquisition Methods

The test field was monitored three times during the vegetation period.

The front and side overlap was 80%. The flight altitude was 25 m, with a speed of 3 m/s and a ground sample distance (GSD) of 1.4 cm/px in every case. The total covered area of the flight mission was 0.12 ha and approx. 5 min. for each flight. In order to ensure repeatability, the recordings were always made between 10 am and 11 am. On all three survey dates, the weather was clear and sunny. The capture mode was hover and capture at point. Capture was used by the time interval. Agricultural UAV (drone): DJI Phantom 4 Multispectral RTK drone, which is equipped with a 6-camera Multispectral imaging system 1 visible light and 5 monochrome sensors. FOV (Field of View): 62.7°. Focal Length: 5.74 mm (35 mm format equivalent: 40 mm), autofocus set at ∞. Aperture: f/2.2 [30]. The camera can take 6 pictures simultaneously on 5 different channels. Band 1 is Red (R) 650 nm ± 16 nm, Band 2 is Green (G) 560 nm ± 16 nm, Band 3 is Blue (B) 450 nm ± 16 nm, Band 4 is Red Edge (RE) 730 nm ± 16 nm and Band 5 is Near-Infra RED (NIR) and captures at 840 nm ± 26 nm wavelengths. For each camera sensor, the effective pixels were 2.02 MP (2.12 in total), and the cameras were equipped with a global shutter and a 3-axis gimbal stabilizer [30]. The UAV control software used was DJ GS Pro on IOS with an iPad Mini.

WebODM, the software used in the analysis and evaluation process, is a free application that was used for UAV image processing. Georeferenced maps, point clouds and textured 3D models can be created from aerial photographs (WebODM) [31].

High-resolution orthomosaics were created from the field captures (high-resolution HR 2 cm/pixel). We used these orthomosaics for further QGIS analysis (The imagery was orthorectified using WebODM). Orthophotos were taken on the 5 different channels (Red R, Green G, Blue B, Near-infrared NIR, Red edge RE) on each survey date 7 days (10 June), 14 days (17 June) and 21 days (24 June) following the infection period.

The digitization footprint and sizes of the captured images are presented in Table 2 (raw image format and post-processed Tiff).

Table 2. The digitization footprint of the experiment.

Date	Overlap Front/Side	GSD	Number of Captured Channels	Size in Mb (Raw)	Size in Mb (Tiff)
7 DAI	80%	1.4 cm/px	6	1640	84.7
14 DAI	80%	1.4 cm/px	6	1250	97.1
21 DAI	80%	1.4 cm/px	6	1620	80.6

Quantum GIS 3.20 was used to create and evaluate the different vegetation index (VI) maps (Figure 5). The vegetation index maps were prepared with the raster analysis function and formulas shown in Table 3 (NDVI, NDRE, GNDVI, LCI and ENDVI).

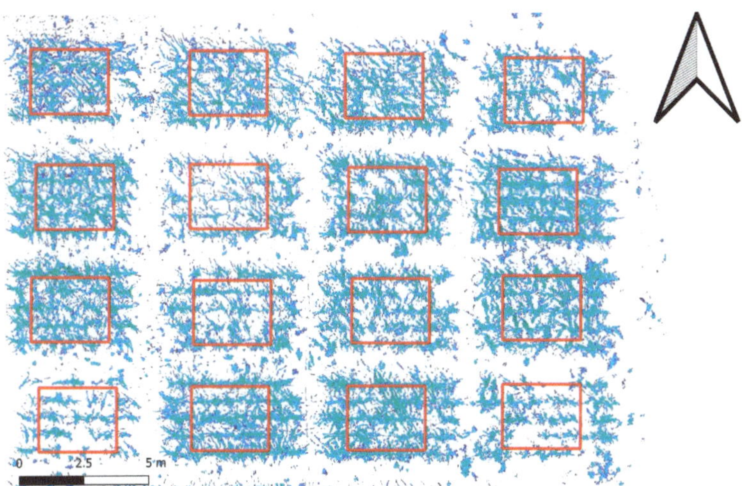

Figure 5. Example of the maps that were created in QGIS 3.20 visualization. NDVI filtered map HR high resolution 2 cm/px containing only maize pixels separated from the background.

Table 3. Vegetation index formulas used in the study.

Abbrev.	Formula	Reference
NDVI	$(R_{NIR} - R_{Red})/(R_{NIR} + R_{Red})$	[32]
GNDVI	$(R_{NIR} - R_{Green})/(R_{NIR} + R_{Green})$	[33]
NDRE	$(R_{NIR} - R_{RedEdge})/(R_{NIR} + R_{RedEdge})$	[34]
LCI	$(R_{NIR} - R_{RedEdge})/(R_{NIR} + R_{Red})$	[35,36]
ENDVI	$(R_{NIR} + R_{Green} - 2 \times R_{Blue})/(R_{NIR} + R_{Green} + 2 \times R_{Blue})$	[37]

In each case, a mask file was created for extracting the maize pixels separated from the background and values of the soils and shadow, i.e., only maize plant pixels were used for further analysis.

Sample values were collected for all the vegetation index maps (ENDVI, LCI, NDRE, etc.) based on the method shown in Figure 6. Each sample box was a 3 × 2.5 m rectangle that covered a multipolygon shape file (Shp) area (NDVI sampling spot). Following the principle of unification, the sample values for each map were collected from the same area.

Guan S. et al. (2019) also collected the sample NDVI values with this method in their experiment [38]. Using these data, statistical mean values were obtained in MS Excel.

Further statistical analyses were performed in R 4.1.2. [39] using the RStudio [40] graphic interface and the "agricolae" [41] packages.

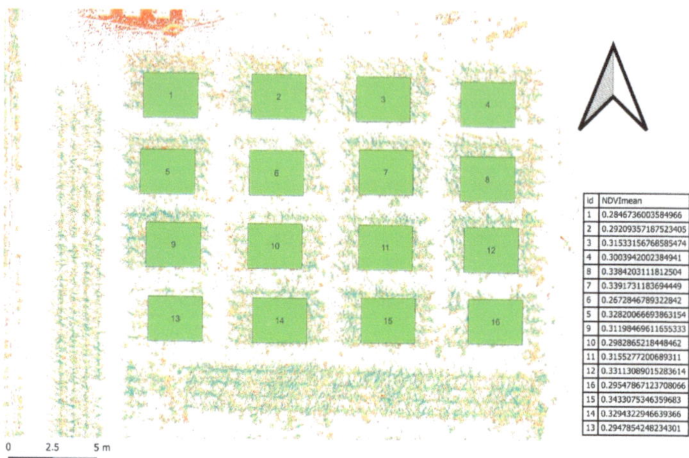

id	NDVImean
1	0.2846736003584966
2	0.29209357187523405
3	0.31533156768585474
4	0.3003942002384941
8	0.3384203111812504
7	0.3391731183694449
6	0.2672846789322842
5	0.32820066693863154
9	0.31198469611655333
10	0.2982865218448462
11	0.3155277200689311
12	0.33113089015283614
16	0.29547867123708066
15	0.3433075346359683
14	0.3294322946639366
13	0.2947854248234301

Figure 6. Sample data collection method. Placement of the experimental area and sampling boxes on the plots, as well as the attribute table of average values from the zonal statistics (NDVI map).

The figures and graphs were created in MS Excel 2016. Based on the example of Huzsvai [42], the repeated measures ANOVA (RMA) model was used to examine the effect of the recording method, dates and infection level and their combined effect, as well as their impact on the vegetation index value changes during the first three weeks of the infection period. The same model was used in each VI map in the case of NDVI, GNDVI, NDRE, LCI and ENDVI and also for each hybrid.

3. Results

The symptoms were fully visible on the experiment field by the third week. The symptoms of the stem (tumors) and leaf corn smut infection were also identified, i.e., discolored brown spots with a yellowish border, necrotic tissues and structural destruction were visible on the surface of the leaves, as shown in Figure 7.

Figure 7. Appearance of visible symptoms (leaf symptom and stalk tumors of *Ustilago maydis* [DC.] Corda) in the treated plots. Third week after infection (21 DAI).

The first recordings were performed 1 week after infection (7 DAI), the second were performed 14 days after infection (14 DAI) and the third were carried out 21 days (DAI) after infection.

The forage maize hybrids P 9025 and Armagnac were more tolerant to the infection than the sweet maize hybrids Dessert R 73 and NOA, as shown also via the symptoms in the field.

The Dessert R 73 results revealed three VIs with a significant effect of the infection, based on the values in Table 4.

Table 4. Significant effect on the values of infection in different VIs in the case of the D73 hybrid.

Infection D73	Df	Sum Sq	Mean Sq	F Value	Pr (>F)
LCI	3	0.0030246	0.0010082	21,059	0.00138 **
NDVI	3	0.003874	0.0012913	7.927	0.0165 *
GNDVI	3	0.010228	0.003409	15.61	0.00307 **

Note: ** indicates that the infection had a significant impact on the values at the 0.01 level; * indicates that the impact was significant at the 0.05 level.

3.1. Results of the Sweet Maize Hybrid Dessert R 73

As shown in the values of the Dessert R 73 hybrid in Table 4, in the case of three VI NDVI, LC, and GNDVI of the five calculated indices, the infection affected the values at various levels.

Based on Figure 8, the values of the control plot had the highest NDVI values at 0.3052, and the values of the plot with a high spore concentration (High 10,000) were the lowest at 0.2560.

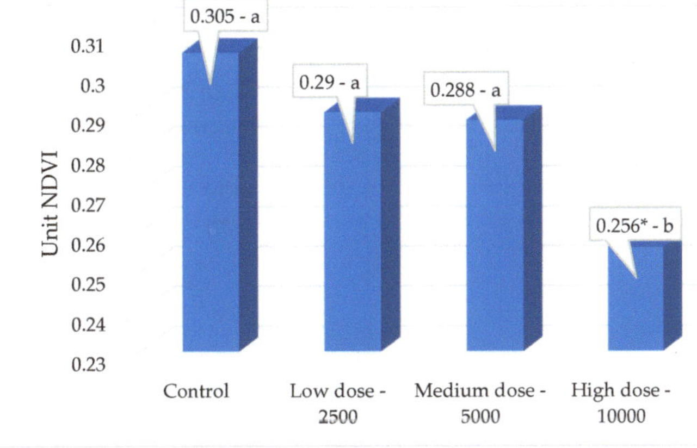

Figure 8. Dessert R 73 Results of NDVI values. Lower case letters (a, b) show the significant differences among treatments based on the ANOVA test ($p < 0.05$). Asterisks mark significant differences compared to control. * $p < 0.05$.

The values of the low-dose plot were 0.2907 (Low 2500), and the values of the plot treated with a medium dose (Medium 5000) showed an NDVI value of 0.2886.

The results of the control plot produced higher values than those of the Low 2500 and Medium 5000 plots. However, they did not differ significantly from each other, and all three belong to Group (a). Based on the NDVI results, the control plot (a) is statistically different from the High 10,000 values (b).

In the case of the high concentration dose 10,000 (b) plot, the NDVI values are lower than both the two treatments and the uninfected control plot. The values of High 10,000 (b) are significantly different from the NDVI values of the other treated plots.

Based on Figure 9, the values of the control plot were the highest with a GNDVI value of 0.2576, and the lowest GNDVI values of the plot with a high spore concentration (High 10,000) were 0.1813.

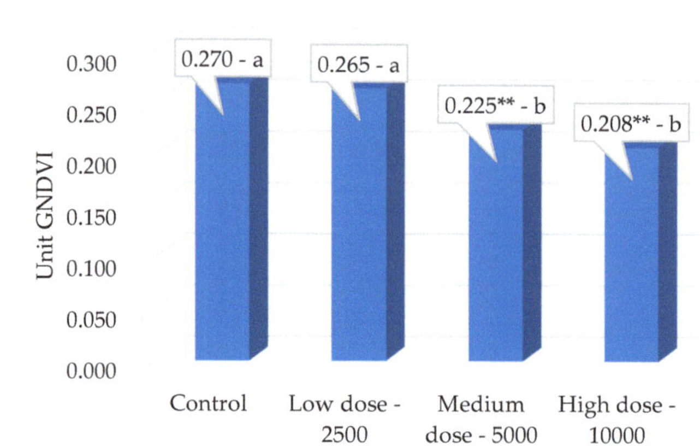

Figure 9. Dessert R 73 results of GNDVI values. Lower case letters (a, b) show the significant differences among treatments based on the ANOVA test ($p < 0.05$). Asterisks mark significant differences compared to control. ** $p < 0.01$. The values of the medium dose plot were 0.2095 (Medium 5000), and the values of the plot treated with a low dose (Low 2500) were 0.2400 GNDVI.

The results of the control plot showed higher GNDVI values than the Low 2500, Medium 5000 and High 10,000 plots and were significantly different from the Medium 5000 (b) and the High 10,000 (b). The Control (a) plot was not significantly different from Low 2500 (a).

In the case of the High 10,000 (b) plot, the GNDVI values were lower than both the two treatments and the uninfected control plot. The values of High 10,000 (b) were significantly different from the GNDVI values of Low 2500 (a) and Control (a). The values of the High 10,000 (b) plot were not significantly different from those of the Medium 2500 (b).

Based on Figure 10, the values of the control plot were the highest with an LCI value of 0.1102, and the values of the plot with a high spore concentration (High 10,000) showed the lowest value of 0.07137.

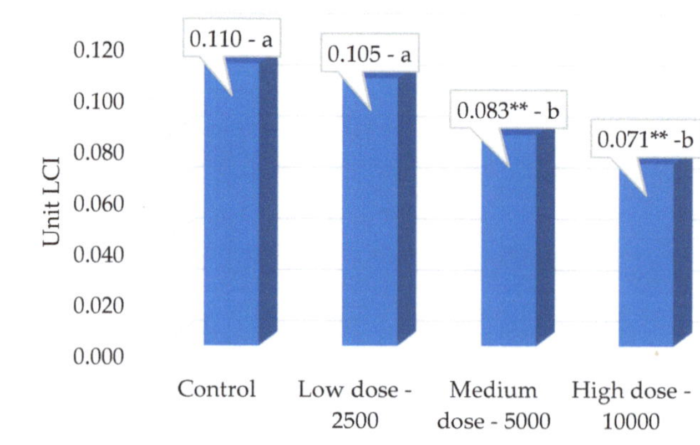

Figure 10. Dessert R 73 results of LCI values. Lower case letters (a, b) show the significant differences among treatments based on the ANOVA test ($p < 0.05$). Asterisks mark significant differences compared to control. ** $p < 0.01$.

The values of the medium dose plot were 0.08258 (Medium 5000), and the values of the low dose treated plot (Low 2500) showed an LCI value of 0.1046.

The results of the control plot showed higher values than the Low 2500 and Medium 5000 plots and also the High 10,000 plot and were significantly different from the Medium 5000 (b) and the High 10,000 (b). The Control (a) plot was not significantly different from the Low 2500 (a).

In the case of the High 10,000 (b) concentration plot, the LCI values were lower than both the two treatments and the uninfected control plot. The values of High 10,000 (b) were significantly different from the LCI values of Low 2500 (a) and Control (a). The values of the High 10,000 (b) plot were not significantly different from those of the Medium 5000 concentration (b).

3.2. Results of the Sweet Maize Hybrid NOA

Based on Figure 11, the values of the control plot were the highest with an ENDVI value of 0.335286, and the values of the plot with a high spore concentration (High 10,000) showed the lowest value of 0.266592. The F value wad 9.173 for the ENDVI, and 11.78 for the GNDV (Table 5).

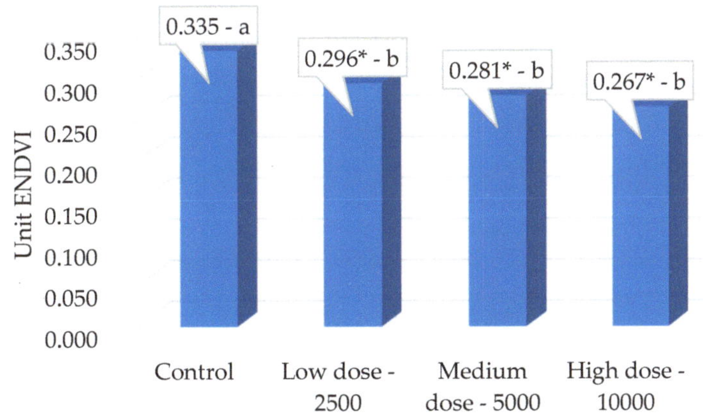

Figure 11. NOA results of ENDVI values. Lower case letters (a, b) show the significant differences among treatments based on the ANOVA test ($p < 0.05$). Asterisks mark significant differences compared to control. * $p < 0.05$.

Table 5. Significant effect on the values of infection in different VIs in the case of the hybrid NOA.

Infection NOA	Df	Sum Sq	Mean Sq	F Value	Pr (>F)
ENDVI	3	0.007891	0.002630	9.73	0.01167 *
GNDVI	3	0.008164	0.002721	11.78	0.00631 **

Note: ** indicates that the infection had a significant impact on the values at the 0.01 level; * indicates that the impact was significant at the 0.05 level.

The values of the medium dose plot were 0.295733 (Medium 5000), and the values of the low dose treated plot (Low 2500) showed an ENDVI value of 0.280846.

The results of the control plot showed higher values than the Low 2500 (b) and Medium 5000 (b) plots and the High 10,000 (b) plot and were significantly different from these three.

In the case of the NOA ENDVI maps, the Medium 5000 (b) was higher than the Low 2500 (b) concentration plot values but was not significantly different.

Based on Figure 12, the values of the control plot were the highest with a GNDVI value of 0.269864, and the values of the plot with a high spore concentration (High 10,000) showed the lowest value of 0.208266.

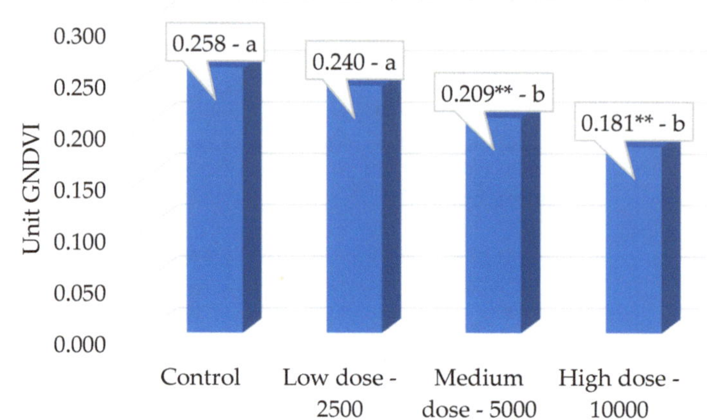

Figure 12. NOA results of GNDVI values. Lower case letters (a, b) show the significant differences among treatments based on the ANOVA test ($p < 0.05$). Asterisks mark significant differences compared to control. ** $p < 0.01$. The values of the medium dose plot were 0.225392 (Medium 5000), and the values of the low dose treated plot (Low 2500) showed an ENDVI value of 0.265049.

The results of the control plot produced higher values than the Low 2500 (a) and Medium 5000 (b) plots and the High 10,000 (b) plot and were significantly different from these two groups.

A summary of the results is shown in Table 6.

Table 6. Summary table of the results.

Infection D73	Control	Low Dose	Medium Dose	High Dose	Pr (>F)
LCI **	0.110 a	0.105 a	0.083 b	0.071 b	0.00138 **
NDVI *	0.305 a	0.290 a	0.288 a	0.256 b	0.0165 *
GNDVI **	0.270 a	0.265 a	0.225 b	0.208 b	0.00307 **
Infection NOA	Control	Low Dose	Medium Dose	High Dose	Pr (>F)
ENDVI *	0.335 a	0.296 b	0.281 b	0.267 b	0.01167 *
GNDVI **	0.258 a	0.240 b	0.209 b	0.181 b	0.00631 **

Groups marked with different letters are significantly different from each other. Alpha: 0.05 Significant levels * $p < 0.05$, ** $p < 0.01$—the marked results is had significant effect).

4. Discussion

The basis of this VIS-SWIR remote sensing technology method lies in the spectral response of plants under different stress factors. Crops that are affected by different diseases show changes in water content, cell structure, pigment synthesis and colorization of the green parts. In this study, four different hybrids were observed: two forage maize hybrids (P 9025 and Armagnac) and two sweet maize hybrids (Dessert R 73 and NOA). In the case of the forage maize hybrids, there were changes in the VI values, but no significant difference was detected between those indices. It is commonly known that forage maize hybrids are much more resilient against smut fungus but are not fully resistant. In the case of the sweet maize hybrid, previous studies showed that after 11 DACSI (days after corn smut infection), the photosynthetic pigments in the plants were significantly reduced [27]. Chlorosis is one of the most typical symptoms of corn smut fungus [22].

Garcia-Ruiz, F. et al. (2013) investigated HLB-infected citrus trees in Florida. They used multispectral images for the detection and classification. The study included seven vegetation indices, and two indices used in this study (NDVI and GNDVI) were also observed for plant pathogen diagnostic purposes.

Their results showed that the AIV (average index values) of the NDVI and GNDVI in the case of HLB-infected trees were both lower than those of the healthy plants. HLB also affected the leaf's photosynthetic pigments, such as chlorophyll and carotenoids. Similar results were observed in this study. In the case of the plots that were infected with the high (10.000) concentration compared to the healthy plots, the average VI values of the infected plots were significantly lower [43].

De Castro et al. (2015) focused on the early detection of laurel wilt disease in avocado plants. For the purpose of evaluation and analysis, multispectral images were taken, and different VIs were calculated. Spectral data analysis showed significant differences between healthy and LW-infected plants in different infection stages. The GNDVI was an optimal vegetation index for this type of survey based on their results [44].

Albetis J. et al. also used the NDVI and GNDVI in their study for the detection of *Flavescence dorée* in red and white grapevine cultivars. The NDVI and GNDVI were also important variables in their multivariate classification and measuring methods [45].

Gennaro and S.F.D monitored grapevine leaf stripe disease (GLSD) with a multispectral UAV. Their results showed that the GLSD grapevine disease affected the values of the NDVI compared to the healthy plants. They also confirmed that GLSD significantly influenced the NDVI values [46].

Abdulridha et al. extracted vegetation indices from hyperspectral images (UAV-based) to detect tomato diseases. According to the authors, they were able to detect bacterial spots in every stage of the disease development using NDVI (850) [47].

Based on the works of Kaur R. et al., the LCI (Leaf Chlorophyll Index) showed $R^2 = 0.73$ regression with chlorophyll content in wheat [48]. In this experiment, in the case of the Dessert R 73 hybrid, the LCI values were able to detect the effect of the corn smut infection, *Ustilago maydis* [DC.] Corda. The plots that were treated with the highest 10.000 spore concentration showed significantly lower values than the healthy ones.

It was detected that the infection had an impact on the values during the infection period. The field was monitored on three different dates 7 days after the infection (7 DAI), 14 days after infection (14 DAI) and 21 days (21 DAI) after infection.

Orthomosaics were created in WebODM with a high spatial resolution (2 cm/pixel). Using raster analysis, five different VIs were calculated (NDVI, GNDVI, ENDVI, NDRE and LCI). Using Quantum GIS 3.20, in addition to creating maps, sample data were also collected with the sample box method, i.e., a rectangle polygon shape file with the same parameters for all the plots. After acquiring the zonal statistics from all the VI maps, a database was created. Statistical analysis was performed in RStudio 4.1.2 [39] with the RStudio [40] graphic interface using the "agricolae" [41] packages.

The figures and graphs were created in Microsoft Excel 2016. Based on the example of Huzsvai [42], a repeated measures ANOVA (RMA) model was used to examine the effect of the recording method and infection level and their combined effect, as well as their impact on the vegetation index value changes.

In the case of the Dessert R 73 hybrid, three different VIs showed that the infection had affected the obtained values, GNDVI, NDVI and LCI. These three VIs performed well based on the study results.

In the case of the NOA sweet maize hybrid, two VIs showed the effect of the corn smut infection (GNDVI and ENDVI).

5. Conclusions

The novelty of our study is that the obtained results show that the GNDVI showed an outstanding performance in the case of the corn smut fungus *Ustilago maydis* [DC.] Corda monitoring. It was shown that the infection had a significant impact on the values at the 0.01 ** level in the two sweet maize hybrids, NOA and Dessert R 73. The NDVI, LCI and ENDVI also showed great results regarding tracking the pathogen *Ustilago maydis* [DC.]. These two hybrids are very popular in the Hungarian sweet maize market, so all research and economic studies are relevant for our producers and also for our researchers. This method can be used to detect the changes in different VIs during the period of infection with the modified RMA model. As the concentration was increased, the values became lower in the plots treated with the fungi, and the dynamic changes were evident. With this survey and sample collection method, the sample values from the VIs can be analyzed and evaluated and can be adopted in every other culture. It is possible to investigate different biotic or abiotic stressors on a field scale. The results showed that the dynamics of the values changed over time and were significantly affected by the different infection levels, i.e., field validation is still required in every case, and the values of severity scales should be determined after evaluation, depending on the culture, vegetation period and on-site validation. In addition, using this multispectral device, which is available on the market, experimental sites or fields can be observed in high quality.

Author Contributions: L.R. (László Radócz 1), data curation, writing of the manuscript and performing measurements; C.B. and A.S. (Adrienn Széles), formal analysis, methodology and conceptualization; A.T. and E.H., writing—review and editing; L.R. (László Radócz 2). and P.R., formal analysis; A.T., A.V. and A.S. (Atala Szabó), supervision and review and editing; Á.I., project administration, corresponding author and review of the manuscript. All authors have read and agreed to the published version of the manuscript.

Funding: Project no. TKP2021-NKTA-32 has been implemented with the support provided by the Ministry of Innovation and Technology of Hungary from the National Research, Development and Innovation Fund, financed under the TKP2021-NKTA funding scheme.

Institutional Review Board Statement: Not applicable.

Informed Consent Statement: Not applicable.

Data Availability Statement: All the data supporting the conclusions of this article are included in this article.

Conflicts of Interest: The authors declare no conflict of interest.

References

1.   Available online: https://www.ksh.hu/docs/hun/xftp/stattukor/vet/20210601/index.html (accessed on 25 January 2023).

2.   Széles, A.; Kovács, K.; Ferencsik, S. The effect of crop years and nitrogen basal and top dressing on the yield of different maize genotypes and marginal revenue. *Időjárás/Q. J. Hung. Meteorol. Serv.* **2019**, *123*, 265–278. [CrossRef]

3.   Széles, A.; Harsányi, E.; Kith, K.; Nagy, J. The effect of fertilisation and weather extremities caused by climate change on maize (*Zea mays* L.) yield in Hungary. *J. Agric. Food Dev.* **2018**, *4*, 1–9. [CrossRef]

4.   Széles, A.; Ragán, P.; Nagy, J. Abiotic stress impacts caused by weather and nutrient replenishment on the yield of maize (*Zea mays* L). *Columella: J. Agric. Environ. Sci.* **2017**, *4*, 39–44.

5.   Rácz, D.; Szőke, L.; Tóth, B.; Kovács, B.; Horváth, É.; Zagyi, P.; Duzs, L.; Széles, A. Examination of the Productivity and Physiological Responses of Maize (*Zea mays* L.) to Nitrapyrin and Foliar Fertilizer Treatments. *Plants* **2021**, *10*, 2426. [CrossRef]

6.   Oerke, E.C. Crop losses to pests. *J. Agric. Sci.* **2006**, *144*, 31–43. [CrossRef]

7.   Christensen, J.J. *Corn Smut Caused by Ustilago Maydis*; Monographs; American Phytopathology Society: Worcester, MA, USA, 1963; Volume 2.

8.   Cao, X.; Luo, Y.; Zhou, Y.; Duan, X.; Cheng, D. Detection of powdery mildew in two winter wheat cultivars using canopy hyperspectral reflectance. *Crop Prot.* **2013**, *45*, 124–131. [CrossRef]

9.   Zhang, J.; Huang, Y.; Pu, R.; Gonzalez-Moreno, P.; Yuan, L.; Wu, K.; Huang, W. Monitoring plant diseases and pests through remote sensing technology: A review. *Comput. Electron. Agric.* **2019**, *165*, 104943. [CrossRef]

10.   Su, J.; Liu, C.; Hu, X.; Xu, X.; Guo, L.; Chen, W.H. Spatio-temporal monitoring of wheat yellow rust using UAV multispectral imagery. *Comput. Electron. Agric.* **2019**, *167*, 105035. [CrossRef]

11.   Su, J.; Liu, C.; Coombes, M.; Hu, X.; Wang, C.; Xu, X.; Chen, W.H. Wheat yellow rust monitoring by learning from multispectral UAV aerial imagery. *Comput. Electron. Agric.* **2018**, *155*, 157–166. [CrossRef]

12.   Calou, V.B.C.; dos Santos Teixeira, A.; Moreira, L.C.J.; Lima, C.S.; de Oliveira, J.B.; de Oliveira, M.R.R. The use of UAVs in monitoring yellow sigatoka in banana. *Biosyst. Eng.* **2020**, *193*, 115–125. [CrossRef]

13. Ye, H.; Huang, W.; Huang, S.; Cui, B.; Dong, Y.; Guo, A.; Ren, Y.; Jin, Y. Recognition of banana fusarium wilt based on UAV remote sensing. *Remote Sens.* **2020**, *12*, 938. [CrossRef]

14. Sankaran, S.; Mishra, A.; Ehsani, R.; Davis, C. A review of advanced techniques for detecting plant diseases. *Comput. Electron. Agric.* **2010**, *72*, 1–13. [CrossRef]

15. Gómez Caro, S. Infection and Spread of *Peronospora sparsa* on *Rosa* sp. (Berk.). Ph.D. Thesis, Universitäts und Landesbibliothek Bonn, Bonn, Germany, 2014.

16. Oerke, E.C.; Froehling, P.; Steiner, U. Thermographic assessment of scab disease on apple leaves. *Precis. Agric.* **2011**, *12*, 699–715. [CrossRef]

17. Penuelas, J.; Filella, I. Visible and near-infrared reflectance techniques for diagnosing plant physiological status. *Trends Plant Sci.* **1998**, *3*, 151–156. [CrossRef]

18. Sishodia, R.P.; Ray, R.L.; Singh, S.K. Applications of remote sensing in precision agriculture: A review. *Remote Sens.* **2020**, *12*, 3136. [CrossRef]

19. Polischuk, V.P.; Shadchina, T.M.; Kompanetz, T.I.; Bi, G.; Sozinov, A.L. Changes in reflectance spectrum characteristic of Nicotiana debneyi plant under the influence of viral infection. *Arch. Phytopathol. Plant Prot.* **1997**, *31*, 115–119. [CrossRef]

20. Kuska, M.; Wahabzada, M.; Leucker, M.; Dehne, H.W.; Kersting, K.; Oerke, E.C.; Steiner, U.; Mahlein, A.K. Hyperspectral phenotyping on the microscopic scale: Towards automated characterization of plant-pathogen interactions. *Plant Methods* **2015**, *11*, 28. [CrossRef]

21. Martinelli, F.; Scalenghe, R.; Davino, S.; Panno, S.; Scuderi, G.; Ruisi, P.; Villa, P.; Stroppiana, D.; Boschetti, M.; Goulart, L.R.; et al. Advanced methods of plant disease detection. A review. *Agron. Sustain. Dev.* **2015**, *35*, 1–25. [CrossRef]

22. Snetselaar, K.M.; Mims, C.W. Light and electron microscopy of Ustilago maydis hyphae in maize. *Mycol. Res.* **1994**, *98*, 347–355. [CrossRef]

23. Frommer, D.; Veres, S.; Radócz, L. Susceptibility of stem infected sweet corn hybrids to common smut disease. *Acta Agrar. Debr.* **2018**, *74*, 55–57. [CrossRef]

24. Morrison, E.N.; Emery, R.J.N.; Saville, B.J. Fungal derived cytokinins are necessary for normal Ustilago maydis infection of maize. *Plant Pathol.* **2017**, *66*, 726–742. [CrossRef]

25. Mills, L.J.; Vanstaden, J. Extraction of cytokinins from maize, smut tumors of maize and Ustilago maydis cultures. *Physiol. Plant Pathol.* **1978**, *13*, 73–80. [CrossRef]

26. Turian, G.; Hamilton, R.H. Chemical detection of 3-indolylacetic acid in Ustilago zeae tumors. *Biochim. Biophy. Acta* **1960**, *41*, 148–150. [CrossRef] [PubMed]

27. Szőke, L.; Moloi, M.J.; Kovács, G.E.; Biró, G.; Radócz, L.; Hájos, M.T.; Kovács, B.; Rácz, D.; Danter, M.; Tóth, B. The application of phytohormones as biostimulants in corn smut infected Hungarian sweet and fodder corn hybrids. *Plants* **2021**, *10*, 1822. [CrossRef]

28. Moura, R.M.; Pedrosa, E.M.; Guimarães, L.M. A rare syndrome of corn smut. *Fitopatol. Bras.* **2001**, *26*, 782. [CrossRef]

29. Király, G.; Rizzo, G.; Tóth, J. Transition to Organic Farming: A Case from Hungary. *Agronomy* **2022**, *12*, 2435. [CrossRef]

30. Available online: https://www.dji.com/hu/p4-multispectral/specs (accessed on 25 January 2023).

31. Available online: https://opendronemap.org/webodm (accessed on 25 January 2023).

32. Tucker, C.J. Red and photographic infrared linear combinations for monitoring vegetation. *Remote Sens. Environ.* **1979**, *8*, 127–150. [CrossRef]

33. Gitelson, A.A.; Merzlyak, M.N. Remote sensing of chlorophyll concentration in higher plant leaves. *Adv. Space Res.* **1998**, *22*, 689–692. [CrossRef]

34. Barnes, E.M.; Clarke, T.R.; Richards, S.E.; Colaizzi, P.D.; Haberland, J.; Kostrzewski, M.; Waller, P.; Choi, C.; Riley, E.; Thompson, T. Coincident detection of crop water stress, nitrogen status and canopy density using ground based multispectral data. In Proceedings of the 5th International Conference on Precision Agriculture and Other Resource Management, Bloomington, MN, USA, 16–19 July 2000.

35. Datt, B. A new reflectance index for remote sensing of chlorophyll content in higher plants: Tests using Eucalyptus leaves. *J. Plant. Physiol.* **1999**, *154*, 30–36. [CrossRef]

36. Thenkabail, P.S.; Smith, R.B.; De Pauw, E. Hyperspectral vegetation indices and their relationships with agricultural crop characteristics. *Remote Sens. Environ.* **2000**, *71*, 158–182. [CrossRef]

37. Rasmussen, J.; Ntakos, G.; Nielsen, J.; Svensgaard, J.; Poulsen, R.N.; Christensen, S. Are vegetation indices derived from consumer-grade cameras mounted on UAVs sufficiently reliable for assessing experimental plots? *Eur. J. Agron.* **2016**, *74*, 75–92. [CrossRef]

38. Guan, S.; Fukami, K.; Matsunaka, H.; Okami, M.; Tanaka, R.; Nakano, H.; Sakai, T.; Nakano, K.; Ohdan, H.; Takahashi, K. Assessing correlation of high-resolution NDVI with fertilizer application level and yield of rice and wheat crops using small UAVs. *Remote Sens.* **2019**, *11*, 112. [CrossRef]

39. R Core Team. *R: A Language and Environment for Statistical Computing*; R Foundation for Statistical Computing: Vienna, Austria, 2022. Available online: http://www.R-project.org/ (accessed on 25 January 2023).

40. RSTUDIO Team. *RStudio: Integrated Development for R*; RStudio, Inc.: Boston, MA, USA, 2022. Available online: http://www.rstudio.com/ (accessed on 25 January 2023).

41. De Mendinburu, F.; Agricolae: Statistical Procedures for Agricultural Research. R Package Version 1. 3-5. 2021. Available online http://CRAN.R-project.org/package=agricolae (accessed on 25 January 2023).

42. Huzsvai, L.; Balogh, P. Lineáris Modellek az R-ben. Seneca Books, Debrecen. 109–124. 2015. Available online: http://seneca-books.hu/doc/Linearis_modellek.pdf (accessed on 25 January 2023).

43. Garcia-Ruiz, F.; Sankaran, S.; Maja, J.M.; Lee, W.S.; Rasmussen, J.; Ehsani, R. Comparison of two aerial imaging platforms for identification of Huanglongbing-infected citrus trees. *Comput. Electron. Agric.* **2013**, *91*, 106–115. [CrossRef]

44. De Castro, A.I.; Ehsani, R.; Ploetz, R.; Crane, J.H.; Abdulridha, J. Optimum spectral and geometric parameters for early detection of laurel wilt disease in avocado. *Remote Sens. Environ.* **2015**, *171*, 33–44. [CrossRef]

45. Albetis, J.; Duthoit, S.; Guttler, F.; Jacquin, A.; Goulard, M.; Poilvé, H.; Féret, J.-B.; Dedieu, G. Detection of Flavescence dorée grapevine disease using unmanned aerial vehicle (UAV) multispectral imagery. *Remote Sens.* **2017**, *9*, 308. [CrossRef]

46. Di Gennaro, S.F.; Battiston, E.; Di Marco, S.; Facini, O.; Matese, A.; Nocentini, M.; Palliotti, A.; Mugnai, L. Unmanned Aerial Vehicle (UAV)-based remote sensing to monitor grapevine leaf stripe disease within a vineyard affected by esca complex. *Phytopathol. Mediterr.* **2016**, *55*, 262–275.

47. Abdulridha, J.; Ampatzidis, Y.; Kakarla, S.C.; Roberts, P. Detection of target spot and bacterial spot diseases in tomato using UAV-based and benchtop-based hyperspectral imaging techniques. *Precis. Agric.* **2020**, *21*, 955–978. [CrossRef]

48. Kaur, R.; Singh, B.; Singh, M.; Thind, S.K. Hyperspectral indices, correlation and regression models for estimating growth parameters of wheat genotypes. *J. Indian Soc. Remote Sens.* **2015**, *43*, 551–558. [CrossRef]

Article

Estimation of the Total Soil Nitrogen Based on a Differential Evolution Algorithm from ZY1-02D Hyperspectral Satellite Imagery

Rongrong Zhang [1], Jian Cui [2,3], Wenge Zhou [1], Dujuan Zhang [4], Wenhao Dai [1], Hengliang Guo [4] and Shan Zhao [1,*]

[1] School of Geoscience and Technology, Zhengzhou University, Zhengzhou 450001, China; 202012562017389@gs.zzu.edu.cn (R.Z.); zhouge2022@gs.zzu.edu.cn (W.Z.); 202022562017409@gs.zzu.edu.cn (W.D.)
[2] Henan Institute of Geological Survey, Zhengzhou 450001, China; 2021811@zzuli.edu.cn
[3] National Engineering Laboratory Geological Remote Sensing Center for Remote Sensing Satellite Application, Zhengzhou 450001, China
[4] National Supercomputing Center in Zhengzhou, Zhengzhou University, Zhengzhou 450001, China; duduzdj@zzu.edu.cn (D.Z.); guohen-gliang@zzu.edu.cn (H.G.)
* Correspondence: zhaoshan4geo@zzu.edu.cn; Tel.: +86-1370-0875-423

Abstract: Precise fertilizer application in agriculture requires accurate and dependable measurements of the soil total nitrogen (TN) concentration. Henan Province is one of the most important grain-producing areas in China. In order to promote the development of precision agriculture in Henan Province, this study took the high-standard basic farmland construction area in central Henan Province as the research area. Using single-phase images acquired from the ZY1-02D satellite hyperspectral sensor on 28 January 2021 (with a spatial resolution of 30 m × 30 m, a spectral range that covered 400–2500 nm, and a revisit period of 3 days) for spectral reflectance transformation and feature spectral band extraction. Based on multiple representation models, such as multiple linear regression, partial least squares regression, and support vector machine (SVM), all bands, feature bands, feature band combinations, and differential evolution (DE) algorithms were used to extract the secondary characteristic variables of the combination of characteristic bands, which were used as model inputs to estimate the content of TN in the study area. It was found that (1) the spectral reflectance transformation could help to improve the accuracy of prediction by reducing the interference from noise in the model, but the optimal spectral transformation method differed between different models and even between the training and test sets of the same model; (2) the estimation accuracy of the TN content model based on the minimum shrinkage and feature selection operator of the feature band was usually better than that of the full band, the feature combination band contained more effective information related to the TN content, and the combination of the DE algorithm and the SVM model achieved a better estimation accuracy for secondary feature extraction and TN content estimation of the feature combination band; and (3) ZY1-02D hyperspectral satellite data have the potential for the dynamic and non-destructive monitoring of regional TN content.

Keywords: soil total nitrogen; ZY1-02D/AHSI hyperspectral; feature selection; model estimation

Citation: Zhang, R.; Cui, J.; Zhou, W.; Zhang, D.; Dai, W.; Guo, H.; Zhao, S. Estimation of the Total Soil Nitrogen Based on a Differential Evolution Algorithm from ZY1-02D Hyperspectral Satellite Imagery. *Agronomy* **2023**, *13*, 1842. https://doi.org/10.3390/agronomy13071842

Academic Editors: Karsten Schmidt and Peng Fu

Received: 2 June 2023
Revised: 8 July 2023
Accepted: 11 July 2023
Published: 12 July 2023

1. Introduction

Soil total nitrogen (TN) is a fundamental indicator of soil fertility and a critical factor in plant growth and development [1,2]. The soil's abundance of or deficiency in nitrogen will directly impact crop growth and yield [3]. Therefore, the dynamic, large-scale, and accurate estimation of TN content is a significant measure that can be used to guide agricultural field fertilization schemes and crop growth status monitoring [4].

The traditional method for determining TN is to gather soil samples from different sites for laboratory chemical analysis, after which the TN distribution is obtained using spatial interpolation. In order to guarantee the precision of the interpolation results, this method frequently calls for collecting numerous soil samples, which is time-consuming and invasive [5,6]. Recent research has demonstrated that satellite-based hyperspectral images offer excellent spectral and spatial resolution, enabling dynamic, effective, and precise predictions of soil components [7–9]. The ZY1-02D satellite carries the Advanced Hyper Spectral Imager (AHSI), which can acquire 166 bands of different wavelengths between 400 and 2500 nm (covering visible to shortwave infrared), with an image strip width of 60 km [10]. Recent research has demonstrated that the ZY1-02D/AHSI satellite, China's first hyperspectral satellite for civil usage, has promising application possibilities, and images obtained using the AHSI sensor can be used to estimate the component composition of the soil [11,12].

Hyperspectral data are rich in spectral information that may be used to identify changes in soil characteristics and offer precise estimations of the elemental content in the soil [9,12,13]. However, there are several problems with using hyperspectral data to estimate soil nutrient contents, such as the fact that unprocessed original hyperspectral data typically contain a lot of redundant spectral information because they have a lot of spectral channels with high spectral resolution [14]. The recent research on hyperspectral-based soil content estimation has mainly included the processing of spectral data through reflectance transformation, spectrum feature selection, and the formulation of estimation models. Reflectance transformation processing can enhance feature bands and facilitate the rejection of noise [15]. Therefore, choosing an appropriate reflectance transformation technique is crucial in guaranteeing the model's precision [16].

Studies have shown that spectral data transformation methods, such as the inverse reflectance [17], natural logarithm of the reflectance [18], and first-order derivative reflectance [19], can enhance the characteristic bands and improve model accuracy [20,21]. In addition, some scholars have studied spectral information and found that the selection of an appropriate spectral feature band can reduce data redundancy, simplify the model, improve model accuracy, and lead to better estimation results [22,23]. Spectral feature extraction has been carried out using methods such as the differential evolutionary (DE) technique and the least absolute shrinkage and selection operator (LASSO) [4,24]. However, most studies on the extraction of characteristic bands of soil composition information are based on original spectral data or single spectral transformation data without considering the application potential of characteristic spectral combination data in soil composition information model estimation. Currently, machine learning models, such as support vector machines (SVMs), back propagation neural networks, and random forests [25–28], as well as linear models, such as multiple linear regression (MLR) and partial least squares regression (PLSR) [29–31], are commonly used to estimate the soil component content. Scholars demonstrated that SVM models produce more precise predictions than the RF and PLSR models [32], and the SVM model has good stability and versatility in solving nonlinear problems.

However, the current hyperspectral data used for TN content estimation are mainly based on hyperspectral data obtained by laboratories and ground platforms [33], and there is still a lack of research on TN content estimation methods based on satellite platform hyperspectral data, especially the new ZY1-02D hyperspectral data. Summarizing the previous research results, this study established a set of TN content estimation processes and methods based on ZY1-02D hyperspectral remote sensing data from spectral reflectance transformation processing, spectral band selection, and model construction methods. In order to make full use of the computational spectral data obtained by spectral reflectance transformation and feature selection, this study also proposed the use of characteristic spectral combination data for the estimation of TN content. Additionally, this study aimed to tackle the problem of possible redundant information in the characteristic spectral combination data, and proposed performing secondary feature selection on the characteristic spectral combination data to remove invalid variables and improve the accuracy of model estimation.

In this work, by comparing previous studies, the ZY1-02D hyperspectral data were transformed using four widely used techniques: the original reflectance, the inverse reflectance, the natural logarithm of the reflectance, and the first-order derivative reflectance. The LASSO technique was used to extract the primary spectral characteristics for individual spectral reflectance transform data, and a first attempt was made at secondary feature extraction using the DE algorithm for LASSO feature spectral band combination data. Finally, based on the chosen characteristics, the MLR, PLSR, and SVM models were used to estimate the TN content based on the selected characteristics, and the best model was selected to map the TN content. These were the key goals: (1) to investigate the best spectral reflectance transformation methods for different models; (2) to compare the estimation of all bands of a single spectral reflectance transformation with the LASSO feature bands; and (3) to investigate the model estimate capabilities of secondary feature selection for a combination of LASSO feature bands using the DE technique.

2. Materials and Methods

2.1. Study Area

In this study, the construction area of high-standard basic farmland in central Henan Province, China, which is located between 34°16′ and 34°40′ N and 113°20′ and 113°54′ E (Figure 1), was used as the study area. The terrain is relatively elevated in the east, with hills in the center and mountains in the west, and with elevations ranging from 78 to 787 m. The area is one of the major grain-producing areas in Henan Province, with an average annual temperature of 14.4 °C and a mild monsoon environment. The primary crops grown there are soybeans, maize, and wheat.

2.2. Soil Sample Acquisition and TN Content Determination

Using the five-point sampling technique (collect soil samples at the center and four corners of a rectangular area of 30 m × 30 m and mix the five samples evenly as the final collected soil sample), 595 soil samples were gathered at depths of 0–20 cm in the research area in March 2021 and their positions were recorded (Figure 1c). The soil samples were sieved to remove crop roots, gravel, and other contaminants and allowed to dry in a chamber, and then the total nitrogen content of the soil samples was determined using an automatic Kjeldahl nitrogen analyzer [34]. At the same time, the amounts of heavy metals (cadmium, mercury, arsenic, lead, chromium, copper, nickel, and zinc) and other soil nutrients (organic matter, total phosphorus, total potassium, and pH value) in the soil samples were also determined.

2.3. ZY1-02D/AHSI Remote Sensing Image Collection and Pre-Processing

Soil sampling was conducted in March 2021, which is when winter wheat was the dominant crop in the study area. This is the time when young winter wheat seedlings turn green and begin to grow. Considering the influence of the soil sampling time and clouds on the image quality, hyperspectral remote sensing images acquired by the ZY1-02D/AHSI sensor on 28 January 2021 (Figure 1c) were the closest high-quality images to the soil sampling time, and they were taken at a time when wheat and other crops were growing slowly in winter and the plants were short and did not cover the ground. The soil surface reflectance obtained with hyperspectral satellites during this period was less affected by crops and was similar to the spectral reflectance of bare soil; therefore, this study was based on the images from this period for the estimation of TN content in the study area. The ZY1-02D/AHSI sensor's parameters are listed in Table 1. The satellite operated in a Sun-synchronous orbit and acquired hyperspectral images in 166 spectral channels over a swath width of 60 km. The spectrum range covered was between 400 and 2500 nm. The resulting images comprised 90 channels between 1055 and 2500 nm with a resolution of 20 nm and 76 spectral channels between 400 and 1040 nm with a spectral resolution of 10 nm.

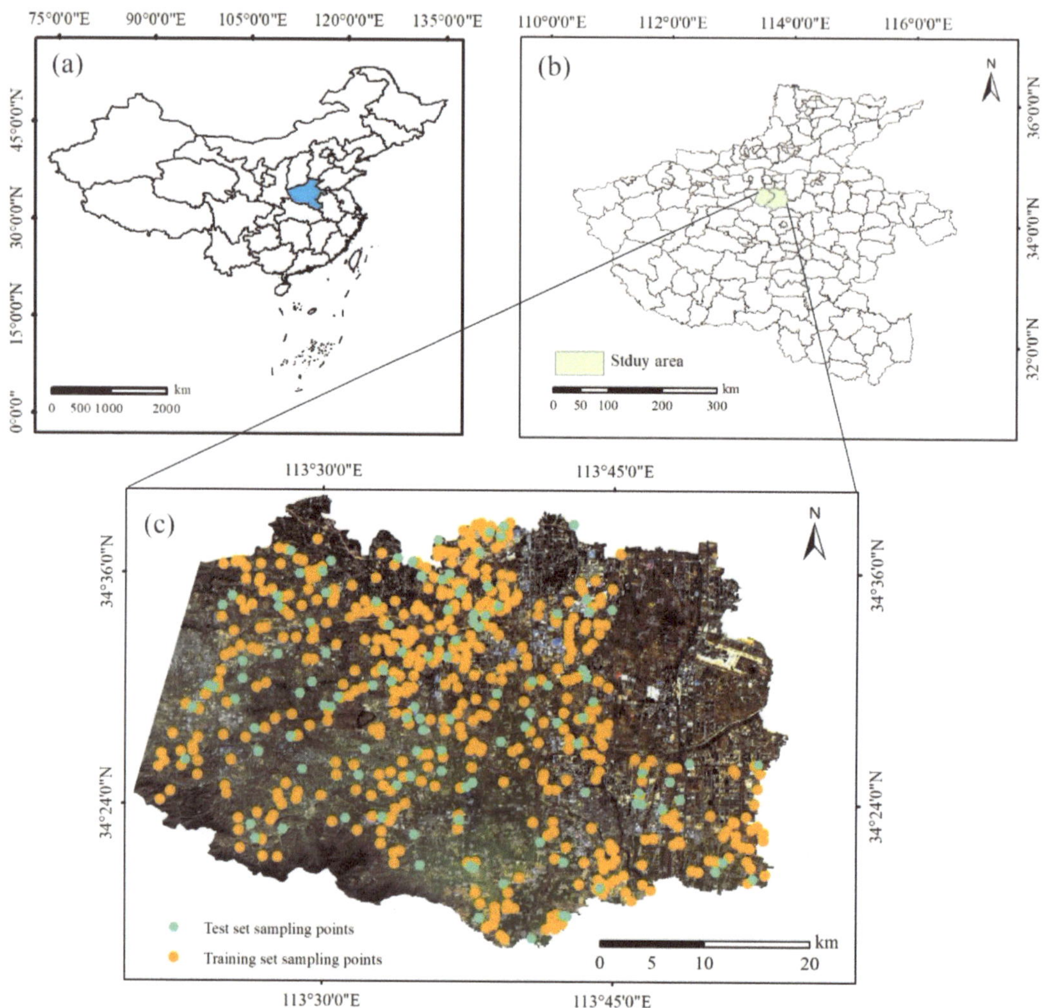

Figure 1. Maps of the research site. (**a**) Map of China, where the blue area is Henan Province; (**b**) map of Henan Province; (**c**) hyperspectral ZY1-02D/AHSI image of the research area, together with the locations of the soil sampling sites (in this study, only the reflectance data of satellite images in cultivated areas were extracted for experimentation, excluding buildings and other areas).

Table 1. Device parameters of ZY1-02D/AHSI.

Items	Parameters
Date of launch	12 September 2019
Spectral bands	76 (VNIR), 90 (SWIR)
Spectral range (nm)	400–2500
Spectral resolution (nm)	10 (VNIR), 20 (SWIR)
Spatial resolution (m)	30
Swath width (km)	60
Revisit cycle (d)	3

The images were pre-processed to improve their quality and reduce the effects caused by weather and atmospheric factors. The images were radiometrically and atmospherically corrected and orthorectified using ENVI v5.3. Subsequently, to determine the reflectance value of the hyperspectral pictures corresponding to the coordinates of the soil sampling location, ArcGIS v10.5 was utilized. In this study, three spectral bands that overlap with VNIR and SWIR and four bands near 1400 nm and 1900 nm that are strongly influenced by atmospheric water absorption (two bands near 1400 nm: B100 and B101 at 1391.65 nm and 1408.49 nm, and two bands near 1900 nm: B130 and B131 at 1896.11 nm and 1912.97 nm) were used, for a total of 159 experimental spectral bands.

2.4. Methodology

In this study, the TN estimation using satellite hyperspectral imagery consisted of the following three main steps: (1) data acquisition and pre-processing, as well as preparation of soil sampling points and remote sensing image data (as described in Sections 2.2 and 2.3 above); (2) data processing, spectral reflectance transformation, and feature extraction; (3) and TN content model estimation and mapping. The experimental approach for this investigation is shown in a flow chart in Figure 2.

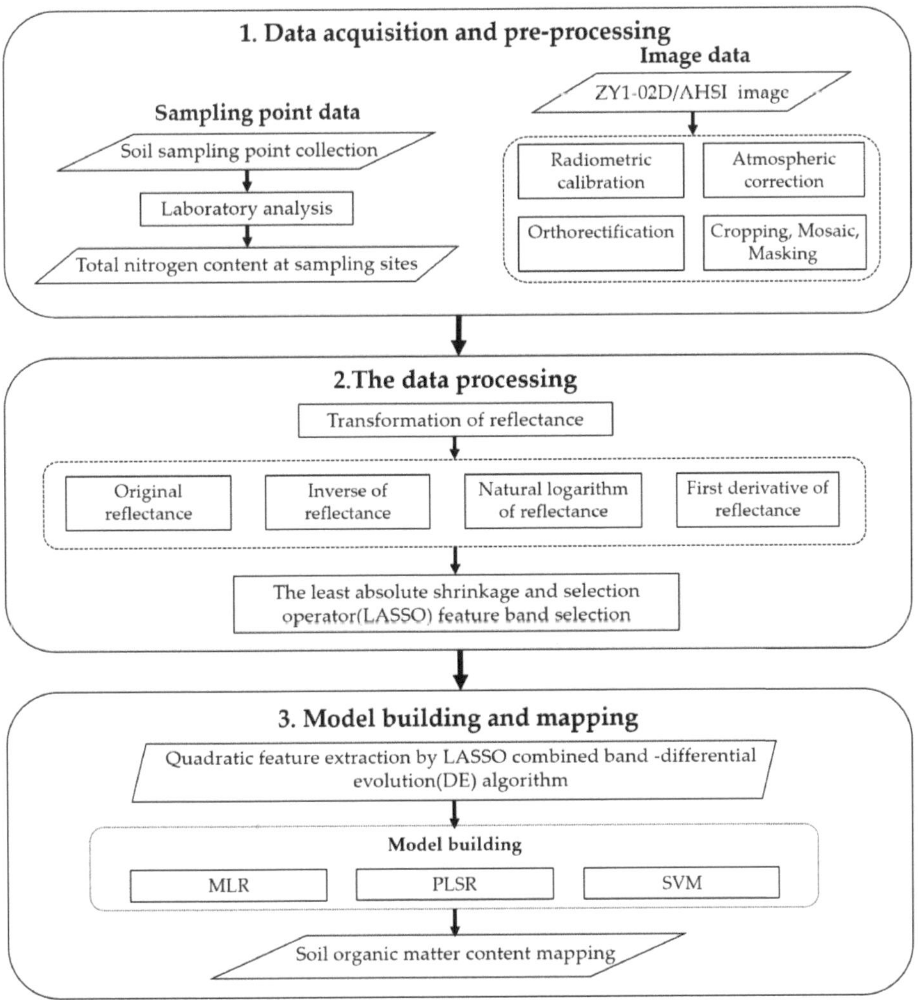

Figure 2. Flow chart of the experimental steps.

2.4.1. Spectral Reflectance Transformation

To identify the sensitive links between TN content and spectral reflectance and emphasize distinctive spectral bands, it is crucial to transform the spectral data. In this study, the inverse reflectance (IR), the natural logarithm of the reflectance (NLR), and the first-order derivative reflectance (FDR) were used to transform the original reflectance (OR) values. Spectral transformation processing can increase spectral sensitivity and aid in increasing models' predictive ability [35]. The equations for the different transformations are shown below:

$$\mathrm{IR}(\lambda_i) = \frac{1}{\lambda_i} \tag{1}$$

$$\mathrm{NLR}(\lambda_i) = \mathrm{LnR}(\lambda_i) \tag{2}$$

$$\mathrm{FDR}(\lambda_i) = \frac{\mathrm{R}(\lambda_{i+1}) - \mathrm{R}(\lambda_{i-1})}{2\Delta\lambda} \tag{3}$$

where λ_{i-1}, λ_i, and λ_{i+1} are the wavelengths of band i and its adjacent bands; $\Delta\lambda$ indicates the interval between two adjacent wavelengths; $\mathrm{IR}(\lambda_i)$ is the inverse reflectance of wavelength λ_i; $\mathrm{NLR}(\lambda_i)$ is the natural logarithmic reflectance of wavelength λ_i; and $\mathrm{FDR}(\lambda_i)$ is the first-order derivative reflectance of wavelength λ_i.

2.4.2. Selection of Spectral Feature Variables

According to previous research, the identification of distinctive spectral characteristics is one of the key techniques for estimating the concentration of soil components using hyperspectral data [36,37]. It is essential to choose the proper feature variables in order to ensure the precision of the model estimation. TN content and spectral reflectance values have a complicated, non-linear relationship. Thus, in this work, the TN content feature bands were selected using non-linear algorithms (LASSO and DE). The LASSO algorithm was used for the initial feature selection of the above four sets of spectral reflectance data. Then, the DE algorithm combined with a predictive model was used to perform secondary feature selection from the feature band combinations selected using LASSO. The LASSO and DE algorithms were implemented using Python 3.8 and are described below.

LASSO is a paradigm-based algorithm in which a 1-paradigm regularization penalty term was introduced on top of the ordinary least squares function to constrain the residual sum of squares [38], which could successfully reduce the dimensionality of the data and resolve sparsity issues with high-dimensional data. To carry out feature selection, the estimation was compressed using a penalty function that compressed the regression coefficients. The coefficients of the less-sensitive variables were adjusted to zero when the total absolute value of the regression coefficients was less than a specific value. The following is a mathematical definition of the LASSO algorithm.

$$\underset{\varrho}{\mathrm{argmin}}\left\{ \sum_{i=1}^{n} \left(p_i - \sum_{j=1}^{m} h_{ij}X_j \right)^2 \right\} \\ \text{subject to} \sum_{j=1}^{m} |X_j| \leq \varepsilon \tag{4}$$

where the numbers n and m denote the samples and variables, respectively; the independent and dependent variables for each sample are denoted by h_{ij} and p_i; and ε and X_j are the critical and specific gravity values.

The DE method, which is a global optimization approach, was initially put forward by R. Storn and K. Price [39]. It is extensively utilized in many domains because of its straightforward structure, implementation, quick convergence, and robustness. Some academics have also employed the method for feature variable selection [40]. The DE algorithm is a population-based heuristic search algorithm. Its fundamental premise is to produce candidates for selection through individual variations within a population, followed by crossover and selection operations to accomplish population evolution. In order to advance to the next generation and determine the optimal model variable selection, the DE algorithm goes through an evolutionary process, which is depicted in Figure 3 and comprises mutation, crossover, and selection processes.

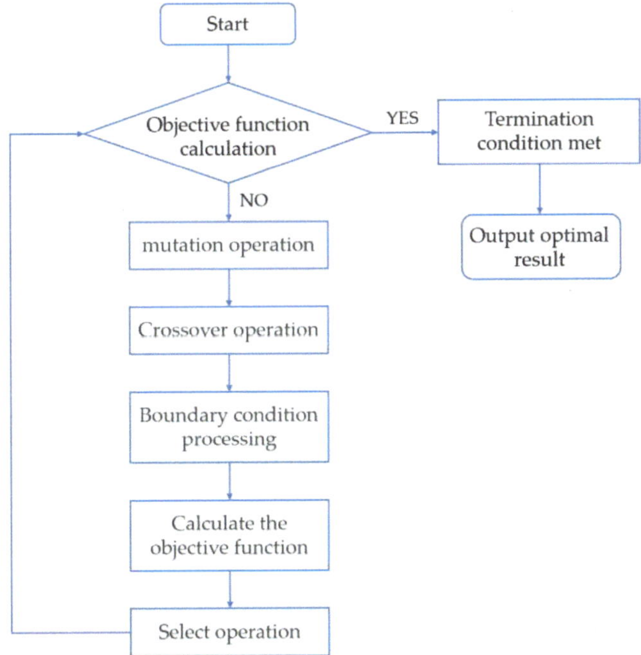

Figure 3. DE algorithm's flow chart.

Some of the expressions for the calculation of the DE algorithm are shown below:

(a)   Initializing the population: M individuals, each made up of an n-dimensional vector, are created uniformly and randomly in the solution space. The vector and the j-th dimensional value assigned to the i-th individual, respectively, are given in Equations (5) and (6):

$$X_i(0) = (x_{i,1}(0), x_{i,2}(0), x_{i,3}(0), \ldots, x_{i,n}(0))$$
$$i = 1,\ 2,\ 3, \ldots,\ M \tag{5}$$

$$X_{i,j}(0) = L_{j\_min} + rand(0,1)\left(L_{j\_max} - L_{j\_min}\right)$$
$$i = 1,\ 2,\ 3, \ldots,\ M; j = 1,\ 2,\ 3 \ldots,\ n \tag{6}$$

where $X_i(0)$ is the i-th individual; j denotes the j-th dimension; M denotes the population size parameter; n denotes the optimization dimension; L_{j_min} and L_{j_max} are the lower and upper bounds of the j-th dimension, respectively; and $rand(0,1)$ denotes a random number on the interval $[0, 1]$.

(b)   Mutation operation: The DE algorithm implements an individual mutation operation through a difference strategy. Equation (7) shows the vector mutation operation for each individual:

$$W_i(G+1) = X_{c1}(G) + Z(X_{c2}(G) - X_{c3}(G)) \tag{7}$$

where c1, c2, and c3 are random numbers; Z is a scaling factor; and G is the index of individual mutation generations.

(c)   Crossover operation: the mutated individuals are subject to the crossover operation shown in formula eight:

$$U_{i,j}(G+1) = \begin{cases} W_{i,j}(G+1) & \text{if rand } (0,1) \leq CR \\ x_{i,j}(G) & \text{otherwise} \end{cases} \tag{8}$$

where CR is the crossover probability.

(d)   The operation for selecting the next generation of individuals is shown in Equation (9):

$$X_i(G+1) = \begin{cases} U_i(G+1) & \text{if } (U_i(G+1) \leq f(X_i(G))) \\ X_i & \text{otherwise} \end{cases} \tag{9}$$

where f is the objective function.

2.4.3. Construction and Evaluation of the TN Content Estimation Model

The TN content was used as the dependent variable in this research, and the independent variables were the spectral band reflectance values of various forms. Three models, namely, MLR, PLSR, and SVM, were used to describe the relationship between the spectral data and TN content. The models' specific characteristics are listed below.

MLR is a type of regression analysis where the dependent variable is predicted or estimated using a combination of several independent factors. MLR is also widely used as a classical prediction model for predicting soil information content [41,42]. The MLR model has the benefit of being straightforward to construct and simple to apply when compared with other models. However, the MLR model cannot produce a precise forecast of the target variables when there is a non-linear connection between the independent factors and the target variables.

PLSR is a popular multivariate statistical technique for forecasting soil element content using hyperspectral data [43,44], which can address the issue of spectral band covariance. The PLSR approach compares several dependent variables to numerous independent variables in a multivariate statistical regression modeling setting. Principal component analysis, traditional correlation analysis, and multiple linear regression analysis are the three fundamental analytical methods combined in PLSR.

The SVM algorithm is a supervised machine learning model that employs non-linear mapping to map data in a high-dimensional data feature space. In a high-dimensional feature space, this enables the formulation of appropriate linear regression characteristics between the independent and dependent variables [45], enabling fitting in the higher-dimensional space and the subsequent return to the initial space. The core of the SVM regression model lies in the selection of the kernel function. In this study, several kernel functions' model prediction capabilities were analyzed before settling on the radial basis function (RBF). The SVM is widely used in the prediction of soil component composition using spectral data due to its high stability and generalization in solving non-linear issues [13,46].

The coefficient of determination (R^2), the mean absolute error (MAE), and the root mean square error (RMSE) were used to evaluate the accuracy of the prediction model. Lower MAE and RMSE values and higher R^2 values correspond to more precise model estimation [47].

Studies have shown that the d-factor can be used to evaluate the uncertainty of an estimation model [19,48]. The degree of uncertainty of an estimation model is proportional to the calculated value of the d-factor; that is, the degree of uncertainty of the estimated model will increase with the increase in the calculated value of the d-factor, and the d-factor calculation formula is as follows:

$$\overline{d_r} = \frac{1}{m}\sum_{i=1}^{m}(P_{Ui} - P_{Li}) \tag{10}$$

$$d - factor = \frac{\overline{d_r}}{\sigma_P} \tag{11}$$

where m is the sample number; P_{Ui} and P_{Li} are upper and lower confidence limits, respectively; $\overline{d_r}$ is the average distance between the upper and lower confidence limits; and σ_P is the standard deviation of the measured value of TN.

3. Results

3.1. Statistical Description of the TN Content of the Sampling Points

The 595 soil samples in this study were used, and for the train_test_split module in the Python programming language the random_state parameter was set to 15, and the soil samples were randomly divided into training and test sets at a ratio of 4:1. The numbers of training set and test set samples were 476 and 119, respectively, and the sample distribution of the training set and test set is shown in Figure 1c. The statistics of the samples that were taken are shown in Table 2.

Table 2. Results of the statistical description of the TN content.

Set	N	Max (g/kg)	Min (g/kg)	Mean (g/kg)	SD (g/kg)	CV
Whole set	595	1.80	0.37	1.01	0.22	0.22
Training set	476	1.80	0.37	1.01	0.22	0.22
Test set	119	1.71	0.43	1.02	0.24	0.23

3.2. TN Content Spectral Features Analysis and Spectral Transformation Processing

3.2.1. Spectral Features Analysis of TN Content

The ZY1-02D/AHSI hyperspectral reflectance spectral patterns of soil samples in the research area with various TN contents are shown in Figure 4. The distribution of the spectral reflectance curves for the various content levels followed a similar pattern, as shown by the original reflectance curves in Figure 4a. The spectral reflectance had two absorption peaks in the wavelength ranges 1100–1400 nm and 1750–2000 nm. According to pertinent research, the primary cause of the absorption peaks at 1100–1400 nm and 1750–2000 nm in soil water is hydroxide ions [49]. Generally, the soil reflectance values decrease as the TN content increases.

3.2.2. Spectral Transformation Processing

As can be seen in Figure 4b, the overall trend of the inverse reflectance spectral curve was the opposite of the original reflectance spectral curve, with a distinct reflection peak in the 1750–2000 nm wavelength range. Although the reflectance trough between 1750 and 2000 nm was also evident in the case of the natural logarithm of the reflectance (Figure 4c), the reflectivity curve was comparable with that of the original reflectance. Compared with the original reflectance, the inverse reflectance and natural logarithm of the reflectance showed a flattening out of the reflectance values in the wavelength range, except for the reflectance and absorption peaks that were highlighted at specific wavelengths, which helped to reduce noise-enhancing characteristic spectral variables. Figure 4d shows the first-order derivative reflectance spectral profile, which had more pronounced absorption peaks at 1100 nm, 1850 nm, and 1950 nm and distinct reflection peaks at 1150 nm, 1380 nm, and 2000 nm compared with the three other kinds of reflectance values.

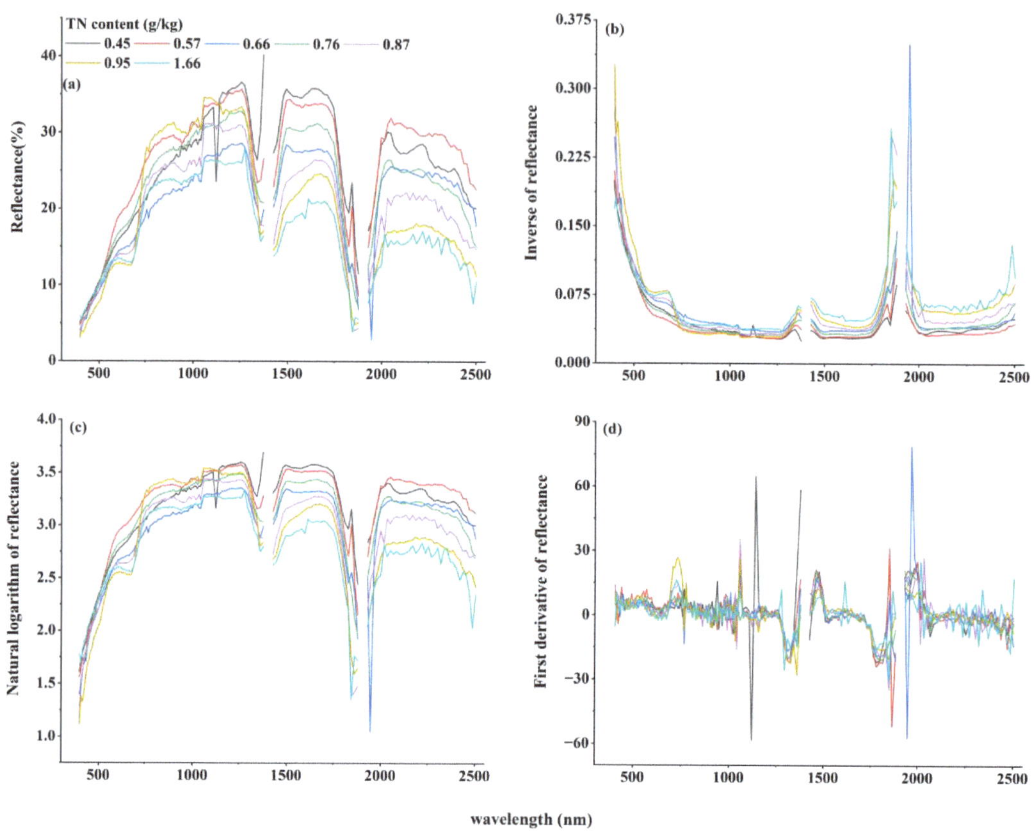

Figure 4. Spectral reflectance profiles for different TN contents: (**a**) original reflectance, (**b**) inverse reflectance, (**c**) natural logarithm of the reflectance, and (**d**) first-order derivative reflectance.

3.3. Selection of TN Spectral Characteristics

The LASSO algorithm was used to select the TN spectral bands based on the original and transformed reflectance readings. It can be seen from the results of the LASSO feature band selection in Figure 5 and Table 3 that the bands were much more distinctive for the original reflectance (OR) and first-order derivative reflectance (FDR) than for the inverse reflectance (IR) and natural logarithm of the reflectance (NLR). The main reason for this phenomenon was that the IR and NLR had prominent reflection and absorption peaks at specific wavelengths compared with the OR and FDR, and they had reflectance values that differed significantly from those in the other wavelength ranges. Therefore, the bands at the reflection and absorption peaks of IR and NLR were given larger coefficients when performing feature selection and were retained as feature variables, while the coefficients in other wavelength ranges were compressed to zero and excluded. The OR and FDR bands were evenly distributed between 400 and 2500 nm, while the IR and NLR bands were more concentrated, with most of the bands in the 400–700 nm and 1500–2000 nm wavelength ranges. The number of bands chosen for the LASSO feature is shown in Table 3 as n.

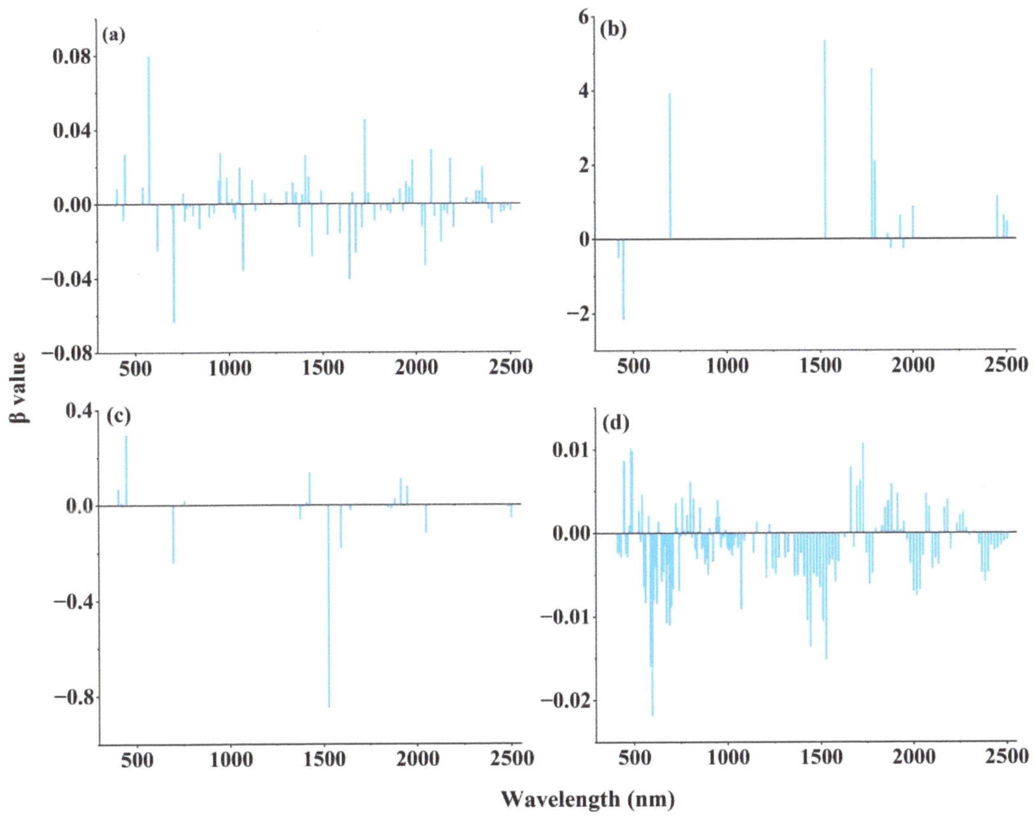

Figure 5. LASSO algorithm TN feature band selection results: (**a**) OR–LASSO, (**b**) IR–LASSO, (**c**) NLR–LASSO, and (**d**) FDR–LASSO.

Table 3. LASSO algorithm TN feature band selection results.

Reflectance Representation	n	Wavelengths (nm)
OR	77	396–405, 439–227, 542, 577, 619, 696–705, 756–774, 791, 808, 842, 894, 920, 945–954, 988–997, 1014–1073, 1123–1139, 1190, 1224, 1308, 1341–1442, 1493, 1526, 1594, 1644–1678, 1711–1745, 1779, 1812–1880, 1930–1981, 2014–2048, 2081–2098, 2132–2199, 2267, 2300–2401, 2450–2501
IR	17	395–404, 422, 447, 697, 1526, 1778–1795, 1845–1880, 1929–1947, 1998, 2451, 2484–2501
NLR	19	404, 422, 447, 697, 757, 1375, 1425, 1526, 1594, 1644, 1845–1880, 1930–1947, 2048, 2199, 2484–2501
FDR	141	404–430, 447–490, 524–559, 576–628, 645–705, 722–1106, 1139–1173, 1207–1274, 1307–1324, 1357–1594, 1627, 1660–1795, 1828–2132, 2165–2199, 2233–2317, 2350–2501
Total	254	

3.4. TN Content Model Estimation Results

3.4.1. Results of All Bands Based on the Individual Spectral Reflectance Transformation

MLR, PLSR, and SVM estimation models were formulated using the TN content as the dependent variable, and all bands of the four spectral transformations were used as independent variables. Table 4 presents the estimation results (OR, IR, NLR, and FDR in the table indicate the original, inverse, natural logarithmic, and first-order derivative reflectances for all bands, respectively). MLR's training set accuracy was higher than that of PLSR and the SVM; however, its performance on the test set model was sub-par. The SVM model's prediction accuracy was better for both the training and test sets than that of PLSR, except in the case of first-order derivative reflectance. The test set performance of the SVM in terms of R^2 ranged from 0.45 to 0.59, the MAE ranged from 0.11 to 0.13 g/kg, and the RMSE ranged from 0.15 to 0.17 g/kg. SVM was the best model for estimating the TN content based on all bands after combining the data from the training and test set models.

Table 4. Estimation results of the TN content for all bands based on individual spectral reflectance transformations.

Model	Reflectance	Training Set			Test Set		
		R^2	MAE (g/kg)	RMSE (g/kg)	R^2	MAE (g/kg)	RMSE (g/kg)
MLR	OR	0.68	0.10	0.13	0.29	0.15	0.19
	IR	0.69	0.10	0.12	0.22	0.16	0.20
	NLR	0.69	0.10	0.12	0.28	0.15	0.19
	FDR	0.68	0.10	0.13	0.27	0.16	0.20
PLSR	OR	0.51	0.12	0.16	0.54	0.12	0.16
	IR	0.54	0.12	0.15	0.55	0.12	0.15
	NLR	0.53	0.12	0.15	0.54	0.12	0.16
	FDR	0.50	0.12	0.16	0.47	0.13	0.17
SVM	OR	0.60	0.10	0.14	0.59	0.11	0.15
	IR	0.61	0.10	0.14	0.58	0.11	0.15
	NLR	0.61	0.10	0.14	0.58	0.11	0.15
	FDR	0.58	0.10	0.14	0.45	0.13	0.17

The MLR model had the best results when the natural logarithm of the reflectance was used as the independent variable. The PLSR model with the natural logarithm of reflectance and inverse reflectance as inputs showed varying degrees of improvement in estimation accuracy compared with the original reflectance, with the natural logarithm of the reflectance having the best predictive power. OR–SVM was the best combination of models for the TN content estimation for all bands based on individual spectral reflectance conversions (Figure 6): R^2 was 0.59, the MAE was 0.11 g/kg, and the RMSE was 0.15 g/kg for the test set.

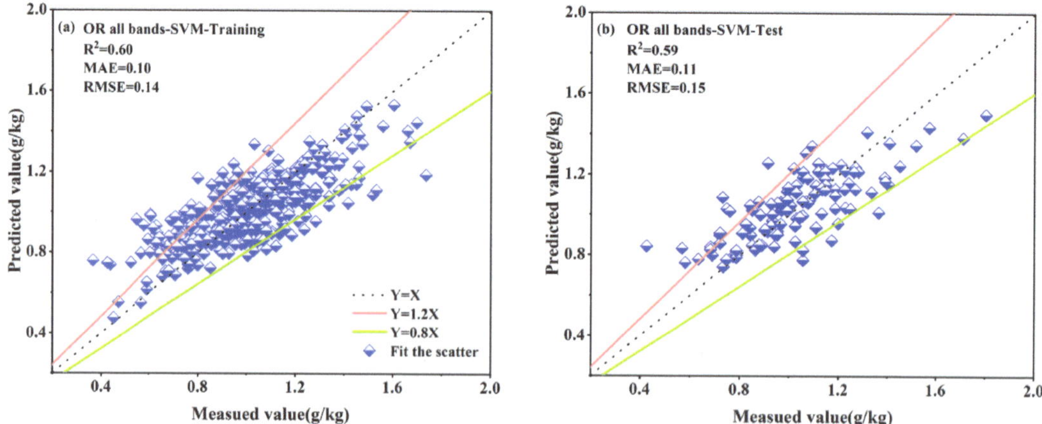

Figure 6. Scatter plots of measured and predicted TN contents based on individual spectral reflectance conversions of the best estimation model for all bands: (**a**) OR–SVM training set model; (**b**) OR–SVM test set model.

3.4.2. Results of the LASSO Feature Selection Based on the Individual Spectral Reflectance Transformations

Table 5 displays the results of the TN content estimate models' predictions, which utilized feature bands chosen with LASSO from the various spectral reflectance transformations. The outcomes of the training set models demonstrate that MLR and SVM had greater prediction accuracies than PLSR. The R^2 value of the MLR training set models ranged from 0.53 to 0.67, and higher R^2 values corresponded to lower MAE and RMSE values. The training set R^2 value for the SVM varied from 0.54 to 0.62, and the lowest values of MAE and RMSE were 0.10 and 0.14 g/kg, respectively. According to the outcomes of the test set models, the SVM model made the best predictions, followed by the PLSR and MLR models. The R^2 of the SVM for the test set ranged from 0.47 to 0.61, the MAE ranged from 0.11 to 0.13 g/kg, and the RMSE ranged from 0.14 to 0.17 g/kg. Combining the training and test set model results, the SVM was the optimal model for TN content estimation based on the LASSO feature selection.

Table 5. Estimation results of the TN content based on the LASSO feature selection.

Model	Reflectance	Training Set			Test Set		
		R^2	MAE (g/kg)	RMSE (g/kg)	R^2	MAE (g/kg)	RMSE (g/kg)
LASSO–MLR	OR	0.62	0.11	0.14	0.49	0.13	0.16
	IR	0.53	0.12	0.15	0.56	0.12	0.15
	NLR	0.54	0.12	0.15	0.55	0.12	0.15
	FDR	0.67	0.10	0.13	0.32	0.15	0.19
LASSO–PLSR	OR	0.55	0.12	0.15	0.54	0.12	0.16
	IR	0.52	0.12	0.15	0.57	0.12	0.15
	NLR	0.53	0.12	0.15	0.55	0.12	0.15
	FDR	0.51	0.12	0.16	0.48	0.13	0.17
LASSO–SVM	OR	0.58	0.11	0.14	0.61	0.11	0.14
	IR	0.54	0.12	0.15	0.56	0.12	0.15
	NLR	0.62	0.10	0.14	0.58	0.11	0.15
	FDR	0.57	0.11	0.15	0.47	0.13	0.17

A comparison of the different spectral reflectance transformations revealed that the inverse reflectance and natural logarithm of reflectance were the best inputs for the MLR and PLSR, respectively. With the SVM models, the natural logarithm of reflectance performed best in the training set, whereas OR performed best in the test set. OR–LASSO–SVM was, overall, the optimal model combination for LASSO feature selection to estimate TN content (Figure 7), with test set model metrics of $R^2 = 0.61$, MAE = 0.11 g/kg, and RMSE = 0.14 g/kg.

3.4.3. LASSO-Selected Spectral Band Combinations for the Four Spectral Reflectance Transformations

The four groups of LASSO characteristic bands of spectral reflectance transformation were combined as the input variables of the TN content prediction model. Table 6 displays the outcomes of the model estimations of TN content based on attributes collected from band combinations chosen by LASSO. The LBC–MLR training set model achieved better TN content estimation, with an R^2 of 0.80; nevertheless, when compared with the other models, the test set model did not perform as well (LBC: LASSO-selected spectral band combination). In terms of the test set accuracy, the LBC–PLSR and LBC–SVM models both outperformed the LBC–MLR, with the LBC–SVM model outperforming the LBC–PLSR model in terms of prediction outcomes. The LBC–SVM model, with test set model metrics of $R^2 = 0.57$, MAE = 0.12 g/kg, and RMSE = 0.15 g/kg, was the best model for TN content estimation among the LASSO-selected feature band combinations. The scatter plots of the LBC–SVM model's TN content measurements fitted to the predicted values are shown in Figure 8.

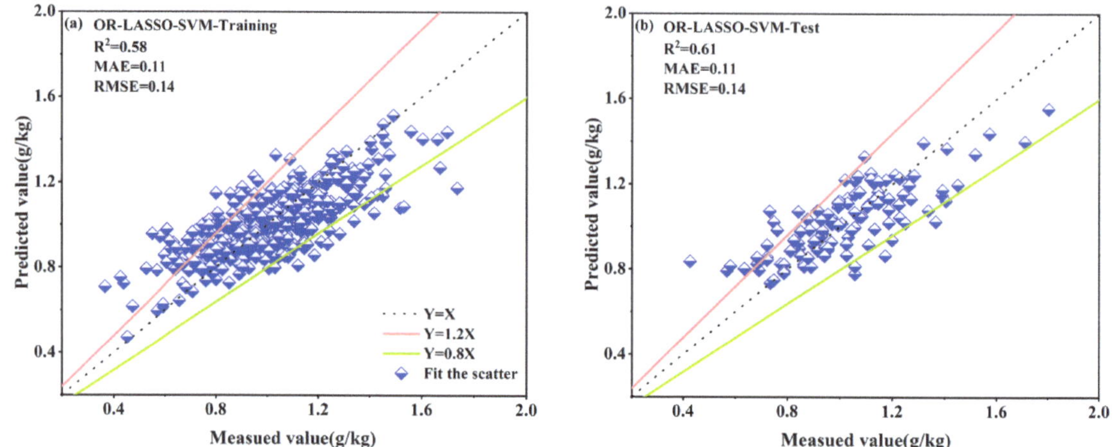

Figure 7. Scatter plots of the measured and predicted values of the best estimation model for TN content based on the LASSO feature selection: (**a**) OR–LASSO–SVM training set model; (**b**) OR–LASSO–SVM test set model.

Table 6. Estimation results of the TN content model based on the LBC.

Model	Training Set			Test Set		
	R^2	MAE (g/kg)	RMSE (g/kg)	R^2	MAE (g/kg)	RMSE (g/kg)
LBC–MLR	0.80	0.08	0.10	0.23	0.19	0.24
LBC–PLSR	0.52	0.12	0.15	0.54	0.12	0.16
LBC–SVM	0.65	0.10	0.13	0.57	0.12	0.15

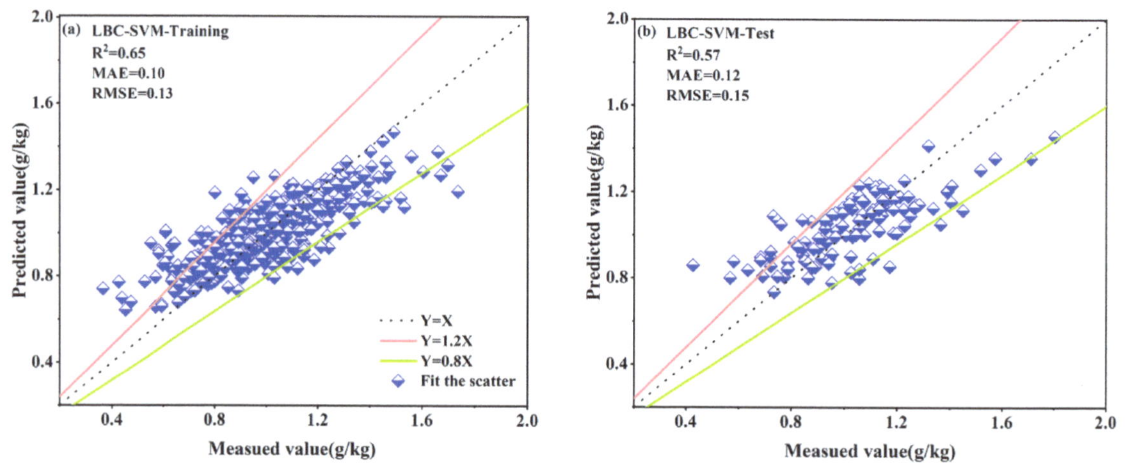

Figure 8. Scatter plots of the measured and predicted values of the best estimation model for TN content based on the LBC: (**a**) LBC–SVM training set model; (**b**) LBC–SVM test set model.

3.4.4. DE Secondary Feature Selection from LASSO-Selected Spectral Band Combinations Based on Four Spectral Reflectance Transformations

In this research, secondary feature selection from the LBC was carried out using the DE algorithm in combination with a prediction model. For the DE algorithm, 100 individuals were used and 1500 iterations were performed. In Table 7, the outcomes of the LBC–DE quadratic feature selection are displayed. LBC–DE–MLR was used to pick 98 variables,

LBC–DE–PLSR to select 61 variables, and LBC–DE–SVM to select 50 variables. N in Table 7 is the number of variables extracted.

Table 7. DE feature selection results for the LBC.

Reflectance Representation	MLR Wavelengths (nm)	N	PLSR Wavelengths (nm)	N	SVM Wavelengths (nm)	N
OR	404, 542, 576, 619, 765–774, 791, 954, 1023, 1056, 1139, 1190, 1308, 1442, 1526, 1644, 1745, 1812–1845, 1880, 1930-1947, 1981, 2031–2048, 2082-2098, 2183–2199, 2267, 2301, 2451	33	705, 757, 1880, 2132	4	447, 954, 1139, 1308, 1341, 1375, 1425, 1812, 2317, 2451	11
IR	396, 697, 1526, 1795, 1880, 1998, 2501	7	1526, 1880, 1930–1947, 2451, 2484–2501	7	404, 422, 1795, 1862, 2484, 2501	6
NLR	697, 757, 1425, 1526, 1880, 1930, 2199, 2484–2501	9	422, 1526, 1593, 1880, 1947	5	404, 4222, 1644, 1880, 1947	5
FDR	413, 490, 551–559, 594–628, 654–662, 679–688, 722–731, 757, 834, 851, 868–877, 894, 1073–1089, 1156–1173, 1257, 1375, 1442–1476, 1526, 1678-1728, 1762, 1795, 1947–1981, 2048, 2081, 2199, 2267, 2367–2417, 2451	49	413, 447–456, 551–559, 576-602, 619, 757, 808–842, 877, 894, 928, 946, 980, 1073, 1139, 1173, 1277, 1241, 1375, 1442, 1510–1526, 1678, 1728, 1829–1846, 1880, 1930–1947, 1981, 2048–2098, 2132, 2183, 2367	45	482, 525, 594–602, 619, 622, 679–688, 705, 748, 834, 877–885, 1073, 1089, 1224, 1526, 1745, 1778–1795, 1880, 1930, 2098, 2183, 2233, 2301, 2501	28
Total		98		61		50

The LBC–DE–SVM model had an excellent ability to predict the TN content; its training set metrics were $R^2 = 0.85$, MAE = 0.08 g/kg, and RMSE = 0.09 g/kg, while the corresponding values for the test set were $R^2 = 0.72$, MAE = 0.08 g/kg, and RMSE = 0.12 g/kg (Table 8). The LBC–DE–MLR and LBC–DE–PLSR models were comparable in predictive ability but inferior to the LBC–DE–SVM model. The LBC–DE–SVM was the best model for TN content estimation based on the LBC–DE quadratic feature selection (Figure 9). Moreover, compared with Figure 6, Figure 7 Figure 8, the fitted scatter plots of both the training set and test set models in Figure 9 were closer to the 1:1 line. The results show that the LBC-DE-SVM model had a better TN content estimation ability than the other models.

Table 8. TN content model estimation results based on LBC–DE.

Model	Training Set			Test Set		
	R^2	MAE (g/kg)	RMSE (g/kg)	R^2	MAE (g/kg)	RMSE (g/kg)
LBC–DE–MLR	0.64	0.10	0.13	0.65	0.10	0.14
LBC–DE–PLSR	0.57	0.11	0.15	0.60	0.11	0.14
LBC–DE–SVM	0.85	0.08	0.09	0.72	0.08	0.12

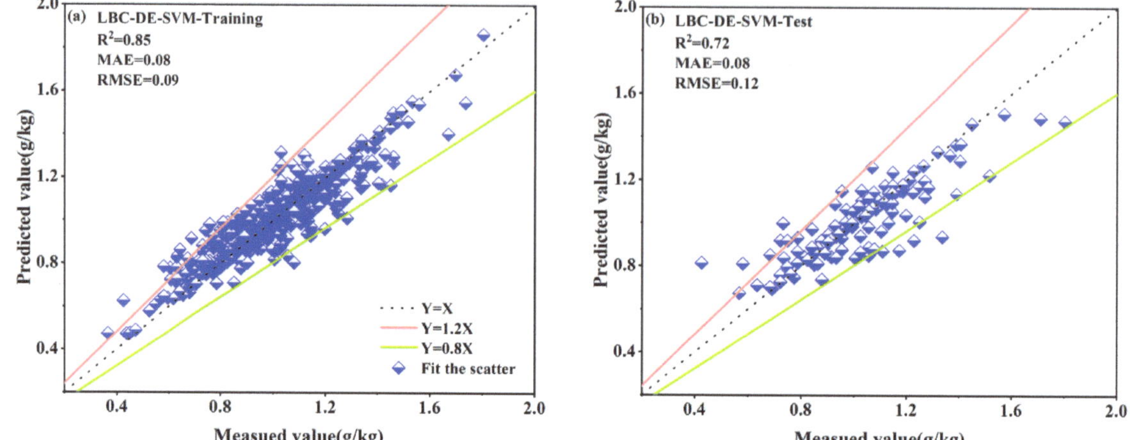

Figure 9. Scatter plot of the measured and predicted values of the best estimation model for TN content based on LBC–DE: (**a**) LBC–DE–SVM training set model; (**b**) LBC–DE–SVM test set model.

4. Discussion

4.1. Role of Spectral Reflectance Transformation Processing

In modeling the soil elemental content using hyperspectral data, an essential step is the mathematical transformation of the data. Spectral reflectance transformation can help to increase the precision of the prediction by lowering the interference from noise in the model [50,51]. In this study, the inverse reflectance and the natural logarithm of the reflectance enhanced the reflection peaks between 1750 and 2500 nm, and the first-order derivative reflectance transformation highlighted the absorption peaks at 1100 nm, 1850 nm, and 1950 nm and the reflection peaks at 1150 nm, 1380 nm, and 2000 nm. However, research demonstrated that not all spectral transformation techniques can produce outcomes superior to those obtained using the original data [16]. Regarding this, it should be mentioned that the best spectrum transformation varied depending on the model and even had an impact on how well a model performed on both the training and test sets. Using the prediction outcomes of the SVM model for LASSO feature selection based on the individual spectral reflectance change as an illustration, the training set model had the best accuracy when the natural logarithm of the reflectance spectral data was used as the input, but the test set model with the best estimation capability was that utilizing original reflectance spectral data. As a result, in this research, DE secondary feature extraction and modeling of the LASSO feature selection band combination (LBC) were carried out for the four spectral reflectance transforms in order to completely utilize the spectral reflectance transformation and to improve the prediction accuracy. Compared with the individual spectral reflectance, the accuracy of all models based on the LBC–DE quadratic feature selection was improved to different degrees, according to the model results (Table 8). This shows that the LASSO feature band combination data contained more essential information than the individual LASSO feature bands, and extracting this information from the combined spectra for model estimation could improve the models' predictive power.

4.2. The Role of Spectral Feature Extraction

Research demonstrated that using feature bands for modeling soil elemental content can lower the dimensionality of the data, make the prediction process simpler, and enhance the prediction outcomes [52,53]. Moreover, in the quantitative inversion of lake water depth and lake mineral content based on multispectral satellite remote sensing data, some scholars used the reflectance transformation combination data and the data after feature extraction as input variables for model estimation and achieved good estimation results [54–56].

Therefore, in this study, the LASSO algorithm was first used to screen the TN-predicting feature variables from the individual spectral reflectance representations to formulate a prediction model. Then, the DE algorithm was used to perform secondary feature selection and TN content estimation for the LASSO feature spectral band combination (LBC). The results from this study's test set model demonstrate that (see Figure 10), except for individual models, the LASSO-feature-selection-based estimation models had higher accuracies than the all-bands-based models (best R^2 value using LASSO feature selection: 0.61; best R^2 value of models using all bands: 0.59). This was mainly because the LASSO feature selection method could extract the characteristic spectral bands of TN according to the subtle relationship between TN and spectral reflectance and then achieve the purpose of eliminating redundant spectral variables and estimating the accuracy using the model. The model prediction results based on LBC–DE were noticeably superior to those based on the LBC, as shown in Figure 9 (best R^2 value for LBC–DE: 0.72; best R^2 value for LBC: 0.57). This implies that the LBC contained a significant amount of redundant information. The DE technique eliminated non-essential variables and retained valuable information, which enhanced the model's estimation capacity.

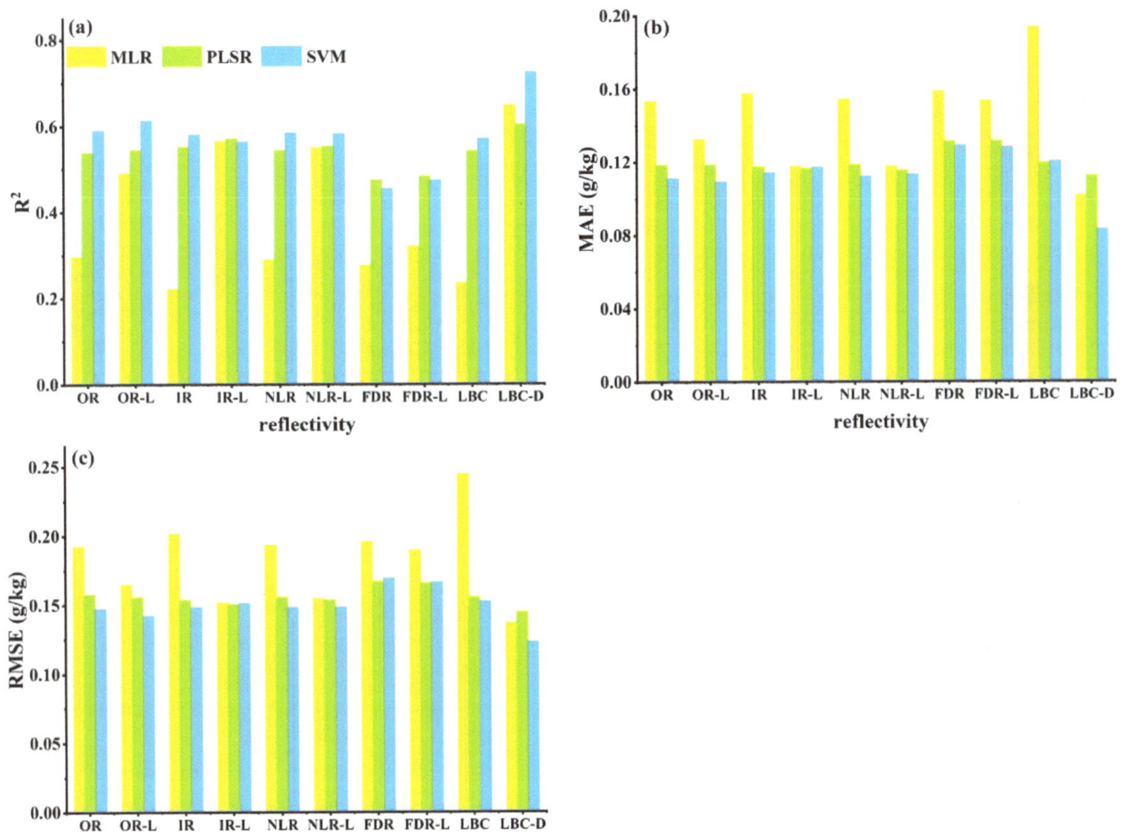

Figure 10. Comparison of the accuracy of the test set models: (**a**) R^2, (**b**) MAE, and (**c**) RMSE.

4.3. Estimated Model Comparison and Best Model TN Content Mapping

According to the comparison of the estimation accuracy and uncertainty evaluation indicators (Table 9) of the MLR, PLSR, and SVM models, the estimation accuracy of the SVM model was better than that of the MLR and PLSR models, and for the uncertainty calculation of the SVM model training set and test set, the index d-factor value was comparable with the MLR and PLSR models. Overall, the SVM model had a better estimation ability for TN content. An SVM is a machine learning model used for solving non-linear problems [13,44] and can accurately estimate TN content and capture the link between TN content and complex spectral data. According to this study's findings, the accuracy of the SVM model was increased when the DE technique was applied to extract the feature variables from the LBC. The LBC–DE–SVM model was the best TN content estimation model in this study. As a result, the LBC–DE–SVM model was utilized to calculate the TN content of the research region's farming regions (Figure 11). There was a total of 692,389 image elements in the arable area of the study area, i.e., a total of 62,311.894 hectares of arable land, and the estimated map raster occupied 45.813 MB of memory per hectare. The TN concentration ranged from 0.57 to 1.52 g/kg in the study region, with a trend of being higher in the west and lower in the east.

Table 9. Calculation value of the estimated model d-factor.

Model	Training Set	Test Set	Model	Training Set	Test Set
OR–MLR	0.014	0.032	FDR–LASSO–MLR	0.014	0.033
IR–MLR	0.014	0.034	OR–LASSO–PLSR	0.013	0.029
NLR–MLR	0.014	0.033	IR–LASSO–PLSR	0.012	0.028
FDR–MLR	0.014	0.033	NLR–LASSO–PLSR	0.012	0.028
OR–PLSR	0.012	0.027	FDR–LASSO–PLSR	0.012	0.025
IR–PLSR	0.012	0.028	OR–LASSO–SVM	0.012	0.027
NLR–PLSR	0.012	0.028	IR–LASSO–SVM	0.012	0.027
FDR–PLSR	0.012	0.025	NLR–LASSO–SVM	0.013	0.028
OR–SVM	0.012	0.028	FDR–LASSO–SVM	0.010	0.021
IR–SVM	0.012	0.028	LBC–MLR	0.015	0.041
NLR–SVM	0.012	0.028	LBC–PLSR	0.012	0.028
FDR–SVM	0.010	0.021	LBC–SVM	0.012	0.025
OR–LASSO–MLR	0.013	0.031	LBC–DE–MLR	0.014	0.031
IR–LASSO–MLR	0.012	0.027	LBC–DE–PLSR	0.013	0.028
NLR–LASSO–MLR	0.013	0.028	LBC–DE–SVM	0.015	0.030

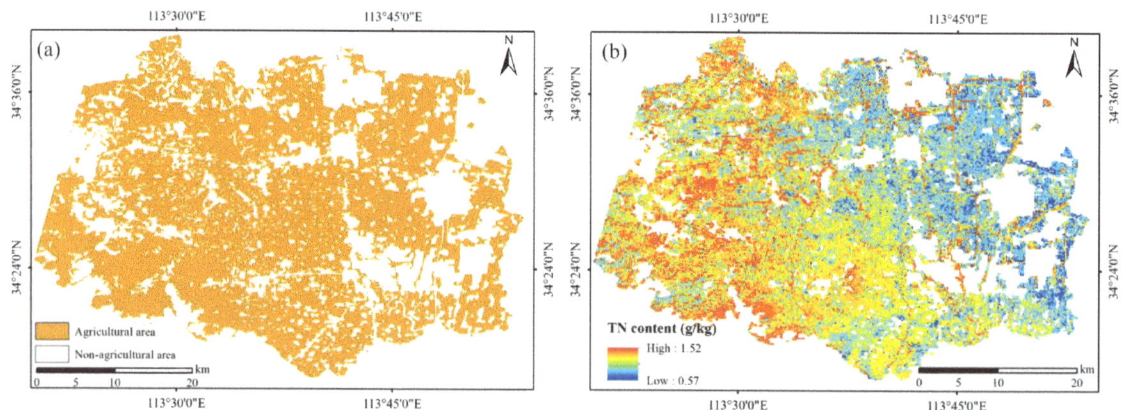

Figure 11. Map of the best model (LBC–DE–SVM) for estimating the TN content in the agricultural areas of the study area: (**a**) agricultural distribution in the study area and (**b**) spatial distribution of the TN content.

5. Conclusions

In this work, the ZY1-02D/AHSI hyperspectral satellite's remote sensing data and the measured soil sample data were used to estimate the TN concentration in the study region. In order to achieve an efficient and precise estimation of the TN content, the inverse reflectance, natural logarithm of the reflectance, and first-order derivative reflectance were selected for the transformation of the original reflectance remote sensing data. The LASSO method was used to select the feature bands for the four sets of spectral reflectance transformations, and the combination of the DE algorithm and the prediction model was proposed to conduct secondary feature extraction and model estimation of the TN content for the LASSO feature band combination (LBC). The key conclusions that can be made include the following:

(1) The transformation of the spectral reflectance data can highlight some of the enhanced spectral information. However, the best spectral data pre-processing methods for different estimation models differ, where even the optimal spectral transformation methods for the training and test sets of the same model are different. Suitable spectral reflectance transformation methods can be selected for different prediction models in the TN content estimation studies in other regions to improve the estimation accuracy.

(2) Using the LASSO method for feature variable selection for full-band data not only reduced the spectral data redundancy and simplified the model but also improved the estimation accuracy of the model. Compared with individual spectral reflectance data, the LBC contained more valid spectral information and concentrated a large amount of noise information. This study used a combination of the DE algorithm and the prediction model to extract feature variables from the LBC, which can achieve the purpose of retaining valid information in the LBC and eliminating invalid information and can provide a reference for future research in making full use of the spectral reflectance transform and feature data for TN content estimation.

(3) Compared with ground-based hyperspectral data and airborne hyperspectral data, ZY1-02D/AHSI hyperspectral satellite image data have the advantages of wide image coverage, the automatic acquisition of hyperspectral remote sensing image data, and a short return cycle, and thus, it can enable the dynamic, rapid, and large area estimation of TN content.

Although the present study achieved satisfactory TN estimation in the central agricultural region of Henan Province, there were still some shortcomings, and further studies are needed to improve the existing results. On the one hand, there is a need for long-time-series monitoring of TN content based on the ZY1-02D hyperspectral data; in this study, only a single period of hyperspectral image data was used to establish a relationship with the soil total nitrogen content for the estimation study, and a study of the change in TN content over time and its correspondence with the change in spectral reflectance values of the images in different periods was missing. Therefore, to achieve long-time-series monitoring of TN content, there is still a need to further investigate the variation pattern of TN content in a long time series, as well as the correspondence between TN content and image spectral data in a long time series. On the other hand, this study was an estimation study based only on the relationship between spectral data and TN content, while TN levels in agricultural areas may be affected by the soil's physical and chemical properties, the temperature, the amount of moisture, the fertilizer application, and other factors. Therefore, various factors related to the TN content should be added to the estimation model in subsequent studies to improve the reliability of the model estimation.

Author Contributions: Writing—review and editing, R.Z. and J.C.; validation, W.Z.; methodology, D.Z.; data curation, W.D.; conceptualization, H.G.; project administration, S.Z. All authors have read and agreed to the published version of the manuscript.

Funding: This research was funded by Major Science and Technology Special Projects in Henan Province (221100210600), Major Science and Technology Special Projects in Henan Province (201400211000), Major Science and Technology Special Projects in Henan Province (201400210100), and the Science and Technology Tackling Plan of Henan Province (222102320220).

Data Availability Statement: Not applicable.

Conflicts of Interest: The authors declare no conflict of interest.

References

1. Li, H.; Wang, J.; Zhang, J.; Liu, T.; Acquah, G.E.; Yuan, H. Combining Variable Selection and Multiple Linear Regression for Soil Organic Matter and Total Nitrogen Estimation by DRIFT-MIR Spectroscopy. *Agronomy* **2022**, *12*, 638. [CrossRef]
2. Otto, R.; Castro, S.A.Q.; Mariano, E.; Castro, S.G.Q.; Franco, H.C.J.; Trivelin, P.C.O. Nitrogen Use Efficiency for Sugarcane-Biofuel Production: What Is Next? *BioEnergy Res.* **2016**, *9*, 1272–1289. [CrossRef]
3. Kaushal, S.S.; Lewis, W.M., Jr.; McCutchan, J.H., Jr. Land Use Change and Nitrogen Enrichment of A Rocky Mountain Watershed. *Ecol. Appl.* **2006**, *16*, 299–312. [CrossRef]
4. Peng, Y.; Wang, L.; Zhao, L.; Liu, Z.; Lin, C.; Hu, Y.; Liu, L. Estimation of Soil Nutrient Content Using Hyperspectral Data. *Agriculture* **2021**, *11*, 1129. [CrossRef]
5. Bao, Y.; Meng, X.; Ustin, S.; Wang, X.; Zhang, X.; Liu, H.; Tang, H. Vis-SWIR spectral prediction model for soil organic matter with different grouping strategies. *CATENA* **2020**, *195*, 104703. [CrossRef]
6. Kuang, B.; Mouazen, A.M. Non-biased prediction of soil organic carbon and total nitrogen with vis–NIR spectroscopy, as affected by soil moisture content and texture. *Biosyst. Eng.* **2013**, *114*, 249–258. [CrossRef]
7. Sun, H.; Zhao, Z.; Zhao, J.; Chen, W. Inversion of Topsoil Organic Matter Content by Hyperspectral Remote Sensing of Zhuhai-1. *Remote Sens. Inf.* **2020**, *35*, 40–46. [CrossRef]
8. Wang, Y.; Zhang, X.; Sun, W.; Wang, J.; Ding, S.; Liu, S. Effects of hyperspectral data with different spectral resolutions on the estimation of soil heavy metal content: From ground-based and airborne data to satellite-simulated data. *Sci. Total Environ.* **2022**, *838*, 156129. [CrossRef]
9. Yin, F.; Wu, M.; Liu, L.; Zhu, Y.; Feng, J.; Yin, D.; Yin, C.; Yin, C. Predicting the abundance of copper in soil using reflectance spectroscopy and GF5 hyperspectral imagery. *Int. J. Appl. Earth Obs. Geoinf.* **2021**, *102*, 102420. [CrossRef]
10. Liu, Y.; Sun, D.; Han, B.; Zhu, H.; Liu, S.; Yuan, J. Development of Advanced Visible and Short-wave Infrared Hyperspectral Imager Onboard ZY-1-02D Satellite. *Spacecr. Eng.* **2020**, *29*, 85–92. [CrossRef]
11. Shang, K.; Gu, H.; Yang, Y. Inversion of Total Copper Content in Mining Soils with Different Spectral Pretreatment Techniques Using AHSI/ZY1-02D Data. In Proceedings of the 2021 IEEE International Geoscience and Remote Sensing Symposium IGARSS 2021, Brussels, Belgium, 11–16 July 2021; pp. 6461–6464. [CrossRef]
12. Yang, Y.; Shang, K.; Xiao, C.; Wang, C.; Tang, H. Spectral Index for Mapping Topsoil Organic Matter Content Based on ZY1-02D Satellite Hyperspectral Data in Jiangsu Province, China. *ISPRS Int. J. Geo-Inf.* **2022**, *11*, 111. [CrossRef]
13. Meng, X.; Bao, Y.; Liu, J.; Liu, H.; Zhang, X.; Zhang, Y.; Wang, P.; Tang, H.; Kong, F. Regional soil organic carbon prediction model based on a discrete wavelet analysis of hyperspectral satellite data. *Int. J. Appl. Earth Obs. Geoinf.* **2020**, *89*, 102111. [CrossRef]
14. Zhang, Y.; Li, M.; Zheng, L.; Qin, Q.; Lee, W.S. Spectral features extraction for estimation of soil total nitrogen content based on modified ant colony optimization algorithm. *Geoderma* **2019**, *333*, 23–34. [CrossRef]
15. Rossel, R.A.V.; Cattle, S.R.; Ortega, A.; Fouad, Y. In situ measurements of soil colour, mineral composition and clay content by vis–NIR spectroscopy. *Geoderma* **2009**, *150*, 253–266. [CrossRef]
16. Xie, S.; Ding, F.; Chen, S.; Wang, X.; Li, Y.; Ma, K. Prediction of soil organic matter content based on characteristic band selection method. *Spectrochim. Acta Part A Mol. Biomol. Spectrosc.* **2022**, *273*, 120949. [CrossRef]
17. Shen, L.; Gao, M.; Yan, J.; Li, Z.; Leng, P.; Yang, Q.; Duan, S. Hyperspectral Estimation of Soil Organic Matter Content using Different Spectral Preprocessing Techniques and PLSR Method. *Remote Sens.* **2020**, *12*, 1206. [CrossRef]
18. Liu, J.; Dong, Z.; Xia, J.; Wang, H.; Meng, T.; Zhang, R.; Han, J.; Wang, N.; Xie, J. Estimation of soil organic matter content based on CARS algorithm coupled with random forest. *Spectrochim. Acta Part A Mol. Biomol. Spectrosc.* **2021**, *258*, 119823. [CrossRef] [PubMed]
19. Yu, Q.; Yao, T.; Lu, H.; Feng, W.; Xue, Y.; Liu, B. Improving estimation of soil organic matter content by combining Landsat 8 OLI images and environmental data: A case study in the river valley of the southern Qinghai-Tibet Plateau. *Comput. Electron. Agric.* **2021**, *185*, 106144. [CrossRef]
20. Vašát, R.; Kodešová, R.; Klement, A.; Borůvka, L. Simple but efficient signal pre-processing in soil organic carbon spectroscopic estimation. *Geoderma* **2017**, *298*, 46–53. [CrossRef]
21. Zhang, Z.; Ding, J.; Zhu, C.; Wang, J. Combination of efficient signal pre-processing and optimal band combination algorithm to predict soil organic matter through visible and near-infrared spectra. *Spectrochim. Acta Part A Mol. Biomol. Spectrosc.* **2020**, *240*, 118553. [CrossRef]
22. Gao, L.; Zhu, X.; Han, Z.; Wang, L.; Zhao, G.; Jiang, Y. Spectroscopy-Based Soil Organic Matter Estimation in Brown Forest Soil Areas of the Shandong Peninsula, China. *Pedosphere* **2019**, *29*, 810–818. [CrossRef]
23. Leardi, R. Application of genetic algorithm-PLS for feature selection in spectral data sets. *J. Chemom.* **2000**, *14*, 643–655. [CrossRef]
24. Zhang, Y.; Chen, W.; Tang, Z.; Gu, J.; Mo, L.; Chen, H. Application of Interval Partial Least Squares with Differential Evolution Algorithm in Wavelength Selection of Near Infrared Spectroscopy for Fishmeal. *FENXI CESHI XUEBAO J. Instrum. Anal.* **2020**, *39*, 1392–1397. [CrossRef]
25. Cao, F.; Yang, Z.; Ren, J.; Jiang, M.; Ling, W. Linear vs. Nonlinear Extreme Learning Machine for Spectral-Spatial Classification of Hyperspectral Images. *Sensors* **2017**, *17*, 2603. [CrossRef] [PubMed]
26. Kooistra, L.; Wehrens, R.; Leuven, R.S.E.W.; Buydens, L.M.C. Possibilities of visible–near-infrared spectroscopy for the assessment of soil contamination in river floodplains. *Anal. Chim. Acta* **2001**, *446*, 97–105. [CrossRef]

27. Zhang, S.; Shen, Q.; Nie, C.; Huang, Y.; Wang, J.; Hu, Q.; Ding, X.; Zhou, Y.; Chen, Y. Hyperspectral inversion of heavy metal content in reclaimed soil from a mining wasteland based on different spectral transformation and modeling methods. *Spectrochim. Acta Part A Mol. Biomol. Spectrosc.* **2019**, *211*, 393–400. [CrossRef]

28. Zhao, H.; Liu, P.; Qiao, B.; Wu, K. The Spatial Distribution and Prediction of Soil Heavy Metals Based on Measured Samples and Multi-Spectral Images in Tai Lake of China. *Land* **2021**, *10*, 1227. [CrossRef]

29. Kinoshita, R.; Roupsard, O.; Chevallier, T.; Albrecht, A.; Taugourdeau, S.; Ahmed, Z.; van Es, H.M. Large topsoil organic carbon variability is controlled by Andisol properties and effectively assessed by VNIR spectroscopy in a coffee agroforestry system of Costa Rica. *Geoderma* **2016**, *262*, 254–265. [CrossRef]

30. Mouazen, A.M.; Kuang, B.; De Baerdemaeker, J.; Ramon, H. Comparison among principal component, partial least squares and back propagation neural network analyses for accuracy of measurement of selected soil properties with visible and near infrared spectroscopy. *Geoderma* **2010**, *158*, 23–31. [CrossRef]

31. Xu, S.; Zhao, Y.; Wang, M.; Shi, X. Comparison of multivariate methods for estimating selected soil properties from intact soil cores of paddy fields by Vis–NIR spectroscopy. *Geoderma* **2018**, *310*, 29–43. [CrossRef]

32. Rossel, R.A.V.; Behrens, T. Using data mining to model and interpret soil diffuse reflectance spectra. *Geoderma* **2010**, *158*, 46–54. [CrossRef]

33. Xu, Z.; Chen, S.; Zhu, B.; Chen, L.; Ye, Y.; Lu, P. Evaluating the Capability of Satellite Hyperspectral Imager, the ZY1–02D, for Topsoil Nitrogen Content Estimation and Mapping of Farmlands in Black Soil Area, China. *Remote Sens.* **2022**, *14*, 1008. [CrossRef]

34. Gao, F.; Chen, S.; Du, X.; Zhou, X.; Wang, X. FOSS Kjeltec 8400 Automatic Kjeldahl Nitrogen Determinator for Determination of Total Nitrogen in Soil. *Tianjin Agric. Sci.* **2022**, *28*, 76–80. [CrossRef]

35. Zhang, Q.; Zhang, H.; Liu, W.; Zhao, S. Inversion of heavy metals content with hyperspectral reflectance in soil of well-facilitied capital farmland construction areas. *Trans. Chin. Soc. Agric. Eng.* **2017**, *33*, 230–239. [CrossRef]

36. Petropoulos, G.P.; Arvanitis, K.; Sigrimis, N. Hyperion hyperspectral imagery analysis combined with machine learning classifiers for land use/cover mapping. *Expert Syst. Appl.* **2012**, *39*, 3800–3809. [CrossRef]

37. Sorol, N.; Arancibia, E.; Bortolato, S.A.; Olivieri, A.C. Visible/near infrared-partial least-squares analysis of Brix in sugar cane juice. *Chemom. Intell. Lab. Syst.* **2010**, *102*, 100–109. [CrossRef]

38. Tibshirani, R. Regression Shrinkage and Selection via the Lasso. *J. R. Stat. Soc. Ser. B* **1996**, *58*, 267–288. [CrossRef]

39. Storn, R. On the usage of differential evolution for function optimization. In Proceedings of the North American Fuzzy Information Processing, Berkeley, CA, USA, 19–22 June 1996. [CrossRef]

40. Chattopadhyay, S.; Mishra, S.; Goswami, S. *Feature Selection Using Differential Evolution with Binary Mutation Scheme*; IEEE: Durgapur, India, 2016. [CrossRef]

41. Minhoni, R.T.D.A.; Scudiero, E.; Zaccaria, D.; Saad, J.C.C. Multitemporal satellite imagery analysis for soil organic carbon assessment in an agricultural farm in southeastern Brazil. *Sci. Total Environ.* **2021**, *784*, 147216. [CrossRef] [PubMed]

42. Wang, F.; Shi, Z.; Biswas, A.; Yang, S.; Ding, J. Multi-algorithm comparison for predicting soil salinity. *Geoderma* **2020**, *365*, 114211. [CrossRef]

43. Demattê, J.A.M.; Ramirez-Lopez, L.; Marques, K.P.P.; Rodella, A.A. Chemometric soil analysis on the determination of specific bands for the detection of magnesium and potassium by spectroscopy. *Geoderma* **2017**, *288*, 8–22. [CrossRef]

44. Jakab, G.; Rieder, Á.; Vancsik, A.V.; Szalai, Z. Soil organic matter characterisation by photometric indices or photon correlation spectroscopy: Are they comparable? *Hung. Geogr. Bull.* **2018**, *67*, 109–120. [CrossRef]

45. Smola, A.; Scholkopf, B. A tutorial on support vector regression. *Stat. Comput.* **2004**, *14*, 199–222. [CrossRef]

46. Peng, X.; Shi, T.; Song, A.; Chen, Y.; Gao, W. Estimating Soil Organic Carbon Using VIS/NIR Spectroscopy with SVMR and SPA Methods. *Remote Sens.* **2014**, *6*, 2699–2717. [CrossRef]

47. Wang, X.; Zhang, F.; Kung, H.; Johnson, V.C. New methods for improving the remote sensing estimation of soil organic matter content (SOMC) in the Ebinur Lake Wetland National Nature Reserve (ELWNNR) in northwest China. *Remote Sens. Environ.* **2018**, *218*, 104–118. [CrossRef]

48. Guo, H.; Zhang, R.; Dai, W.; Zhou, X.; Zhang, D.; Yang, Y.; Cui, J. Mapping Soil Organic Matter Content Based on Feature Band Selection with ZY1-02D Hyperspectral Satellite Data in the Agricultural Region. *Agronomy* **2022**, *12*, 2111. [CrossRef]

49. Zornoza, R.; Guerrero, C.; Mataix-Solera, J.; Scow, K.M.; Arcenegui, V.; Mataix-Beneyto, J. Near infrared spectroscopy for determination of various physical, chemical and biochemical properties in Mediterranean soils. *Soil Biol. Biochem.* **2008**, *40*, 1923–1930. [CrossRef]

50. Buddenbaum, H.; Steffens, M. The Effects of Spectral Pretreatments on Chemometric Analyses of Soil Profiles Using Laboratory Imaging Spectroscopy. *Appl. Environ. Soil Sci.* **2012**, *2012*, 274903. [CrossRef]

51. Li, B.; Xie, W. Adaptive fractional differential approach and its application to medical image enhancement. *Comput. Electr. Eng.* **2015**, *45*, 324–335. [CrossRef]

52. Mehmood, T.; Liland, K.H.; Snipen, L.; Sæbø, S. A review of variable selection methods in Partial Least Squares Regression. *Chemom. Intell. Lab. Syst.* **2012**, *118*, 62–69. [CrossRef]

53. Vohland, M.; Ludwig, M.; Thiele-Bruhn, S.; Ludwig, B. Quantification of Soil Properties with Hyperspectral Data: Selecting Spectral Variables with Different Methods to Improve Accuracies and Analyze Prediction Mechanisms. *Remote Sens.* **2017**, *9*, 1103. [CrossRef]

54. Yang, H.; Ju, J.; Guo, H.; Qiao, B.; Nie, B.; Zhu, L. Bathymetric Inversion and Mapping of Two Shallow Lakes Using Sentinel-2 Imagery and Bathymetry Data in the Central Tibetan Plateau. *IEEE J. Sel. Top. Appl. Earth Obs. Remote Sens.* **2022**, *15*, 4279–4296 [CrossRef]

55. Yang, H.; Guo, H.; Dai, W.; Nie, B.; Qiao, B.; Zhu, L. Bathymetric mapping and estimation of water storage in a shallow lake using a remote sensing inversion method based on machine learning. *Int. J. Digit. Earth* **2022**, *15*, 789–812. [CrossRef]

56. Guo, H.; Dai, W.; Zhang, R.; Zhang, D.; Qiao, B.; Zhang, G.; Zhao, S.; Shang, J. Mineral content estimation for salt lakes on the Tibetan plateau based on the genetic algorithm-based feature selection method using Sentinel-2 imagery: A case study of the Bieruoze Co and Guopu Co lakes. *Front. Earth Sci.* **2023**, *11*, 1118118. [CrossRef]

Article

Semi-Supervised Semantic Segmentation-Based Remote Sensing Identification Method for Winter Wheat Planting Area Extraction

Mingmei Zhang [1], Yongan Xue [2], Yuanyuan Zhan [3] and Jinling Zhao [3,*]

[1] Department of Geological and Surveying Engineering, Shanxi Institute of Energy, Jinzhong 030600, China; zhangmm@sxie.edu.cn
[2] College of Mining Engineering, Taiyuan University of Technology, Taiyuan 030024, China; xueyongan@tyut.edu.cn
[3] National Engineering Research Center for Agro-Ecological Big Data Analysis & Application, Anhui University, Hefei 230601, China; p20201094@stu.ahu.edu.cn
* Correspondence: zhaojl@ahu.edu.cn

Abstract: To address the cost issue associated with pixel-level image annotation in fully supervised semantic segmentation, a method based on semi-supervised semantic segmentation is proposed for extracting winter wheat planting areas. This approach utilizes self-training with pseudo-labels to learn from a small set of images with pixel-level annotations and a large set of unlabeled images, thereby achieving the extraction. In the constructed initial dataset, a random sampling strategy is employed to select 1/16, 1/8, 1/4, and 1/2 proportions of labeled data. Furthermore, in conjunction with the concept of consistency regularization, strong data augmentation techniques are applied to the unlabeled images, surpassing classical methods such as cropping and rotation to construct a semi-supervised model. This effectively alleviates overfitting caused by noisy labels. By comparing the prediction results of different proportions of labeled data using SegNet, DeepLabv3+, and U-Net, it is determined that the U-Net network model yields the best extraction performance. Moreover, the evaluation metrics MPA and MIoU demonstrate varying degrees of improvement for semi-supervised semantic segmentation compared to fully supervised semantic segmentation. Notably, the U-Net model trained with 1/16 labeled data outperforms the models trained with 1/8, 1/4, and 1/2 labeled data, achieving MPA and MIoU scores of 81.63%, 73.31%, 82.50%, and 76.01%, respectively. This method provides valuable insights for extracting winter wheat planting areas in scenarios with limited labeled data.

Keywords: semi-supervised classification; sematic segmentation; winter wheat; self-training; data augmentation

Citation: Zhang, M.; Xue, Y.; Zhan, Y.; Zhao, J. Semi-Supervised Semantic Segmentation-Based Remote Sensing Identification Method for Winter Wheat Planting Area Extraction. *Agronomy* **2023**, *13*, 2868. https://doi.org/10.3390/agronomy13122868

Academic Editor: Mario Cunha

Received: 27 October 2023
Revised: 19 November 2023
Accepted: 20 November 2023
Published: 22 November 2023

1. Introduction

Semantic segmentation is a fundamental task in the field of computer vision, and it has made significant progress in many application areas [1–3]. Fully supervised semantic segmentation learns to assign pixel-level semantic labels by generalizing from a large number of densely annotated images. Despite rapid progress, the cost of manually annotating pixels is much higher compared to other visual tasks such as image classification and object detection. Reliable pixel-wise segmentation annotations are typically only available for a few classes and images, making fully supervised semantic segmentation challenging for tasks involving diverse objects. To further simplify the process of acquiring high-quality data, semi-supervised semantic segmentation has been proposed, which learns models from a small number of labeled images and a large number of unlabeled images to achieve the final segmentation goal. Many researchers have applied semi-supervised semantic segmentation to crop extraction. Casado-García et al. [4] used a pseudo-label semi-supervised learning method combined with a semantic segmentation network to segment natural color images of vineyards captured by cameras for plant yield monitoring. Zheng et al. [5] improved segmentation accuracy with a small number of labeled data by using a semi-supervised adversarial semantic segmentation network for building extraction on three different resolution datasets (WBD, MBD, and GID). Mukhtar et al. [6] proposed a semi-supervised method based on cross-consistency for semantic segmentation of RGB images captured by drones and further counted the extracted plant clusters, providing insights for studying other crops (such as rice and maize) using a small number of labeled images. In the field of semi-supervised semantic segmentation, two widely used forms of semi-supervised methods are entropy minimization and consistency regularization [7].

Self-training [8,9], also known as self-supervised learning, is a semi-supervised learning technique where a model iteratively generates pseudo-labels for unlabeled data and then uses these pseudo-labeled samples to retrain itself. It aims to improve the model's accuracy by iteratively refining its own predictions, often by considering high-confidence predictions as pseudo-labels. Zhu et al. [10] achieved optimal performance with less supervision using the self-training semi-supervised method on the Cityscape, CamVid, and KITTI datasets. Feng et al. [11] proposed a dynamic self-training and class-balanced curriculum algorithm for semi-supervised semantic segmentation. They achieved robustness to pseudo-label noise in self-training by weighting the pixel-wise loss based on predicted confidences and providing a pseudo-label curriculum to gradually label all unlabeled data, demonstrating good performance on PASCAL VOC 2012 and Cityscapes. Zoph et al. [12] compared self-training with ImageNet pre-training and revealed the universality and flexibility of self-training. These studies serve as typical examples of self-training methods. However, this approach also has certain limitations. If the teacher model trained based on labeled data contains errors, it can lead to erroneous results in the subsequent student model.

Recently, methods based on consistency regularization have been proposed, which rely on the fact that unlabeled and labeled data share the same distribution. Therefore, these methods aim to train models that have consistent and reliable predictions for both labeled and unlabeled data. One widely researched approach is to use different unlabeled images under different transformations as inputs to the model and enforce consistency loss on the predicted masks. In consistency regularization methods [13,14], consistency is enhanced by augmenting input images, feature representations, and networks to ensure consistency with various perturbations, such as input perturbations, which involve perturbing unlabeled data [15,16]. Based on the clustering assumption, data points with different class labels are separated from each other, while similar data points have similar outputs. Therefore, if an unlabeled data point is perturbed and consistency constraints are enforced on the predicted masks of the perturbed images, the predicted results should not change significantly. Common transformations include data augmentation techniques such as adding noise, cropping, and scaling [17], as well as methods like CutMix [18] and ClassMix [19]. These form the basis of consistency-based methods and many self-supervised learning methods, all of which focus on leveraging unlabeled data. For example, the Π-model parameterizes input samples with different noises and adds a regularization term to reduce the difference between the outputs of the perturbed samples and the original inputs. Temporal ensembling [20] and mean teacher [21] involve ensemble learning, using exponentially moving average (EMA) weights to improve the quality of labels for perturbed samples. CCT (cross-consistency training) [22] is a semi-supervised semantic segmentation method based on cross-consistency, which applies perturbations to the input of the encoder instead of directly perturbing the input, exploring the application of perturbations at different network layers in the segmentation network. Interpolation consistency training (ICT) achieves a modification in the perturbation method by using a mixture with another unlabeled sample instead of adding random noise, which is considered more effective in dealing with low-margin unlabeled points. Berthelot et al. [23] further propose sharpening artificial labels for unlabeled data and using MixUp to mix labeled and unlabeled data.

Generative adversarial networks (GANs) are generally utilized for tasks related to generative modeling, where they generate data that are similar to the training data. They indirectly might benefit semi-supervised learning but not as a direct supervision signal, are not easy to optimize, and may encounter mode collapse issues [24]. Pseudo-labeling [25] is a combination of entropy minimization and consistency regularization, which relies on the assumption that pseudo-labels generated by a teacher model can benefit the training of a new model. FixMatch [26] achieves consistency training by applying strong data augmentation (Cutout, CTAugment, and RandAugment) to the unlabeled loss and input for pseudo-label generation, encouraging the outputs of two augmented inputs to remain consistent. As an extension of FixMatch, PseudoSeg [27] uses weak and strong augmentations on an unlabeled image and uses the predictions of the weakly augmented image as pseudo-labels for the strongly augmented image. This method generates pseudo-labels using only the image itself.

In conclusion, semi-supervised learning offers a significant reduction in the annotation process, resulting in resource savings in terms of manpower and materials. Nevertheless, it can be found that the performance of these semi-supervised semantic methods can be limited by the quality and quantity of the labeled data available. Inaccurate or inconsistent annotations can hinder the learning process and result in suboptimal segmentation performance. In addition, some methods may require complex architectures or training procedures, which can make them computationally expensive and difficult to implement in real-world applications. It is necessary to reduce the reliance of segmentation methods on high-quality pseudo-labels while simultaneously minimizing the number of iterative rounds required to generate such labels. Currently, most related studies have focused on publicly available datasets, with limited research on remote sensing crop extraction. This paper aims to extract winter wheat planting areas using a semi-supervised semantic segmentation method that combines entropy minimization and consistency regularization. The proposed approach utilizes a deep learning semantic segmentation network to achieve the desired results. By eliminating the need to filter high-quality pseudo-labels and generating them only once for unannotated images, significant reductions in training time can be achieved.

2. Materials and Methods

2.1. Study Area

The selected study area (Figure 1) is Zhengding County and Zengcun Town, Gaocheng District, Shijiazhuang City, Hebei Province, China, located at approximately 37°51′~38°18′ N, 114°39′~114°59′ E. The study area has a temperate semi-humid to semi-arid continental monsoon climate, with distinct seasons in most regions. The annual average temperature is 12.9 °C, and the annual average precipitation is 550 mm. Cultivated land is the predominant land use type in this region, with winter wheat being one of the major food crops.

2.2. Data Preprocessing and Label Generation

Based on the phenological characteristics of winter wheat, during the grain filling stage, its growth is more advanced compared to other crops, which are either not yet sown or have just been sown. At this stage, there is a significant difference between winter wheat and other land cover types, enabling the high-precision segmentation and extraction of wheat planting areas. Therefore, in this study, remote sensing imagery data were acquired during the milk-ripe stage of winter wheat, around mid-to-late May. A Landsat-8 OLI (operational land imager) image was acquired, and data preprocessing was also performed. After finishing the geometric and radiometric corrections, the image fusion with 15 m panchromatic and 30 m multispectral bands was carried out using the high-pass filtering fusion.

Subsequently, the labeled data were generated using the preprocessed image. Firstly, the remote sensing imagery was opened using ArcGIS Desktop 10.8 software. The wheat area was delineated on the original image, and a vector was generated to create labels in PNG format. In the labels, wheat areas were represented by white color with a pixel value of 1, while non-wheat areas (background) were represented by black color with a pixel value of 0. Simultaneously, the original remote sensing imagery was saved as JPG format output. Both the original imagery data and label data were randomly cropped to 256 × 256 pixels. The file names of the original imagery data corresponded to the label data. Data augmentation operations were performed, including rotation, mirror operations, brightness adjustment, and adding noises (Gaussian noise, salt and pepper noise).

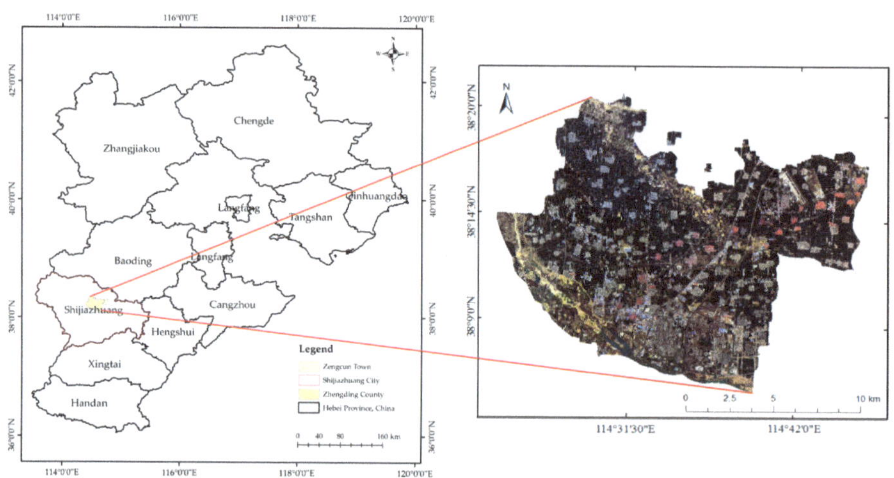

Figure 1. Geographical location of the study area.

2.3. Data Augmentation

Self-training, relying on an initial model trained with labeled data, does not attempt to address the negative impact of noisy pseudo-labels and is not well-suited for the sparse labeling mechanism in semi-supervised learning. Specifically, when iteratively overfitting to incorrect supervision, errors in pseudo-labels accumulate and significantly degrade the performance of the student model. Additionally, in this self-learning process, the introduced information during training is insufficient, resulting in a severe coupling problem between the teacher model and the retrained student model. As the predicted pseudo-labels in the second stage may still contain a considerable amount of noise, directly retraining on these images and labels containing noise can easily lead to overfitting to the noisy labels. Specifically, the student model is forced to learn the pseudo-labels from the teacher model in a supervised manner during retraining. However, due to the same network structure and similar initialization of the teacher and student models, they tend to make similar true and false predictions on unlabeled images, making it difficult for the student model to learn additional information beyond minimizing entropy during training.

To address the above issues, namely overfitting to noisy labels and the prediction coupling between the student and teacher models, data augmentation is applied to the unlabeled images during the retraining stage to propose more challenging optimization objectives for the student model. Data augmentation is an effective regularization technique, with basic strategies such as random flipping and cropping commonly used in training visual models. Classic augmentation perturbations, such as cropping, scaling, and rotation, were also used on the original images in the data set construction, and these basic perturbations as weak data augmentation are difficult to confuse output categories. Therefore, stronger and more diverse augmentations were used on the unlabeled images in semi-supervised semantic segmentation. Color transformation data augmentation was used, including Grayscale, Colorjitter, Blur, and RandomInvert, as well as spatial transformation using Cutout [28] (Figure 2).

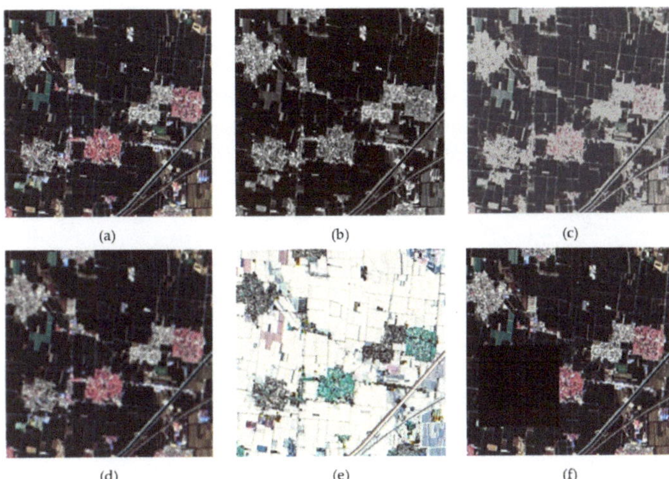

Figure 2. Comparison of different data augmentation methods: (**a**) Original image; (**b**) Grayscale; (**c**) Colorjitter; (**d**) Blur; (**e**) RandomInvert; and (**f**) Cutout.

2.4. Experimental Environment and Parameter Setting

The experimental setup consisted of an Intel Xeon Gold 6248R processor, 192 GB of memory, an NVIDIA Quadro P4000 graphics card, and the GPU acceleration library CUDA 10.0 with the PyTorch deep learning framework. For model training, the SGD optimizer was chosen as the parameter optimizer, with a base learning rate of 0.001, 80 training iterations, and a step size of 8. Weak data augmentation techniques, such as rotation, mirroring, and noise addition, were applied as discussed in the previous section. Strong data augmentation techniques on unlabeled images were implemented with the following settings: Grayscale random probability p was set to 0.5; Colorjitter had brightness, contrast, and saturation values of 0.5 and a hue value of 0.25; Blur had a sigma value range of (0.1, 2.0); RandomInvert had a random probability p of 0.5; and Cutout had a size value range of (0.02, 0.4).

2.5. Methodology

2.5.1. Entropy Minimization

Entropy minimization [29] is an effective semi-supervised approach for achieving clustering assumptions, as it encourages more confident predictions for unlabeled data by forcing the classifier to make low-entropy predictions. Under the assumption that the majority of data points are far from the decision boundary, this method prevents the decision boundary of the network from being close to the data points; otherwise, it would be forced to produce less reliable predictions. This approach can be achieved by adding a loss term for predictions, as follows:

$$L = -\sum_{k=1}^{c} f_\theta(x)_k \log f_\theta(x)_k \tag{1}$$

where k is the number of classes and $f_\theta(x)_k$ is the model's confidence in predicting whether x belongs to class k. The sum of confidence for all classes is 1, meaning that when the prediction value for one class is close to 1, the prediction values for other classes are close to 0, resulting in minimized entropy.

Self-training is a form of entropy minimization in semi-supervised learning. Its main idea is to collect both labeled and unlabeled data, but only use the labeled data to train an initial teacher model. This model is then used to construct pseudo-labels for the unlabeled data, which are combined with the labeled data for joint training to obtain a new student prediction model. However, self-training also has certain limitations. For example, the pseudo-labels generated based on the teacher model can be erroneous and unreliable, leading to a deterioration in the performance of the student model during iterative training. Therefore, in recent years, researchers have been exploring ways to improve the feasibility of self-training in order to enhance the accuracy and reliability of the model. One approach is to set a probability threshold, where pseudo-labels are only generated for unlabeled data when the prediction probability exceeds the threshold. Another approach is to use iterative training, where, after generating pseudo-labels for unlabeled data, a new supervised model is trained by combining the pseudo-labels and labeled data. This new model is then used for further predictions, allowing for iterative updates to correct previous erroneous pseudo-labels. Alternatively, training multiple teacher models can be employed, where only pseudo-labels recognized by the majority of teacher models are successfully generated.

2.5.2. Consistency Regularization

In the realm of semi-supervised semantic segmentation algorithms, the concept of consistency regularization has been widely employed to enhance segmentation accuracy [30]. By perturbing unlabeled data, the current optimization model is compelled to produce stable and consistent predictions for the same unlabeled data under different perturbations, such as variations in shape and color. Leveraging the assumption of clustering, the addition of perturbations does not significantly affect the actual output results. This approach does not require specific label information, making it suitable for semi-supervised learning. By constructing an unsupervised regularization loss term between the predictions obtained from perturbed and unperturbed versions of unlabeled data, the model's generalization ability is improved. Mathematically, the formulation is as follows:

$$D[p_{\text{model}}(y|\text{Augment}(x), \theta), p_{\text{model}}(y|\text{Augment}(x), \theta)] \tag{2}$$

where D represents a metric function, typically utilizing KL (Kullback–Leibler) divergence or JS (Jensen–Shannon) divergence, and can employ cross-entropy or mean squared error; Augment($\cdot$) denotes a data augmentation function that introduces noise perturbations, and θ represents the model parameters.

To mitigate the potential errors and unreliability of pseudo-labels generated by the teacher model, methods to improve the quality of generated labels can be employed. There are two approaches to enhance label quality: rational selection of the teacher model and appropriate choice of input perturbations. In the study [21], the mean teacher architecture was utilized, which is a consistency regularization method based on the assumption of smoothness. By perturbing unlabeled data using CutMix, the overfitting problem of neural networks is reduced. This allows the model to consistently predict both the given unlabeled data and its perturbed versions. By combining labeled data and pseudo-labels, the student model's parameters are updated through backpropagation. The student model's parameters are exponentially moving and averaged to serve as the teacher model's parameters, continuously iterating to shape an improved teacher model. Consequently, the quality of the student model is continuously enhanced, resulting in improved segmentation accuracy.

CCT is a semi-supervised semantic segmentation method based on cross-consistency that achieves prediction consistency through various forms of perturbations applied to the encoder outputs [22]. By training a shared encoder and a main decoder with labeled data, multiple auxiliary decoders are trained to leverage unlabeled data. The different perturbed versions of the shared encoder's output are used as inputs, and consistency is enforced between the predictions of the main decoder and the auxiliary decoders. The encoder's representation is strengthened by training signals extracted from unlabeled data. Perturbations such as Cutout are applied to unlabeled images, resulting in higher segmentation accuracy on the PASCAL VOC 2012 dataset.

PseudoSeg continues the exploration of consistency regularization by applying different perturbations to images [27]. PseudoSeg employs two types of data augmentation on input images: weak augmentation (random cropping and flipping, for instance) and strong augmentation (color jittering). Both augmented images are fed into the same network, producing two distinct outputs. Due to the stability of training under weak augmentation, confidence vectors are constructed from the outputs generated by weakly augmented images to generate pseudo-labels. Finally, the loss is computed using the pseudo-labels and the vectors obtained from strongly augmented images.

2.5.3. Self-Training Algorithm

Semi-supervised semantic segmentation [31] aims to generalize from a combination of pixel-labeled images of $D^l = \{(x_i, y_i)\}_{i=1}^{M}$ and unlabeled images of $D^u = \{u_i\}_{i=1}^{N}$, where, in most cases, $N \gg M$, and the overall optimization objective is formalized as follows:

$$L = L^s + \lambda L^u \tag{3}$$

Here, λ is the weight between labeled and unlabeled data. The unsupervised loss is represented as L^u, while the supervised loss L^s is the cross-entropy loss between the predicted and manually annotated masks.

It involves three steps and does not require iterative training (Algorithm 1):

(1) Supervised learning: Train a teacher model T using the cross-entropy loss on a labeled dataset of D^l.
(2) Pseudo-labeling: Use the trained teacher model T to predict one-hot pseudo-labels on an unlabeled dataset of D^u, resulting in $\hat{D}^u = \{(u_i, T(u_i))\}_{i=1}^{N}$.
(3) Retraining: Combine the labeled and pseudo-labeled data of $D^l \cup \hat{D}^u$ and retrain a student model S for final testing.

Here, the unsupervised loss L^u can be expressed as follows:

$$L^u = H(T(x), S(A^w(x))) \tag{4}$$

Here, x represents the input image, and T and S map the image x to the output space. A^w applies random, weak data augmentation to the original image. H minimizes the entropy between the student and teacher.

Due to the random, strong data augmentation applied to each unlabeled image before feeding it into the network model, it is highly beneficial for decoupling predictions of the same input and mitigating overfitting to noise pseudo-labels. In other words, although the images are the same, the inputs from different training batches are constantly changing. In this scenario, the model is less prone to overfitting the noise from pseudo-labels. Additionally, compared to the teacher model, the student model can obtain more comprehensive representations. Moreover, the randomly augmented images are supervised by the same pseudo-label, which means that consistency regularization is applied to the same unlabeled image across different training batches. Therefore, the self-training approach with strong data augmentation combines two methods: entropy minimization and consistency regularization, commonly used in semi-supervised learning. At this point, the unsupervised objective can be formalized as follows:

$$L^u = H(T(x), S(A^s(A^w(x)))) \tag{5}$$

where A^s denotes the application of strong data augmentation to the unlabeled data.

The approach employed in this study eliminates the need to select high-quality pseudo-labels based on a threshold. Moreover, it only requires the construction of pseudo-labels once for the unlabeled images, without the need for iterative regeneration of pseudo-labels by the student model. This enables training to proceed in a fully supervised manner, saving training time. The pseudocode for the semi-supervised self-training is presented in Algorithm 1 [31], while the segmentation process is illustrated in Figure 3.

Algorithm 1 Self-training pseudocode

Input: Labeled training set $D^l = \{(x_i, y_i)\}_{i=1}^{M}$
      Unlabeled training set $D^u = \{u_i\}_{i=1}^{N}$
      Weak/Strong data augmentations A^w / A^s
      Teacher/Student model T/S
Output: Student model S
Train T on D^l with cross-entropy loss L_{ce}
Obtain pseudo labeled $\hat{D}^u = \{(u_i, T(u_i))\}_{i=1}^{N}$
Over-sample D^l to around the sized of $\hat{D}^u$
for *minbatch* $\{(x_k, y_k)\}_{k=1}^{B} \subset (D^l \cup \hat{D}^u)$ **do**
    for $k \in \{1, \cdots, B\}$ **do**
      if $k \in \{1, \cdots, B\}$ **then**
        $x_k, y_k \leftarrow A^s(A^w(x_k, y_k))$
      else
        $x_k, y_k \leftarrow A^w(x_k, y_k)$
      $\hat{y}_k = S(x_k)$
    Update S to minimize L_{ce} of $\{(\hat{y}_k, y_k)\}_{k=1}^{B}$
return S

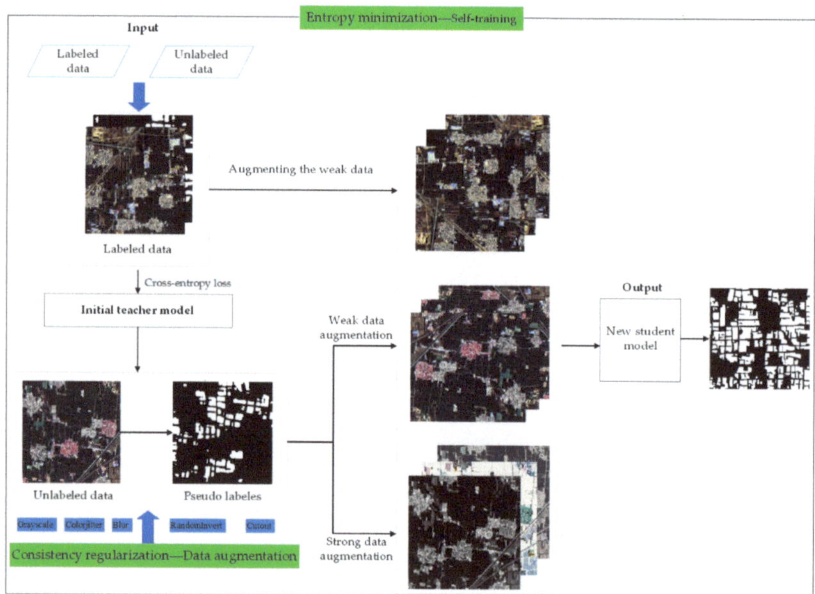

Figure 3. Workflow chart of the semi-supervised classification.

2.6. Evaluation Metrics and Comparative Methods

Semantic segmentation is considered a multi-classification problem, with mean pixel accuracy (MPA) and mean intersection over union (MIoU) serving as the evaluation metrics for wheat segmentation [32]. A higher value for these metrics indicates better segmentation performance for the model. Let k denote the number of object categories that can be segmented in the dataset, with a total of $k + 1$ categories, where 1 represents the background. This experiment includes two categories: wheat and background. Let p_{ij} denote the number of pixels belonging to category i that are incorrectly classified as category j, p_{ii} denote the number of pixels correctly classified as category i, and p_{jj} denote the number of pixels correctly classified as category j.

Pixel accuracy (PA) is the ratio of correctly classified wheat and background pixels to the total number of pixels in the image. Mean PA (MPA) is an improved version of PA that calculates the prediction accuracy of wheat and background pixels separately, taking the average of the two to better reflect the overall semantic segmentation accuracy of the model. It is defined as follows:

$$\text{MPA} = \frac{1}{k+1} \times \sum_{i=0}^{k} \frac{p_{ii}}{\sum\limits_{j=0}^{k} p_{ij}} \tag{6}$$

MIoU is a commonly used metric for evaluating the effectiveness of image semantic segmentation. It calculates the ratio between the intersection and union of the true and predicted values, measuring the degree of similarity between the predicted result and the ground truth. A value closer to 1 indicates a more accurate prediction. It is defined as follows:

$$\text{MIoU} = \frac{1}{k+1} \sum_{i=0}^{k} \frac{P_{ii}}{\sum\limits_{j=0}^{k} p_{ij} + \sum\limits_{j=0}^{k} p_{ji} - p_{ii}} \tag{7}$$

To provide a comparative analysis with fully supervised semantic segmentation, three segmentation networks, namely SegNet [33], U-Net [34], and DeepLabv3+ [35], were selected for evaluating the results. SegNet is a deep, fully convolutional neural network architecture for semantic pixel-wise segmentation, which consists of an encoder network and a corresponding decoder network, followed by a pixel-wise classification layer. The fundamental concept of U-Net lies in the integration of skip connections, resulting in a substantial enhancement in image segmentation accuracy, which comprises three primary components: the decoder, encoder, and bottleneck layer. In contrast to SegNet and U-Net, the primary distinguishing feature of DeeplabV3+ lies in its incorporation of dilated convolutions. This augmentation, without compromising information integrity, expands the receptive field, enabling each convolutional output to encompass a broader range of information. This facilitates the extraction of multiscale information, thus enhancing the model's semantic segmentation capabilities. All three models were trained and tested using the same data, and the evaluation metrics employed were MPA and MIoU.

3. Results

3.1. Comparison of Segmentation Accuracies

The experimental training data consisted of remote sensing images from Zhengding County captured by Landsat-8 OLI. A random sampling strategy was employed to extract 1/16, 1/8, 1/4, and 1/2 proportions of labeled data, resulting in datasets with 312, 625, 1250, and 2500 original images, respectively. The remaining 4688, 4375, 3750, and 2500 images were used as unlabeled data. Initially, the labeled data in different proportions were fed into the segmentation networks to train the teacher models. Subsequently, pseudo-labels were generated for the unlabeled data using the teacher models, and a dataset combining pseudo-labeled and labeled data was created for training the student models. This process was repeated for each of the four datasets with different proportions of labeled data. Finally, the student models were trained using the semi-supervised approach. To reduce the error caused by manual labeling, random sampling of 1/16, 1/8, 1/4, and 1/2 proportions of the labeled data was performed three times on the original 5000 images, and the average results of the three runs were reported. The test set for all experiments was the Zengcun town image in Gaocun, Dingxing County, captured by Landsat-8 OLI. The evaluation metrics for the three segmentation networks in the semi-supervised setting are presented in Table 1.

Table 1. Comparison of the semi-supervised classification effects of different methods.

Method	Indicator	1/16(312)	1/8(625)	1/4(1250)	1/2(2500)
Fully supervised SegNet with only the labeled data	MPA/%	61.82	68.87	76.95	78.17
	MIoU/%	50.26	62.52	68.19	70.18
Semi-supervised SegNet	MPA/%	66.01	73.77	79.66	80.51
	MIoU/%	54.81	66.46	72.28	73.48
Fully supervised DeepLabv3+ with only the labeled data	MPA/%	78.29	79.33	81.14	82.27
	MIoU/%	71.82	72.48	73.39	74.49
Semi-supervised DeepLabv3+	MPA/%	79.25	80.60	82.13	82.50
	MIoU/%	73.45	74.29	75.30	75.74
Fully supervised U-Net with only the labeled data	MPA/%	81.63	82.88	83.99	84.62
	MIoU/%	73.31	74.11	75.73	76.45
Semi-supervised U-Net	MPA/%	82.50	84.60	84.27	85.52
	MIoU/%	76.01	76.84	77.52	77.83

3.2. Visualization Effects Using Different Models

This paper presents a semi-supervised approach that combines self-training with unlabeled data augmentation to analyze the experimental results of three semantic segmentation networks: SegNet, DeepLabv3+, and U-Net. Initially, only 1/16, 1/8, 1/4, and 1/2 of the labeled data are used to train the SegNet, DeepLabv3+, and U-Net segmentation networks in a fully supervised manner, resulting in the experimental results shown in Figure 4, Figure 6 and Figure 8, and Table 1 (annotated as the fully supervised section). The remaining unlabeled data are then used for self-training in a semi-supervised manner using the SegNet, DeepLabv3+, and U-Net segmentation networks. Two representative result images are selected: one with a higher density of wheat and another with a lower density, along with a higher presence of buildings and bare soil areas. The fully supervised and semi-supervised prediction results of SegNet with different proportions of labeled data are shown in Figures 4 and 5, respectively. The fully supervised and semi-supervised prediction results of DeepLabv3+ with different proportions of labeled data are shown in Figures 6 and 7, respectively. The fully supervised and semi-supervised prediction results of U-Net with different proportions of labeled data are shown in Figures 8 and 9, respectively. From the prediction results, it can be observed that as the proportion of labeled data increases, the model learns better, and the predicted result images show superior performance.

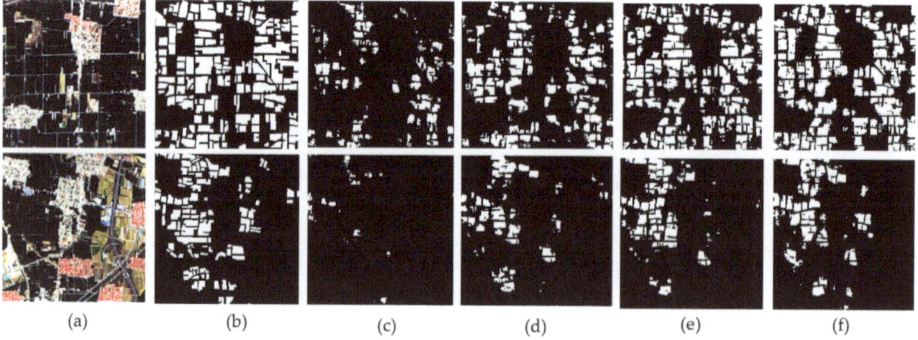

(a)   (b)   (c)   (d)   (e)   (f)

Figure 4. Comparison of extraction effects for the fully supervised SegNet under different ratios of labeled data: (**a**) Original image; (**b**) labeled data; (**c**) 1/16; (**d**) 1/8; (**e**) 1/4; and (**f**) 1/2.

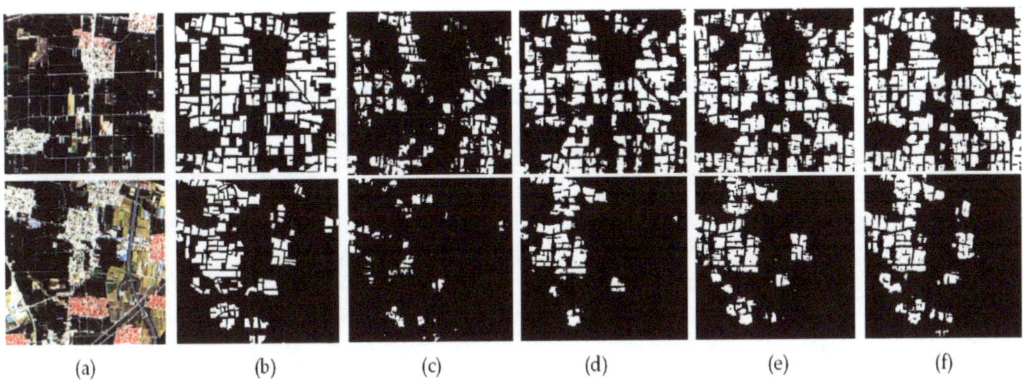

Figure 5. Comparison of extraction effects for the semi-supervised SegNet under different ratios of labeled data: (**a**) Original image; (**b**) labeled data; (**c**) 1/16; (**d**) 1/8; (**e**) 1/4; and (**f**) 1/2.

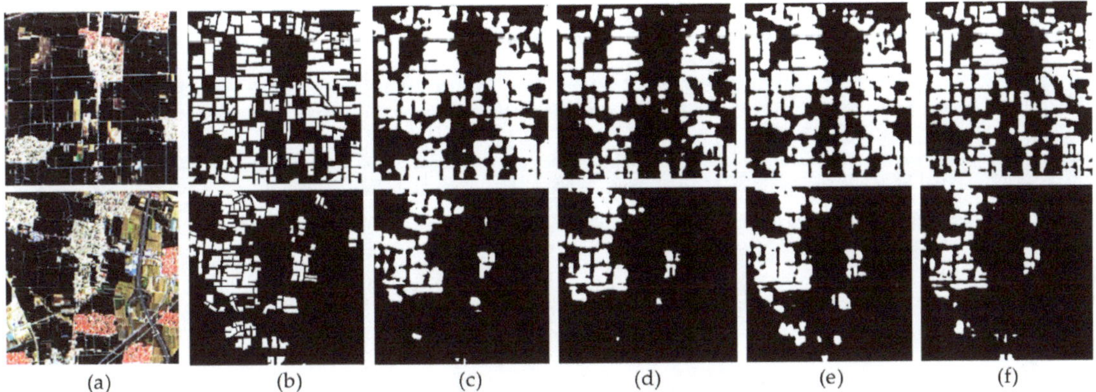

Figure 6. Comparison of extraction effects for the fully supervised DeepLabv3+ under different ratios of labeled data: (**a**) Original image; (**b**) labeled data; (**c**) 1/16; (**d**) 1/8; (**e**) 1/4; and (**f**) 1/2.

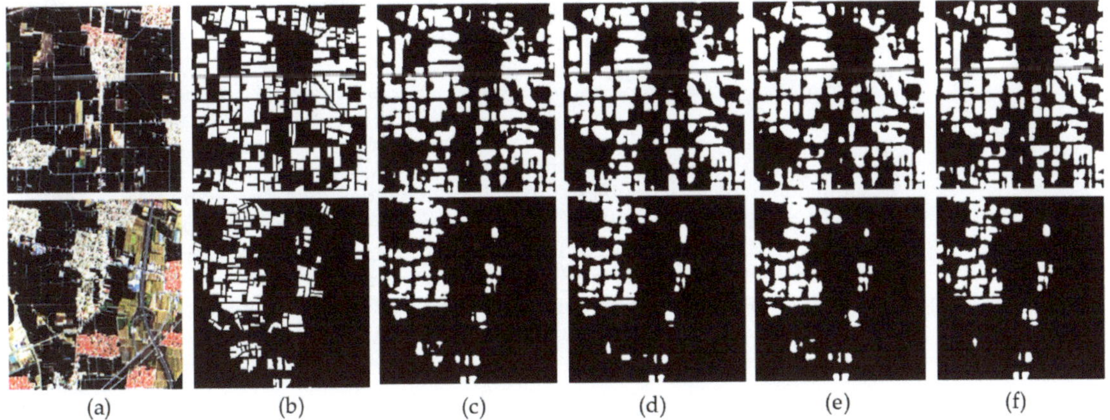

Figure 7. Comparison of extraction effects for the semi-supervised DeepLabv3+ under different ratios of labeled data: (**a**) Original image; (**b**) labeled data; (**c**) 1/16; (**d**) 1/8; (**e**) 1/4; and (**f**) 1/2.

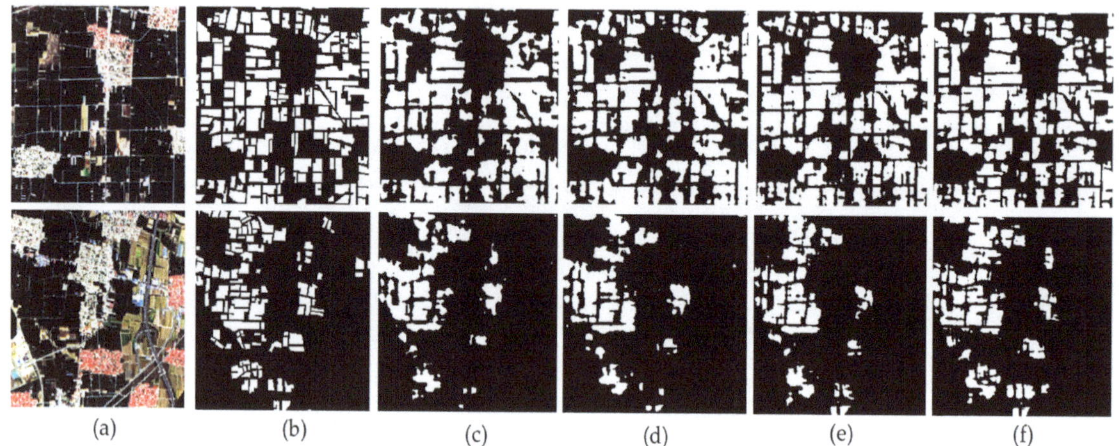

Figure 8. Comparison of extraction effects for the fully supervised U-Net under different ratios of labeled data: (**a**) Original image; (**b**) labeled data; (**c**) 1/16; (**d**) 1/8; (**e**) 1/4; and (**f**) 1/2.

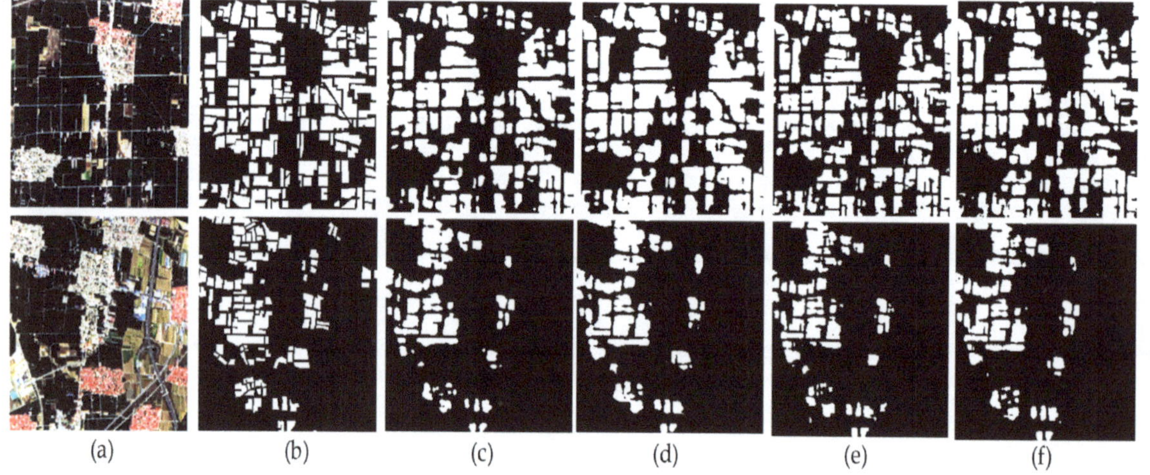

Figure 9. Comparison of extraction effects for the semi-supervised U-Net under different ratios of labeled data: (**a**) Original image; (**b**) labeled data; (**c**) 1/16; (**d**) 1/8; (**e**) 1/4; and (**f**) 1/2.

Overall, this study demonstrates the superiority of the semi-supervised self-training method, which incorporates unlabeled data with strong data augmentation, over fully supervised learning methods when training models with different proportions of labeled data. Particularly, the effectiveness of the semi-supervised approach is more pronounced in scenarios with limited labeled data, such as when only 1/16 of the data are labeled. For SegNet, the fully supervised prediction model achieves an MPA of 61.82% and an MIoU of 50.26%, while the semi-supervised prediction model achieves an MPA of 66.01% and an MIoU of 54.81%. For DeepLabv3+, the fully supervised prediction model achieves an MPA of 78.29% and an MIoU of 71.82%, while the semi-supervised prediction model achieves an MPA of 79.25% and an MIoU of 73.45%. For U-Net, the fully supervised prediction model achieves an MPA of 81.63% and an MIoU of 73.31%, while the semi-supervised prediction model achieves an MPA of 82.50% and an MIoU of 76.01%. Comparing the semi-supervised results with the fully supervised results, both the MPA and MIoU show improvement, enabling the attainment of superior solutions even with insufficient labeled data.

3.3. Mapping Wheat Planting Areas Using the Semi-Supervised U-Net

Based on the analysis of experimental results, the semi-supervised method employed in this study demonstrates improvements in training with labeled data proportions of 1/16, 1/8, 1/4, and 1/2. Furthermore, this method is applicable to three different semantic segmentation networks: SegNet, DeepLabv3+, and U-Net. Among these networks, U-Net exhibits superior overall prediction performance in semantic segmentation. Figure 10 illustrates the overall results of semi-supervised prediction using different proportions of labeled data for the Landsat-8 OLI image in the rural area. It can be observed from the figure that when only 1/16 of the data are labeled for fully supervised segmentation, a significant portion of the wheat areas remain unidentified. However, with the application of semi-supervised semantic segmentation, there is a noticeable improvement in wheat segmentation.

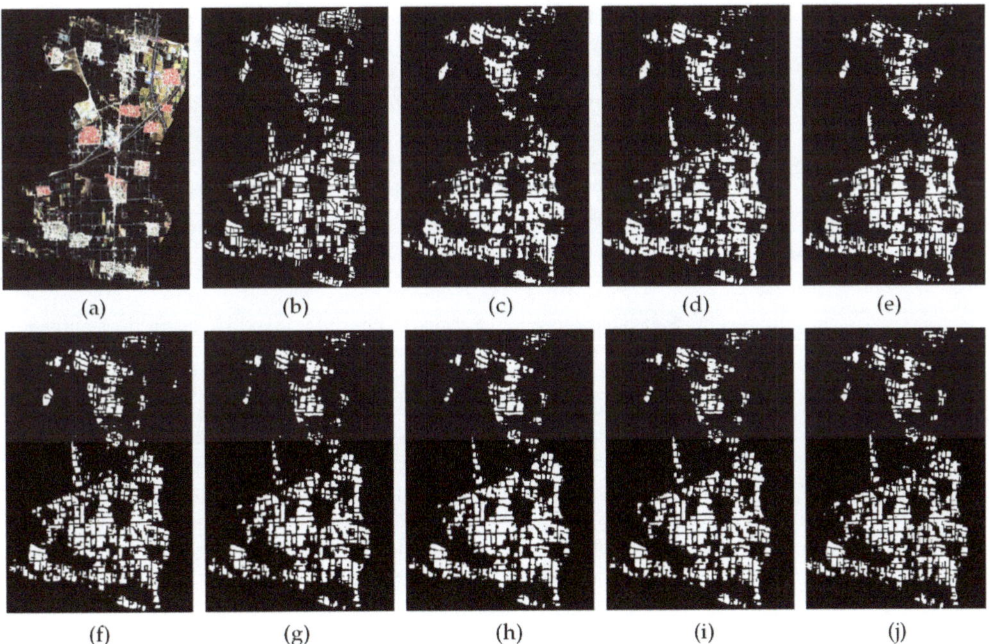

Figure 10. Comparison of extraction effects for the semi-supervised U-Net under different ratios of labeled data: (**a**) Original image; (**b**) labeled data; (**c**) 1/16 fully supervised; (**d**) 1/8 fully supervised; (**e**) 1/4 fully supervised; (**f**) 1/2 fully supervised; (**g**) 1/16 semi-supervised; (**h**) 1/8 semi-supervised; (**i**) 1/4 semi-supervised; and (**j**) 1/2 semi-supervised.

4. Discussion

4.1. Analysis of Training Ratios of Labeled Data for Different Models

In the case of SegNet, when the labeled data are scarce, the segmentation results are poor, with most wheat areas remaining unrecognized, especially in areas with fewer wheat crops. Even with only 1/16 of the labeled data used for training, the generated prediction model is almost unable to identify the wheat areas. However, as the proportion of labeled data increases, the segmentation improves. From the experimental results, it can be seen that the semi-supervised approach improves the segmentation results for different proportions of labeled data, with a significant increase in the predicted area. In the case of DeepLabv3+, when the labeled data are scarce, the wheat areas are often connected in large clusters, making it difficult to distinguish them from other non-wheat areas and resulting in misclassifications. By comparing the fully supervised prediction results of 1/16 and 1/2 labeled data, it can be observed that increasing the amount of labeled data partially alleviates the issue of connected prediction regions. In contrast, the semi-supervised approach, based on the fusion of labeled and unlabeled data, effectively mitigates the problem of connected prediction regions caused by limited labeled data and performs better in terms of prediction accuracy compared to using only labeled data [36] U-Net performs better than SegNet and DeepLabv3+ in predicting wheat planting areas. Comparing the results with fully supervised learning, a smaller amount of training data leads to less information learned by the model. The semi-supervised self-training method used in this paper improves performance in various proportions of labeled data compared to fully supervised learning.

4.2. Influence of Spatial Resolution on Remote Imagery

The spatial resolution of remote sensing imagery plays a crucial role in object segmentation and extraction [37]. Higher spatial resolution allows for the identification of smaller objects in an image. Fine details and intricate features of objects can be captured more accurately with higher spatial resolution imagery. This is particularly important when dealing with small and closely spaced objects, such as individual trees in a forest or vehicles in an urban area. Spatial resolution also affects the ability to accurately delineate the boundaries of objects. Higher spatial resolution imagery provides sharper and better-defined edges, enabling more precise and accurate object segmentation. This is particularly important when dealing with objects that have complex or irregular shapes, such as rivers, coastlines, or land-use boundaries.

Nevertheless, it is not necessarily better to have a higher spatial resolution in remote sensing imagery. On the contrary, it is important to consider the distribution range, spatial aggregation characteristics, area coverage, and contrast with the background features of the objects to be segmented and extracted in the image [38]. Sometimes, it is necessary to consider the efficiency of object extraction as well as the cost of acquiring the imagery. Since the selected experimental data are based on Landsat-8 OLI imagery with a resolution of 15 m, the wheat labeling map reveals that manually annotated wheat areas often appear as small, fragmented patches, especially in areas where wheat is interspersed with buildings or bare soil. For instance, in the upper right corner of Figure 10b, the labeling is not well defined, which may lead to biased predictions and instances of missed segmentation. Nevertheless, this approach helps to reduce the cost of acquiring high-resolution remote sensing imagery. Overall, the wheat areas show minimal deviation in terms of their spatial distribution, thereby validating the reliability of the prediction results.

5. Conclusions

The study combines entropy minimization and consistency regularization to perform semi-supervised learning. Three mainstream segmentation networks, namely SegNet, DeepLabv3+, and U-Net, are employed to comparatively identify wheat planting areas using semi-supervised semantic segmentation. The labeled data are randomly sampled at proportions of 1/16, 1/8, 1/4, and 1/2, and both fully supervised and self-training semi-supervised methods are employed to train models. A comparative analysis is conducted on the same test set, validating the effectiveness of the semi-supervised approach for different proportions of labeled data and different semantic segmentation networks. The segmentation evaluation metrics, MPA and MIoU, demonstrate varying degrees of improvement, mitigating to some extent the challenge of inadequate labeled data leading to subpar segmentation results. In future research, we will delve into the realm of unsupervised semantic segmentation methods, aiming to automatically extract label data and classification features from the original input imagery for model training purposes.

Author Contributions: J.Z. and Y.Z. conceived and designed the experiments; M.Z. and Y.X. performed the experiments; M.Z. and Y.Z. analyzed the data; J.Z. wrote and proofread the paper. All authors have read and agreed to the published version of the manuscript.

Funding: This research was funded by the National Natural Science Foundation of China (42301103) and the Industry-University Collaborative Education Project of the Ministry of Education (220802313205840, 202102245009, 22087106262449).

Data Availability Statement: Data are contained within the article.

Acknowledgments: We gratefully acknowledge the anonymous reviewers for their valuable comments that helped to considerably improve the manuscript.

Conflicts of Interest: The authors declare no conflict of interest.

References

1. Mo, Y.; Wu, Y.; Yang, X.; Liu, F.; Liao, Y. Review the state-of-the-art technologies of semantic segmentation based on deep learning. *Neurocomputing* **2022**, *493*, 626–646. [CrossRef]
2. Yuan, X.; Shi, J.; Gu, L. A review of deep learning methods for semantic segmentation of remote sensing imagery. *Expert Syst. Appl.* **2021**, *169*, 114417. [CrossRef]
3. Yang, E.; Zhou, W.; Qian, X.; Lei, J.; Yu, L. DRNet: Dual-stage refinement network with boundary inference for RGB-D semantic segmentation of indoor scenes. *Eng. Appl. Artif. Intell.* **2023**, *125*, 106729. [CrossRef]
4. Casado-García, A.; Heras, J.; Milella, A.; Marani, R. Semi-supervised deep learning and low-cost cameras for the semantic segmentation of natural images in viticulture. *Precis. Agric.* **2022**, *23*, 2001–2026. [CrossRef]
5. Zheng, Y.; Yang, M.; Wang, M.; Qian, X.; Yang, R.; Zhang, X.; Dong, W. Semi-Supervised Adversarial Semantic Segmentation Network Using Transformer and Multiscale Convolution for High-Resolution Remote Sensing Imagery. *Remote Sens.* **2022**, *14*, 1786. [CrossRef]
6. Mukhtar, H.; Khan, M.U.G.; Saba, T.; Latif, R. Wheat Plant Counting Using UAV Images Based on Semi-supervised Semantic Segmentation. In Proceedings of the 2021 1st International Conference on Artificial Intelligence and Data Analytics, Riyadh, Saudi Arabia, 6–7 April 2021; pp. 257–261.
7. Oliver, A.; Odena, A.; Raffel, C.A.; Cubuk, E.D.; Goodfellow, I. Realistic evaluation of deep semi-supervised learning algorithms. In Proceedings of the 32nd annual Conference on Neural Information Processing Systems (NeurIPS 2018), Montréal, QC, Canada, 3–8 December 2018; pp. 3235–3246.
8. Xie, Q.; Luong, M.T.; Hovy, E.; Le, Q.V. Self-training with noisy student improves imagenet classification. In Proceedings of the IEEE/CVF Conference on Computer Vision and Pattern Recognition, Seattle, WA, USA, 13–19 June 2020; pp. 10687–10698.
9. Wei, C.; Sohn, K.; Mellina, C.; Yuille, A.; Yang, F. CReST: A Class-Rebalancing Self-Training Framework for Imbalanced Semi-Supervised Learning. In Proceedings of the IEEE/CVF Conference on Computer Vision and Pattern Recognition, Nashville, TN, USA, 20–25 June 2021; pp. 10857–10866.
10. Zhu, Y.; Zhang, Z.; Wu, C.; Zhang, Z.; He, T.; Zhang, H.; Manmatha, R.; Li, M.; Smola, A.J. Improving Semantic Segmentation via Efficient Self-Training. *arXiv* **2020**, arXiv:2004.14960. [CrossRef] [PubMed]
11. Feng, Z.; Zhou, Q.; Cheng, G.; Tan, X.; Shi, J.; Ma, L. Semi-supervised semantic segmentation via dynamic self-training and class-balanced curriculum. *arXiv* **2020**, arXiv:2004.08514.
12. Zoph, B.; Ghiasi, G.; Lin, T.Y.; Cui, Y.; Liu, H.; Cubuk, E.D.; Le, Q. Rethinking pre-training and self-training. In Proceedings of the 34th Conference on Neural Information Processing Systems (NeurIPS 2020), Vancouver, BC, Canada, 6–12 December 2020; pp. 3833–3845.
13. Luo, Y.; Zhu, J.; Li, M.; Ren, Y.; Zhang, B. Smooth neighbors on teacher graphs for semi-supervised learning. In Proceedings of the IEEE Conference on Computer Vision and Pattern Recognition, Salt Lake City, UT, USA, 18–21 June 2018; pp. 8896–8905.
14. Jeong, J.; Shin, J. Consistency regularization for certified robustness of smoothed classifiers. In Proceedings of the 34th Conference on Neural Information Processing Systems (NeurIPS 2020), Vancouver, BC, Canada, 6–12 December 2020; pp. 10558–10570.
15. French, G.; Laine, S.; Aila, T.; Mackiewicz, M.; Finlayson, G. Semi-supervised semantic segmentation needs strong, varied perturbations. *arXiv* **2019**, arXiv:1906.01916.
16. Kim, J.; Jang, J.; Park, H.; Jeong, S. Structured consistency loss for semi-supervised semantic segmentation. *arXiv* **2020**, arXiv:2001.04647.
17. Chen, T.; Kornblith, S.; Norouzi, M.; Hinton, G. A simple framework for contrastive learning of visual representations. In Proceedings of the 37th International Conference on Machine Learning, Vienna, Austria, 12–18 July 2020; pp. 1597–1607.
18. Yun, S.; Han, D.; Oh, S.J.; Chun, S.; Choe, J.; Yoo, Y. CutMix: Regularization Strategy to Train Strong Classifiers with Localizable Features. In Proceedings of the IEEE/CVF International Conference on Computer Vision (ICCV), Seoul, Republic of Korea, 27 October–2 November 2019; pp. 6023–6032.
19. Olsson, V.; Tranheden, W.; Pinto, J.; Svensson, L. Classmix: Segmentation-based data augmentation for semi-supervised learning. In Proceedings of the IEEE/CVF Winter Conference on Applications of Computer Vision, Waikoloa, HI, USA, 3–8 January 2021; pp. 1369–1378.
20. Laine, S.; Aila, T. Temporal ensembling for semi-supervised learning. *arXiv* **2016**, arXiv:1610.02242.

21. Tarvainen, A.; Valpola, H. Mean teachers are better role models: Weight-averaged consistency targets improve semi-supervised deep learning results. In Proceedings of the 31st International Conference on Neural Information Processing Systems (NIPS 2017), Long Beach, CA, USA, 4–8 December 2017; pp. 1195–1204.
22. Ouali, Y.; Hudelot, C.; Tami, M. Semi-supervised semantic segmentation with cross-consistency training. In Proceedings of the IEEE/CVF Conference on Computer Vision and Pattern Recognition, Seattle, WA, USA, 13–19 June 2020; pp. 12674–12684.
23. Berthelot, D.; Carlini, N.; Goodfellow, I.; Papernot, N.; Oliver, A.; Raffel, C.A. Mixmatch: A holistic approach to semi supervised learning. In Proceedings of the 33rd International Conference on Neural Information Processing Systems (NIPS 2019), Vancouver, BC, Canada, 8–14 December 2019; pp. 5049–5059.
24. Mittal, S.; Tatarchenko, M.; Brox, T. Semi-Supervised Semantic Segmentation with High- and Low-Level Consistency. *IEEE Trans Pattern Anal. Mach. Intell.* **2019**, *43*, 1369–1379. [CrossRef] [PubMed]
25. Lee, D.H. Pseudo-label: The simple and efficient semi-supervised learning method for deep neural networks. In Proceedings of the Workshop on Challenges in Representation Learning, ICML (2013), Atlanta, GA, USA, 16–21 June 2013; pp. 1–6.
26. Sohn, K.; Berthelot, D.; Carlini, N.; Zhang, Z.; Zhang, H.; Raffel, C.A.; Cubuk, E.D.; Kurakin, A.; Li, C.L. Fixmatch: Simplifying semi-supervised learning with consistency and confidence. In Proceedings of the 34th Conference on Neural Information Processing Systems (NeurIPS 2020), Vancouver, BC, Canada, 6–12 December 2020; pp. 596–608.
27. Zou, Y.; Zhang, Z.; Zhang, H.; Li, C.L.; Bian, X.; Huang, J.B.; Pfister, T. Pseudoseg: Designing pseudo labels for semantic segmentation. *arXiv* **2020**, arXiv:2010.09713.
28. Santos, C.F.G.D.; Papa, J.P. Avoiding overfitting: A survey on regularization methods for convolutional neural networks. *ACM Comput. Surv.* **2022**, *54*, 1–25. [CrossRef]
29. Grandvalet, Y.; Bengio, Y. Semi-supervised learning by entropy minimization. In Proceedings of the 17th International Conference on Neural Information Processing Systems, Vancouver, BC, Canada, 13–18 December 2004; pp. 529–536.
30. Wang, J.; Ding, C.H.Q.; Chen, S.; He, C.; Luo, B. Semi-Supervised Remote Sensing Image Semantic Segmentation via Consistency Regularization and Average Update of Pseudo-Label. *Remote Sens.* **2020**, *12*, 3603. [CrossRef]
31. Yang, L.; Zhuo, W.; Qi, L.; Shi, Y.; Gao, Y. St++: Make self-training work better for semi-supervised semantic segmentation. In Proceedings of the IEEE/CVF Conference on Computer Vision and Pattern Recognition, New Orleans, LA, USA, 18–24 June 2022, pp. 4268–4277.
32. Alberto, G.G.; Sergio, O.E.; Sergiu, O.; Victor, V.M.; Jose, G.R. A review on deep learning techniques applied to semantic segmentation. *arXiv* **2017**, arXiv:1704.06857.
33. Badrinarayanan, V.; Kendall, A.; Cipolla, R. SegNet: A deep convolutional encoder-decoder architecture for image segmen-tation. *IEEE Trans. Pattern Anal. Mach. Intell.* **2017**, *39*, 2481–2495. [CrossRef] [PubMed]
34. Ronneberger, O.; Fischer, P.; Brox, T. U-net: Convolutional networks for biomedical image segmentation. In Proceedings of the Medical Image Computing and Computer-Assisted Intervention–MICCAI 2015: 18th International Conference, Munich, Germany, 5–9 October 2015; pp. 234–241.
35. Chen, L.C.; Zhu, Y.; Papandreou, G.; Schroff, F.; Adam, H. Encoder-decoder with atrous separable convolution for semantic image segmentation. In Proceedings of the European Conference on Computer Vision (ECCV), Munich, Germany, 8–14 September 2018; pp. 801–818.
36. Song, E.; Huang, D.; Ma, G.; Hung, C.-C. Semi-supervised multi-class Adaboost by exploiting unlabeled data. *Expert Syst. Appl.* **2011**, *38*, 6720–6726. [CrossRef]
37. Wang, M.; Dong, Z.; Cheng, Y.; Li, D. Optimal Segmentation of High-Resolution Remote Sensing Image by Combining Superpixels with the Minimum Spanning Tree. *IEEE Trans. Geosci. Remote Sens.* **2018**, *56*, 228–238. [CrossRef]
38. Yang, R.; Xu, Q.; Long, H. Spatial distribution characteristics and optimized reconstruction analysis of China's rural settlements during the process of rapid urbanization. *J. Rural Stud.* **2016**, *47*, 413–424. [CrossRef]

Article

Monitoring Indicators for Comprehensive Growth of Summer Maize Based on UAV Remote Sensing

Hao Ma [1,2], Xue Li [1], Jiangtao Ji [1,2], Hongwei Cui [1,2,*], Yi Shi [1], Nana Li [3] and Ce Yang [4]

1   College of Agricultural Equipment Engineering, Henan University of Science and Technology, Luoyang 471023, China
2   Longmen Laboratory, Luoyang 471000, China
3   Caterpillar Mechanical Components, Inc., Wuxi 214000, China
4   Department of Bioproducts and Biosystems Engineering, University of Minnesota, Minneapolis, MN 55455-0213, USA
*   Correspondence: hongwei21@haust.edu.cn

Abstract: Maize is one of the important grain crops grown globally, and growth will directly affect its yield and quality, so it is important to monitor maize growth efficiently and non-destructively. To facilitate the use of unmanned aerial vehicles (UAVs) for maize growth monitoring, comprehensive growth indicators for maize monitoring based on multispectral remote sensing imagery were established. First of all, multispectral image data of summer maize canopy were collected at the jointing stage, and meanwhile, leaf area index (LAI), relative chlorophyll content (SPAD), and plant height (VH) were measured. Then, the comprehensive growth monitoring indicators $CGMI_{CV}$ and $CGMI_{CR}$ for summer maize were constructed by the coefficient of variation method and the CRITIC weighting method. After that, the $CGMI_{CV}$ and $CGMI_{CR}$ prediction models were established by the partial least-squares (PLSR) and sparrow search optimization kernel extremum learning machine (SSA-KELM) using eight typical vegetation indices selected. Finally, a comparative analysis was performed using ground-truthing data, and the results show: (1) For $CGMI_{CV}$, the R^2 and RMSE of the model built by SSA-KELM are 0.865 and 0.040, respectively. Compared to the model built by PLSR, R^2 increased by 4.5%, while RMSE decreased by 0.3%. For $CGMI_{CR}$, the R^2 and RMSE of the model built by SSA-KELM are 0.885 and 0.056, respectively. Compared to the other model, R^2 increased by 4.6%, and RMSE decreased by 2.8%. (2) Compared to the models by single indicator, among the models constructed based on PLSR, the $CGMI_{CR}$ model had the highest R^2. In the models constructed based on SSA-KELM, the R^2 of models by the $CGMI_{CR}$ and $CGMI_{CV}$ were larger than that of the models by SPAD ($R^2 = 0.837$), while smaller than that of the models by LAI ($R^2 = 0.906$) and models by VH ($R^2 = 0.902$). In summary, the comprehensive growth monitoring indicators prediction model established in this paper is effective and can provide technical support for maize growth monitoring.

Keywords: multispectral remote sensing; summer maize; comprehensive growth; SSA-KELM; vegetation index

Citation: Ma, H.; Li, X.; Ji, J.; Cui, H.; Shi, Y.; Li, N.; Yang, C. Monitoring Indicators for Comprehensive Growth of Summer Maize Based on UAV Remote Sensing. *Agronomy* **2023**, *13*, 2888. https://doi.org/10.3390/agronomy13122888

Academic Editor: Francisco Manzano Agugliaro

Received: 30 October 2023
Revised: 21 November 2023
Accepted: 22 November 2023
Published: 24 November 2023

1. Introduction

Maize is one of the most important crops grown globally and accounts for the highest proportion of the global cereal market. Ensuring maize production is of great significance in guaranteeing global economic development and food security [1]. The growth of maize is an important data source for early yield estimation, and its final yield can be predicted to a certain extent. Sujan Sapkota et al. [2] used a drone equipped with a multispectral camera to obtain the canopy spectral information of maize plants, and successfully monitored the growth and yield estimation of maize by using multiple growth parameters such as maize plant height, biomass, and leaf area index combined with vegetation index. Therefore, real-time and effective monitoring of field corn growth is of great significance in guiding field management [3]. In maize growth monitoring research, many parameters characterize maize growth [4–7], such as aboveground biomass, chlorophyll content, leaf area index, plant height, and leaf nitrogen content. In the study, three growth indicators, leaf area index (LAI, total plant leaf area per unit land area as a multiple of land area), relative chlorophyll content (SPAD, relative proportions of chlorophyll content in different samples), and plant height (VH, distance from the base of the plant to the top of the main stem, i.e., between the growing points of the main stem), were selected to initiate research and analysis of summer maize growth monitoring.

The leaf area index (LAI) is an important basis for characterizing the canopy structure and growth of crops, and is one of the most important parameters for evaluating crop growth conditions [8]. Ma JW et al. [9] constructed and validated a winter wheat LAI prediction model by combining four methods (GPR, artificial neural network (ANN), partial least-squares regression (PLSR), and spectral index (SI), with hyperspectral data), and the results showed the best results for ANN and GPR. Hang YH et al. [10] explored the use of spectral features, texture index, and crop cover to construct LAI estimation models through UAV remote sensing data, and the results showed that the combination of the three types of indices to construct multiple stepwise regression and artificial neural network models estimated LAI with the best accuracy. Tao HL et al. [11] used a UAV with a hyperspectral camera, and used stepwise regression analysis (SWR) and PLSR methods to predict AGB and LAI using vegetation index, red edge parameters, and their combinations, respectively. The results showed that the PLSR algorithm had the best effect in predicting AGB and LAI by combining the vegetation index with the red edge parameters. Chlorophyll (SPAD) is the main pigment that converts light energy into chemical energy and is the driving force behind photosynthesis in plants [12]. Qiao L et al. [13] utilized multispectral remote sensing images of maize captured by drones during the tasseling stage. They observed a significant linear correlation between near-ground vegetation indices and maize canopy chlorophyll content under medium and low crop coverage. For high coverage, a noticeable nonlinear correlation emerged, and the model for monitoring chlorophyll content, established based on partial least-squares regression (PLSR), demonstrated the best performance. Guo Y [14] and team identified optimal combinations of drone spectral indices and texture indices using the SRM model. Applying support vector machine (SVM) and random forest (RF) models, they estimated maize SPAD values, with the SVM model yielding the most optimal predictive results. Plant height (VH), as an important indicator of crop growth, can be used to indirectly obtain crop biomass [15], and can also be used to predict crop yield in conjunction with vegetation index and canopy cover [16]. Bending et al. [17] obtained visible-light data of barley based on a UAV-mounted digital camera. By constructing a CMS, they realized the accurate extraction of crop height and established a barley plant height estimation model. Xu YF et al. [18] constructed 22 multispectral vegetation indices with image spectral reflectance features, and selected three different machine learning algorithms to establish a plant height monitoring model for winter wheat. The results showed that the optimal prediction model for plant height was the MLR-VH model. Under normal circumstances, a single growth indicator characterizes the physiological information status of crops in a certain aspect. The data belong to the 'point' data in essence, and to a certain extent, they cannot represent the overall growth situation of crops, the use of multiple indicators, and the relationship between them to establish a comprehensive growth monitoring indicator (CGMI), which is of practical significance for remote sensing monitoring of crop growth.

Comprehensive growth monitoring indicators can be modeled according to different assignment methods, and the commonly used assignment method is the equalization method. Pei HJ et al. [19] synthesized five indicators reflecting wheat growth, including leaf area index, leaf chlorophyll content, plant nitrogen content, plant water content, and biomass, into a single growth indicator according to equal weights, and combined wheat growth by combining with partial least-squares regression, and achieved a high-precision model. This method did not take into account the contribution of a single indicator to the integrated growth monitoring indicator and simply constructed each indicator into a composite indicator according to equal weights. Although the above studies have good predictive capabilities, most of them are limited to using a single growth indicator to predict crop growth, or simply constructing a comprehensive growth indicator without considering the contribution of each indicator to crop growth, which makes them constructed from the prediction model has certain limitations. Considering the different degrees of importance of each indicator in the comprehensive growth monitoring indicators and the non-uniformity of the scale of each indicator, the coefficient of variation method and the CRITIC weighting method were used to try to construct the comprehensive growth monitoring indicators (CGMI).

In the study, we used summer maize multispectral image data to establish a crop growth monitoring model, and comparatively analyzed the correlation between comprehensive growth monitoring indices and individual indices and vegetation index. The main objectives include: (1) use of the coefficient of variation method and CRITIC weighting method to weight chlorophyll content, leaf area index, and plant height according to their contribution to the overall growth, and construct comprehensive growth monitoring indices respectively; (2) analysis of the correlation of comprehensive growth monitoring indicators and individual indicators with vegetation indices; (3) use of the PLSR and SSA-KELM algorithms to construct a comprehensive growth monitoring indicator prediction model, followed by analysis in comparison with a single growth indicator prediction model.

2. Materials and Methods

2.1. Overview of the Experimental Area

The experimental area was situated in China's first tractor company limited intelligent agriculture demonstration farm in Yiyang County, Luoyang City, Henan Province, China, at coordinates 112°37′11.72″ E, 34°47′79.03″ N, belonging to the temperate continental monsoon-type climate. The area where the geographical coordinates are located is shown in Figure 1. The terrain in the experimental field was flat. The maize variety in this experiment was Zheng Dan 958, which was planted in rotation with wheat. The planting density is 67,000 plants m^{-2}, the row spacing is 0.6 m, and the plant spacing is 0.2 m. Its management methods, such as irrigation, fertilization, pest, and weed control, were the same as those of local conventional farmland.

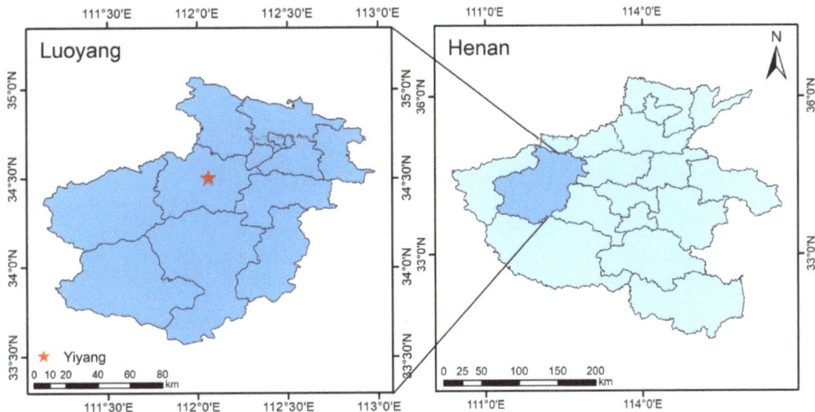

Figure 1. Geographical location of China's first tractor company limited intelligent agriculture demonstration farm.

2.2. UAV Data Acquisition and Preprocessing

This study utilized a DJI Phantom 4 drone (DJI, Shenzhen, China) to collect multispectral remote sensing data of summer maize, equipped with an integrated multispectral imaging system consisting of one visible-light camera and five multispectral cameras (blue, green, red, red edge, and near-infrared). The experiment took place on 22 July 2022, from 10:00 to 12:00 during the summer maize jointing stage, with stable light intensity, clear weather, and no wind.

Before image acquisition, flight routes were planned using mapping software (DJI Terra, V3.9.4, DJI, Shenzhen, China), and spectral image collection followed the planned routes. The drone flew at a height of 70 m, with the sensor lens pointing vertically downward and an 85% overlap rate in both longitudinal and lateral directions. The flight direction was north-south, with a speed of 3.5 m/s.

For reflectance calibration, a calibration reflectance panel was placed on the ground in the experimental area before and after image acquisition. The drone, manually controlled, hovered 2 m directly above the calibration panel to capture spectral images, obtaining standard reflectance values for the test.

Import the gathered multispectral images into Pix4D Mapper software (from Pix4D, V4.4.12, Lausanne, Switzerland) for preprocessing to derive the reflectance spectrum of summer corn within the sampling point ROI region (illustrated as the box in Figure 2). The primary steps encompass (1) ortho-image processing; (2) calibration using reflectance board DN values, generating stitched images; (3) geometric correction with concurrently obtained high-definition digital images as references; (4) choosing the necessary ROI region pertinent to this study; (5) deriving the average reflectance spectrum within the ROI, representing the summer corn's reflectance spectrum within that specific ROI region. For this study, 20 experimental ROI regions, each measuring 10 m × 5 m, are chosen based on growth gradients. Of the data, 70% is designated as the training set (70 plants of maize), while the remaining 30% forms the validation set (30 plants of maize).

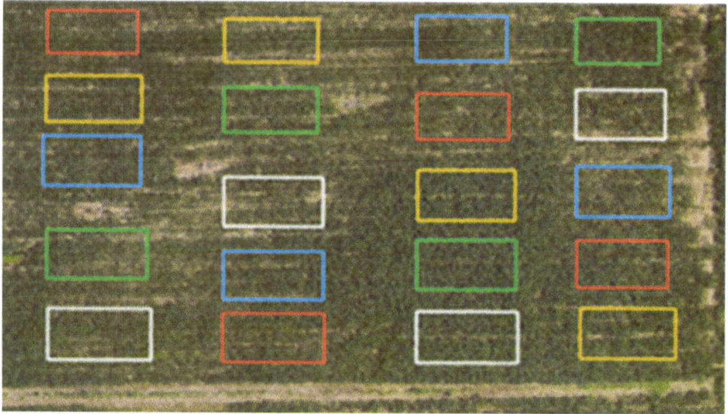

Figure 2. ROI regional distribution map [20].

2.3. Field Data Acquisition

Growth indicators are time-sensitive [21]. Ground-truthing data, mainly including relative chlorophyll content, leaf area index, and plant height, were synchronously collected on the same day that summer maize multispectral remote sensing data were acquired.

(1)   Measurement of Relative Chlorophyll Content

In this experiment, a portable SPAD-502Plus chlorophyll meter (Konica Minolta, Tokyo, Japan) was used for SPAD determination. Five maize plants were selected in each experimental area according to the five-point sampling method, avoiding leaf veins to collect the SPAD values of the three parts of the ear leaves of each maize plant, the root, the middle of the leaf, and the tip of the leaf. Each part was repeated twice, and the average of the six times' data was taken as the SPAD value of the plant [22]. The mean SPAD value of 5 plants in the experimental area was calculated as the relative chlorophyll content of maize plants in that area.

(2) Determination of Leaf Area Index

In this experiment, leaf area index data were collected using the aspect ratio method. Measure summer maize leaf length L_{ij} and maximum leaf width B_{ij} with a tape measure. Five maize plants were selected from each experimental area according to the five-point sampling method, and the L_{ij} and B_{ij} of their leaves were measured, and each part was repeated three times to take the average value [23]. The mean LAI value of five plants in the experimental area was calculated as the LAI of maize plants in the area. The LAI calculation formula is as follows:

$$\text{LAI} = \alpha \times \rho \frac{\sum_{i=1}^{m} \sum_{j=1}^{n} \left(L_{ij} \times B_{ij} \right)}{m} \tag{1}$$

where α is the maize leaf area conversion factor with a value of 0.75, ρ is the planting density (plants m^{-2}), n is the total number of leaves (pieces) of the jth plant, and m is the number of plants measured.

(3) Maize Height Measurement

Five maize plants were selected in each ROI area according to the five-point sampling method, and the vertical distance from the ground to the highest point of the plant was measured physically using a tape measure with the plant's leaves naturally spread. Measurements were repeated three times for each plant to take the average value. The average heights of the five maize plants in the experimental area were taken as the plant height of the maize in that area.

2.4. Comprehensive Growth Indicator Construction

Based on the importance of relative chlorophyll content, leaf area index, and plant height in the comprehensive growth monitoring indicator of summer maize, the Comprehensive growth monitoring indicator (CGMI) of maize was constructed. For this reason, in this study, the CRITIC weighting method was used in this study to determine the weights of the three indicators, and the weights of LAI, SPAD, and VH were A, B, and C. This led to the establishment of the comprehensive growth monitoring indicator CGMI$_{CR}$.

The CRITIC weighting method is an objective method of assigning weights based on the volatility of data [24]. Its basic idea lies in two indicators, which are volatility (contrast strength) and conflict (correlation) indicators. Volatility is expressed using the standard deviation. If the larger the standard deviation of the data indicates greater volatility, the higher the weight will be. Conflict is expressed using the correlation coefficient table. If the value of the correlation coefficient between the indicators is larger, it means that there is more similarity between the indicators, and the smaller the conflict, then its weight will be lower.

(1) Data Normalization

To eliminate the influence of different scales on the evaluation results, it is necessary to normalize the indicators. The normalization formula is as follows:

$$X'_{ij} = \frac{X_{ij} - X_{jmin}}{X_{jmax} - X_{jmin}} \tag{2}$$

$$U_j = \left(X'_{1j}, X'_{2j}, X'_{3j} \ldots \ldots X'_{ij} \right) \tag{3}$$

where X_{ij} denotes the value of the jth evaluation indicator for the ith sample, X_{jmin} denotes the minimum value in the jth indicator, X_{jmax} denotes the maximum value in the jth indicator, and U_j denotes the data normalized to the jth indicator.

(2) Calculation of Indicator Volatility

In the CRITIC weighting method, a larger standard deviation indicates that the indicator is more volatile and more informative, and more weight should be given to the indicator.

$$\overline{X}_j = \frac{1}{n}\sum_{i=1}^{n} X_{ij} \tag{4}$$

$$S_j = \sqrt{\frac{\sum_{i=1}^{n}\left(X_{ij} - \overline{X}_j\right)^2}{n-1}} \tag{5}$$

where $\overline{X}_j$ denotes the mean of the jth indicator, and S_j denotes the standard deviation of the jth indicator is the volatility of that indicator.

(3) Calculation of Conflicting Indicators

The conflict between indicators is expressed using the correlation coefficient; the stronger the correlation between an indicator and other indicators, the less conflict there is between the indicator and other indicators, reflecting more of the same information, and the more repetitive the content of the evaluation that can be embodied, which to a certain extent weakens the intensity of the evaluation of the indicator, and the weight assigned to the indicator should be reduced.

$$R = \frac{\sum_{j,k=1}^{P}\left(x_{ij} - \overline{x}_j\right)\left(x_{ik} - \overline{x}_k\right)}{\sqrt{\sum_{j=1}^{P}\left(x_{ij} - \overline{x}_j\right)^2 \sum_{k=1}^{P}\left(x_{ik} - \overline{x}_k\right)^2}} \tag{6}$$

$$A_j = \sum_{i=1}^{P}\left(1 - r_{ij}\right) \tag{7}$$

where R denotes the correlation matrix of the indicator, j, k denote the jth indicator and the kth indicator, r_{ij} denotes the correlation coefficient between the ith indicator and the jth indicator, and P denotes the number of indicators.

(4) Calculation of the Information Content of the Indicator

$$C_j = S_j \times A_j \tag{8}$$

(5) Calculation of Objective Weights for Indicators

$$W_j = \frac{C_j}{\sum_{j=1}^{P} C_j} \tag{9}$$

where W_j denotes the weight of the jth indicator in the whole evaluation system, and P denotes the number of indicators.

(6) Comprehensive Growth Monitoring Indicator $\mathrm{CGMI_{CR}}$

The CRITIC weight method was used to derive A, B, and C as 0.331, 0.390, and 0.279, respectively, and the comprehensive growth monitoring indicator $\mathrm{CGMI_{CR}}$ was constructed. The expression of $\mathrm{CGMI_{CR}}$ is as follows:

$$\mathrm{CGMI_{CR}} = 0.331U_1 + 0.390U_2 + 0.279U_3 \tag{10}$$

where U_1 is the normalized LAI value, U_2 is the normalized SPAD value, and U_3 is the normalized VH value.

2.5. Selection of Vegetation Indices

In the initial stage of the study, the correlation analysis between a single band and the maize growth indicator showed that the correlation between the single band and the maize growth indicator was low, which was similar to the results of previous studies [25,26].

Therefore, the vegetation index was used to predict the growth of maize. The vegetation index (VI) changes through the combination of reflectance in different bands, which weakens the interference of background and other factors on the spectral characteristics of vegetation to a certain extent, and helps to improve the accuracy of remote sensing data to express crop growth. [27]. To better reflect the information contained in CGMI, this study extracted the spectral reflectance of 20 ROI regions from the spliced raw images and constructed vegetation indices by linear or nonlinear combinations. Eight vegetation indices with high relevance and wide application were selected: Green Ratio Vegetation Index (GRVI), Green Optimal Soil-Adjusted Vegetation Index (GOSAVI), Optimal Vegetation Index (VIopt), Normalized Difference Vegetation Index (NVI), Green Difference Vegetation Index (GDVI), Ratio Vegetation Index (RVI), Normalized Difference Vegetation Index (NDVI), Vegetation Index (RVI), Green Normalized Difference Vegetation Index (GNDVI), and Canopy Chlorophyll Content Index (CCCI). The formula for calculating the vegetation indices is shown in Table 1.

Table 1. Vegetation indices and its calculation formula [20].

Vegetation Indices	Equation	References
GOSAVI	$(1 + 0.16) \times (NIR - RE)/(NIR + RE + 0.16)$	Marin D B et al. [28]
GDVI	$NIR - G$	Zhou X F et al. [29]
RVI	NIR/R	Jiang J et al. [30]
GNDVI	$(NIR - G)/(NIR + G)$	Jiang J et al. [30]
GRVI	NIR/G	Motohka T et al. [31]
VIopt	$(1 + 0.45) \times (2NIR + 1)/(R + 0.45)$	Motohka T et al. [31]
NDVI	$(NIR - R)/(NIR + R)$	Deng L et al. [32,33]
CCCI	$(NIR - RE)/(NIR + RE)$	Shu M Y et al. [34]

Note: R is the red band, G is the green band, NIR is the near-infrared band, and RE is the red-edge band.

2.6. Model Construction

In this study, kernel-based extreme learning machine based on sparrow search algorithm optimization (SSA-KELM) was used to construct a comprehensive growth monitoring model.

Kernel-based extreme learning machine (KELM) is an improved algorithm based on ELM, and combined with the kernel function, KELM guarantees good generalization ability and faster learning speed based on retaining the advantages of ELM to improve the prediction performance of the model [35]. The predictive performance of KELM is significantly influenced by the regularization coefficient C and kernel function parameter S. Inadequate parameter optimization ability and poor local search capabilities can result in convergence on local optima, leading to low prediction accuracy. The SSA-KELM model optimizes the regularization coefficient C and kernel function parameter S using the sparrow search algorithm, enhancing the model's predictive capability. To further select the regularization coefficient C and kernel function parameter S, the fitness function is designed as the training set's mean squared error (MSE):

$$MSE = \frac{1}{n}\sum_{i=1}^{n}(y - x)^2 \tag{11}$$

$$fitness = argmin\left(MSE_{pridect}\right) \tag{12}$$

where x is the estimated value, y is the measured value, and n is the number of samples.

The fitness function selects the post-training MSE. A smaller MSE indicates a higher alignment between predicted and original data. The final optimization yields the best regularization coefficient C and kernel function parameter S. Using these optimized values, the network is trained on the testing dataset [20]. The flowchart of the SSA-KELM algorithm is shown in Figure 3.

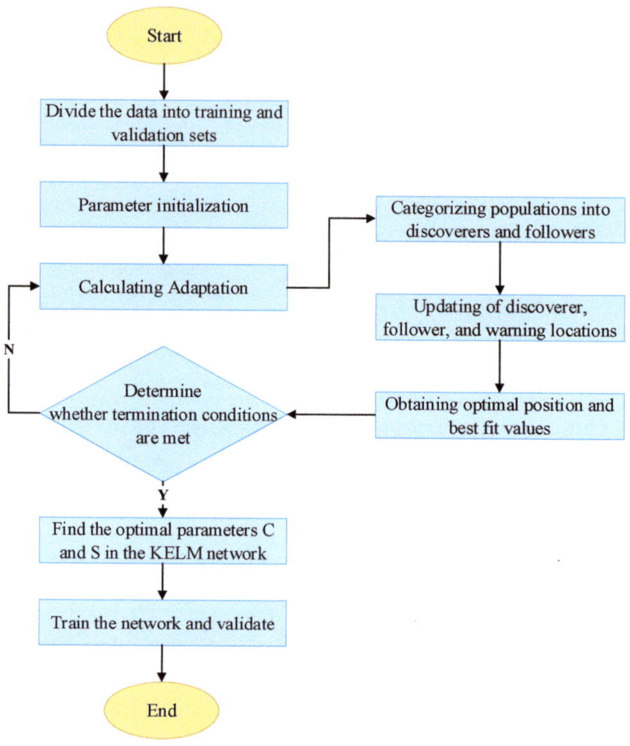

Figure 3. Flow chart of SSA-optimized KELM algorithm.

2.7. Experimental Comparison Methods

2.7.1. Coefficient of Variation Method

The coefficient of variation method was selected as the comparison method of the CRITIC weight method. The coefficient of variation method is based on the evaluation of the degree of variation of the indicator value reflected in the amount of information to determine the weight of the indicator. That is, the weight of the indicator with the changes in the value of the indicator changes is a dynamic and objective method of assigning weights, and the results of its ascertainment of the weight are not subject to the influence of the outline [36]. In this study, the coefficient of variation method was used as the comparison method of the CRITIC weight method, and the weights of LAI, SPAD, and VH were determined to be A, B, and C, respectively, and a comprehensive growth monitoring indicator CGMI$_{CV}$ was established.

(1)   The data normalization process is the same as the CRITIC weight method;
(2)   Calculation of the coefficient of variation.

$$\overline{X}_j = \frac{1}{n}\sum_{i=1}^{n} X_{ij} \tag{13}$$

$$S_j = \sqrt{\frac{\sum_{i=1}^{n}\left(X_{ij} - \overline{X}_j\right)^2}{n-1}} \tag{14}$$

$$V_j = \frac{S_j}{\overline{X}_j} \tag{15}$$

where $\overline{X}_j$ denotes the mean value of the jth indicator, S_j denotes the standard deviation (mean square error) of the jth indicator, and V_j denotes the coefficient of variation of the jth indicator.

(3) Calculation of Indicator Weights

$$W_j = \frac{V_j}{\sum_{i=1}^{P} V_j} \tag{16}$$

where W_j denotes the weight of the jth indicator in the whole evaluation system and P denotes the number of indicators.

(4) Comprehensive Growth Monitoring Indicator $CGMI_{CV}$

The coefficient of variation method was used to derive A, B, and C as 0.440, 0.202, and 0.358, respectively, and the comprehensive growth monitoring indicator $CGMI_{CV}$ was constructed, and the expression of $CGMI_{CV}$ is as follows:

$$CGMI_{CV} = 0.440U_1 + 0.202U_2 + 0.358U_3 \tag{17}$$

where U_1 is the normalized LAI value, U_2 is the normalized SPAD value, and U_3 is the normalized VH value.

2.7.2. Partial Least-Squares Regression (PLSR) Method

PLSR is a technique aimed at identifying the optimal fitting function for a dataset by minimizing the sum of squared errors. It blends statistical methods like correlation analysis, principal component analysis, and multiple linear regression, effectively addressing issues related to data covariance [37]. Xie P et al. [38] used PLSR and PSO-SVM algorithms to compare and establish a hyperspectral quantitative inversion model of soil selenium content. Fu HY et al. [39] used PLSR, SVM, RF, and other modeling methods to construct the estimation models of relative chlorophyll content, leaf area index, and leaf relative water content of ramie, respectively, and proposed a more suitable dynamic monitoring method for physical and chemical traits of ramie in the field. In comparison to multivariate linear regression algorithms, PLSR can establish a multivariate linear model when the independent variables have multicollinearity and a limited number of sample points, ensuring predictive accuracy. In contrast to principal component analysis, partial least squares not only absorb the idea of extracting information from the explanatory variables in principal component analysis but also pay attention to the problem of explaining the dependent variable by the independent variables, which is ignored in the principal component analysis [40].

2.8. Model Evaluation Methodology

In the study, three indicators, coefficient of determination (R^2), root mean square error (RMSE), and mean relative error (MRE) were used to judge the prediction effect of the model, and the specific formulas are shown below. R^2 indicates the degree of fit between the predicted value and the measured value, and RMSE reflects the degree of deviation between the predicted value and the measured value. The more R^2 tends to be close to 1, the smaller RMSE is, indicating that the model predicts better. MRE is used to describe the error between the model prediction result and the actual value to evaluate the model stability. The smaller the MRE is, the more stable the model is.

$$R^2 = \frac{\left[\sum_{i=1}^{n}(x - \overline{x})(y - \overline{y})\right]^2}{\sum_{i=1}^{n}(x - \overline{x})^2 \sum_{i=1}^{n}(y - \overline{y})^2} \tag{18}$$

$$RMSE = \sqrt{\frac{\sum_{i=1}^{n}(x - y)^2}{n}} \tag{19}$$

$$\text{MRE} = \frac{\sum_{i=1}^{n}|x - y|}{ny} \tag{20}$$

where x denotes the predicted value, $\bar{x}$ denotes the mean of the predicted value, y denotes the measured value, $\bar{y}$ denotes the mean of the measured value, and n denotes the number of samples.

3. Results and Discussion

3.1. Correlation Analysis of Different Vegetation Indices with Combined Growth Indicators and Single Growth Indicators

The Pearson significance test was chosen to correlate the composite growth monitoring indicators CGMI_{CV}, CGMI_{CR}, and the individual growth indicators that make up the composite growth monitoring indicators with the eight vegetation indices in turn. The correlations are shown in Table 2.

Table 2. The correlation of vegetation indices with growth indicators.

Vegetation Indices	LAI	SPAD	VH	CGMI_{CV}	CGMI_{CR}
GRVI	0.630 **	0.540 *	0.701 **	0.552 *	0.656 **
GOSAVI	0.652 **	0.567 **	0.704 **	0.575 **	0.674 **
VIopt	0.648 **	0.564 **	0.707 **	0.573 **	0.673 **
NDVI	0.776 **	0.590 *	0.735 **	0.605 **	0.704 **
GDVI	0.608 **	0.536 *	0.654 **	0.528 *	0.626 **
RVI	0.589 **	0.517 *	0.663 **	0.514 *	0.616 **
GNDVI	0.687 **	0.589 **	0.798 **	0.612 **	0.714 **
CCCI	0.630 **	0.601 **	0.743 **	0.616 **	0.710 **

Note: ** indicates a significant correlation at the 0.01 level; * indicates a significant correlation at the 0.05 level.

All the growth indicators showed a positive correlation with each vegetation index. Comparing and analyzing the single growth indicators, the correlation between the growth indicator VH and the vegetation index was the strongest. All of them were significantly correlated at the 0.01 level, and the correlation coefficients r were all greater than 0.6. The correlation between the SPAD indicator and the eight vegetation indices was generally low, and the correlation with the vegetation index CCCI was the highest, reaching 0.601.

In constructing the comprehensive growth monitoring indicators, the correlation between the comprehensive growth monitoring indicators CGMI_{CV} and CGMI_{CR} and the vegetation index was generally lower than that of the VH indicators due to the simultaneous consideration of the three single growth indicators and the assignment of weights to each of them, with the lower correlation between the SPAD indicators and the vegetation index. For the composite growth monitoring indicator CGMI_{CV}, the vegetation indices GOSAVI, VIopt, NDVI, GNDVI, CCCI, and CGMI_{CV} were significantly correlated at the 0.01 level, whereas GRVI, GDVI, RVI, and CGMI_{CV} were significantly correlated at the 0.05 level, and the correlation coefficients r were all greater than 0.5. For the comprehensive growth monitoring indicator CGMI_{CR}, the correlation coefficients between the eight selected vegetation indices and CGMI_{CR} were all greater than 0.6, which is a strong correlation, and all of them were significantly correlated at the 0.01 level.

3.2. Construction and Testing of the CGMI Model for Comprehensive Growth Monitoring Indicators

3.2.1. Construction and Testing of the CGMI_{CV} Monitoring Model

The PLSR and SSA-KELM algorithms combined with eight vegetation indices were used for model construction and prediction analysis of the comprehensive growth monitoring indicators CGMI_{CV} and CGMI_{CR}. The effect of the prediction model of CGMI_{CV} built based on PLSR and SSA-KELM is shown in Table 3, and the linear fitting comparison between its predicted and measured values is shown in Figure 4.

Table 3. Comparison of results from CGMI$_{CV}$ prediction models based on PLSR and SSA-KELM.

Algorithm Model	Training		Validate	
	R^2	RMSE	R^2	RMSE
PLSR	0.820	0.043	0.816	0.048
SSA-KELM	0.865	0.040	0.871	0.045

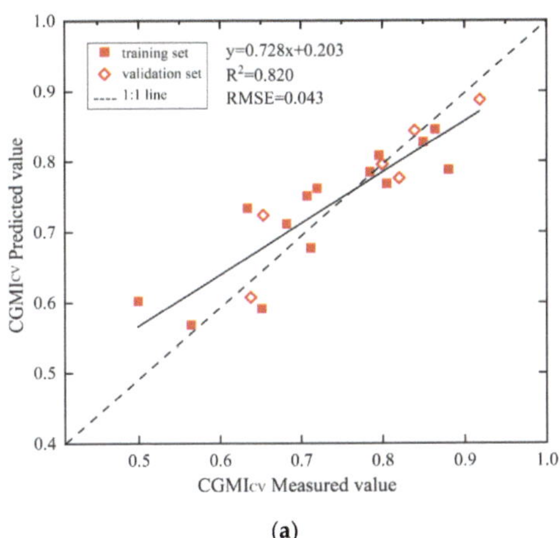

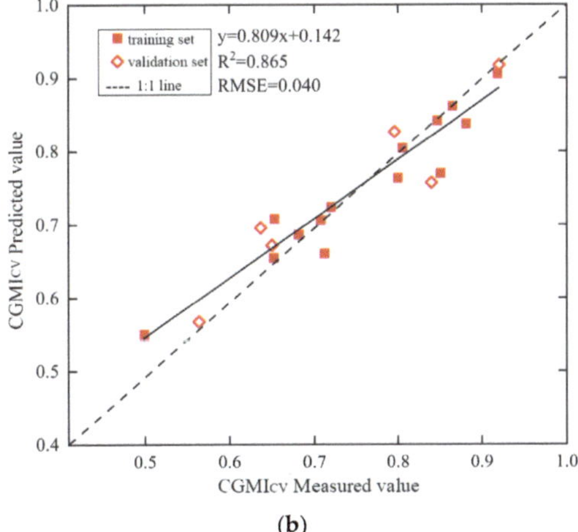

Figure 4. Relationship between predicted and measured values of the CGMI$_{CV}$ models based on PLSR and SSA-KELM: (**a**) PLSR-CGMI$_{CV}$ predictive model; (**b**) SSA-KELM-CGMI$_{CV}$ predictive model.

For the PLSR model, the R^2 of the training set was 0.820, the RMSE was 0.043, the validation set decreased by 0.004, the RMSE increased by 0.005 from the prediction set R^2, and the prediction effect decreased slightly. For the SSA-KELM model, the training set has an R^2 of 0.865 and an RMSE of 0.040, and the validation set has an increase of 0.006 and an increase of 0.005 in the RMSE over the prediction set, with a small increase in accuracy but an increased in prediction error. Overall, the SSA-KELM model increased the R^2 of the training set by 0.045 and decreased the RMSE by 0.003 compared to the PLSR model, while the validation set increased the R^2 by 0.055 and decreased the RMSE by 0.003. The prediction of both the training set and validation set of the SSA-KELM model is better than that of the PLSR model.

3.2.2. Construction and Testing of the CGMI$_{CR}$ Monitoring Model

The prediction results of the CGMI$_{CR}$ prediction model based on PLSR and SSA-KELM are shown in Table 4, and the linear fit comparison of its predicted and measured values is shown in Figure 5.

Table 4. Comparison of results from CGMI$_{CR}$ prediction models based on PLSR and SSA-KELM.

Algorithm Model	Training		Validate	
	R^2	RMSE	R^2	RMSE
PLSR	0.839	0.084	0.840	0.084
SSA-KELM	0.885	0.056	0.889	0.058

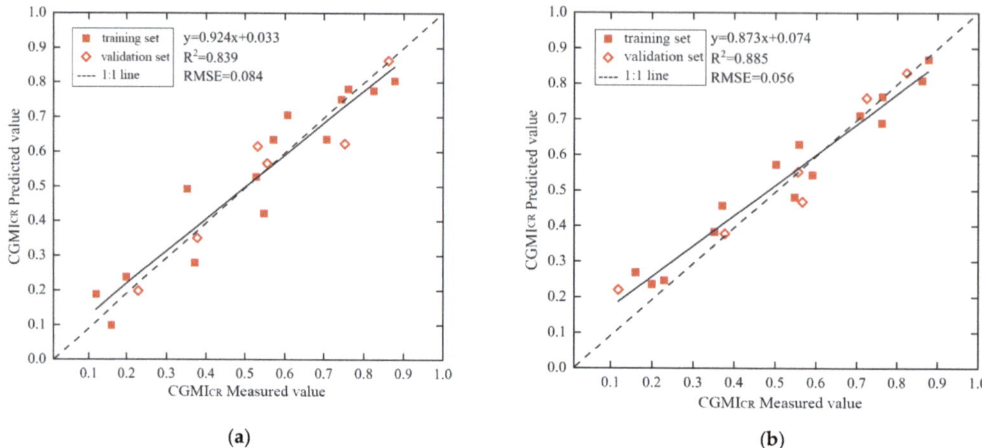

Figure 5. Relationship between predicted and measured values of the $CGMI_{CR}$ models based on PLSR and SSA-KELM: (**a**) PLSR-$CGMI_{CR}$ predictive model; (**b**) SSA-KELM-$CGMI_{CR}$ predictive model.

For the PLSR model, the RMSE was 0.084 for both the training and validation sets, and the validation set had an increase of 0.001 in R^2 over the training set, with essentially the same prediction effect. For the SSA-KELM model, the R^2 of the training set was 0.885, the RMSE was 0.056, and the validation set had an increase of 0.004 in R^2 and an increase of 0.002 in RMSE compared to the training set, with a decreased in the fitting effect but a smaller overall difference. In summary, the model built by SSA-KELM increased R^2 by 0.046 and decreased RMSE by 0.028 for the training set and increased R^2 by 0.049 and decreased RMSE by 0.026 for the validation set compared to the PLSR model. The prediction effect of both training and validation sets of the SSA-KELM model is better than the PLSR model.

Comparing and analyzing the comprehensive growth monitoring indicators $CGMI_{CV}$ and $CGMI_{CR}$, the prediction effect of the $CGMI_{CR}$ model is better than that of the $CGMI_{CV}$ model because of the CRITIC weighting method, compared with the coefficient of variation method, not only takes into account the fluctuation of each evaluation index, but also involves the conflict between different indexes, and it can more realistically reflect the crop's growth situation.

3.3. Stability Analysis of the Comprehensive Growth Monitoring Indicator Model

When constructing a nonlinear model, the presence of multicollinearity among independent variables can significantly undermine the reliability of model testing and yield unstable analytical outcomes. To assess the stability of the CGMI monitoring model, training and validation sets mirroring the modeling proportions were meticulously chosen. Subsequently, the $CGMI_{CV}$ and $CGMI_{CR}$ regression models, constructed using PLSR and SSA-KELM methodologies, respectively, underwent rigorous testing in 100 random runs. The average mean relative error (MRE) across the 100 prediction results were calculated. The results are shown in Figure 6.

The fluctuation range of the average relative errors of the $CGMI_{CV}$ models constructed by different algorithms varies less, and the fluctuation range of the average relative error of the PLSR-$CGMI_{CV}$ model is from 8.2% to 24.0%, with a median of 17.4% and a mean of 17.3%. The fluctuation range of the average relative error of the SSA-KELM-$CGMI_{CV}$ model is from 6.5% to 22.1%, with a median of 12.7% and a mean of 12.9%. The overall average relative error is smaller than PLSR-$CGMI_{CV}$, which indicates that the SSA-KELM-$CGMI_{CV}$ model is more stable.

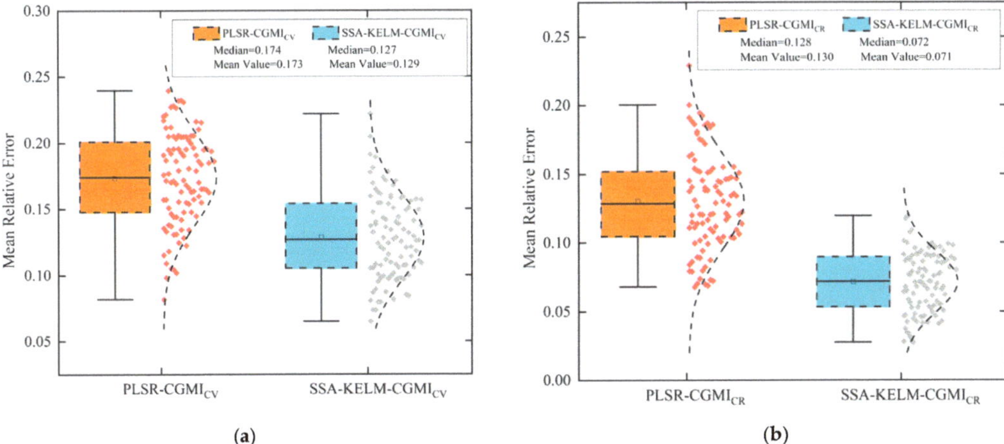

Figure 6. Predictive effects of CGMI models constructed based on PLSR and SSA-KELM: (**a**) prediction effects of $CGMI_{CV}$ models; (**b**) prediction effects of $CGMI_{CR}$ models.

The PLSR-$CGMI_{CR}$ model's average relative error fluctuates from 6.8% to 20.0%, with a median of 12.8% and a mean of 13.0%. The SSA-KELM-$CGMI_{CR}$ model's average relative error fluctuates from 2.7% to 11.9%, which is reduced compared to that of PLSR-$CGMI_{CR}$, with a median of 7.2% and a mean of 7.1%, which likewise indicates the greater stability of modeling using SSA-KELM.

3.4. Discussions

3.4.1. Comparative Analysis of the Predictive Effects of Different Growth Indicators

To verify the effect of the prediction model for the comprehensive growth monitoring indicators, the prediction model was constructed using PLSR and SSA-KELM methods for leaf area index (LAI), relative chlorophyll content (SPAD), and plant height (VH) sequentially, and the results of the model prediction are shown in Table 5.

Table 5. Comparison of all growth indicator prediction models based on PLSR and SSA-KELM.

Growth Indicator	PLSR	SSA-KELM
LAI	0.821	0.906
SPAD	0.723	0.837
VH	0.834	0.902
$CGMI_{CV}$	0.820	0.865
$CGMI_{CR}$	0.839	0.885

For all growth indicators, the R^2 of the SSA-KELM model is greater than that of the PLSR model. Among all the models constructed using a single growth indicator, the SSA-KELM-LAI model had the best prediction effect, with an R^2 of 0.906, which is 8.5% higher than that of PLSR-LAI. The R^2 of the SSA-KELM-VH model is 0.902, which is 6.8% higher than that predicted by PLSR-VH. The accuracy of the SSA-KELM-SPAD model is low, and the R^2 is 0837, which is 10.5% higher than that predicted by PLSR-VH. Among all the models constructed using the comprehensive growth monitoring indicators, the $CGMI_{CR}$ monitoring model was better than the $CGMI_{CV}$ monitoring model, with SSA-KELM-$CGMI_{CR}$ achieving the best monitoring effect ($R^2 = 0.885$). The reason is that when constructing the comprehensive growth monitoring indicators, the CRITIC weighting method, compared with the coefficient of variation method, not only takes into account the influence of the volatility of the evaluation indicators on the weights but also involves the conflicting nature between different indicators. Therefore, the weights of LAI and

VH are appropriately reduced, and the weight of SPAD is increased, which improved the prediction effect of the constructed $CGMI_{CR}$ model compared with that of $CGMI_{CV}$. However, the sensitivity of SPAD to multispectral information is relatively weak in this stage, and increasing the weights of SPAD also weakened the monitoring ability of $CGMI_{CR}$.

3.4.2. Stability Analysis of Predictive Models for Different Growth Indicators

The PLSR and SSA-KELM methods were used to test the regression model of a single growth indicator for 100 random runs in sequence, and the average relative error (MRE) of the 100 prediction results was counted and analyzed in comparison with the stability of the CGMI monitoring model. Since the modeling effect of all indicators using the PLSR algorithm is lower than the modeling effect of the SSA-KELM algorithm, only the stability of SSA-KELM algorithm modeling is discussed this time. The stability results of the predictive models constructed based on SSA-KELM for all the growth indicators are shown in Table 6.

Table 6. Stability comparison of all growth indicator prediction models constructed based on SSA-KELM.

Growth Indicator	Error Fluctuation Range	Median Error	Mean Value of Error
$CGMI_{CV}$	6.5~22.1%	12.7%	12.9%
$CGMI_{CR}$	2.7~11.9%	7.2%	7.1%
LAI	9.7~42.5%	22%	23.4%
SPAD	1.3~5.9%	3.9%	3.8%
VH	4.8~22.2%	14.8%	14.6%

Compared with the monitoring models of single growth indicators constituting CGMI, the stability of the $CGMI_{CV}$ and $CGMI_{CR}$ models is smaller than that of the SPAD model, mainly because the error fluctuation range of the LAI indicator is large, which leads to the larger error fluctuation range of $CGMI_{CV}$ and $CGMI_{CR}$. The stability of the $CGMI_{CV}$ and $CGMI_{CR}$ models was greater than that of the LAI and VH models because a single growth indicator has limitations and can only simply reflect a certain physiological information of the crop, which is essentially a 'point' of data and cannot uniformly reflect the overall characteristics of the crop. Zhai LT et al. [41] inverted the nitrogen content, chlorophyll content, and water content of winter wheat based on the PLSR model, the R^2 of single index inversion was 0.72, 0.31, 0.61, respectively, and the R^2 of comprehensive index inversion was 0.75. The results showed that the inversion effect of the comprehensive index model was better than that of a single index, and the results of this study were similar to the present study. In summary, compared with the single growth indicator, the comprehensive growth monitoring indicator has a better prediction effect. Compared with the comprehensive growth monitoring indicator $CGMI_{CV}$ and $CGMI_{CR}$, the error fluctuation range, median error and mean value of the error of the $CGMI_{CR}$ model are smaller than those of the $CGMI_{CV}$ model, which indicates that the comprehensive growth monitoring indicator $CGMI_{CR}$ has better stability.

However, this experiment only discussed the growth of maize at the jointing stage, which is a more vigorous stage of maize growth and development, and also the stage of low crop cover, which will have a certain impact on the results of the experiment. As the growth cycle progresses, further research and analysis are needed to monitor the growth of maize plants during different growth stages, such as the staminate stage, silking stage, and grouting stage after the jointing stage, to achieve more accurate monitoring of maize plant growth.

4. Conclusions

In the study, modeling prediction analysis of comprehensive summer maize growth was conducted based on UAV multispectral image data, and the main conclusions are as follows:

(1) The comprehensive growth monitoring indicators $CGMI_{CV}$ and $CGMI_{CR}$ constructed using the coefficient of variation method and the CRITIC weighting method were both positively correlated with the vegetation index. The correlation coefficients between $CGMI_{CR}$ and the vegetation index were both greater than the correlation coefficients between $CGMI_{CV}$ and $CGMI_{CR}$.

(2) A comprehensive growth monitoring indicator model based on SSA-KELM was established. For $CGMI_{CV}$, the model R^2 was 0.871, and the RMSE was 0.045; for $CGMI_{CR}$, the model R^2 was 0.889, and the RMSE was 0.058, which shows that the model built by SSA-KELM-$CGMI_{CR}$ is more stable and more effective.

(3) Based on the $CGMI_{CV}$ and $CGMI_{CR}$ monitoring models built by PLSR, the monitoring effect of the PLSR-$CGMI_{CV}$ model is lower than that of the PLSR-$CGMI_{CR}$ model, and both of them are lower than that of the model constructed by SSA-KELM. In summary, the model constructed based on SSA-KELM has a better monitoring effect.

Author Contributions: Conceptualization, H.M., J.J. and H.C.; Methodology, H.M., X.L., H.C. and N.L.; Software, X.L. and H.C.; Validation, H.M., X.L. and H.C.; Formal analysis, H.M., X.L. and H.C.; Investigation, H.M. and X.L.; Resources, H.M., X.L. and N.L.; Data curation, H.M. and X.L.; Writing—Original Draft Preparation, H.M.; Writing—Review and Editing, H.M., X.L. and H.C.; Visualization, H.M.; Supervision, Y.S. and C.Y.; Project Administration, J.J., H.C. and Y.S.; Funding Acquisition, J.J. and H.C. All authors have read and agreed to the published version of the manuscript.

Funding: Major Science and Technology Project of Henan Province (No. 221100110800), Longmen Laboratory Major Projects (No. 231100220200); Postgraduate Education Reform Project of Henan Province (No. 2021SJGLX138Y). The present research was supported by the 2020 Training Plan for Young Backbone Teachers in Colleges and Universities of Henan Province (No. 2020GGJS075); Henan Provincial University Science and Technology Innovation Talent Support Program (No. 23HASTIT020); Henan Provincial Science and Technology Research Project (No. 222102110067).

Data Availability Statement: The data that support the findings of this study are available on request from the corresponding author.

Conflicts of Interest: Author Nana Li was employed by the company Caterpillar Mechanical Components. The remaining authors declare that the research was conducted in the absence of any commercial or financial relationships that could be construed as a potential conflict of interest.

References

1. Yu, Y.; Jiang, Z.H.; Wang, G.J.; Kattel, G.R.; Chuai, X.W.; Shang, Y.; Zou, Y.F.; Miao, L.J. Disintegrating the impact of climate change on maize yield from human management practices in China. *Agric. For. Meteorol.* **2022**, *327*, 109235. [CrossRef]

2. Sapkota, S.; Paudyal, D.R. Growth Monitoring and Yield Estimation of Maize Plant Using Unmanned Aerial Vehicle (UAV) in a Hilly Region. *Sensors* **2023**, *23*, 5432. [CrossRef] [PubMed]

3. Wang, X.Y.; Yang, H.; Li, X.X.; Zheng, Y.J.; Yan, H.J.; Li, N. Corn growth monitoring based on visible spectrum remote sensing by unmanned aerial vehicle (UAV). *Spectrosc. Spectr. Anal.* **2021**, *41*, 265–270.

4. Wu, B.F.; Zhang, F.; Liu, C.L.; Zhang, L.; Luo, Z.M. An integrated method for crop condition monitoring. *J. Remote Sens.* **2004**, *8*, 498–514.

5. Li, Z.N. Study on indicators for remote sensing monitoring of winter wheat growth. *Chin. Acad. Agric. Sci.* **2010**, *33*, 74–82.

6. Meng, J.H. Study on Indicators for Remote Sensing Monitoring of Crop Growth. Ph.D. Thesis, Graduate School of the Chinese Academy of Sciences (Institute of Remote Sensing Applications), Beijing, China, 2006; pp. 1–12.

7. Zhao, J.; Pan, F.J.; Xiao, X.; Hu, L.B.; Wang, X.L.; Yan, Y.; Zhang, S.L.; Tian, B.Q.; Yu, H.L.; Lan, Y.B. Summer Maize Growth Estimation Based on Near-Surface Multi-Source Data. *Agronomy* **2023**, *13*, 532. [CrossRef]

8. He, J.; Wang, L.G.; Guo, Y.; Zhang, Y.; Yang, X.Z.; Liu, T.; Zhang, H.L. Estimation of maize LAI based on multi-spectral remote sensing by unmanned aerial vehicle (UAV). *J. Big Data Agric.* **2021**, *3*, 20–28.

9. Ma, J.W.; Wang, L.J.; Chen, P.F. Comparing Different Methods for Wheat LAI Inversion Based on Hyperspectral Data. *Agriculture* **2022**, *12*, 1353. [CrossRef]

10. Hang, Y.H.; Su, H.; Yu, Z.Y.; Liu, H.J.; Guan, H.X.; Kong, F.C. Estimation of Rice Leaf Area Index Combining Spectral and Textural Characteristics and Coverage of Unmanned Aerial Vehicles. *J. Agric. Eng.* **2021**, *37*, 64–71.

11. Tao, H.; Feng, H.; Xu, L.; Miao, M.; Long, H.; Yue, J.; Li, Z.; Yang, G.; Yang, X.; Fan, L. Estimation of crop growth parameters using UAV-based hyperspectral remote sensing data. *Sensors* **2020**, *20*, 1296. [CrossRef]

12. Trawczynski, C. Assessment of the nutrition of potato plants with nitrogen according to the NNI test and SPAD indicator. *J. Elem.* **2019**, *24*, 687–700. [CrossRef]

13. Qiao, L.; Tang, W.J.; Gao, D.H.; Zhao, R.M.; An, L.L.; Li, M.Z.; Sun, H.; Song, D. UAV-based chlorophyll content estimation by evaluating vegetation index responses under different crop coverages. *Comput. Electron. Agric.* **2022**, *196*, 106775. [CrossRef]

14. Guo, Y.H.; Chen, S.Z.; Li, X.X.; Cunha, M.; Jayavelu, S.; Cammarano, D.; Fu, Y.S. Machine Learning-Based Approaches for Predicting SPAD Values of Maize Using Multi-Spectral Images. *Remote Sens.* **2022**, *14*, 1337. [CrossRef]

15. Gil-Docampo, M.L.; Arza-García, M.; Ortiz-Sanz, J.; Martínez-Rodríguez, S.; Marcos-Robles, J.L.; Sánchez-Sastre, L.F. Above-ground biomass estimation of arable crops using UAV-based SfM photogrammetry. *Geocarto International.* **2020**, *35*, 687–699. [CrossRef]

16. Belton, D.; Helmholz, P.; Long, J.; Zerihun, A. Crop Height Monitoring Using a Consumer-Grade Camera and UAV Technology. *PFG-J. Photogramm. Remote Sens. Geoinf. Sci.* **2019**, *87*, 249–262. [CrossRef]

17. Bendig, J.; Yu, K.; Aasen, H.; Bolten, A.; Bennertz, S.; Broscheit, J.; Gnyp, M.L.; Bareth, G. Combining UAV-based plant height from crop surface models, visible, and near-infrared vegetation indices for biomass monitoring in barley. *Int. J. Appl. Earth Obs. Geoinf.* **2015**, *39*, 79–87. [CrossRef]

18. Xu, Y.F. Parameter Inversion and Integrated Growth Monitoring of Winter Wheat Based on Multi-Spectral Remote Sensing by Unmanned Aerial Vehicle (UAV). Master's Thesis, Anhui University of Science and Technology, Huainan, China, 2022; pp. 39–52.

19. Pei, H.J.; Feng, H.K.; Li, C.C.; Jin, X.L.; Li, Z.H.; Yang, G.J. Unmanned aerial remote sensing monitoring of winter wheat growth based on comprehensive indicators. *J. Agric. Eng.* **2017**, *33*, 74–82.

20. Ji, J.T.; Li, N.N.; Cui, H.W.; Li, Y.C.; Zhao, X.B.; Zhang, H.L.; Ma, H. Study on Monitoring SPAD Values for Multispatial Spatial Vertical Scales of Summer Maize Based on UAV Multispectral Remote Sensing. *Agriculture* **2023**, *13*, 1004. [CrossRef]

21. Brewer, K.; Clulow, A.; Sibanda, M.; Gokool, S.; Naiken, V.; Mabhaudhi, T. Predicting the Chlorophyll Content of Maize over Phenotyping as a Proxy for Crop Health in Smallholder Farming Systems. *Remote Sens.* **2022**, *14*, 518. [CrossRef]

22. Zhang, S.M.; Zhao, G.X.; Lang, K.; Su, B.W.; Chen, X.N.; Xi, X.; Zhang, H.B. Integrated Satellite, Unmanned Aerial Vehicle (UAV) and Ground Inversion of the SPAD of Winter Wheat in the Reviving Stage. *Sensors* **2019**, *19*, 1485. [CrossRef]

23. Zhang, X.; Tao, S.Y.; Zhang, M. Inversion of leaf area index of spring maize from FY-3B satellite based on LSTM algorithm. *J. Jilin Univ.* **2022**, *52*, 2071–2080.

24. Zhang, L.J.; Zhang, X. Weighted clustering method based on improved CRITIC method. *Stat. Decis.-Mak.* **2015**, *22*, 65–68.

25. Zhou, Q.; Wang, J.J.; Huo, Z.Y.; Liu, C.; Wang, W.; Ding, L. UAV multispectral remote sensing estimation of SPAD values in wheat canopy at different growth stages. *Spectrosc. Spectr. Anal.* **2023**, *43*, 1912–1920.

26. Huang, W.; Guan, Q.; Luo, J.; Zhang, J.; Zhao, J.; Liang, D.; Huang, L.; Zhang, D. New optimized spectral indices for identifying and monitoring winter wheat diseases. *IEEE J. Sel. Top. Appl. Earth Obs. Remote Sens.* **2014**, *7*, 2516–2524. [CrossRef]

27. Hirooka, Y.; Homma, K.; Shiraiwa, T. Parameterization of the vertical distribution of leaf area index (LAI) in rice (*Oryza sativa* L.) using a plant canopy analyzer. *Sci. Rep.* **2018**, *8*, 6387. [CrossRef]

28. Marin, D.B.; Ferraz, G.A.E.S.; Guimarães, P.H.S.; Schwerz, F.; Santana, L.S.; Barbosa, B.D.S.; Barata, R.A.P.; Faria, R.D.; Dias, J.E.L.; Conti, L.; et al. Remotely Piloted Aircraft and Random Forest in the Evaluation of the Spatial Variability of Foliar Nitrogen in Coffee Crop. *Remote Sens.* **2021**, *13*, 1471. [CrossRef]

29. Zhou, X.F.; Zhang, J.C.; Chen, D.M.; Huang, Y.B.; Kong, W.P.; Yuan, L.; Ye, H.C.; Huang, W.J. Assessment of leaf chlorophyll content models for winter wheat using Landsat-8 multispectral remote sensing data. *Remote Sens.* **2020**, *12*, 2574. [CrossRef]

30. Jiang, J.; Johansen, K.; Stanschewski, C.S.; Wellman, G.; Mousa, M.A.A.; Fiene, G.M.; Asiry, K.A.; Tester, M.; McCabe, M.F. Phenotyping a diversity panel of quinoa using UAV-retrieved leaf area index, SPAD-based chlorophyll and a random forest approach. *Precis. Agric.* **2022**, *23*, 961–983. [CrossRef]

31. Motohka, T.; Nasahara, K.N.; Oguma, H.; Tsuchida, S. Applicability of Green-Red Vegetation Index for Remote Sensing of Vegetation Phenology. *Remote Sens.* **2010**, *2*, 2369–2387. [CrossRef]

32. Deng, L.; Mao, Z.H.; Li, X.J.; Hu, Z.W.; Duan, F.Z.; Yan, Y.N. UAV-based multispectral remote sensing for precision agriculture: A comparison between different cameras. *ISPRS J. Photogramm. Remote Sens.* **2018**, *146*, 124–136. [CrossRef]

33. Zhang, L.; Zhang, H.; Niu, Y.; Han, W. Mapping maize water stress based on UAV multispectral remote sensing. *Remote Sens.* **2019**, *11*, 605. [CrossRef]

34. Shu, M.Y.; Fei, S.P.; Zhang, B.Y.; Yang, X.H.; Guo, Y.; Li, B.G.; Ma, Y.T. Application of UAV Multisensor Data and Ensemble Approach for High-Throughput Estimation of Maize Phenotyping Traits. *Plant Phenomics* **2022**, *2022*, 11. [CrossRef]

35. Zhao, L.; Zhao, X.; Li, Y.; Shi, Y.; Zhou, H.; Li, X.; Wang, X.; Xing, X. Applicability of hybrid bionic optimization models with kernel-based extreme learning machine algorithm for predicting daily reference evapotranspiration: A case study in arid and semiarid regions, China. *Environ. Sci. Pollut. Res. Int.* **2022**, *30*, 22396–22412. [CrossRef] [PubMed]

36. Tao, Z.F.; Ge, L.L.; Chen, H.Y. A class of non-negative variable weight combination prediction methods based on sliding window. *Control. Decis.-Mak.* **2020**, *35*, 1446–1452.

37. Wu, Q.; Sun, H.; Li, M.Z.; Song, Y.Y.; Zhang, Y.E. Research on accurate segmentation and chlorophyll diagnosis method for multispectral images of maize crops. *Spectrosc. Spectr. Anal.* **2015**, *35*, 178–183.

38. Xie, P.; Wang, Z.H.; Xiao, B.; Cao, H.-l.; Huang, Y.; Su, W.-l. Hyperspectral quantitative inversion of soil selenium content based on sCARS-PSO-SVM. *Spectrosc. Spectr. Anal.* **2023**, *43*, 3599–3606.

39. Fu, H.Y.; Wang, W.; Lu, J.N.; Yue, Y.K.; Cui, G.X.; Yu, W. Estimation of ramie physicochemical property based on UAV multispectral remote sensing and machine learning. *Trans. Chin. Soc. Agric. Mach.* **2023**, *54*, 194–200.

40. Liu, T.; Zhang, H.; Wang, Z.Y.; He, C.; Zhang, Q.G.; Jiao, Y.Z. Estimation of leaf area index and chlorophyll content in wheat using unmanned aerial vehicle multispectral estimation. *J. Agric. Eng.* **2021**, *37*, 65–72.
41. Zhai, L.T.; Wei, F.Y.; Feng, H.K.; Li, C.C.; Yang, G.J. Winter wheat growth monitoring based on integrated indicators. *Jiangsu Agric. Sci.* **2020**, *48*, 244–249.

agronomy

MDPI

Article

Estimation of Soil Moisture Content Based on Fractional Differential and Optimal Spectral Index

Wangyang Li, Youzhen Xiang *, Xiaochi Liu, Zijun Tang, Xin Wang, Xiangyang Huang, Hongzhao Shi, Mingjie Chen, Yujie Duan, Liaoyuan Ma, Shiyun Wang, Yifang Zhao, Zhijun Li and Fucang Zhang

The Key Laboratory of Agricultural Soil and Water Engineering in Arid Areas, Ministry of China, Northwest A&F University, Xianyang 712100, China; lwy1222@nwafu.edu.cn (W.L.); 2023055903@nwsuaf.edu.cn (X.L.); tangzijun@nwsuaf.edu.cn (Z.T.); wx123@nwsuaf.edu.cn (X.W.); 2023055900@nwsuaf.edu.cn (X.H.); shihongzhao7@nwafu.edu.cn (H.S.); 2023012631@nwafu.edu.cn (M.C.); 2023012614@nwafu.edu.cn (Y.D.); 2023012608@nwafu.edu.cn (L.M.); wsydyxhs2023@nwafu.edu.cn (S.W.); 2023012616zyf@nwafu.edu.cn (Y.Z.); lizhij@nwsuaf.edu.cn (Z.L.); zhangfc@nwsuaf.edu.cn (F.Z.)
* Correspondence: youzhenxiang@nwsuaf.edu.cn

Abstract: Applying hyperspectral remote sensing technology to the prediction of soil moisture content (SMC) during the growth stage of soybean emerges as an effective approach, imperative for advancing the development of modern precision agriculture. This investigation focuses on SMC during the flowering stage under varying nitrogen application levels and film mulching treatments. The soybean canopy's original hyperspectral data, acquired at the flowering stage, underwent 0–2-order differential transformation (with a step size of 0.5). Five spectral indices exhibiting the highest correlation with SMC were identified as optimal inputs. Three machine learning methods, namely support vector machine (SVM), random forest (RF), and back propagation neural network (BPNN), were employed to formulate the SMC prediction model. The results indicate the following: (1) The correlation between the optimal spectral index of each order, obtained after fractional differential transformation, and SMC significantly improved compared to the original hyperspectral reflectance data. The average correlation coefficient between each spectral index and SMC under the 1.5-order treatment was 0.380% higher than that of the original spectral index, with mNDI showing the highest correlation coefficient at 0.766. (2) In instances of utilizing the same modeling method with different input variables, the SMC prediction model's accuracy follows the order: 1.5 order > 2.0 order > 1.0 order > 0.5 order > original order. Conversely, with consistent input variables and a change in the modeling method, the accuracy order becomes RF > SVM > BPNN. When comprehensively assessing model evaluation indicators, the 1.5-order differential method and RF method emerge as the preferred order differential method and model construction method, respectively. The R^2 for the optimal SMC estimation model in the modeling set and validation set were 0.912 and 0.792, RMSEs were 0.005 and 0.004, and MREs were 2.390% and 2.380%, respectively. This study lays the groundwork for future applications of hyperspectral remote sensing technology in developing soil moisture content estimation models for various crop growth stages and sparks discussions on enhancing the accuracy of these different soil moisture content estimation models.

Keywords: soybean; hyperspectrum; fractional order differentiation; optimal spectral index; soil moisture content

Citation: Li, W.; Xiang, Y.; Liu, X.; Tang, Z.; Wang, X.; Huang, X.; Shi, H.; Chen, M.; Duan, Y.; Ma, L.; et al. Estimation of Soil Moisture Content Based on Fractional Differential and Optimal Spectral Index. *Agronomy* **2024**, *14*, 184. https://doi.org/10.3390/agronomy14010184

Academic Editor: Peng Fu

Received: 14 December 2023
Revised: 31 December 2023
Accepted: 11 January 2024
Published: 15 January 2024

1. Introduction

Soybean is one of the most essential oil products in the world agricultural trade [1]. China, a significant soybean producer, has consistently ranked fourth globally in total output in recent years [2]. In the past, China was one of the major consumers of soybean; China used to purchase around 62% of the soybean that was traded internationally, and only 14.3% of its soybean consumption was self-sufficient. The self-sufficiency rate of soybean is less than 15%, which strains the global food supply [3]. The production and quality of soybeans hold a crucial strategic role in constructing a modern product with distinct Chinese characteristics. Soybean output and quality directly impact China's food security level [4]. The developmental condition of soybeans during the flowering stage significantly affects their subsequent reproductive growth and ultimate yield formation. A soil moisture content (SMC) that is too high or too low will directly affect the flowering quality of soybean, thus affecting the yield and quality of soybean [5,6]. A scientific and efficient acquisition of SMC is significant for growth status evaluation and yield prediction of soybean at the flowering stage [7].

SMC plays a crucial role in the creation, alteration, and utilization of surface water resources [8]. SMC stands as a crucial indicator for assessing crop soil drought conditions, while also serving as a pivotal determinant of crop yield and overall crop quality [9]. Monitoring soil moisture holds substantial significance in advancing water-efficient irrigation practices and optimizing water resource utilization [5]. Many conventional drying methods of determining SMC persist, as the drying method is characterized by its high precision and accuracy [10]. Notably, it has been formally adopted as both a national and international standard. Nevertheless, in situ methodologies for measuring SMC face difficulties in consistently attaining continuous observations and require considerable time and labor [11]. Zai Songmei's research has revealed the feasibility of using a spectrophotometer to determine soil moisture content, providing evidence for the rapid assessment of soil moisture content through soil spectral characteristics [12]. Cao Qi's study found that the radar ground wave method can accurately invert soil volumetric water content, but inversion accuracy is influenced by land use types [13]. Gamma-ray transmission methods have been precisely employed in the study of agricultural soil properties [14]. A study by Medhat discovered that the gamma-ray transmission method, utilizing a portable cadmium telluride detector, offers practical, cost-effective, nondestructive, and rapid analytical advantages in measuring soil density and volumetric water content. However, it places high demands on instrument accuracy [15,16].

Researchers have utilized remote sensing data to explore the intricate relationship between soil spectral properties and moisture levels [17]. However, the direct evaluation of soil moisture through remote sensing technology necessitates extensive field sampling periods. Hence, the trend towards indirect SMC detection, established by relating crop reflectance to soil moisture, is on the rise. The direct link between reflectance and vegetation water content forms the cornerstone of SMC monitoring techniques [18]. Numerous studies have underscored the reflectance–water content relationship across various crops, wheat included [19–22]. Additionally, Gouvea et al. [23] probed the impact of soil water stress on physiological traits like photosynthesis, stomatal conductance, transpiration, and CO_2 levels in soybean plants. Thus, the utilization of remote sensing technology and its correlation with crop canopy reflectance provides a viable avenue for SMC monitoring.

While numerous studies have investigated the derivation of crop water status via spectral measurements, the direct assessment of SMC through crop reflectance remains relatively rare. Sobrino et al. [24] estimated soil moisture content across diverse crops and growth stages using both airborne hyperspectral scanners (AHSs) and satellite imagery. Panigrahi and Das employed ground-based hyperspectral measurements to simulate soil water potential in paddy fields during multiple growth stages [25]. However, the effectiveness of soil moisture prediction models utilizing canopy spectral data can be impeded by factors such as canopy structure, leaf area, angles, positions, shadows, and soil backgrounds [26,27]. These complexities hinder the establishment of a robust quantitative relationship between soil moisture and canopy reflectance, consequently limiting the generalizability of developed models when applied to novel agricultural regions [28].

Certain researchers have leveraged the integer differential transformation method to preprocess raw hyperspectral reflectance data, thus attenuating background noise to some degree and bolstering modeling precision [29]. Nevertheless, it is worth noting that the application of first-order, second-order, and even higher-order integer-order differential transformation techniques, while dampening background noise, has been observed to disregard the continuity and gradient of spectral information, ultimately leading to a loss of spectral characteristics [30,31]. This scenario prompts the emergence of fractional differential transform, a mathematical extension of the integer-order differential approach. This novel approach holds the capability to accentuate subtle shifts within spectral information, enhancing weaker spectral features and amplifying the signal-to-noise ratio inherent in spectral reflectance [32]. Tang et al. [33] demonstrated the estimation of soil salinity through a machine learning framework grounded in remote sensing fractional derivative. Despite these advancements, limited research has been dedicated to investigating the nuanced connection between crop canopy reflectance and soil moisture subsequent to fractional differential transformation.

This study investigates the flowering stage SMC across various nitrogen application rates and film mulching treatments. The original hyperspectral reflectance data underwent fractional differential transformation of 0–2 orders with a step size of 0.5. Employing the correlation matrix method, spectral bands exhibiting the highest correlation with SMC within the 350 to 1830 nm range were identified, leading to the construction of five sets of 25 optimal spectral indices. Building upon this foundation, a predictive model for flowering SMC was established using support vector machine (SVM), random forest (RF), and back propagation neural network (BPNN) algorithms. This study further scrutinized the influence of diverse differential orders and machine learning methods on the predictive accuracy of the SMC model. By doing so, this investigation aims to furnish a theoretical framework that contributes to a more precise and rapid determination of flowering SMC. It also provides a theoretical basis for artificial intelligence applications to quickly and accurately analyze large-scale hyperspectral satellite remote sensing information [34].

2. Materials and Methods

2.1. Research Area and Test Design

In this study, a two-year (2021–2022) soybean field experiment was conducted at the Institute of Water-Saving Agriculture (34°18′ N, 108°24′ E, 524.7 m a.s.l.) within the Key Laboratory of Agricultural Water and Soil Engineering of the Ministry of Education at Northwest A & F University (Figure 1). The mean maximum temperature for June to October 2021 was 30.3 °C, with a minimum of 20.0 °C, while in 2022, it was 31.3 °C and 21.2 °C, respectively. Precipitation during the sowing period in 2021 and 2022 was 432.6 mm and 279.5 mm, respectively. In comparison to the 30-year average rainfall of 345 mm during the soybean season (1991–2020), 2021 was categorized as a wet year, whereas 2022 was considered a dry year.

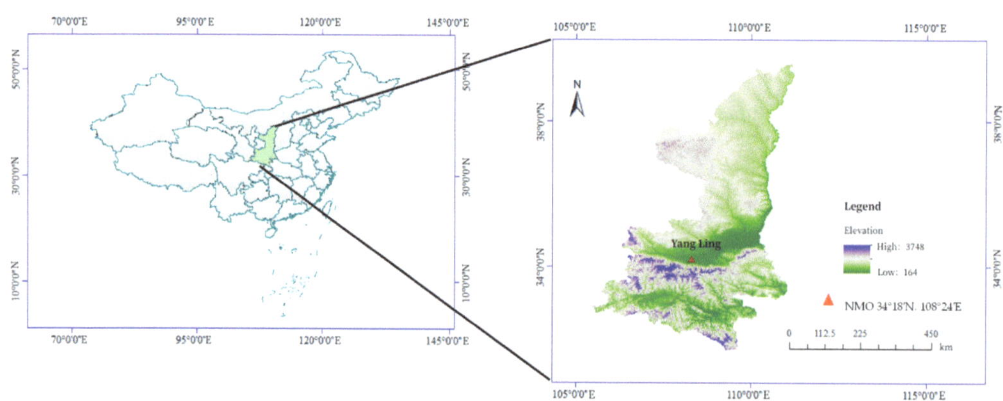

Figure 1. Study area.

In this experiment, three nitrogen application levels were set: N1: 60 kg/hm^2, N2: 120 kg/hm^2, N3: 180 kg/hm^2. Four mulching treatments were also set: NM: no mulching treatment, SM: straw mulching, FM: agricultural film mulching, SFM: straw + agricultural film mulching. The straw mulching amount was 6000 kg/hm^2. The FM and SFM treatments were carried out by ridging and covering the film side. Two ridges with a width of 40 cm and a height of 25 cm were raised in the experimental plot, and the ridge surface was covered with a 60 cm wide plastic film. Soybeans were sown at 5 cm on the film side with a row spacing of 50 cm. This study designed 12 treatments, 2 replicates, and a total of 24 plots. The experimental plot area was 2.5 m × 6 m = 15 m^2. Slow-release nitrogen fertilizer (SNF), potassium fertilizer (K$_2$O, 30 kg/hm^2), and phosphorus fertilizer (P$_2$O$_5$, 30 kg/hm^2) were applied as base fertilizer before sowing. The soybean variety used in the experiment was Shanning 17. It was sown on 17 June 2021 and 9 June 2022 and harvested on 30 September 2021 and 28 September 2022, respectively. The soybean flowering periods were 28 July 2021–24 August 2021 and 23 July 2022–20 August 2022, respectively. There were no obvious diseases and insect pests during the soybean growth period. The experimental scheme design of this study is shown in Table 1.

Table 1. Test scheme design.

SFM	NM	SFM	SM	SM	FM
N1	N3	N2	N3	N2	N1
NM	SFM	SM	FM	FM	NM
N1	N3	N1	N3	N2	N1
SFM	SFM	SFM	NM	NM	FM
N2	N1	N3	N3	N2	N3
SM	SM	SM	NM	FM	FM
N2	N1	N3	N2	N2	N1

2.2. Measurements and Methods

2.2.1. Data Acquisition

This study used the most basic drying method to measure SMC during the soybean flowering period (6 August 2021 and 10 August 2022) in the two-year experiment. At the same time, the spectral data were collected. The weather was sunny, and the light was stable. During the flowering period of soybean, soil samples were taken from the middle of two plants, at a position of 15 cm from each plant and at a horizontal position in the middle of bare land. Soil samples were taken at intervals of 20 cm in each soil layer from 0 cm to 60 cm of soil depth. The soil samples were mixed thoroughly and dried in an oven at 105 °C for 8 h to determine SMC. In this study, six sites were randomly selected in each plot during the flowering period of soybean to determine soil moisture content. The mean value of the six sampling points was the soil moisture content of the plot for a total of 24 plots, and the corresponding hyperspectral remote sensing information was obtained at the same time. A total of 48 groups of SMC and hyperspectral reflectance samples were obtained and tested in the two-year experiment (Table 2).

Table 2. Statistics of soil moisture content of soybean at flowering stage.

Statistical Indicators	Soil Moisture Content
Sample Size	48.00
Maximum Value	0.16
Minimum Value	0.11
Mean Value	0.08
Standard Deviation	0.01
Coefficient of Variation/%	12.5

This study selected the original hyperspectral information under the SFMN3, SMN3, FMN3, and NMN3 treatments and plotted its spectral characteristic curves, as shown in Figure 2.

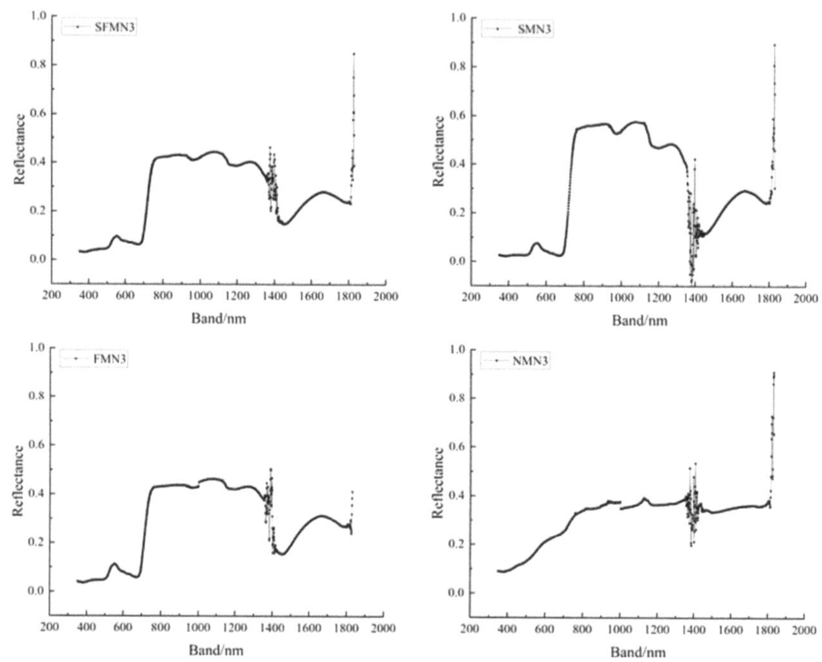

Figure 2. Hyperspectral reflectance characteristic curve graph.

2.2.2. Spectral Data Acquisition of Soybean Crown Height

This study collected the hyperspectral reflectance data of soybean canopy in the experimental area from 10:40 a.m. to 1:00 p.m. on 6 August 2021 and 10 August 2022, respectively. During this period, the light was sufficient, the spectral information was measured, and the error was small. The hyperspectral measurement instrument for the experiment was the Field Spec4 visibl/near-infrared portable ground object hyperspectral spectrometer produced by the American company ASD (Analytical Spectral Devices, Inc., Boulder, CO, USA). The instrument's band range is 350~1830 nm. The spectral resolution of 350~1000 nm is 3 nm, and the sampling interval is 1.4 nm. The 1000~1830 nm resolution is 10 nm, and the sampling interval is 2 nm. The instrument automatically interpolates the sampling data to 1 nm interval output, and the field of view is 25°.

Prior to obtaining hyperspectral data, the spectrometer underwent preheating and optimization, with the reference plate test and comparison completed within 1 min. Following the acquisition of hyperspectral reflectance data for the initial test area, reference plate correction was conducted before acquiring hyperspectral reflectance data for the subsequent test area. In each experimental plot, a crop canopy with balanced growth was selected. The testers held a spectral sensor probe and collected data vertically downwards 75 cm from the top of the crop canopy. A total of 10 hyperspectral reflectance data were collected and recorded each time. According to the '3σ' principle [33], each plot's final hyperspectral reflectance data were used as the average value of the remaining spectral bands.

2.2.3. Soil Moisture Content Data Collection

In this study, the most basic drying method was used to determine the SMC in the field. Three sites were randomly selected in 48 experimental plots to collect soil samples. After drying, the mean value was calculated as the measured value of each experimental field, and the data were collected on the day of hyperspectral reflectance data image acquisition.

2.3. Spectral Data Processing

In this study, the Savitzky–Golay (SG) smoothing method was used to denoise the spectral data. The G-L fractional differential algorithm executed a 0–2-order (step size 0.5) fractional differential transformation on the hyperspectral reflectance data. This algorithm has the capability to extend the conventional integer order differential to any order differential, enabling a more comprehensive representation of nuanced changes and overall data information. The formula for the α-order differential of hyperspectral reflectance data in this experiment is as follows [29]:

$$\frac{d^{\alpha} f(x)}{dx^{\alpha}} \approx f(x) + (-\alpha)f(x-1) + \frac{(-\alpha)(-\alpha+1)}{2}f(x-2) + \ldots + \frac{\Gamma(-\alpha+1)}{n!\Gamma(-\alpha+n1)}f(x-n) \quad (1)$$

In the formula, x is the value of the corresponding point; α is the fractional differential order; Γ is the Gamma function; and n is the difference between the upper and lower limits of differentiation. The order is 0, indicating no preprocessing.

Spectral data preprocessing and spectral index calculation were completed in MAT-LAB2021 (MathWorks, Inc., Natick, MA, USA), and Origin2021 (Origin Lab Corp., Northampton, MA, USA) was used for graphic drawing.

2.4. Selection and Construction of Spectral Index

In this study, 10 spectral indices were selected. The ratio vegetation index (RI) and triangular vegetation index (TVI) have a strong correlation with chlorophyll content and LAI of plants, but when vegetation is dense, their sensitivity will be reduced [33]. The modified simple ratio (mSR) and modified normalized difference index (mNDI) can optimize the specular emission effect of leaves and are sensitive to changes in leaves [35]. The difference vegetation index (DI), normalized difference vegetation index (NDVI), and soil-adjusted vegetation index (SAVI) can reflect the background influence of the plant canopy and eliminate some radiation errors [36,37]. The correlation between the three-band index (TBI-1, TBI-2, TBI-3) and SMC is more stable [38]. The specific spectral index formula is shown in Table 3.

Table 3. Spectral index and construction formula.

Select Index	Computing Formula	Reference
RI	R_i / R_j	[33]
TVI	$0.5\left[120(R_i - R_{550}) - 200\left(R_j - R_{550}\right)\right]$	[33]
DI	$R_i - R_j$	[36]
NDVI	$R_i - R_j / R_i + R_j$	[36]
SAVI	$(1 + 0.16)\frac{R_i - R_j}{R_i + R_j + 0.16}$	[37]
mSR	$R_i - R_{455} / R_j - R_{455}$	[35]
mNDI	$R_i - R_i / R_j + R_j - 2R_{455}$	[35]
TBI-1	$R_{1400} / (R_i + R_j)$	[38]
TBI-2	$(R_{1400} - R_i) / (R_{1400} + R_j)$	[38]
TBI-3	$(R_{1400} - R_i) - (R_{\lambda 1} - R_j)$	[38]

R_i (i = 1, 2, 3) is reflectance at any band; R_j (j = 1, 2, 3) is reflectance at any band; R_{455} and R_{1400} are hyperspectral reflectance at 455 nm and 1400 nm wavelengths.

2.5. Model Construction

In this research study, the most effective combination of spectral indices of various orders served as the input variable, and three machine learning techniques, namely BPNN, SVM, and RF, were employed for modeling and predicting SMC during the flowering stage.

The training-set-to-validation-set ratio was set at 2:1, and the final model fitting result in this experiment was determined by averaging multiple prediction fitting outcomes from the machine learning models.

BPNN is a multi-layer network that propagates forward according to error and is mostly used to solve difficult nonlinear problems [39]. The optimal BPNN combination forecasting model uses m prediction methods to obtain the predicted results as the input of the network and the actual historical data value as the network's output. The weights of various prediction methods in the prediction are obtained according to the self-learning of the network [40].

SVM is a binary classification machine learning algorithm utilizing a Gaussian kernel and polynomial kernel as the foundational kernel function. The weight coefficient is optimized using a gradient descent algorithm. SVM demonstrates an excellent generalization ability and robustness, without the issue of overfitting. Its widespread applications include pattern recognition, classification, and small-sample regression analysis, guided by the principle of minimizing cross-validation error [41,42].

RF is a composite model founded on the 'Bagging' model. Due to its simplicity and convenience, it finds extensive application in diverse regression and prediction challenges. As the RF model employs weighted averages of each tree's results to attain the final output, its implementation involves constructing numerous decision trees. The model builds a set of decision trees through the exchange and alteration of covariates to enhance prediction performance [43,44]. This study determined the number of decision trees in the RF model to be 100 after multiple training and error analyses.

2.6. Data Processing

(1)   Model evaluation index

The model fitting results were evaluated by a determination coefficient (R^2), root mean square error (RMSE), and mean relative error (MRE) [45,46]. The closer R^2 is to 1, the higher the model's prediction accuracy is. The smaller the MRE, the more stable the performance of the model and the more concentrated the prediction results. The calculation formula is as follows:

$$R^2 = \frac{\sum_{i=1}^{n}(\hat{y}_i - \overline{y})^2}{\sum_{i=1}^{n}(y_i - \overline{y})^2} \tag{2}$$

$$\text{RMSE} = \sqrt{\frac{\sum_{i=1}^{n}(y_i - \overline{y})^2}{n}} \tag{3}$$

$$\text{MRE} = \frac{1}{n}\sum_{i=1}^{n}\frac{|\hat{y}_i - y_i|}{y_i} \times 100\% \tag{4}$$

In the formula, $\hat{y}_i$ is model prediction; y_i is actual sample value; $\overline{y}$ is average; and n is number of samples.

(2)   Significance test

Concerning the autocorrelation coefficient test table, when the degree of freedom (i.e., sample size) is 48 and the correlation coefficient is greater than 0.361, it reaches an extremely significant correlation level ($p < 0.01$). When the degree of freedom is 32 and the correlation coefficient value is greater than 0.436, it reaches an extremely significant correlation level ($p < 0.01$). When the degree of freedom is 16 and the correlation coefficient is greater than 0.590, it reaches an extremely significant correlation level ($p < 0.01$).

3. Results and Analysis

3.1. Spectral Index Construction and Optimal Spectral Index Band Combination Extraction

To maximize the utilization of information within hyperspectral reflectance data, this study selected 10 representative spectral indices. Firstly, the spectral indices of all bands of hyperspectral reflectance after 0–2-order fractional differential treatment were calculated by band-by-band spectral indices. Then, the correlation matrix method was used to analyze the correlation between spectral indices and SMC. The i and j wavelengths with the largest correlation coefficient were used to construct different-order spectral indices. From the set of ten spectral indices, five with the strongest correlation to SMC were chosen to form the optimal spectral index combination. The correlation matrix diagram, depicted below, illustrates the correlation between the spectral indexes and SMC, ranging from a high negative correlation in blue to a high positive correlation in red.

Figures 3–7 show the correlation matrix diagrams of 0–2-order fractional differential hyperspectral spectral indexes and SMC after fractional differential treatment. The correlation of SMC with each order spectral index is greater than 0.361 ($p < 0.01$), reaching a very significant correlation level, indicating that the 10 spectral indexes selected in this study can be used to predict SMC at the flowering stage. The average values of the correlation coefficients between the 0–2-order spectral index and SMC were 0.523, 0.688, 0.728, 0.721, and 0.718, respectively. The correlation between the optimal spectral index and SMC calculated by the fractional differential treatment was significantly improved compared with the original spectral index. Under the 1.5-order differential treatment, the highest correlation coefficient with SMC was mNDI, the correlation coefficient value was 0.763, and the wavelength combination coordinate was (740, 696). The correlation coefficient values of each spectral index and SMC ranked from high to low are mNDI > TBI-1 > NDVI > TVI > TBI-3 > DI = SAVI = mSR > RI > TBI-2. From the aforementioned ten spectral indices, five, namely mNDI, TBI-1, NDVI, TVI, and TBI-3, exhibiting the highest correlation coefficients, were chosen to compose the optimal spectral index combinations. The associated bands, (740, 696), (726, 700), (688, 729), (754, 708), and (726, 700), were identified as the optimal spectral index band combinations. Table 4 displays the corresponding bands for both the remaining fractional differential optimal spectral index combination and the optimal spectral index combination.

3.2. Construction and Comparison of Soil Moisture Content Prediction Model

The optimal spectral index combination of each order was used as the independent variable, and SMC was used as the response variable. SVM, RF, and BPNN were used to construct the SMC estimation model of the soybean flowering stage. The accuracy of the model was comprehensively evaluated in terms of three aspects, namely R^2, RMSE, and MRE. The prediction results of different modeling methods for SMC are shown in Table 5. The results show that the R^2 of each SMC estimation model under different order differential transformation treatments is 1.5 order > 2 order > 1 order > 0.5 order > 0 order. The MRE performance of each model is 1.5 order < 2 order < 1 order < 0.5 order < 0 order. The RMSEs of the validation set of each model of 1.5 order are 0.01977, 0.00507, and 0.00579, which are smaller than the corresponding models of other orders. The validation set R^2 of the SVM, RF, and BPNN prediction models of SMC constructed by a 1.5-order differential spectral index are 0.90576, 0.91233, and 0.87778, respectively, which are higher than 0.539, reaching a very significant correlation level and having an excellent linear fitting effect. Under the same order differential transformation processing, the accuracy of the modeling set and verification set of the SMC estimation model constructed by the three modeling methods is as follows: RF > SVM > BPNN. Under each order differential treatment, the validation set R^2 of the SMC prediction model based on RF was 0.025–0.287 higher than that of SVM and GA-BP. The MRE decreased by 11.17–45.81%. In summary, the 1.5-order differential treatment and the RF model are the optimal differential order and the optimal model in this study, respectively. The R^2 of the modeling set and the verification set of the optimal SMC estimation model are 0.912 and 0.792, and the RMSEs are 0.912 and 0.792, respectively.

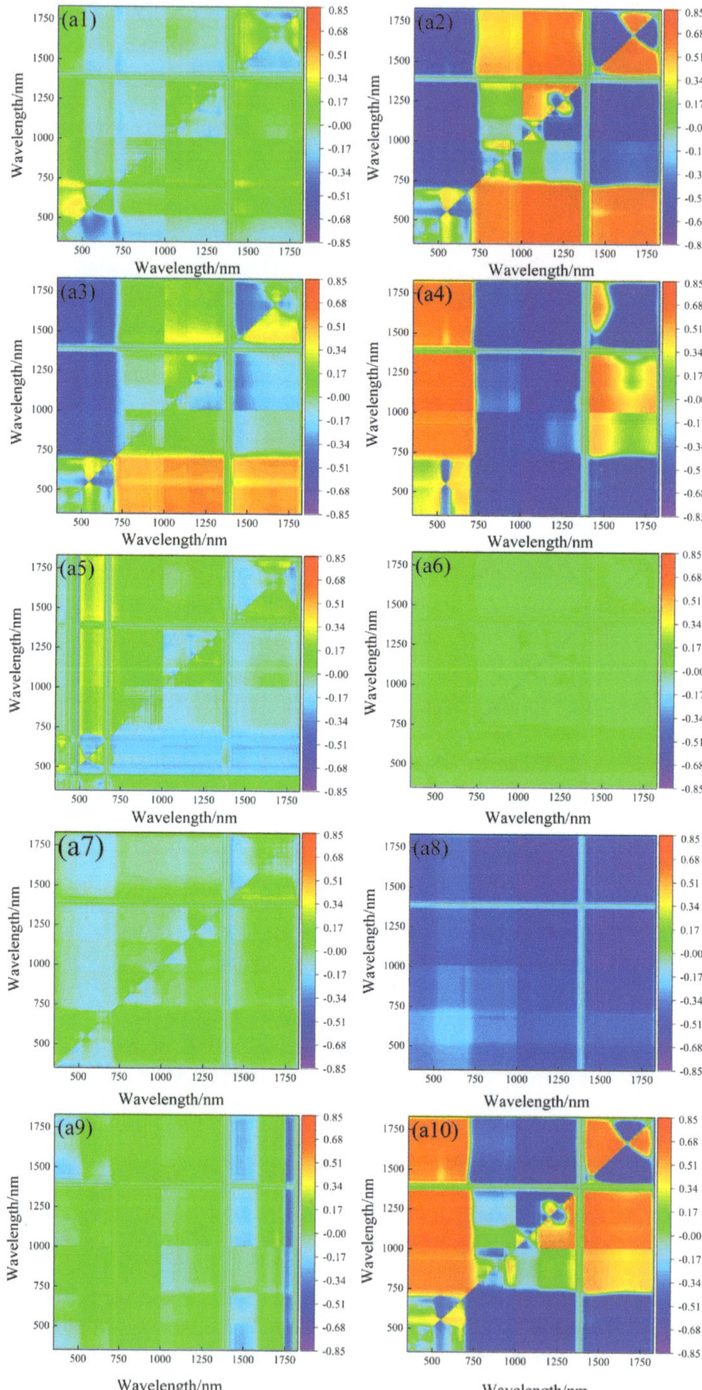

Figure 3. (a1–a10): The correlation matrix of RI, DI, SAVI, TVI, mSR, mNDI, NDVI, TBI-1, TBI-2, TBI-3, and soil moisture content under 0—order differential.

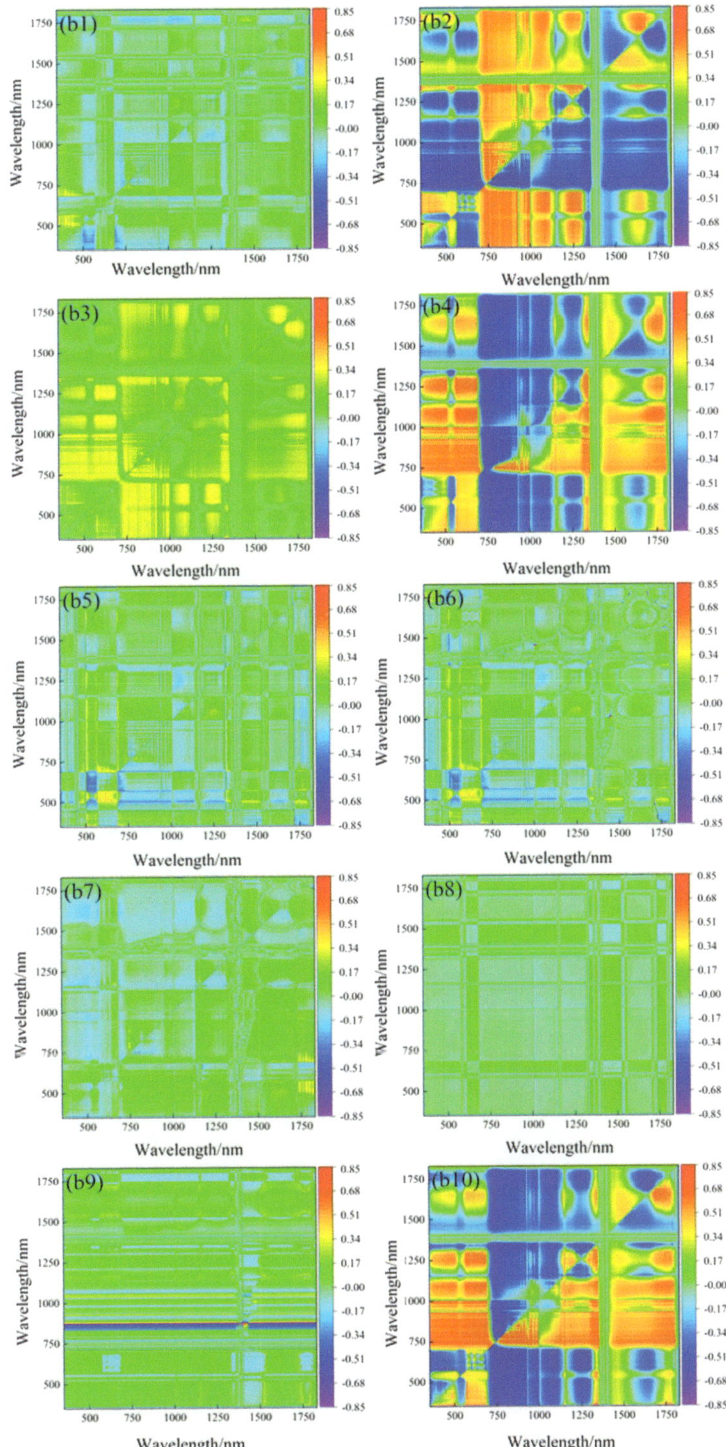

Figure 4. (**b1–b10**): The correlation matrix of RI, DI, SAVI, TVI, mSR, mNDI, NDVI, TBI-1, TBI-2, TBI-3, and soil moisture content under 0.5—order differential.

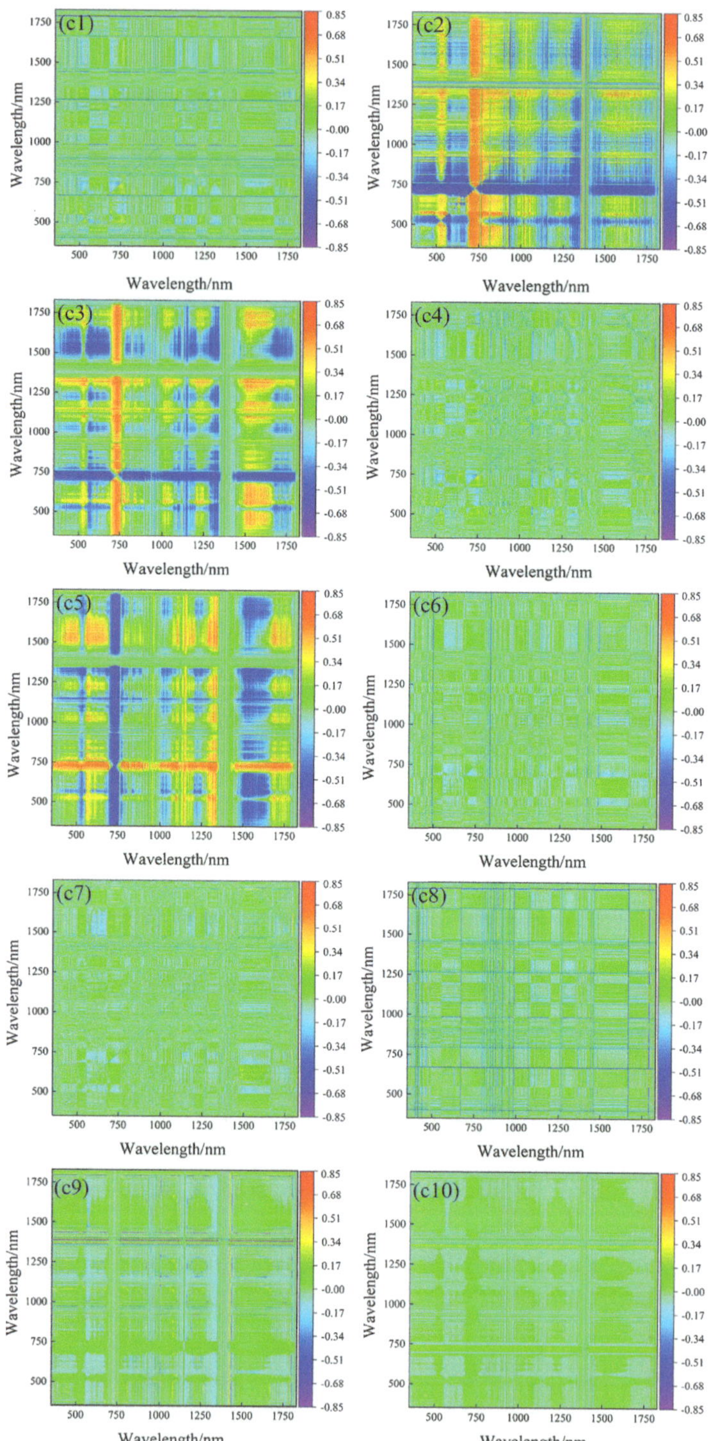

Figure 5. (**c1–c10**): The correlation matrix of RI, DI, SAVI, TVI, mSR, mNDI, NDVI, TBI-1, TBI-2, TBI-3, and soil moisture content under 1—order differential.

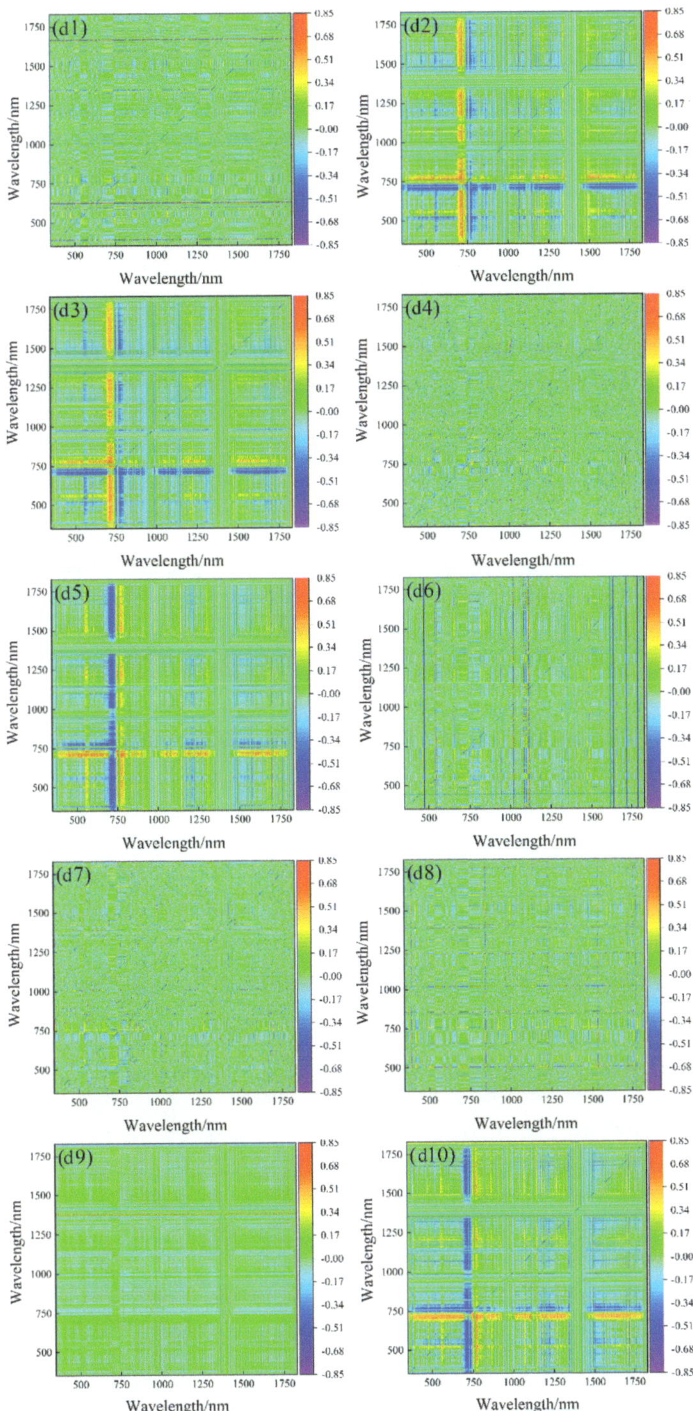

Figure 6. (**d1–d10**): The correlation matrix of RI, DI, SAVI, TVI, mSR, mNDI, NDVI, TBI-1, TBI-2, TBI-3, and soil moisture content under 1.5—order differential.

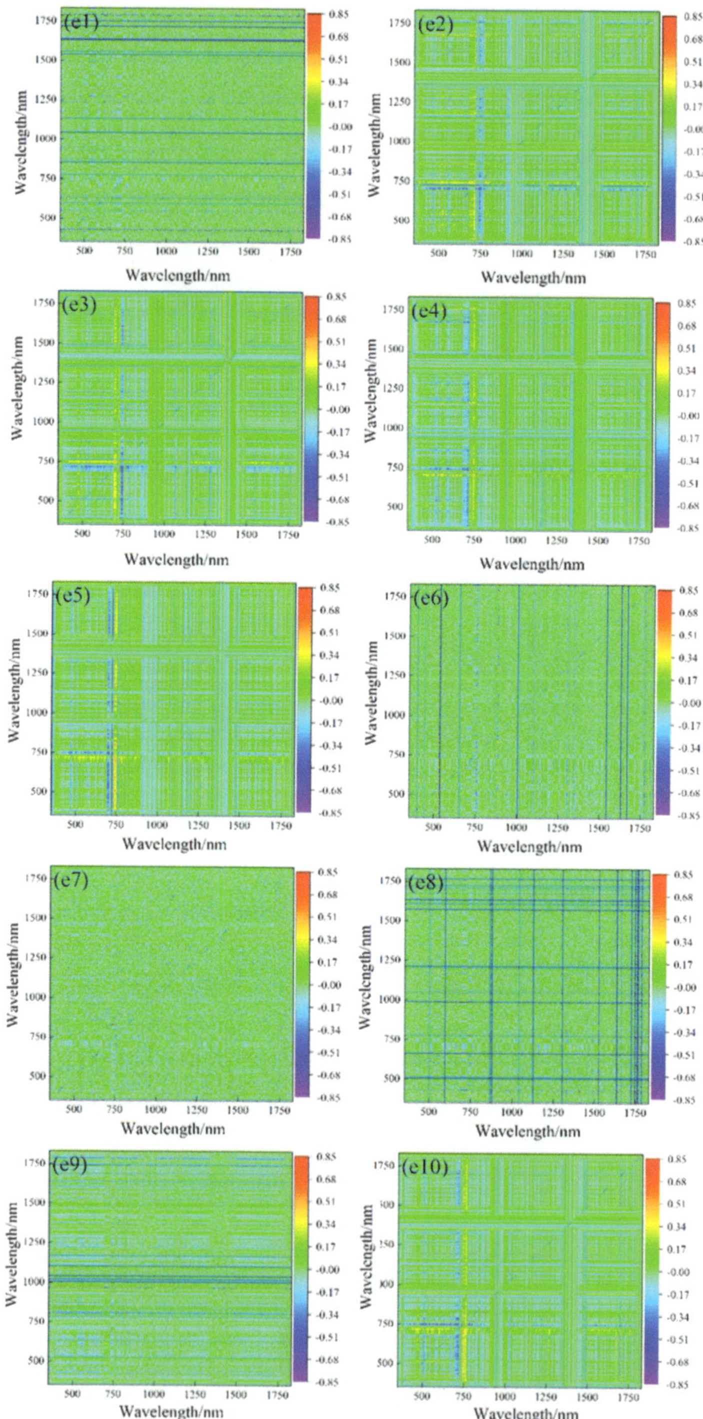

Figure 7. (**e1–e10**): The correlation matrix of RI, DI, SAVI, TVI, mSR, mNDI, NDVI, TBI-1, TBI-2, TBI-3, and soil moisture content under 2—order differential.

Table 4. Preferred spectral index wavelength combinations under various differential orders.

Differential Order	Spectral Index	Correlation Coefficient	Position of Wavelength (i, j)/(nm)	Optimal Spectral Index Combination
0	DI	0.647 **	747, 745	DI, SAVI, TVI, mSR, TBI-3
	RI	0.411 **	717, 720	
	NDVI	0.412 **	717, 720	
	SAVI	0.589 **	717, 720	
	TVI	0.659 **	759, 670	
	mSR	0.619 **	719, 718	
	mNDI	0.447 **	719, 718	
	TBI-1	0.376 **	758, 759	
	TBI-2	0.368 **	757, 717	
	TBI-3	0.711 **	747, 745	
0.5	DI	0.726 **	690, 720	RI, SAVI, TBI-3, mNDI, NDVI
	RI	0.651 **	737, 748	
	NDVI	0.698 **	745, 737	
	SAVI	0.731 **	748, 737	
	TVI	0.682 **	757, 692	
	mSR	0.653 **	719, 717	
	mNDI	0.690 **	717, 720	
	TBI-1	0.648 **	720, 753	
	TBI-2	0.668 **	753, 696	
	TBI-3	0.737 **	737, 748	
1	DI	0.719 **	731, 721	DI, NDVI, TBI-3, mNDI, TVI
	RI	0.688 **	673, 726	
	NDVI	0.722 **	711, 734	
	SAVI	0.716 **	726, 673	
	TVI	0.722 **	737, 696	
	mSR	0.713 **	721, 720	
	mNDI	0.763 **	757, 681	
	TBI-1	0.700 **	729, 690	
	TBI-2	0.628 **	700, 724	
	TBI-3	0.719 **	673, 726	
1.5	DI	0.721 **	726, 700	mNDI, TVI, TBI-1, NDVI, TBI-3
	RI	0.689 **	676, 738	
	NDVI	0.736 **	688, 729	
	SAVI	0.721 **	726, 700	
	TVI	0.726 **	754, 708	
	mSR	0.721 **	692, 726	
	mNDI	0.766 **	740, 696	
	TBI-1	0.737 **	726, 700	
	TBI-2	0.671 **	726, 677	
	TBI-3	0.722 **	726, 700	
2	DI	0.685 **	700, 726	TVI, NDVI, TBI-1, mSR, mNDI
	RI	0.669 **	694, 746	
	NDVI	0.729 **	727, 682	
	SAVI	0.685 **	726, 700	
	TVI	0.703 **	723, 714	
	mSR	0.718 **	700, 726	
	mNDI	0.762 **	755, 720	
	TBI-1	0.736 **	759, 694	
	TBI-2	0.698 **	675, 676	
	TBI-3	0.687 **	747, 737	

'**' indicates that the correlation is significant at 0.01.

Table 5. Comparison of model prediction accuracy evaluation under different differential order.

Differential Order	Evaluating Indicator	BPNN		RF		SVM	
		Training Sets	Validation Set	Training Sets	Validation Sets	Training Set	Validation Set
0	R^2	0.629	0.652	0.714	0.722	0.642	0.679
	RMSE (g/kg)	0.007	0.027	0.005	0.006	0.006	0.006
	MRE (%)	4.31	4.477	3.467	4.622	3.903	4.565
0.5	R^2	0.652	0.693	0.838	0.725	0.678	0.695
	RMSE (g/kg)	0.007	0.031	0.005	0.005	0.008	0.007
	MRE (%)	3.42	11.76	3.765	3.116	4.552	4.887
1	R^2	0.794	0.707	0.842	0.737	0.799	0.719
	RMSE (g/kg)	0.007	0.029	0.006	0.007	0.02	0.029
	MRE (%)	4.428	5.471	3.164	4.831	4.444	6.4
1.5	R^2	0.878	0.759	0.912	0.792	0.906	0.772
	RMSE (g/kg)	0.006	0.027	0.005	0.004	0.02	0.027
	MRE (%)	3.674	3.176	2.891	2.780	2.988	3.974
2	R^2	0.863	0.745	0.889	0.762	0.867	0.757
	RMSE (g/kg)	0.02	0.007	0.006	0.004	0.02	0.026
	MRE (%)	3.918	5.189	2.448	2.542	2.838	5.574

4. Discussion

In this study, the differential transform method is introduced for processing vegetation canopy spectra, providing an advantageous technique for analyzing reflectance spectra. This approach effectively addresses the challenge of multicollinearity inherent in high-dimensional spectral data [47,48]. The employment of differential transform technology significantly impacts peak extraction in finely detailed spectra, thereby enhancing sensitivity to spectral features and curves. This method facilitates baseline correction [49,50], thereby intensifying the correlation between hyperspectral reflectance and SMC, subsequently enhancing inversion model accuracy [32,51]. As depicted in Figure 8, the highest accuracy is observed with 1.5-order processing. Consequently, the fractional order differential algorithm demonstrates superior capability in extracting SMC-relevant spectral data from hyperspectral spectrometers when compared to integer-order derivatives. Nevertheless, with increasing differential orders, background noise gradually diminishes while high-frequency noise progressively amplifies. This phenomenon concurrently leads to a reduction in potential sensitive information within reflectivity data, consequently lowering the signal-to-noise ratio of spectral information, thereby affecting model accuracy [52]. Table 4 displays the model results, indicating that the accuracy of certain fractional differential models surpasses that of raw spectral data, as well as first and second derivatives. Notably, accurate prediction models often cannot be established solely with raw spectral reflectance data. This underscores the foundational rationale behind employing spectral preprocessing for robust spectral data analysis.

The optimal band combination algorithm effectively addresses wavelength interactions within band combinations and handles the overlapping absorption of soil components [53]. This method has found application in numerous studies [54,55]. In the context of this study, five optimal spectral indices were selected from a pool of ten spectral indices. The varying maximum R^2 values derived from these optimal spectral indices under different fractional order differential transformations (Table 4) reveal dissimilar correlations between SMC and the spectral indexes. This disparity indicates variations in the ten spectral indices propensity to correlate with SMC. Within the 1.5-order reflectivity context, the mNDI index excels, exhibiting a peak R^2 value of 0.766. Selecting the ideal spectral index and processing all spectral data are typically formidable tasks [23]. This underscores the advantage of spectral index combinations. Additionally, the methodology enhances modeling accuracy by extracting information-rich bands while eliminating irrelevant predictors, a contrast to full-spectrum data. The optimal spectral indices, constructed through band screening across the entire spectrum, encapsulate more meaningful information linked to SMC. Among the 50 spectral indices spanning 0–2 orders, 25 corresponding bands are positioned within the red or near-infrared domains. As SMC considerably influences chlorophyll content, canopy chlorophyll indirectly reflects SMC [11]. The red edge band exhibits robust chlorophyll absorption and leaf reflection into the near-infrared, thus rendering it the fastest-growing region in green plant reflectance and a pivotal indicator of plant physiological traits [52]. The red edge band accommodates over 80% of plant physiochemical parameters' spectral information [56]. Hierarchical differential processing not only filters background noise but also preserves the red edge band's capacity to describe plant physiochemical parameters. As such, the band combinations sieved under each differential order in this study demonstrate a substantial correlation with SMC, primarily inhabiting the 670–760 nm range, corroborating prior research outcomes [57].

In this study, the optimal spectral index combinations for each order were chosen as input data, and three machine learning methods—SVM, RF and BPNN—were employed to construct SMC prediction models. Among these methods, the RF-based SMC prediction model exhibited the highest accuracy. This outcome underscores RF's robust capability to extract canopy chlorophyll-associated information from spectral reflectance data, consequently enhancing SMC inversion accuracy. The robustness of the RF algorithm, marked by its strong anti-interference and anti-overfitting attributes, alongside its high tolerance to background noise and outliers, renders it particularly adept at addressing nonlinear problems [58]. This conclusion aligns with the findings of Eyo et al. in their SMC inversion study [59]. BPNN stands as one of the most extensively employed neural network architectures. Through iterative adjustments of the network's interneuron weights, the algorithm minimizes the discrepancy between final output and anticipated results [60]. Yet, the inherent limitations of the BPNN algorithm, such as convergence into local extremes, weight converging to local minima, and sluggish convergence speed, tend to impede the accuracy and generalization capacity of neural network models [42]. In this study, the accuracy of the BPNN model was inferior to that of RF. This discrepancy might have arisen from the relatively limited sample size and the extensive model training iterations, contributing to a reduction in model precision and generalization capability [44]. When compared to RF and BPNN, the predictive efficacy of the SVM model on SMC was less impressive. This could potentially be attributed to SVM's limited anti-interference capacity and constraints stemming from parameter selection, including kernel functions and penalty factors [61].

Currently, substantial progress has been achieved in developing models for crop attributes such as leaf area index (LAI), biomass, nitrogen content, and chlorophyll content employing hyperspectral data [62–65]. However, the commonly employed ground-based hyperspectral data can solely be collected at specific locations, limiting their widespread application. The findings of this investigation reveal that optimal accuracy can be attained by utilizing a 1.5-order optimal spectral index combination as input variables and employing RF to construct a soil moisture content (SMC) prediction model. These findings necessitate validation and refinement through experimentation across varied regions, scales, and crop varieties. Such extensive validation ensures the model's adaptability and estimation precision, serving as a foundation for SMC prediction through diverse remote sensing techniques, including multispectral and UAV hyperspectral data. Furthermore, there exists an avenue for exploring fractional differential transformation of hyperspectral reflectance.

This study employed a step size of 0.5 for fractional order differentiation, which may result in relatively lower accuracy in processing high-spectral information and weaker details in data handling compared to a smaller step size, with a relatively modest compensatory effect. Additionally, the selection of a fixed band in the construction of the three-dimensional spectral index had a certain impact on modeling accuracy. This step lays the groundwork for subsequent fractional order differentiation with a smaller step size. Moreover, this study provides a scientific basis for the rapid and accurate determination of soil moisture content during the crop growing season.

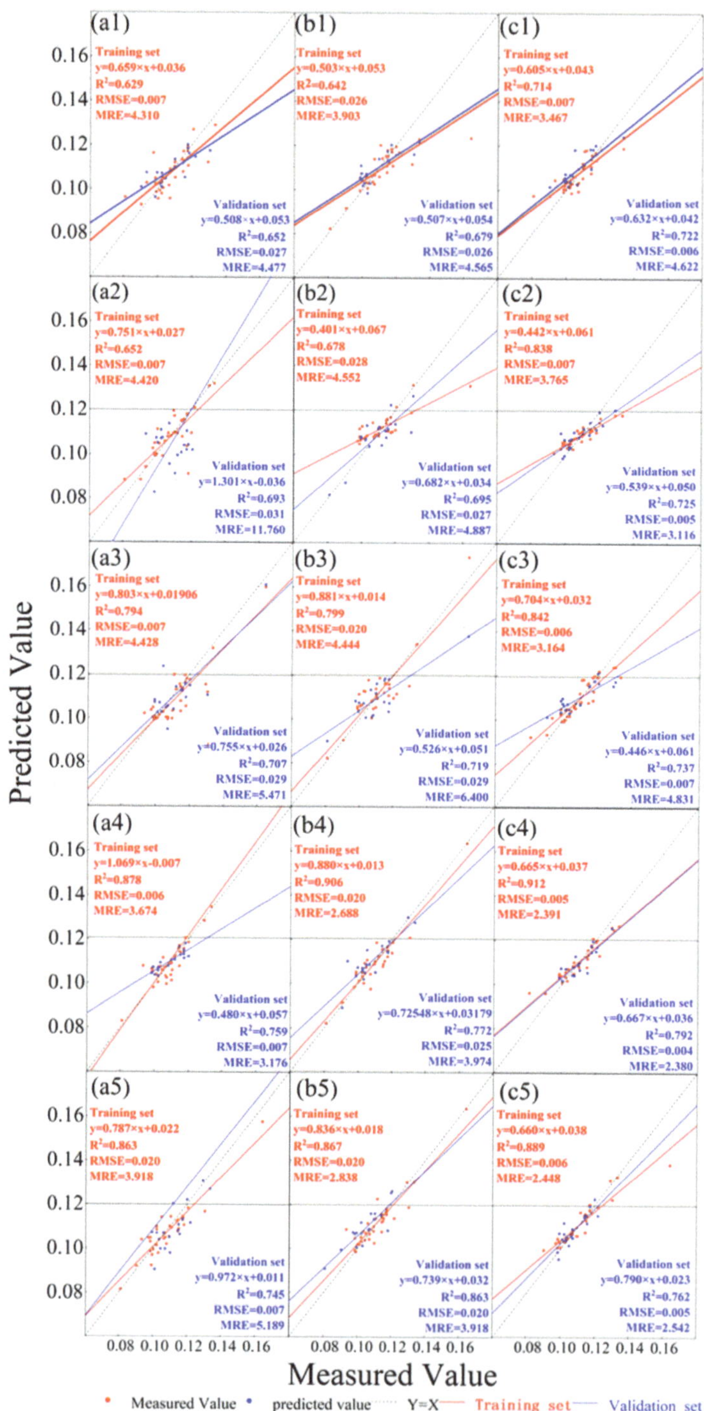

Figure 8. Model evaluation results. (**a1–a5**) is the 0–2−order soil moisture content prediction model constructed by the BPNN method; (**b1–b5**) are the results of the 0–2−order soil moisture content prediction model constructed by the SVM method. (**c1–c5**) are the results of the 0–2−order soil moisture content prediction model constructed by the RF method.

5. Conclusions

In this study, the SMC at the flowering stage was taken as the research object, and the SMC and canopy hyperspectral reflectance data were measured. The original hyperspectral reflectance data were processed by 0–2-order fractional differential, and the band-by-band spectral index was calculated. The correlation matrix method was used to extract the optimal wavelength combination for different order spectral index construction. Finally, based on the optimal spectral index of each order and three machine learning models, namely SVM, RF and BPNN, the SMC prediction model was constructed. The conclusions are as follows:

(1) Compared with the original hyperspectral reflectance data, the correlation between the optimal spectral index of each order extracted after fractional differential transformation and SMC was significantly improved. The average value of the correlation coefficient between each spectral index and SMC under the 1.5-order treatment was 0.380% higher than that of the original spectral index. Among them, mNDI showed the highest correlation, with a correlation coefficient of 0.766.

(2) When the modeling method is the same, and the input variables are different, the accuracy of the SMC prediction model is as follows: 1.5 order > 2.0 order > 1.0 order > 0.5 order > original order. When the input variables are the same and the modeling method changes, the accuracy of the SMC prediction model is as follows: RF > SVM > BPNN. A comprehensive comparison of the model's evaluation indicators shows that the 1.5-order differential and RF methods are the optimal differential order and optimal model construction methods in this study, respectively. The R^2 of the optimal SMC estimation model modeling set and validation set are 0.912 and 0.792, the RMSE is 0.00507 and 0.00393, and the MRE is 2.3901% and 2.3802%.

Author Contributions: W.L.: Methodology, Formal analysis, Writing—original draft, Visualization. X.L. and Z.T.: Methodology, Software. X.W. and X.H.: Data curation, Writing—review and editing, Formal analysis. Y.X.: Supervision, Writing—review and editing, Funding acquisition. M.C., Y.D., L.M., S.W., H.S., Y.Z. and Z.L.: Data curation. F.Z.: Supervision, Project administration. All authors have read and agreed to the published version of the manuscript.

Funding: This work was supported by the National Natural Science Foundation of China: No. 52179045.

Data Availability Statement: Data are contained within the article.

References

1. Wu, F.; Geng, Y.; Zhang, Y.Q.; Ji, C.X.; Chen, Y.F.; Sun, L.; Xie, W.; Ali, T.; Fujita, T. Assessing sustainability of soybean supply in China: Evidence from provincial production and trade data. *J. Clean. Prod.* **2020**, *244*, 119006. [CrossRef]
2. Sfechis, S.; Vidican, R.; Rotar, I.; Stoian, V. Influence of mineral fertilization and zeolite on soybean productivity elements in climatic conditions from ARDS Turda. *Bull. Univ. Agric. Sci. Vet. Med. Cluj-Napoca Agric.* **2014**, *71*, 108771. [CrossRef] [PubMed]
3. Wu, Y.S.; Wang, E.L.; Gong, W.Z.; Xu, L.; Zhao, Z.G.; He, D.; Yang, F.; Wang, X.C.; Yong, T.W.; Liu, J.; et al. Soybean yield variations and the potential of intercropping to increase production in China. *Field Crops Res.* **2023**, *291*, 108771. [CrossRef]
4. Jirayucharoensak, R.; Khuenpet, K.; Jittanit, W.; Sirisansaneeyakul, S. Physical and chemical properties of powder produced from spray drying of inulin component extracted from Jerusalem artichoke tuber powder. *Dry. Technol.* **2019**, *37*, 1215–1227. [CrossRef]
5. Guo, S.; Xu, J. Development Status and Existing Problems of Soybean Food Industry in China. *J. Food Sci. Technol.* **2023**, *41*, 1–8. [CrossRef]
6. Liu, J.; Zhu, L.J.; Zhang, K.; Wang, X.M.; Wang, L.W.; Gao, X.N. Effects of Drought Stress/Rewatering on Photosynthetic Characteristics and Yieldof Soybean at Different Growth Stages. *Ecol. Environ. Sci.* **2022**, *31*, 286–296. [CrossRef]
7. Poudel, S.; Vennam, R.R.; Shrestha, A.; Reddy, K.R.; Wijewardane, N.K.; Reddy, K.N.; Bheemanahalli, R. Resilience of soybean cultivars to drought stress during flowering and early-seed setting stages. *Sci. Rep.* **2023**, *13*, 1277. [CrossRef]
8. Li, S.N.; Zhang, L.; Ma, R.; Yan, M.; Tian, X.J. Improved ET assimilation through incorporating SMAP soil moisture observations using a coupled process model: A study of U.S. arid and semiarid regions. *J. Hydrol.* **2020**, *590*, 125402. [CrossRef]
9. Zhou, Q.W.; Zhu, A.X.; Yan, W.H.; Sun, Z.Y. Impacts of forestland vegetation restoration on soil moisture content in humid karst region: A case study on a limestone slope. *Ecol. Eng.* **2022**, *180*, 106648. [CrossRef]

10. Kelly, B.C.O.; Sivakumar, V. Water Content Determinations for Peat and Other Organic Soils Using the Oven-Drying Method. *Dry. Technol.* **2014**, *32*, 631–643. [CrossRef]

11. Crusiol, L.G.T.; Nanni, M.R.; Furlanetto, R.H.; Sibaldelli, R.N.R.; Sun, L.; Gonçalves, S.L.; Foloni, J.S.S.; Mertz-Henning, L.M.; Nepomuceno, A.L.; Neumaier, N.; et al. Assessing the sensitive spectral bands for soybean water status monitoring and soil moisture prediction using leaf-based hyperspectral reflectance. *Agric. Water Manag.* **2023**, *277*, 108089. [CrossRef]

12. Zai, S.M.; Gao, Y.N.; Wu, F.; Chu, Y.C.; Liu, S.D. Rapid determination of soil moisture based on the principle of spectrophotometry. *Water Sav. Irrig.* **2021**, *1*, 1–6.

13. Cao, Q.; Song, X.D.; Wu, H.Y.; Zhang, G.L.; Yang, S.H. Determination of Basic Parameters of Soil Moisture by Using Ground-Penetrating Radar Ground Wave Method in Red Soil Region of South China. *Chin. J. Soil Sci.* **2020**, *51*, 332–342.

14. Zreda, M.; Shuttleworth, W.J.; Zeng, X.; Zweck, C.; Desilets, D.; Franz, T.; Rosolem, R. COSMOS: The cosmic-ray soil moisture observing system. *Hydrol. Earth Syst. Sci.* **2012**, *16*, 4079–4099. [CrossRef]

15. Medhat, M.E. Application of gamma-ray transmission method for study the properties of cultivated soil. *Ann. Nucl. Energy* **2012**, *40*, 53–59. [CrossRef]

16. Dai, J.J.; Zhang, X.P.; Lv, D.Q.; Luo, Z.D.; He, X.G. Dynamics of Soil Water in *Cinnamomum camphora* Forest in the Red Soil Hilly Region of South China. *Res. Soil Water Conserv.* **2019**, *26*, 123–131.

17. Yang, C.C.H.; Yang, L.; Zhang, L.; Zhou, C.H. Soil organic matter mapping using INLA-SPDE with remote sensing based soil moisture indices and Fourier transforms decomposed variables. *Geoderma* **2023**, *437*, 116571. [CrossRef]

18. Feng, S.J.; Qiu, J.X.; Crow, W.T.; Mo, X.G.; Liu, S.X.; Wang, S.; Gao, L.; Wang, X.H.; Chen, S. Improved estimation of vegetation water content and its impact on L-band soil moisture retrieval over cropland. *J. Hydrol.* **2023**, *617*, 129015. [CrossRef]

19. Solgi, S.; Ahmadi, S.H.; Seidel, S.J. Remote sensing of canopy water status of the irrigated winter wheat fields and the paired anomaly analyses on the spectral vegetation indices and grain yields. *Agric. Water Manag.* **2023**, *280*, 108226. [CrossRef]

20. Li, W.; Li, D.; Liu, S.Y.; Baret, F.; Ma, Z.Y.; He, C.; Warner, T.A.; Guo, C.L.; Cheng, T.; Zhu, Y.; et al. RSARE: A physically-based vegetation index for estimating wheat green LAI to mitigate the impact of leaf chlorophyll content and residue-soil background. *ISPRS J. Photogramm. Remote Sens.* **2023**, *200*, 138–152. [CrossRef]

21. Wen, P.F.; Meng, Y.; Gao, C.K.; Guan, X.K.; Wang, T.C.; Feng, W. Field Identification of Drought Tolerant Wheat Genotypes Using Canopy Vegetation Indices Instead of Plant Physiological and Biochemical Traits. *SSRN Electron. J.* **2022**, *154*, 110781. [CrossRef]

22. Jahromi, M.N.; Zand-Parsa, S.; Razzaghi, F.; Jamshidi, S.; Didari, S.; Doosthosseini, A.; Pourghasemi, H.R. Developing machine learning models for wheat yield prediction using ground-based data; satellite-based actual evapotranspiration and vegetation indices. *Eur. J. Agron.* **2023**, *146*, 126820. [CrossRef]

23. Gouvêa, D.S.; Marenco, R.A. Is a reduction in stomatal conductance the main strategy of *Garcinia brasiliensis* (Clusiaceae) to deal with water stress? *Theor. Exp. Plant Physiol.* **2018**, *30*, 321–333. [CrossRef]

24. Sobrino, J.A.; Franch, B.; Mattar, C.; Jiménez-Muñoz, J.C.; Corbari, C. A method to estimate soil moisture from Airborne Hyperspectral Scanner (AHS) and ASTER data: Application to SEN2FLEX and SEN3EXP campaigns. *Remote Sens. Environ.* **2012**, *117*, 415–428. [CrossRef]

25. Panigrahi, N.; Das, B.S. Canopy Spectral Reflectance as a Predictor of Soil Water Potential in Rice. *Water Resour. Res.* **2018**, *54*, 2544–2560. [CrossRef]

26. He, L.; Liu, M.R.; Zhang, S.H.; Guan, H.W.; Wang, C.Y.; Feng, W.; Guo, T.C. Remote estimation of leaf water concentration in winter wheat under different nitrogen treatments and plant growth stages. *Precis. Agric.* **2023**, *24*, 986–1013. [CrossRef]

27. Gutierrez, M.; Reynolds, M.P.; Klatt, A.R. Association of water spectral indices with plant and soil water relations in contrasting wheat genotypes. *J. Exp. Bot.* **2010**, *61*, 3291–3303. [CrossRef] [PubMed]

28. Wei, Y.L.; Li, X.L.; He, Y. Generalisation of tea moisture content models based on VNIR spectra subjected to fractional differential treatment. *Biosyst. Eng.* **2021**, *205*, 174–186. [CrossRef]

29. Leyden, K.; Goodwine, B. Fractional-order system identification for health monitoring. *Nonlinear Dyn.* **2018**, *92*, 1317–1334. [CrossRef]

30. Shiri, B.; Baleanu, D. System of fractional differential algebraic equations with applications. *Chaos Soliton Fractals* **2019**, *120*, 203–212. [CrossRef]

31. Hong, Y.S.; Liu, Y.L.; Chen, Y.Y.; Liu, Y.F.; Yu, L.; Liu, Y.; Cheng, H. Application of fractional-order derivative in the quantitative estimation of soil organic matter content through visible and near-infrared spectroscopy. *Geoderma* **2019**, *337*, 758–769. [CrossRef]

32. Wang, Z.; Zhang, X.; Zhang, F.; Chan, N.W.; Kung, H.; Liu, S.H.; Deng, L.F. Estimation of soil salt content using machine learning techniques based on remote-sensing fractional derivatives; a case study in the Ebinur Lake Wetland National Nature Reserve, Northwest China. *Ecol. Indic.* **2020**, *119*, 106869. [CrossRef]

33. Tang, Z.J.; Guo, J.J.; Xiang, Y.Z.; Lu, X.H.; Wang, Q.; Wang, H.D.; Cheng, M.H.; Wang, H.; Wang, X.; An, J.Q.; et al. Estimation of Leaf Area Index and Above-Ground Biomass of Winter Wheat Based on Optimal Spectral Index. *Agronomy* **2022**, *12*, 1729. [CrossRef]

34. Wijata, A.M.; Foulon, M.-F.; Bobichon, Y.; Vitulli, R.; Celesti, M.; Camarero, R.; Di Cosimo, G.; Gascon, F.; Longépé, N.; Nieke, J.; et al. Taking Artificial Intelligence Into Space Through Objective Selection of Hyperspectral Earth Observation Applications: To bring the "brain" close to the "eyes" of satellite missions. *IEEE Geosci. Remote Sens. Mag.* **2023**, *11*, 10–39. [CrossRef]

35. Sinhal, A.; Verma, B.A. Distribution Based Approach of Outlier Removal for Software Effort Data. *Int. J. Comput. Appl.* **2013**, *74*, 24–28. [CrossRef]

36. Kumar, V.; Sharma, A.; Bhardwaj, R.; Thukral, A.K. Comparison of different reflectance indices for vegetation analysis using Landsat-TM data. *Remote Sens. Appl. Soc. Environ.* **2018**, *12*, 70–77. [CrossRef]

37. Liang, T.G.; Yang, S.X.; Feng, Q.S.; Liu, B.K.; Zhang, R.P.; Huang, X.D.; Xie, H.J. Multi-factor modeling of above-ground biomass in alpine grassland: A case study in the Three-River Headwaters Region. China. *Remote Sens. Environ.* **2016**, *186*, 164–172. [CrossRef]

38. Lao, C.C.; Chen, J.Y.; Zhang, Z.T.; Chen, Y.W.; Ma, Y.; Chen, H.R.; Gu, X.B.; Ning, J.F.; Jin, J.M.; Li, X.W. Predicting the contents of soil salt and major water-soluble ions with fractional-order derivative spectral indices and variable selection. *Comput. Electron. Agric.* **2021**, *182*, 106031. [CrossRef]

39. Bala, M.; Singh, M. Non destructive estimation of total phenol and crude fiber content in intact seeds of rapeseed-mustard using FTNIR. *Ind. Crops Prod.* **2013**, *42*, 357–362. [CrossRef]

40. Wei, Y.; Bao, Y. Nondestructive Detection Method in Soybean Moisture Content Based on BP Neural Network. *J. Agr. Mech. Res.* **2007**, *2*, 126–129.

41. Lin, H.; Liang, L.; Zhang, L.P.; Du, P. Wheat leaf area index inversion with hyperspectral remote sensing based on support vector regression algorithm. *Nongye Gongcheng Xuebao/Trans. Chin. Soc. Agric. Eng.* **2013**, *29*, 139–146.

42. Atteh, E. The Nature of Mathematics Education; The Issue of Learning Theories and Classroom Practice. *Asian J. Educ. Soc. Stud.* **2020**, *10*, 42–49. [CrossRef]

43. Velis, C.A.; Wilson, D.C.; Gavish, Y.; Grimes, S.M.; Whiteman, A. Socio-economic development drives solid waste management performance in cities: A global analysis using machine learning. *Sci. Total Environ.* **2023**, *872*, 161913. [CrossRef] [PubMed]

44. Fang, K.N.; Wu, J.B.; Zhu, J.P.; Bang, C.N. A review of technologies on random forests. *J. Stat. Inf.* **2011**, *26*, 32–38. [CrossRef]

45. Chen, D.S.; Zhang, F.; Tan, M.L.; Chan, N.W.; Shi, J.C.; Liu, C.J.; Wang, W.W. Improved Na+ estimation from hyperspectral data of saline vegetation by machine learning. *Comput. Electron. Agric.* **2022**, *196*, 106862. [CrossRef]

46. Liu, N.; Xing, Z.Z.; Zhao, R.M.; Qiao, L.; Li, M.N.; Liu, G.; Sun, H. Analysis of chlorophyll concentration in potato crop by coupling continuous wavelet transform and spectral variable optimization. *Remote Sens.* **2020**, *12*, 2826. [CrossRef]

47. Song, S.; Song, X.N.; Tejado, B.I. Adaptive projective synchronization for time-delayed fractional-order neural networks with uncertain parameters and its application in secure communications. *Trans. Inst. Meas. Control.* **2018**, *40*, 3078–3087. [CrossRef]

48. Guo, L.; Chen, Y.Y.; Shi, T.Z.; Zhao, C.; Liu, Y.L.; Wang, S.Q.; Zhang, H.T. Exploring the role of the spatial characteristics of visible and near-infrared reflectance in predicting soil organic carbon density. *ISPRS Int. J. Geo-Inf.* **2017**, *6*, 308. [CrossRef]

49. Zhang, Z.P.; Ding, J.L.; Wang, J.Z.; Ge, X.Y. Prediction of soil organic matter in northwestern China using fractional-order derivative spectroscopy and modified normalized difference indices. *Catena* **2020**, *185*, 104257. [CrossRef]

50. Zhang, Z.P.; Ding, J.L.; Zhu, C.M.; Wang, J.Z.; Ma, G.L.; Ge, X.Y.; Li, Z.S.; Han, L.J. Strategies for the efficient estimation of soil organic matter in salt-affected soils through Vis-NIR spectroscopy: Optimal band combination algorithm and spectral degradation. *Geoderma* **2021**, *382*, 114729. [CrossRef]

51. Hong, Y.S.; Guo, L.; Chen, S.C.; Linderman, M.; Mouazen, A.M.; Yu, L.; Chen, Y.Y.; Liu, Y.L.; Liu, Y.F.; Cheng, H.; et al. Exploring the potential of airborne hyperspectral image for estimating topsoil organic carbon: Effects of fractional-order derivative and optimal band combination algorithm. *Geoderma* **2020**, *365*, 114228. [CrossRef]

52. Berger, K.; Verrelst, J.; Féret, J.B.; Wang, Z.H.; Wocher, M.; Strathmann, M.; Danner, M.; Mauser, W.; Hank, T. Crop nitrogen monitoring: Recent progress and principal developments in the context of imaging spectroscopy missions. *Remote Sens. Environ.* **2020**, *242*, 111758. [CrossRef]

53. Norris, C.E.; Quideau, S.A.; Landhäusser, S.M.; Drozdowski, B.; Hogg, K.E.; Oh, S.W. Assessing structural and functional indicators of soil nitrogen availability in reclaimed forest ecosystems using15N-labelled aspen litter. *Can. J. Soil Sci.* **2018**, *98*, 357–368. [CrossRef]

54. Taghizadeh-mehrjardi, R.; Toomanian, N.; Khavaninzadeh, A.R.; Jafari, A.; Triantafilis, J. Predicting and mapping of soil particle-size fractions with adaptive neuro-fuzzy inference and ant colony optimization in central Iran. *Eur. J. Soil Sci.* **2016**, *67*, 707–725. [CrossRef]

55. Li, X.H.; Li, J.G. Comparative Analysis for Grey Relation Estimation Models of Soil Organic Matter based on Hyperspectral Data. *IOP Conf. Ser. Earth Environ. Sci.* **2021**, *820*, 012002. [CrossRef]

56. Lu, J.S.; Eitel, J.U.H.; Engels, M.; Zhu, J.; Ma, Y.; Liao, F.; Zheng, H.B.; Wang, X.; Yao, X.; Cheng, T.; et al. Improving Unmanned Aerial Vehicle (UAV) remote sensing of rice plant potassium accumulation by fusing spectral and textural information. *Int. J. Appl. Earth Obs. Geoinf.* **2021**, *104*, 102592. [CrossRef]

57. Luo, Y.; Lyu, M.-Z.; Chen, J.-B.; Spanos, P.D. Equation governing the probability density evolution of multi-dimensional linear fractional differential systems subject to Gaussian white noise. *Theor. Appl. Mech. Lett.* **2023**, *13*, 100436. [CrossRef]

58. Fu, Z.P.; Jiang, J.; Gao, Y.; Krienke, B.; Wang, M.; Zhong, K.T.; Cao, Q.; Tian, Y.C.; Zhu, Y.; Cao, W.X.; et al. Wheat growth monitoring and yield estimation based on multi-rotor unmanned aerial vehicle. *Remote Sens.* **2020**, *12*, 508. [CrossRef]

59. Eyo, E.U.; Abbey, S.J.; Lawrence, T.T.; Tetteh, F.K. Improved prediction of clay soil expansion using machine learning algorithms and meta-heuristic dichotomous ensemble classifiers. *Geosci. Front.* **2022**, *13*, 101296. [CrossRef]

60. Chen, Y.F.; Hou, F.H.; Dong, S.H.; Guo, L.Y.; Xia, T.Y.; He, G.Y. Reliability evaluation of corroded pipeline under combined loadings based on back propagation neural network method. *Ocean Eng.* **2022**, *262*, 111910. [CrossRef]

61. Dai, Y.M.; Zhao, P. A hybrid load forecasting model based on support vector machine with intelligent methods for feature selection and parameter optimization. *Appl. Energy* **2020**, *279*, 115332. [CrossRef]

62. Croft, H.; Arabian, J.; Chen, J.M.; Shang, J.L.; Liu, J.G. Mapping within-field leaf chlorophyll content in agricultural crops for nitrogen management using Landsat-8 imagery. *Precis. Agric.* **2020**, *21*, 856–880. [CrossRef]
63. Raya-Sereno, M.D.; Ortiz-Monasterio, J.I.; Alonso-Ayuso, M.; Rodrigues, F.A.; Rodríguez, A.A.; González-Perez, L.; Quemada, M. High-resolution airborne hyperspectral imagery for assessing yield; biomass; grain N concentration; and N output in spring wheat. *Remote Sens.* **2021**, *13*, 1373. [CrossRef]
64. Moharana, S.; Dutta, S. Spatial variability of chlorophyll and nitrogen content of rice from hyperspectral imagery. *ISPRS J. Photogramm. Remote Sens.* **2016**, *122*, 17–29. [CrossRef]
65. Wang, C.Y.; Fu, J.X.; Yao, K.; Xue, K.; Xu, K.X.; Pang, X. Acridine-based fluorescence chemosensors for selective sensing of Fe^{3+} and Ni^{2+} ions. *Spectrochim. Acta-Part A Mol. Biomol. Spectrosc.* **2018**, *199*, 403–411. [CrossRef]

Article

Early Detection of Rice Leaf Blast Disease Using Unmanned Aerial Vehicle Remote Sensing: A Novel Approach Integrating a New Spectral Vegetation Index and Machine Learning

Dongxue Zhao [1], Yingli Cao [1,2], Jinpeng Li [1], Qiang Cao [1], Jinxuan Li [1], Fuxu Guo [1], Shuai Feng [1,2] and Tongyu Xu [1,2,*]

[1] College of Information and Electrical Engineering, Shenyang Agricultural University, Shenyang 110866, China; zhaodx@stu.syau.edu.cn (D.Z.); caoyingli@syau.edu.cn (Y.C.); lijipeng@stu.syau.edu.cn (J.L.); caoqiang@stu.syau.edu.cn (Q.C.); lijinxuan@stu.syau.edu.cn (J.L.); guofuxu@stu.syau.edu.cn (F.G.); fengshuai@syau.edu.cn (S.F.)

[2] Liaoning Key Laboratory of Intelligent Agricultural Technology, Shenyang 110866, China

* Correspondence: xutongyu@syau.edu.cn

Abstract: Leaf blast is recognized as one of the most devastating diseases affecting rice production in the world, seriously threatening rice yield. Therefore, early detection of leaf blast is extremely important to limit the spread and propagation of the disease. In this study, a leaf blast-specific spectral vegetation index RBVI = $9.78(R_{816} - R_{724}) - 2.08(\rho_{736}/R_{724})$ was designed to qualitatively detect the level of leaf blast disease in the canopy of a field and to improve the accuracy of early detection of leaf blast by remote sensing by unmanned aerial vehicle. Stacking integrated learning, AdaBoost, and SVM were used to compare and analyze the performance of the RBVI and traditional vegetation index for early detection of leaf blast. The results showed that the stacking model constructed based on the RBVI spectral index had the highest detection accuracy (OA: 95.9%, Kappa: 93.8%). Compared to stacking, the detection accuracy of the SVM and AdaBoost models constructed based on the RBVI is slightly degraded. Compared with conventional SVIs, the RBVI had higher accuracy in its ability to qualitatively detect leaf blast in the field. The leaf blast-specific spectral index RBVI proposed in this study can more effectively improve the accuracy of UAV remote sensing for early detection of rice leaf blast in the field and make up for the shortcomings of UAV hyperspectral detection, which is susceptible to interference by environmental factors. The results of this study can provide a simple and effective method for field management and timely control of the disease.

Keywords: rice; leaf blast; UAV remote sensing; hyperspectral; spectral vegetation index

Citation: Zhao, D.; Cao, Y.; Li, J.; Cao, Q.; Li, J.; Guo, F.; Feng, S.; Xu, T. Early Detection of Rice Leaf Blast Disease Using Unmanned Aerial Vehicle Remote Sensing: A Novel Approach Integrating a New Spectral Vegetation Index and Machine Learning. *Agronomy* **2024**, *14*, 602. https://doi.org/10.3390/agronomy14030602

Academic Editor: Yanbo Huang

Received: 17 February 2024
Revised: 8 March 2024
Accepted: 15 March 2024
Published: 17 March 2024

1. Introduction

Rice is one of the most important food crops in the world and is the main food ration for nearly half of the world's population [1,2]. However, with the change in global climate, the outbreak of rice diseases has been exacerbated by the frequent occurrence of abnormal weather, such as typhoons and heavy rains. Leaf blast, caused by the rice blast fungus, is considered the most infectious and destructive fungal disease in the major rice-growing regions of the world [3]. Under cloudy and rainy weather conditions, leaf blast can break out throughout rice's growth and development period and spread quickly, and pathogenicity is strong. At the same time, leaf blast disease develops in rice leaves, causing serious damage to leaf tissue cells, reducing photosynthesis in rice plants, and affecting rice growth and development. Disease fungi can form conidia between 10 and 35 °C (25–28 °C is optimal) and spread rapidly between plants by air and water currents, invading rice plants and spreading to neighboring cell tissues, causing disease in rice plants [4]. Under normal circumstances, leaf blast reduces rice yield by 10–30%. And under favorable climatic conditions, it will lead to 90–100% yield loss in rice [5]. In addition to causing yield loss, leaf blast can also jeopardize the natural environment. The application of deleterious agents after leaf blast is the main measure to control the further spread of the disease. However, excessive or indiscriminate application of disease control agents will not only lead to plant damage but also more likely to lead to the pollution of the natural environment (e.g., water and soil) in the field, and the misuse of agents will also increase the production cost of farmers. Therefore, there is an urgent need for a rapid, efficient, and large-scale detection tool for the precise detection and control of rice leaf blast.

Accurate early detection of diseases in the field environment is an important measure to prevent and control large-scale outbreaks of rice leaf blast, as well as a key link to reducing the abuse of pharmaceuticals and avoiding environmental pollution. Up to now, the early detection of crop diseases has been mainly carried out by visual hand inspection. This method relies heavily on experienced plant protection staff to conduct visual field site inspections or field sampling. This method has high accuracy and reliability. However, it needs to consume a lot of human and material resources. Another method with high accuracy is the biochemical detection (polymerase chain reaction) method. However, this in house method requires more specialized knowledge and sophisticated testing instruments. It is difficult to meet the current requirements for efficient and accurate early detection and control of rice diseases. At the same time, both of these methods suffer from the inability to detect large crop areas rapidly. In recent years, many researchers have gradually favored remote sensing technology as an emerging crop detection technology [6,7]. At the same time, the technology is one of the few detection means that can rapidly obtain crop growth and development information over a large area [8]. Many studies have shown that remote sensing technology has good potential for application in several crop information detection fields, such as crop nutrition detection, growth detection, and disease detection. In disease detection, hyperspectral remote sensing technology can effectively distinguish crop diseases due to its rich narrowband and high resolution to obtain subtle changes in crop physiological and biochemical characteristics caused by various stresses [9,10]. However, although hyperspectral remote sensing data have shown great advantages in disease detection, the large number of narrow bands and spectral information in the data bring more redundant feature information for the use of the data. On the other hand, the high cost of hyperspectral instrumentation has also greatly limited the popularization and application of this technology.

Data dimension reduction is the main way of hyperspectral data processing and the key to obtaining effective disease features. Existing data downscaling methods can be categorized into feature compression, feature wavelength screening, and vegetation indices (VIs) extraction. Feature compression can compress, and dimension reduction reduces the dimensionality of the original spectral data and decreases the influence of invalid information [11,12]. However, this method destroys the physical properties of the original spectra to some extent during the linear or nonlinear compression process. At the same time, the practicality of data compression may be limited by the method's computerized system (the original spectral data still need to be acquired during the application process), which greatly restricts its practical popularization and application for crop disease detection at different scales [2,13]. Different from feature compression, to extract disease feature information, feature wavelength screening does not change the original spectral data and only screens a few or dozens of spectral wavelength data containing the main key disease features from the original spectral data, which reduces the data dimensionality and has strong interpretability [14,15]. However, the influence of various factors, such as lighting conditions, results in a large instability of spectral feature wavelengths, affecting the reliability and generalization ability of the detection model. At the same time, more methods do not consider the inter-band correlation in the screening process, resulting in the introduction of useless features or the omission of important information [16,17].

In recent years, as an effective tool capable of characterizing physiological and biochemical changes in crops, spectral vegetation indices (SVIs) have gained more and more researchers' attention and studies in crop detection [18,19]. Compared with spectral feature wavelength screening, SVIs can improve the sensitivity of the information contained in spectra through specific relational expressions and, to a certain extent, can reduce the impact of environmental factors, such as changes in illumination, with better stability [20,21]. It has been shown that SVIs have good potential for application in crop disease detection. For example, Abdulridha et al. [22] successfully detected and differentiated target spot and bacterial spot in tomatoes by constructing vegetation indices and concluded that the photochemical reflectance index (PRI) and chlorophyll index (CHl green) were effective in differentiating between the two tomato diseases under indoor and outdoor conditions. Su et al. [23] studied and analyzed winter wheat yellow rust using UAV multispectral imagery and concluded that ratio vegetation index (RVI), normalized difference vegetation index (NDVI), and optimized soil adjusted vegetation index (OSAVI) are the best SVIs for distinguishing between healthy and diseased plants. It was also shown that NIR and red wavelengths are the best spectral regions for detecting the disease. Abdulridha et al. [24] detected canker-infected citrus by UAV hyperspectral remote sensing, and comparative analysis concluded that the anthocyanin reflectance index (ARI) and transform chlorophyll absorption in reflectance index (TCARI) were the most promising vegetation indices for detecting canker-infected citrus among the 31 available vegetation indices. However, based on the fact that common conventional SVIs, although having variable responses to crop physiological and biochemical properties, are unable to characterize the specific spectral responses of host–pathogen interactions, they cannot quantitatively or qualitatively describe the response mechanism to a particular kind of disease, which, in turn, leads to a large error in large-area disease detection [4,25,26]. Therefore, developing specific SVIs applicable to a certain crop disease can be expected to quantify the disease detection process. And it can effectively simplify the spectral sensor.

In view of this, this study used UAV hyperspectral remote sensing to acquire hyperspectral images of rice leaf blast at different disease levels, including healthy, mild, and severe, under field conditions to analyze the variability of spectral responses to leaf blast infection at different disease levels and to develop targeted specific SVIs to improve the accuracy of remote sensing detection of rice diseases.

2. Materials and Methods

2.1. Experimental Setup

The rice leaf blast experiment was conducted in 2021–2022 at the experimental base of Shenyang Agricultural University in Haicheng City, Liaoning Province (122°43′32.39″ N, 40°58′42.24″ E), as shown in Figure 1. The area shown in the test plot belongs to the temperate monsoon climate zone, with an average annual temperature of 10.4 °C, rainfall of 721.3 mm, mild temperature, abundant rainfall, and a climate suitable for crop growth.

Artificial dyeing test plot (2021): the test plot area was 0.39 hm^2, and Shennong 9816, which is sensitive to leaf blast, was selected as the test variety. Rice transplanting occurred on 25 May 2021 at a planting density of 30 cm × 17 cm, and nitrogen, phosphorus, and potash fertilizers were applied according to the local fertilization standards, which were 89, 40, and 53 kg/hm^2, respectively. Leaf blast fungi cultivated indoors were mixed with water in a certain proportion to configure a spore suspension at a concentration of 9 mg/100 mL, and an artificial disease test was conducted at 5:00 p.m. on 6 July. In the field, the configured suspension was sprayed evenly onto the rice leaves until the rest spread over the leaves. Meanwhile, to promote rice disease, moist black plastic bags were used to wrap the rice until they were removed at 7 a.m. on the second day. We used manual spraying of disease control agents for prevention and control to prevent and control the disease in healthy rice plots, and other field management was normal. Five days after inoculation with the disease (11 July), rice in the inoculated area showed symptoms of leaf blast. With the assistance of plant protection staff, we began canopy hyperspectral image collection and ground survey of the extent of the disease.

Natural incidence area (2022): the test plot area was 0.39 hm^2, and the planting rice variety was Yanfeng 47, producing a density of 30 cm × 20 cm. On 12 July 2022, under the detection of the plant protection staff, the field appeared to have a leaf blast disease infestation.

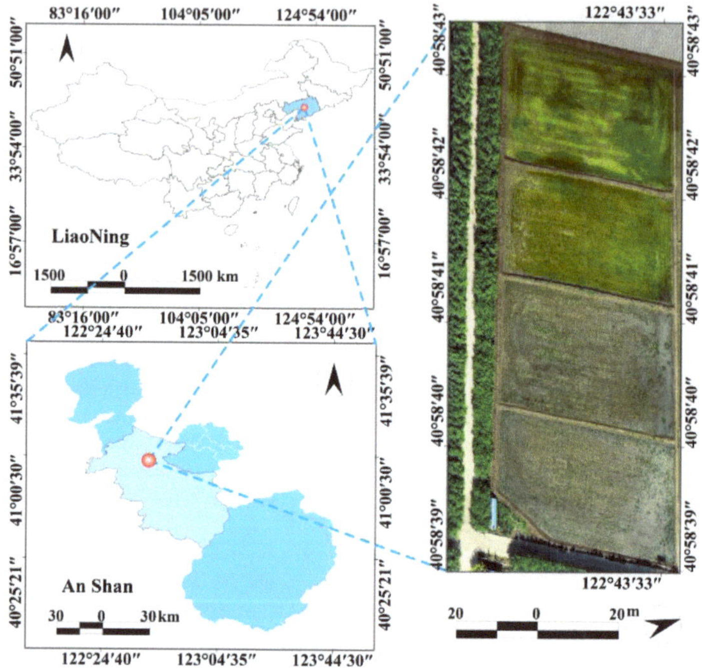

Figure 1. Rice leaf blast test plot.

2.2. Data Acquisition

2.2.1. UAV Hyperspectral Remote Sensing Data Acquisition

A GaiaSky-mini2-VN hyperspectral imager from Sichuan Shuangli Hepu, China, was used to acquire hyperspectral images of rice at critical fertility stages (nodulation and pre-sprouting) in the field canopy using an M600 PRO six-rotor UAV from DJI, China (Figure 2). The spectral range is 400–1000 nm, with a resolution of 3.5 nm and a pixel pitch of 4.54 μm. The flight altitude of the UAV in the experiment was set at 100 m, and the hyperspectral image data acquisition was carried out when the weather was clear and cloudless. At the same time, data acquisition was concentrated between 11 a.m. and 1 p.m. to reduce the influence of external factors such as weather and lighting conditions. A black-and-white standard version of the hyperspectral imager must be calibrated before each acquisition of hyperspectral images.

Meanwhile, the calibrated whiteboard was arranged next to the rice field and photographed along with the area during the UAV flight to be used for atmospheric correction later. After hyperspectral image acquisition, each pixel point in the hyperspectral image was labeled using the classical normalized vegetation index (NDVI) to distinguish between the vegetation index region and non-vegetation region (water, soil, etc.), which served to remove the effect of water, soil, etc., on the extraction of the spectral reflectance of the vegetation. We used ENVI 5.3 software to create regions of interest (ROIs) in the healthy and diseased areas, and the mean value of the spectral reflectance of the pixels occupied by rice in each ROI was used as the spectral reflectance for that degree of disease. At the same time, the spectral reflectance using the SG smoothing method removed the effect of noise due to the environment, instrumentation, etc.

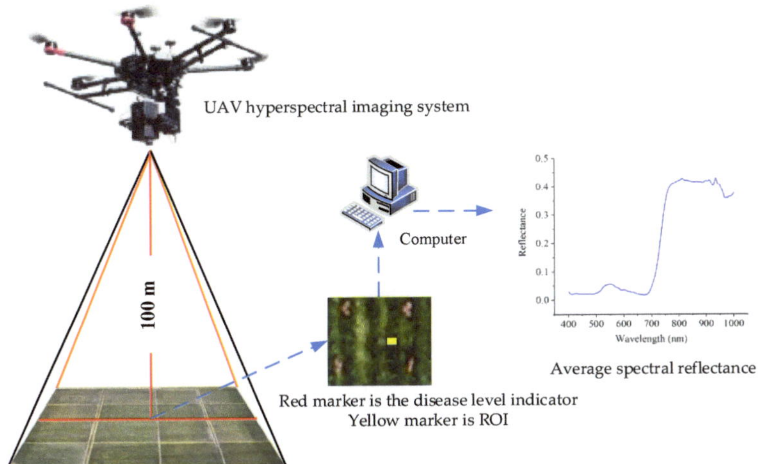

Figure 2. UAV hyperspectral remote sensing data acquisition.

2.2.2. Ground Truth Data Survey

To obtain the disease levels of rice leaf blast in the canopy of the field, during the survey, we marked the survey area of early leaf blast in the field by GPS positioning and field markers (e.g., the marked area in Figure 2). Thirty rice leaves were randomly selected from different plants in the surveyed area. The surveyed leaves were evaluated for the level of disease by visual survey (classified into leaf disease levels of 0–5 based on the percentage of diseased spot areas on the leaves). Then, according to the "GTB 15790-2009 rules of investigation and forecast of the rice blast [27]", the rice leaf disease level and the number of leaves with different levels in the region were counted, and the disease index (DI) was calculated. The formula of DI is shown below.

$$\mathrm{DI} = \frac{\sum_n x_i \times D_i}{n \times D_M},\tag{1}$$

where x_i is the number of diseases at all levels of rice leaves in the disease investigation area, D_i is the representative value of disease degree at all levels, n investigates the total number of rice leaves, and D_M is a representative value of the highest level of disease.

At the same time, we classified the disease level of leaf blast at the canopy scale according to the Chinese standard (GTB 15790-2009) [27] based on the disease index obtained from the calculation, and the specific disease level classification is shown in Table 1.

Table 1. Classification of disease level of rice leaf blast.

Rank	Disease Severity	Disease Index
A	Asymptomatic	DI = 0
M	Mild symptoms	0 < DI < 10
S	Severe symptoms	10 < DI < 30

Ultimately, to ensure the validity and balance of the data for different disease levels, we obtained 720 data entries for the two years 2021 and 2022 (501 acquired in 2021 and 219 in 2022). Data on 167 asymptomatic, mild, and severe symptoms of infestation were obtained in 2021, respectively. And the data of 73 asymptomatic, mild, and severe symptom infestations were obtained in 2022, respectively.

2.3. Experimental Methods

2.3.1. Constructing the Leaf Blast Spectral Vegetation Index

SVIs can provide spectral reflectance changes caused by stress infestation in a highly intuitive and interpretable expression. Therefore, this study developed a leaf blast SVIS based on healthy, light, and heavy canopy-scale rice leaf blast spectral data to provide theoretical support for detecting early rice leaf blast in the field. In this study, the development of rice leaf blast SVIs at the low-altitude canopy scale consisted of three main steps: determining the leaf blast-sensitive spectral region, designing a computational expression for leaf blast SVIs, and screening optimal spectral features and analyzing the detection performance of SVIs.

First, we analyzed spectral data for different disease levels of leaf blast. As shown in Figure 3, under disease infestation, the average spectral reflectance curves of rice canopies with varying levels of disease showed some changes, and the spectral reflectance curves showed a flattening trend, especially in the red-edge region and near-infrared region. In the red-edge region, the slopes of the spectral reflectance curves of different degrees of disease showed a reduced performance compared to the healthy spectrum (Figure 3B). In the near-infrared region, on the other hand, the spectral reflectance gradually decreased as the degree of disease increased. As elaborated by Zhao et al. [28], due to the infestation of the disease, the cellular structure of the canopy leaves is damaged, multiple reflectance is reduced, the content of biochemical components such as chlorophyll is reduced, and anisotropy is not obvious, which in turn contributes to the significant changes in the spectra. The red-light region showed an opposite trend compared to the near-infrared region. As the degree of disease increases, the reflectance values in the red-light region show an increasing trend. At the same time, the change in the red-light region contributes to the gradual flattening of the spectral curve. Therefore, in the subsequent determination of the computational expressions and screening of the optimal spectral features of the SVIs of leaf blast, we mainly analyze the comparison from the red-light, red-edge, and near-infrared regions. At the same time, a comparative analysis of the spectral reflectance of different disease levels based on the spectral coefficient of variation (CV) showed that the CV for asymptomatic, mild, and severe symptoms was 0.1176, 0.1072, and 0.1045, respectively (the CVs for the different disease levels were all less than 0.15). The results show a variability in spectral reflectance for different disease levels and that the spectral reflectance is relatively stable.

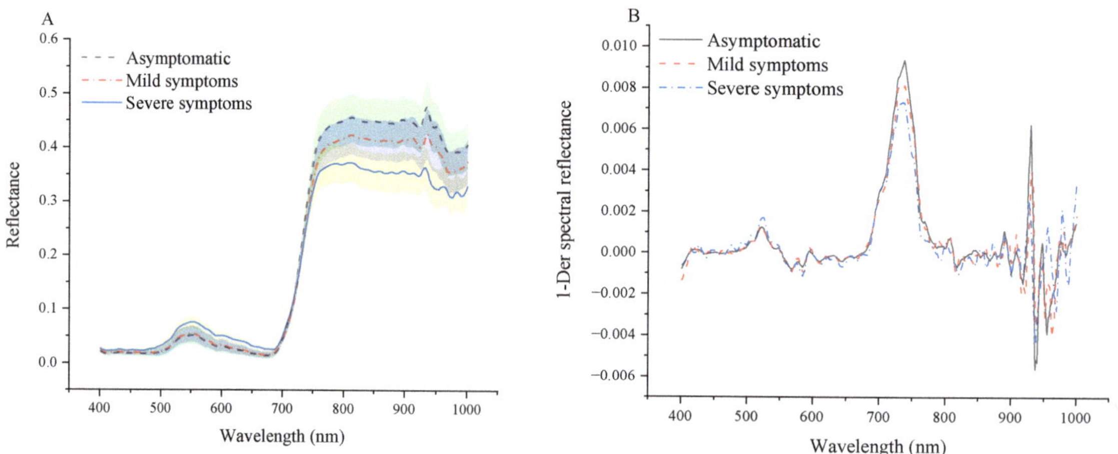

Figure 3. Variation of spectral reflectance and differential spectral reflectance for different disease levels. (**A**) Spectral reflectance. (**B**) Differential spectral reflectance.

Secondly, it can be seen from the above analysis of the changes in spectral reflectance of different disease levels that there is a significant difference between the visible region (390–760 nm) and the near-infrared region (760–1000 nm) in the hyperspectral spectrum of crop canopies. The spectral reflectance shows a trend of first low and then high changes, respectively. Because of the strong absorption in the visible region and high reflection in the near-infrared region, crop spectra are often used to enhance or characterize the implied crop information by performing various linear or nonlinear combinations of ratios, differences, normalization, etc., to form significant contrasts. Examples include difference vegetation index (DVI) and ratio vegetation index (RVI). DVI and RVI showed a gradual decrease with the increase in disease level. The reason for this is that when the leaf blast fungus interacts with rice, the color change at the spot with the disruption of the cellular structure leads to an increase in the reflectance in the red band and a decrease in the spectral reflectance in the near-infrared band [29]. As for the normalized vegetation index (NDVI), it can effectively reflect the health and growth of crops. In disease detection, NDVI can effectively detect the late symptoms of crop diseases, but the effect is not obvious for the early signs of diseases [30].

Meanwhile, in addition to the above combinations, spectral differential transformation processing is often used for the construction of SVIs, and spectral differentiation can significantly increase the feature differentiation between different disease levels in the red-edge region (680–760 nm), especially at the early stage of disease development [31,32]. The SVIs developed in this study were designed to detect rice leaf blast at different disease levels qualitatively. Hence, a combination of wavelength difference, wavelength ratio, and spectral differential transformations seems appropriate. We also performed normalized ratios for any two wavelengths, but the correlation with disease degree was poor.

Finally, combining the above analyses, we found that R_{NIR} and $\rho_{Red-edge}$ had the same trend for different disease levels, while R_{NIR} showed the opposite. At the same time, the two trends mentioned above can increase with the increase in disease severity (Red, Red-edge, and NIR represent the spectral wavelengths in the red, red-edge, and near-infrared regions, respectively; R and ρ represent the spectral reflectance and differential spectral reflectance, respectively). Therefore, from the above trends, we can conclude that as the severity of the disease advances, the value of $R_{NIR} - R_{Red}$ will gradually decrease, showing a monotonically reducing change. Similarly, $\frac{\rho_{Red-edge}}{R_{Red}}$ will have lower rates than healthy samples as the level of disease increases. Meanwhile, since the theoretical value of hyperspectral reflectance obtained by UAV hyperspectral remote sensing is between 0 and 1 (the hyperspectral reflectance in this study is between 0 and 0.6), there is an inconspicuous phenomenon of the degree of change in the values of $R_{NIR} - R_{Red}$ and $\frac{\rho_{Red-edge}}{R_{Red}}$ between health and mild symptoms, which can result in overlapping portions of the symptoms between health and mild symptoms. Therefore, we used the coefficient method to adjust the weight coefficient relationship between $R_{NIR} - R_{Red}$ and $\frac{\rho_{Red-edge}}{R_{Red}}$, which in turn further improves the sensitivity and variability of vegetation indices to leaf blast disease infestation.

Therefore, based on the above forms of combining SVIs and the spectral response laws of different disease degrees, we constructed one leaf blast-specific SVI to increase the variability of the spectra of varying disease degrees, as shown in Equation (2). (α and β are the weighting coefficients).

$$\text{RBVI} = \alpha(R_{NIR} - R_{Red}) + \beta\frac{\rho_{Red-edge}}{R_{Red}} \tag{2}$$

To improve the robustness and transferability of the SVIs proposed in this study, we used Fisher's discriminant analysis to screen the optimal and differential spectral wavelengths in the RBVI, respectively. The Fisher discriminant analysis method determines the differential spectral wavelengths with better separability. This method can classify points in high dimensional space by selecting the optimal direction and projecting them into low dimensional space, ensuring that the selected features have large interclass differences and small intraclass variations [33]. Meanwhile, an exhaustive algorithm was used to arbitrarily combine the spectral wavelengths in the visible and near-infrared regions to screen the spectral wavelengths with the best correlation. To determine the ability of the proposed RBVI to detect diseases, we analyzed the proposed RBVI in comparison with conventional SVIs (shown in Table S1).

2.3.2. Stacking Integrated Learning Detection Models

The stacking integrated learning model consists of a base learner and a meta-learner, which combine multiple base learner models and a meta-learner to obtain a strong learner with high stability and accuracy. To improve the accuracy of UAV hyperspectral remote sensing for detecting rice leaf blast, each base learner applies its machine learning algorithm to process the training data separately. Then, multiple machine learning algorithms are fused according to the stacking integrated learning fusion strategy to obtain a strong learner with a better detection effect, as shown in Figure 4. At the same time, stacking integrated learning can gather the advantages of individual machine learning algorithms and make up for the shortcomings of some algorithms in terms of accuracy and stability, improving the robustness and generalization of the detection model. Given this, we choose five detection models, including random forest (RF) [34], support vector machine (SVM) [35], AdaBoost [36], Lightgbm [37], and Xgboost [38], which have strong learning ability and good generalization, as the base learners. The selection of simple and stable machine learning algorithms for the meta-learner can effectively improve the accuracy of disease detection to avoid the risk of overfitting, so we chose linear discriminant analysis (LDA) as the meta-learner. At the same time, the combination of the prediction results of all base learners was used to train the meta-learner, and the model results of the meta-learner were used as the final detection results and compared and analyzed with the RF, SVM, AdaBoost, Lightgbm, and XgBoost models.

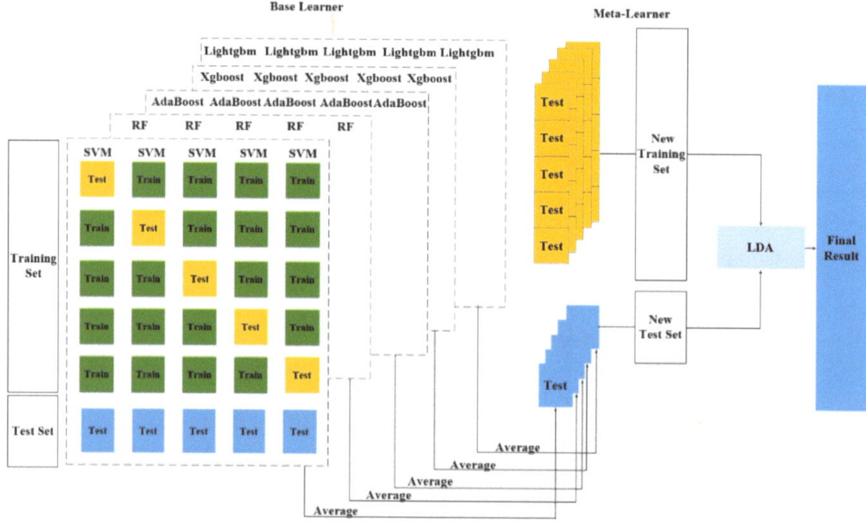

Figure 4. Stacking model structure.

The specific steps of the algorithm are as follows: (1) the original data are divided into training and test sets; (2) each base learner is trained and learned using 5-fold cross-validation, and the prediction results are used as the training set for the meta-learner; (3) the divided test set is input into the base learner for testing, and the test results are averaged as the new test set; (4) the newly constructed training and test sets are inputted into the meta-learner for training, and the obtained test results are the final test results.

2.3.3. Evaluation Indicators

To evaluate the accuracy of the constructed model, overall accuracy (OA), Kappa coefficient, user's accuracy (UA), and producer's accuracy (PA) were used as the evaluation metrics of the model (As shown in Table 2). The above metrics were calculated based on true positive (TP), true negative (TN), false positive (FP), and false negative (FN) values. The specific calculation formula is as follows:

$$OA = \frac{TP + TN}{TP + TN + FP + FN} \tag{3}$$

$$p_e = \sum_{i=1}^{m} \frac{(TP_i + FN_i)(TP_i + FP_i)}{N^2} \tag{4}$$

$$Kappa = \frac{OA - p_e}{1 - p_e} \tag{5}$$

Table 2. Definition of the confusion matrix.

	Positive	Negative	UA (%)
Positive	True Positive (TP)	False Negative (FN)	TP/(TP + FN) × 100%
Negative	False Positive (FP)	True Negative (TN)	TN/(TN + FP) × 100%
PA (%)	TP/(TP + FP) × 100%	TN/(TN + FN) × 100%	

3. Results and Analysis

3.1. Construction of a New Spectral Vegetation Index

The process of determining the red wavelengths, near-infrared wavelengths, and red-edge wavelengths of the differential spectra in the newly constructed RBVI can be divided into three steps. First, Fisher discriminant analysis is used to determine the separability of each wavelength in the first-order differential spectrum to filter the best $\rho_{Red-edge}$, as shown in Figure 5. In Figure 5A, we can tentatively determine that the first-order differential spectra of different disease levels have significant variability in the 720–750 nm interval. However, from Figure 5A, we cannot determine exactly the spectral wavelength corresponding to the peaks with the best separability. The reason is that the average first-order differential spectroscopy curves of different disease levels can only present rough features and not characterize all curve features. Therefore, the first-order differential spectra were statistically analyzed using Fisher discriminant analysis to screen the optimal first-order differential spectral wavelengths. As shown in Figure 5B, the Fisher discriminant analysis was modeled over the entire differential spectral region to obtain the weighting coefficients at every other differential spectral wavelength. The results show that the weighting coefficient curves of all differential spectral wavelengths are highly similar to the average differential spectral curves, especially in the red-edge and near-infrared regions. Among them, the highest peak value is presented at 736 nm in the red-edge region (Fisher score = 1). Therefore, ρ_{736} was determined to have the best separability and was screened as the best feature of the differential spectrum for calculating the RBVI.

Secondly, an exhaustive algorithm was used to determine the optimal wavelength combinations of R_{Red} and R_{NIR} in the ranges of 620–760 nm and 760–1000 nm, respectively, to obtain SVIs that are sensitive to leaf blast at different disease levels, as shown in Figure 6.

As can be seen in Figure 6, the sum of $R_{NIR} - R_{Red}$ and $\frac{\rho_{Red-edge}}{R_{Red}}$ constructed from R_{Red} and R_{NIR} acquired in the spectral ranges of 698–756 nm vs. 760–1000 nm had a high correlation with the degree of disease (correlation: 0.8138–0.930). The highest correlation was achieved when R_{Red} and R_{NIR} were chosen as 724 nm and 816 nm, respectively, with a correlation of 0.9257. At the same time, to further validate and select 724 nm and 816 nm, respectively, for the separability in recognizing different disease levels of leaf blast, we constructed the LDA model by using the exhaustive method in the red and near-infrared regions to judge the best wavelength selection, as shown in Figure 7. We obtained high accuracy (OA > 88.75%) in recognizing leaf blast in the field with different disease levels by obtaining R_{NIR} and R_{NIR} in the 710–747 nm and 760–865 nm range, respectively. The best detection accuracy (OA = 92.22%) was obtained when R_{NIR} and R_{NIR} were chosen as 724 nm and 816 nm, respectively. After verifying all possible spectral wavelengths, R_{NIR} and R_{NIR} were determined to be 724 nm and 816 nm, respectively.

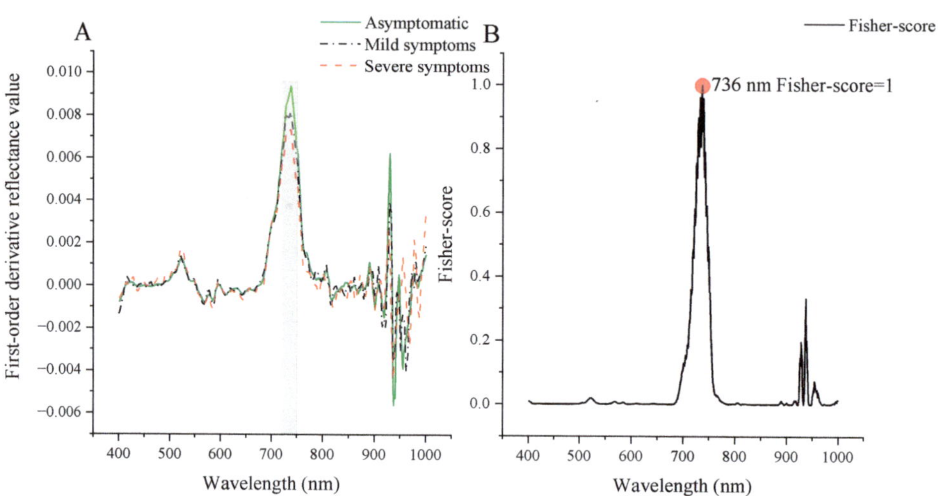

Figure 5. First-order differential spectra with Fisher weighting coefficients. (**A**) Differential spectral reflectance. (**B**) The Fisher score for differential spectral wavelengths of 400–1000 nm.

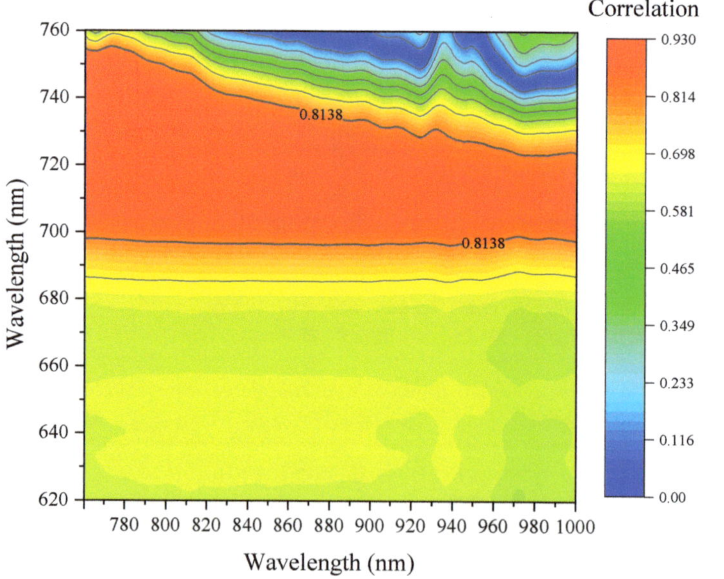

Figure 6. Contour map of the correlation between RBVI and different disease levels.

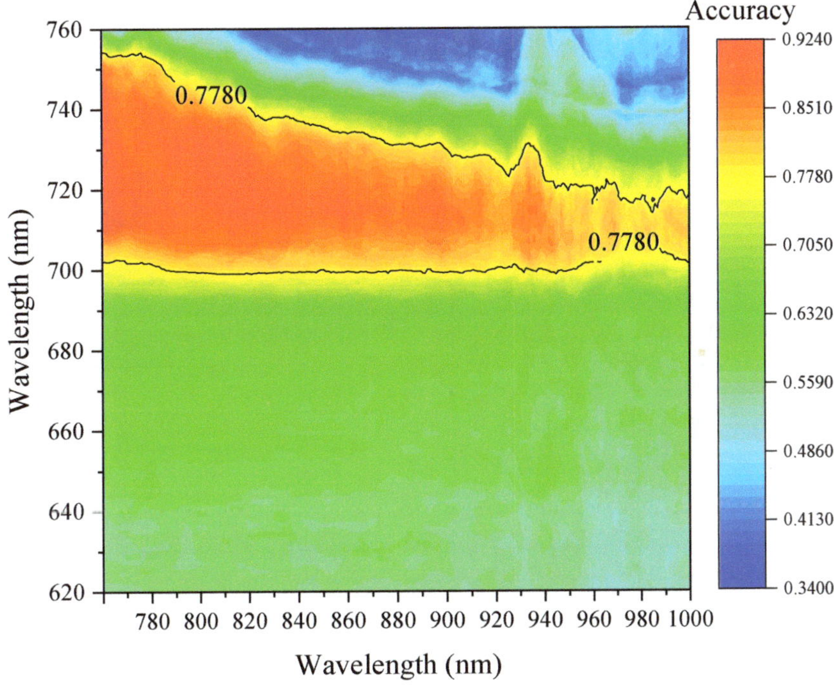

Figure 7. Contour maps visualizing the separability of the red and near-infrared spectral regions at different disease levels.

Finally, we determined the weight coefficients by fitting Fisher's linear discriminant to the training dataset. We concluded that the RBVI achieved the best correlation (correlation: 0.9381) when α and β were 9.78 and -2.08, respectively. The RBVI was finally determined, as shown in Equation (6).

$$\text{RBVI} = 9.78(R_{816} - R_{724}) - 2.08\frac{\rho_{736}}{R_{724}} \tag{6}$$

3.2. Comparative Analysis of RBVI and Traditional SVIs

To preliminarily verify the validity of the SVIs proposed in this study, we first conducted a comparative analysis of 20 SVIs (including the SVIs proposed in this study) using Spearman's correlation analysis, and the results are shown in Figure 8. As a whole, different types of SVIs showed large variability in their correlation with disease levels. A portion of SVIs showed strong correlations with disease levels, such as MTVI-2, $\text{DVI}_{753,665}$, RDVI, MCARI-2, EVI, G, and our proposed RBVI. These seven SVIs were significantly correlated with disease class at the 0.01 level ($|R|$: 0.80–0.94). Among them, our proposed RBVI showed better correlations of -0.9381. In contrast, six SVIs, such as CI, TCARI, and PRI $\times$ CI, showed strong correlations ($|R|$: 0.50–0.72) with the disease levels, and the rest of the SVIs showed poor correlations ($|R|$: 0.09–0.50). Therefore, it was concluded that the existing 13 SVIs and the RBVI proposed in this study have a good correlation with the severity of leaf blast and have a high application value for remote sensing detection of leaf blast.

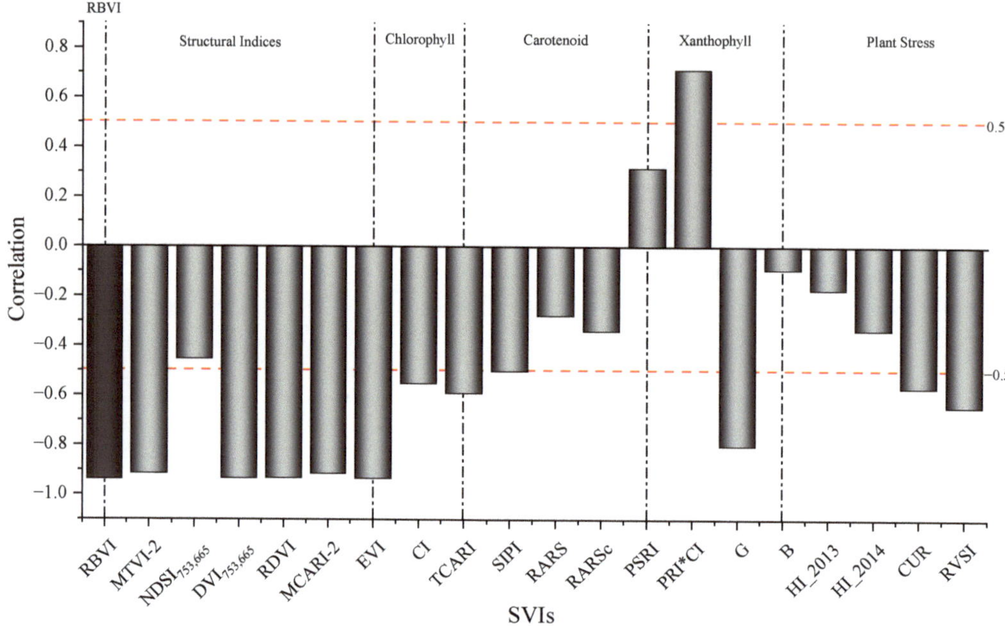

Figure 8. Correlation statistics between different SVIs and the extent of leaf blast disease.

To further screen the discriminative ability of each SVI in leaf blast detection, the mutual information (MI) method, which is computationally small and highly interpretive, was used to rank the feature validity. The value of MI can be interpreted as the amount of information shared between the variables. The higher the value, the greater the interdependence between the two variables. Mutual information captures not only linear relationships between variables but also nonlinear relationships. The MI values between different SVIs and leaf blast disease levels were calculated, and the results are shown in Figure 9.

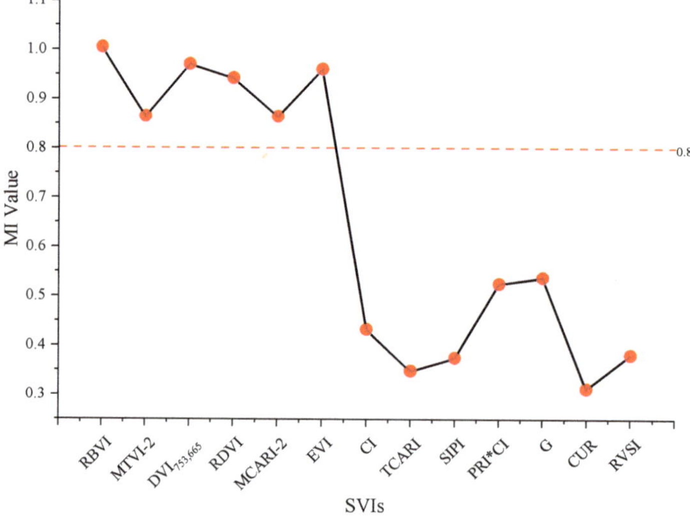

Figure 9. Results of MI values for 21 spectral vegetation indices.

As seen from Figure 9, the 13 SVIs with good correlation had different responses to the detection ability of leaf blast. From the overall determination results, eight SVIs had good discriminatory power with MI values greater than 0.5. In contrast, six SVIs, RBVI, MTVI-2, $DVI_{753,665}$, RDVI, MCARI-2, and EVI, had significant discriminatory ability with MI values greater than 0.80. The RBVI had the highest MI value (MI = 1.00), indicating that our proposed SVIs have a strong early discrimination ability for leaf blast. Therefore, this study used the six SVIs above with significant discriminatory ability for separate modeling to determine the best SVIs for leaf blast detection.

3.3. Results and Analysis of Leaf Blast Detection Based on Stacking Integrated Learning Models

To validate the effectiveness of our proposed SVIs and screen for the best SVIs, a stacking integrated learning leaf blast early detection model was constructed by combining multiple machine learning models (training model for 2022 data, testing model for 2023 data). The modeling results are shown in Table 3.

Table 3. Stacking model detection results for different SVIs.

SVIs		A	M	S	UA (%)	OA (%)	Kappa (%)
RBVI	A	73	0	0	100.0		
	M	5	67	1	91.8		
	S	0	3	70	95.9	95.9	93.8
	PA (%)	93.6	95.7	98.6			
EVI	A	72	1	0	98.6		
	M	4	64	5	87.7		
	S	0	4	69	94.5	93.6	90.4
	PA (%)	94.7	92.8	93.2			
RDVI	A	69	4	0	94.5		
	M	2	65	6	89.0		
	S	0	3	70	95.9	93.2	89.7
	PA (%)	97.2	90.3	92.1			
$DVI_{753,665}$	A	71	2	0	97.3		
	M	9	59	5	80.8		
	S	0	3	70	95.9	91.3	87.0
	PA (%)	88.8	92.2	93.3			
MTVI-2	A	65	8	0	89.0		
	M	11	56	6	76.7		
	S	0	1	72	98.6	88.1	82.2
	PA (%)	85.5	86.2	92.3			
MCARI-2	A	65	8	0	89.0		
	M	11	56	6	76.7		
	S	0	2	71	97.3	87.7	81.5
	PA (%)	85.5	84.9	92.3			

Note: overall accuracy (OA); producer's accuracy (PA); user's accuracy (UA), asymptomatic (A); mild symptoms (M); severe symptoms (S).

As can be seen in Table 3, all six SVIs after preliminary screening had good detection accuracy, with an OA greater than 87.6% and Kappa greater than 81.5%. Compared with the common SVIs (RDVI, etc.), the newly developed RBVI had the highest detection accuracy with an OA and Kappa of 95.9% and 93.8%, respectively. Compared with the second-best EVI, it had further improved leaf blast detection accuracy with a 2.3% improvement in OA and 3.4% in Kappa. These results indicate that the proposed RBVI has good detection accuracy in rice leaf blast detection and can better characterize leaf blast at different disease levels.

In addition, a comparison of UA and PA for different disease levels is also shown in Table 3. The results showed consistency with the overall detection accuracy. Compared with other SVIs, the precision of the RBVI in detecting UA and PA at different disease levels was relatively good, especially showing that the RBVI had better sensitivity and specificity on mild symptoms.

3.4. Comparison and Analysis of the Results of Leaf Blast Detection with Other Models

Based on the different SVIs, our models were compared and analyzed with AdaBoost SVM, RF, Lightgbm, and XgBoost models. The detection results of the RF, Lightgbm, and XgBoost models are shown in Table S2, and the detection results of the AdaBoost and SVM models are shown in Table 4. We selected the better two models (AdaBoost and SVM) from five models, including AdaBoost, for further comparative analysis. The results show that AdaBoost with SVM constructed based on the RBVI has the highest detection accuracy from the perspective of different SVIs. Meanwhile, the modeling accuracy of the RBVI was slightly higher than that of EVI, RDVI, and $DVI_{753,665}$ and much higher than that of MTVI-2 and MCARI-2. This further indicates that the RBVI is significantly sensitive to detecting leaf blast and can characterize the disease more adequately. While analyzed from the perspective of different models, better detection accuracy (OA > 85.8% and Kappa > 78.7%) was achieved in both SVM and AdaBoost modeling. Among them, the SVM leaf blast detection modeling with the RBVI had higher accuracy with an OA and Kappa of 95.4% and 93.2%, respectively. However, compared with the stacking integrated learning model, the detection accuracy of SVM still needs to be improved. Because of this, it can be seen in the comparative analysis of different SVIs and different modeling approaches that the RBVI can better characterize the disease characteristics of varying disease levels of rice early leaf blast at the canopy scale. Using the stacking integrated learning model can help improve the accuracy of early detection of rice leaf blast in the field.

Table 4. AdaBoost and SVM model detection results for different SVIs.

Methods	SVIs		A	M	S	UA (%)	OA (%)	Kappa (%)
AdaBoost	RBVI	A	73	0	0	100.0		
		M	6	67	0	91.8		
		S	0	4	69	94.5	95.0	92.5
		PA (%)	93.6	94.4	98.6			
	EVI	A	72	1	0	98.6		
		M	5	66	2	90.4		
		S	0	7	66	90.4	93.2	89.7
		PA (%)	88.6	90.8	93.3			
	RDVI	A	71	2	0	97.3		
		M	7	60	6	82.2		
		S	0	3	70	95.9	91.8	87.7
		PA (%)	91.0	92.3	92.1			
	$DVI_{753,665}$	A	70	3	0	95.9		
		M	9	59	5	80.8		
		S	0	3	70	95.9	90.9	86.3
		PA (%)	88.6	90.8	93.3			
	MTVI-2	A	61	12	0	83.6		
		M	8	62	3	84.9		
		S	0	5	68	93.2	87.2	80.8
		PA (%)	88.4	78.5	95.8			
	MCARI-2	A	61	12	0	83.6		
		M	8	56	9	76.7		
		S	0	1	72	98.6	86.3	79.5
		PA (%)	88.4	81.2	88.9			

Table 4. *Cont.*

Methods	SVIs		A	M	S	UA (%)	OA (%)	Kappa (%)
SVM	RBVI	A	73	0	0	100.0		
		M	5	67	1	91.8	95.4	93.2
		S	0	4	69	94.5		
		PA (%)	88.4	78.4	90.8			
	EVI	A	70	3	0	95.9		
		M	4	65	4	89.0	93.2	89.7
		S	0	4	69	94.5		
		PA (%)	94.6	90.3	94.5			
	RDVI	A	68	5	0	93.2		
		M	4	62	7	84.9	91.3	87.0
		S	0	3	70	95.9		
		PA (%)	94.4	89.9	90.9			
	$DVI_{753,665}$	A	69	4	0	94.5		
		M	8	59	6	80.8	90.4	85.6
		S	0	3	70	95.9		
		PA (%)	89.6	89.4	92.1			
	MTVI-2	A	61	12	0	83.5		
		M	8	58	7	79.5	85.8	78.8
		S	0	4	69	94.5		
		PA (%)	88.4	78.4	90.8			
	MCARI-2	A	61	12	0	83.5		
		M	8	58	7	79.5	85.8	78.8
		S	0	4	69	94.5		
		PA (%)	88.4	78.4	90.8			

Note: overall accuracy (OA); producer's accuracy (PA); user's accuracy (UA), asymptomatic (A); mild symptoms (M); severe symptoms (S).

4. Discussion

Early detection of rice leaf blast is vital for scientific management and early control of rice fields. In the existing research, most researchers have used spectroscopic technology to carry out an in-depth analysis of early disease detection at the leaf scale of rice, explored the spectral response mechanism and change rule of early disease stress on rice leaves, and achieved better research results. However, this cannot be used and promoted on a large scale outdoors. At the canopy scale, especially in low-altitude remote sensing, it has been documented that using UAVs with spectral sensors for disease detection of crop diseases has a high potential for application [39,40]. However, there is still inapplicability in the early detection of diseases [41]. Therefore, in this study, a rice leaf blast spectral vegetation index (RBVI) was proposed based on UAV hyperspectral remote sensing data to improve the accuracy of early detection of rice leaf blast in the field.

The high accuracy and separability of the RBVI are due to the structure of the spectral vegetation index, the determination of spectral and differential spectral wavelengths, and the weighting coefficients. For the vegetation index structure, it has been shown in existing studies that the use of three-wavelength or feature construction of SVIs can further improve the ability of spectral detection of crops [42]. The reason is that, in general, compared with two-wavelength vegetation indices, three-wavelength vegetation indices can obtain richer crop information with better stability and accuracy—for example, plant senescence reflectance index (PSRI) and healthy index (HI) for monitoring crop senescence and stress. Also, how the three wavelengths are combined for computing is particularly important, which is key to further improving the variability among different substances (different disease levels). At the same time, this combination operation also needs to be determined by combining the special spectral response change law of other detection substances. For example, in this study, by analyzing the spectral response law of healthy, mildly infected, and severely infected rice leaf blast, it was concluded that in the red and near-infrared spectral region, the spectral reflectance showed the change rule of increasing and decreasing with the increase in disease level, respectively.

Moreover, after the spectral reflectance curve was processed by first-order differentiation, it showed a gradual decrease in differential spectral reflectance with increased disease level in the red-edge region. It is the combination of the spectral response law and the use of combinatorial operations that further highlights this variability. As for the determination of the optimal spectral wavelength and differential spectral wavelength, it has been shown in previous studies that the correlation between diseases and crops can show a specific spectral response [43], and this response can be used for crop disease detection. And in this study, Fisher's discriminant analysis and exhaustive algorithm methods were used to screen the spectral wavelengths and differential spectral wavelengths in the vegetation index of RBVIs that were best correlated with different disease levels of leaf blast, respectively. In this way, it was determined that the selected spectral wavelengths had a higher correlation with the differential spectral wavelengths for leaf blast. Finally, based on the training data, a fitting analysis was performed by Fisher's linear discriminant method to determine the better weighting coefficients in the RBVI, which were used to improve the differentiability of the RBVI for early diseases.

To show the ability of the RBVI to detect leaf blast in canopy rice, we used JM distance [2] and the probability histogram [23] for further demonstration and comparative analysis with EVI and MCARI-2. As shown in Table 5, EVI, MCARI-2, and RBVI were distinguishable (JM distance > 1.0) among different leaf blast levels. Among them, our proposed RBVI performed best. The JM distances between A and M and M and S were greater than 1.5, which did not reach the best standard threshold for interclass separability (JM distance > 1.8) but still had better separability than EVI and MCARI-2. Also, the low interclass separability of neighboring disease levels was expected. The reason is a greater similarity in the early spectral data of diseases acquired by UAV hyperspectral remote sensing, especially between neighboring disease levels. And a better interclass separability was presented between A and S with a JM distance of 2.0. Meanwhile, we constructed a probability histogram to demonstrate more intuitively the discriminatory ability of the EVI, MCARI-2, and RBVI in the early detection of leaf blast (Figure 10). It can be seen that the probability distribution of the RBVI has less overlap between different leaf blast levels, followed by the EVI, and the worst is MCARI-2. This indicates that the RBVI has better discriminatory ability compared to the EVI and MCARI-2. Therefore, as a whole, the proposed RBVI has a stronger discriminative ability for early leaf blast levels.

Table 5. JM distances for different leaf blast levels for RBVI, EVI, and MCARI-2.

SVIs	[A, M]	[A, S]	[M, S]
EVI	1.59	1.99	1.39
MCARI-2	1.03	1.93	1.37
RBVI	1.65	2.00	1.52

Note: JM distance between 0 and 1 indicates that SVI has interclass separability, JM distance between 1 and 1.8 means that SVI has some interclass separability, and JM distance between 1.8 and 2.0 suggests that SVI has good interclass separability. A: asymptomatic, M: mild symptoms, S: severe symptoms.

Qualitative studies can generally be processed and analyzed using traditional statistical analysis methods. However, UAV hyperspectral remote sensing for detecting rice leaf blast in the field is susceptible to the external environment and manual investigation factors, so more complex processing and analyzing methods, such as machine learning (ML) algorithms, are needed. Recently, ML algorithms have been successfully applied in disease detection in various food crops, such as wheat, corn, peanut, and rice.

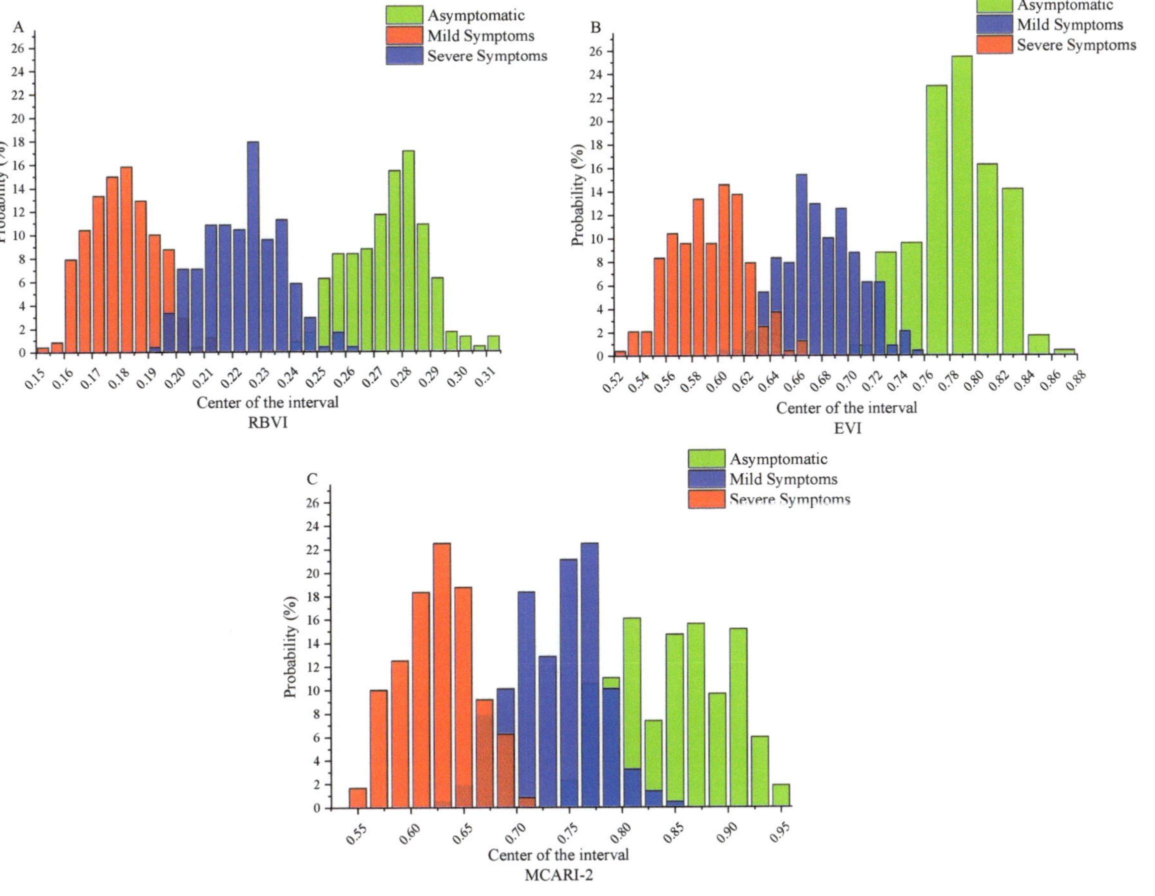

Figure 10. Histogram of the probability of different leaf blast levels for RBVI, EVI, and MCARI-2. (**A**) Histogram of the probability of different leaf blast levels for RBVI. (**B**) Histogram of the probability of different leaf blast levels for EVI. (**C**) Histogram of the probability of different leaf blast levels for MCARI-2.

SVM is a more commonly used ML algorithm in qualitative research and is widely used in disease detection. Compared to other classification algorithms, SVM's excellent correctness rate is due to using the structural rather than empirical principle of risk-minimizing induction [44]. However, in our study, SVM did not achieve the best disease detection results, and the model accuracy was second only to the stacking integrated model because SVM overly pursues the transformation of nonlinear classification in low-dimensional space into linear variety in the high-dimensional area, which drives the model into overfitting [45]. As an iteratively optimized ML classification algorithm, AdaBoost can train different classifiers for the same training set and assemble these classifiers to form a stronger final classifier [46]. However, this study found that AdaBoost is not the best ML algorithm in remote sensing for detecting early leaf blast disease in rice. The possible reason for this is the presence of the external environment (light, wind speed) on the hyperspectral data acquired by UAV remote sensing, affecting AdaBoost's modeling accuracy. Although the overall detection accuracy of the SVM and AdaBoost models did not meet expectations, the modeling results from different SVIs showed that our proposed RBVI could obtain disease characteristics of varying leaf blast disease levels more accurately than other SVIs.

At the same time, under certain circumstances, the use of unique ML algorithms (e.g. SVM) is more prone to the risk of overfitting phenomena [47]. Therefore, to solve this problem, we use an integrated learning algorithm that fuses multiple ML algorithms to obtain a strong learner with better detection results. In this study, five ML algorithms, such as RF, SVM, AdaBoost, Lightgbm, and Xgboost, were used to construct a strong learner based on the principle of stacking integration, and LDA was used as the final classifier of the model. Previous studies reported stacking models in different research areas, such as crop nutrition detection, image recognition, and medical treatment. At the same time, it had not been used for crop disease detection. In our study, the stacking model achieved the best detection accuracy compared to the SVM and AdaBoost models. The stacking model constructed based on the RBVI obtained the highest early detection accuracy (95.9% for OA and 93.8% for Kappa). The possible reason is that the organic stacking of multiple ML algorithms can draw the advantages of different algorithms and improve the generalization and accuracy of the model. It also shows that using stacking integrated learning for crop disease detection is successful.

To show the performance of the RBVI with the stacking model in detecting early leaf blast in rice in the field, Figure 11 shows a graph of the results of leaf blast detection in rice in the area at different times. Also, to highlight the effects of early disease detection, the water bodies in the field were defined as the same color as the bare soil. As shown in Figure 11a, we calibrated the degree of disease in the ground survey, with blue markers for mild and other features for severe. Our comparative analysis revealed great consistency between the areas of different disease levels of leaf blast detected by the RBVI stacking model on 18 July 2023 and the areas of the ground survey. At the same time, the site of leaf blast disease increased with increasing time of infection (without control), which is consistent with the law of disease spread. The large area of disease in the lower right corner of the image is due to the fact that this is the location of the water source of the field, where the disease occurs more severely due to the effects of water temperature and water flow.

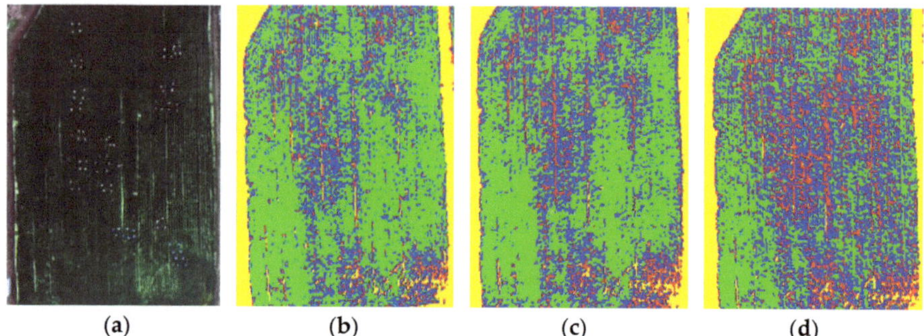

(a)   (b)   (c)   (d)

Figure 11. Graphs showing the results of early detection of rice leaf blast in the field at different times. (**a**) Original image; (**b**) classification result on 18 July; (**c**) classification result on 22 July; (**d**) classification result on 26 July.

In summary, the early detection models for rice leaf blast proposed in this study have achieved relatively promising research results. However, here, we have to elaborate on the study's limitations. The UAV hyperspectral images were all chosen to be acquired when the weather was clear and cloudless to reduce the interference of weather and other external environments on the data and models. In addition, a single rice variety was used for the experiment to more accurately assess the susceptibility of RBVI to leaf blast in the field. In our future research, we will consider acquiring UAV hyperspectral images from multiple regions and in different weather environments to enhance the diversity of data and thus further improve the model's generalization ability. Meanwhile, combining rice growth environment information with remote sensing spectral information for research prediction and forecasting of rice blast disease will also be the focus of our future research.

5. Conclusions

In this study, a leaf blast-specific vegetation index, RBVI, was developed by extracting the information on the spectral reflectance of a rice canopy from hyperspectral images acquired by UAV remote sensing to achieve qualitative detection of asymptomatic, mildly symptomatic, and severely symptomatic leaf blast-infected rice in the field. In order to more significantly highlight the characteristics of different disease levels of leaf blast, we determined the index structure of the RBVI by combining the characteristics of spectral reflectance variation with monotonicity analysis. Then, the best spectral features for calculating the RBVI were screened by correlation analysis and exhaustive search to ensure that the screened spectral features had better sensitivity and specificity for different disease degrees of leaf blast. The results showed that the stacking model for early detection of leaf blast constructed based on the RBVI had the highest detection accuracy, with an OA and Kappa of 95.9% and 93.8%, respectively. The results were also slightly improved compared to the SVM model constructed based on the RBVI with the AdaBoost model. In addition, in order to further indicate the separability of the RBVI for leaf blast, JM distance and a probability histogram were used for comparative analysis. The analysis results showed that the RBVI had a stronger ability to discriminate and differentiate early leaf blast classes compared to other SVIs. Therefore, combining the RBVI with the stacking model can help to improve the detection accuracy of early leaf blast in the field and detect the early signs of rice blast in time. The application of this method can not only reduce the influence of environmental factors such as light changes to a certain extent and improve the accuracy of early detection of rice leaf blast in the field, but it can also reduce the use of chemical pesticides before the spread of the disease by early intervention, thus reducing the negative impact on the environment.

Supplementary Materials: The following supporting information can be downloaded at: https://www.mdpi.com/article/10.3390/agronomy14030602/s1, Table S1: Spectral vegetation indices used in this study and their formulations; Table S2: RF, Lightgbm and XgBoost model detection results; Table S3: Stacking model parameter [48–59].

Author Contributions: D.Z.: methodology, software, validation, formal analysis, visualization, investigation, data curation, resources, writing—original draft, writing—review and editing. Y.C.: formal analysis, validation, data curation, investigation, resources. J.L. (Jinpeng Li): data curation, software. Q.C.: data curation, software. J.L. (Jinxuan Li): data curation, visualization. F.G.: data curation, investigation. T.X.: methodology, validation, investigation, funding acquisition, project administration, supervision. S.F.: data curation, resources, validation, software. All authors have read and agreed to the published version of the manuscript.

Funding: This research was supported by a grant from the Liaoning Province Applied Basic Research Program Project (2023JH2/101300120) and the National Natural Science Foundation of China Youth Program (32201652).

Data Availability Statement: The raw data supporting the conclusions of this article will be made available by the authors on request.

Conflicts of Interest: The authors declare no conflict of interest.

References

1. Tian, L.; Xue, B.; Wang, Z.; Li, D.; Yao, X.; Cao, Q.; Zhu, Y.; Cao, W.; Cheng, T. Spectroscopic detection of rice leaf blast infection from asymptomatic to mild stages with integrated machine learning and feature selection. *Remote Sens. Environ.* **2021**, *257*, 112350. [CrossRef]
2. Feng, S.; Zhao, D.; Guan, Q.; Li, J.; Liu, Z.; Jin, Z.; Li, G.; Xu, T. A deep convolutional neural network-based wavelength selection method for spectral characteristics of rice blast disease. *Comput. Electron. Agric.* **2022**, *199*, 107199. [CrossRef]

3. Deng, Y.W.; Zhai, K.R.; Xie, Z.; Yang, D.Y.; Zhu, X.D.; Liu, J.Z.; Wang, X.; Qin, P.; Yang, Y.Z.; Zhang, G.M.; et al. Epigenetic regulation of antagonistic receptors confers rice blast resistance with yield balance. *Science* **2017**, *355*, 962–965. [CrossRef] [PubMed]

4. Tian, L.; Wang, Z.; Xue, B.; Li, D.; Zheng, H.; Yao, X.; Zhu, Y.; Cao, W.; Cheng, T. A disease-specific spectral index tracks Magnaporthe oryzae infection in paddy rice from ground to space. *Remote Sens. Environ.* **2023**, *285*, 113384. [CrossRef]

5. Samanta, S.; Nayak, S.; Dhua, U.; Mukherjee, A.K. Genotypic-phenotypic diversity and distinctiveness among Magnaporthe grisea isolates from rice and weed Echinochloa colonum. *J. Phytopathol.* **2021**, *169*, 581–596. [CrossRef]

6. Hutt, C.; Bolten, A.; Huging, H.; Bareth, G. UAV LiDAR Metrics for Monitoring Crop Height, Biomass and Nitrogen Uptake: A Case Study on a Winter Wheat Field Trial. *PFG-J. Photogramm. Remote Sens. Geoinf. Sci.* **2023**, *91*, 65–76. [CrossRef]

7. Yang, B.; Zhu, W.X.; Rezaei, E.E.; Li, J.; Sun, Z.G.; Zhang, J.Q. The Optimal Phenological Phase of Maize for Yield Prediction with High-Frequency UAV Remote Sensing. *Remote Sens.* **2022**, *14*, 1559. [CrossRef]

8. Zheng, Q.; Huang, W.J.; Cui, X.M.; Shi, Y.; Liu, L.Y. New Spectral Index for Detecting Wheat Yellow Rust Using Sentinel-2 Multispectral Imagery. *Sensors* **2018**, *18*, 868. [CrossRef]

9. Gorlich, F.; Marks, E.; Mahlein, A.K.; Konig, K.; Lottes, P.; Stachniss, C. UAV-Based Classification of Cercospora Leaf Spot Using RGB Images. *Drones* **2021**, *5*, 34. [CrossRef]

10. Barreto Alcantara, A.A.; Ispizua Yamati, F.R.; Varrelmann, M.; Paulus, S.; Mahlein, A.-K. Disease incidence and severity of Cercospora leaf spot in sugar beet assessed by multispectral unmanned aerial images and machine learning. *Plant Dis.* **2022**, *107*, 188–200. [CrossRef]

11. Sonobe, R.; Sugimoto, Y.; Kondo, R.; Seki, H.; Sugiyama, E.; Kiriiwa, Y.; Suzuki, K. Hyperspectral wavelength selection for estimating chlorophyll content of muskmelon leaves. *Eur. J. Remote Sens.* **2021**, *54*, 512–523. [CrossRef]

12. Lapajne, J.; Knapic, M.; Zibrat, U. Comparison of Selected Dimensionality Reduction Methods for Detection of Root-Knot Nematode Infestations in Potato Tubers Using Hyperspectral Imaging. *Sensors* **2022**, *22*, 367. [CrossRef] [PubMed]

13. Zhao, L.; Zeng, Y.; Liu, P.; He, G.J. Band Selection via Explanations From Convolutional Neural Networks. *IEEE Access* **2020**, *8*, 56000–56014. [CrossRef]

14. Al-Saddik, H.; Simon, J.C.; Cointault, F. Assessment of the optimal spectral bands for designing a sensor for vineyard disease detection: The case of "Flavescence doree". *Precis. Agric.* **2019**, *20*, 398–422. [CrossRef]

15. Pham, Q.T.; Liou, N.S. The development of on-line surface defect detection system for jujubes based on hyperspectral images. *Comput. Electron. Agric.* **2022**, *194*, 106743. [CrossRef]

16. Zhang, C.; Kovacs, J.M. The application of small unmanned aerial systems for precision agriculture: A review. *Precis. Agric.* **2012**, *13*, 693–712. [CrossRef]

17. Yang, S.; Zhang, H.Q.; Fan, W.M. Characteristic wavelengths selection of rice spectrum based on adaptive sliding window permutation entropy. *Food Sci. Technol.* **2022**, *42*, e38922. [CrossRef]

18. Marang, I.J.; Filippi, P.; Weaver, T.B.; Evans, B.J.; Whelan, B.M.; Bishop, T.F.A.; Murad, M.O.F.; Al-Shammari, D.; Roth, G. Machine Learning Optimised Hyperspectral Remote Sensing Retrieves Cotton Nitrogen Status. *Remote Sens.* **2021**, *13*, 1428. [CrossRef]

19. Crusiol, L.G.T.; Sun, L.; Sun, Z.; Chen, R.Q.; Wu, Y.F.; Ma, J.C.; Song, C.X. In-Season Monitoring of Maize Leaf Water Content Using Ground-Based and UAV-Based Hyperspectral Data. *Sustainability* **2022**, *14*, 9039. [CrossRef]

20. Anderegg, J.; Hund, A.; Karisto, P.; Mikaberidze, A. In-Field Detection and Quantification of Septoria Tritici Blotch in Diverse Wheat Germplasm Using Spectral-Temporal Features. *Front. Plant Sci.* **2019**, *10*, 473679. [CrossRef]

21. Domingues Franceschini, M.H.; Bartholomeus, H.; van Apeldoorn, D.; Suomalainen, J.; Kooistra, L. Intercomparison of Unmanned Aerial Vehicle and Ground-Based Narrow Band Spectrometers Applied to Crop Trait Monitoring in Organic Potato Production (vol 17, 1428, 2017). *Sensors* **2017**, *17*, 2265. [CrossRef] [PubMed]

22. Abdulridha, J.; Ampatzidis, Y.; Kakarla, S.C.; Roberts, P. Detection of target spot and bacterial spot diseases in tomato using UAV-based and benchtop-based hyperspectral imaging techniques. *Precis. Agric.* **2020**, *21*, 955–978. [CrossRef]

23. Su, J.Y.; Liu, C.J.; Coombes, M.; Hu, X.P.; Wang, C.H.; Xu, X.M.; Li, Q.D.; Guo, L.; Chen, W.H. Wheat yellow rust monitoring by learning from multispectral UAV aerial imagery. *Comput. Electron. Agric.* **2018**, *155*, 157–166. [CrossRef]

24. Abdulridha, J.; Batuman, O.; Ampatzidis, Y. UAV-Based Remote Sensing Technique to Detect Citrus Canker Disease Utilizing Hyperspectral Imaging and Machine Learning. *Remote Sens.* **2019**, *11*, 1373. [CrossRef]

25. Huo, L.N.; Persson, H.J.; Lindberg, E. Early detection of forest stress from European spruce bark beetle attack, and a new vegetation index: Normalized distance red & SWIR (NDRS). *Remote Sens. Environ.* **2021**, *255*, 112240. [CrossRef]

26. Liu, L.Y.; Dong, Y.Y.; Huang, W.J.; Du, X.P.; Ren, B.Y.; Huang, L.S.; Zheng, Q.; Ma, H.Q. A Disease Index for Efficiently Detecting Wheat Fusarium Head Blight Using Sentinel-2 Multispectral Imagery. *IEEE Access* **2020**, *8*, 52181–52191. [CrossRef]

27. *GBT 15790-2009*; Rules of Investigation and Forecast of the Rice Blast. 2009. Available online: https://www.chinesestandard.net/ AMP/Related.amp.aspx/GBT15790-2009 (accessed on 16 February 2024).

28. Zhao, D.X.; Feng, S.; Cao, Y.L.; Yu, F.H.; Guan, Q.; Li, J.P.; Zhang, G.S.; Xu, T.Y. Study on the Classification Method of Rice Leaf Blast Levels Based on Fusion Features and Adaptive-Weight Immune Particle Swarm Optimization Extreme Learning Machine Algorithm. *Front. Plant Sci.* **2022**, *13*, 879668. [CrossRef] [PubMed]

29. Guan, Q.; Song, K.; Feng, S.; Yu, F.H.; Xu, T.Y. Detection of Peanut Leaf Spot Disease Based on Leaf-, Plant-, and Field-Scale Hyperspectral Reflectance. *Remote Sens.* **2022**, *14*, 4988. [CrossRef]

0. Lee, C.C.; Koo, V.C.; Lim, T.S.; Lee, Y.P.; Abidin, H. A multi-layer perceptron-based approach for early detection of BSR disease in oil palm trees using hyperspectral images. *Heliyon* **2022**, *8*, e09252. [CrossRef]

1. Huang, L.S.; Wu, Z.C.; Huang, W.J.; Ma, H.Q.; Zhao, J.L. Identification of Fusarium Head Blight in Winter Wheat Ears Based on Fisher's Linear Discriminant Analysis and a Support Vector Machine. *Appl. Sci.* **2019**, *9*, 3894. [CrossRef]

2. Ahmadi, P.; Muharam, F.M.; Ahmad, K.; Mansor, S.; Abu Seman, I. Early Detection of Ganoderma Basal Stem Rot of Oil Palms Using Artificial Neural Network Spectral Analysis. *Plant Dis.* **2017**, *101*, 1009–1016. [CrossRef]

3. Fisher, M.C.; Henk, D.A.; Briggs, C.J.; Brownstein, J.S.; Madoff, L.C.; McCraw, S.L.; Gurr, S.J. Emerging fungal threats to animal, plant and ecosystem health. *Nature* **2012**, *484*, 186–194. [CrossRef]

4. Ham, J.; Yangchi, C.; Crawford, M.M.; Ghosh, J. Investigation of the random forest framework for classification of hyperspectral data. *IEEE Trans. Geosci. Remote Sens.* **2005**, *43*, 492–501. [CrossRef]

5. Platt, J. Sequential Minimal Optimization: A Fast Algorithm for Training Support Vector Machines. In *Advances in Kernel Methods-Support Vector Learning*; MIT Press: Cambridge, MA, USA, 1998; Volume 208.

6. Freund, Y.; Schapire, R.E. A Decision-Theoretic Generalization of On-Line Learning and an Application to Boosting. *J. Comput. Syst. Sci.* **1997**, *55*, 119–139. [CrossRef]

7. Meng, Q. LightGBM: A Highly Efficient Gradient Boosting Decision Tree. In Proceedings of the 31st Conference on Neural Information Processing Systems (NIPS 2017), Long Beach, CA, USA, 4–9 December 2017.

8. Chen, T.Q.; Guestrin, C.; Assoc Comp, M. XGBoost: A Scalable Tree Boosting System. In Proceedings of the KDD'16: Proceedings of the 22nd ACM SIGKDD International Conference on Knowledge Discovery and Data Mining, San Francisco, CA, USA, 13–17 August 2016; pp. 785–794.

9. Abdulridha, J.; Ampatzidis, Y.; Qureshi, J.; Roberts, P. Laboratory and UAV-Based Identification and Classification of Tomato Yellow Leaf Curl, Bacterial Spot, and Target Spot Diseases in Tomato Utilizing Hyperspectral Imaging and Machine Learning. *Remote Sens.* **2020**, *12*, 2732. [CrossRef]

40. Zhang, H.; Huang, L.; Huang, W.; Dong, Y.; Weng, S.; Zhao, J.; Ma, H.; Liu, L. Detection of wheat Fusarium head blight using UAV-based spectral and image feature fusion. *Front. Plant Sci.* **2022**, *13*, 1004427. [CrossRef]

41. Guo, A.; Huang, W.; Dong, Y.; Ye, H.; Ma, H.; Liu, B.; Wu, W.; Ren, Y.; Ruan, C.; Geng, Y. Wheat Yellow Rust Detection Using UAV-Based Hyperspectral Technology. *Remote Sens.* **2021**, *13*, 123. [CrossRef]

42. Cao, Z.S.; Cheng, T.; Ma, X.; Tian, Y.C.; Zhu, Y.; Yao, X.; Chen, Q.; Liu, S.Y.; Guo, Z.Y.; Zhen, Q.M.; et al. A new three-band spectral index for mitigating the saturation in the estimation of leaf area index in wheat. *Int. J. Remote Sens.* **2017**, *38*, 3865–3885. [CrossRef]

43. Huang, W.; Guan, Q.; Luo, J.; Zhang, J.; Zhao, J.; Liang, D.; Huang, L.; Zhang, D. New Optimized Spectral Indices for Identifying and Monitoring Winter Wheat Diseases. *IEEE J. Sel. Top. Appl. Earth Obs. Remote Sens.* **2014**, *7*, 2516–2524. [CrossRef]

44. Hesami, M.; Naderi, R.; Tohidfar, M.; Yoosefzadeh-Najafabadi, M. Development of support vector machine-based model and comparative analysis with artificial neural network for modeling the plant tissue culture procedures: Effect of plant growth regulators on somatic embryogenesis of chrysanthemum, as a case study. *Plant Methods* **2020**, *16*, 112. [CrossRef] [PubMed]

45. Guo, Y.; Wang, X.L.; Huang, Y.M.; Xu, L. Collaborative driving style classification method enabled by majority voting ensemble learning for enhancing classification performance. *PLoS ONE* **2021**, *16*, e0254047. [CrossRef]

46. Zhang, Q.W.; Wang, Z.C.; Duan, S.H.; Cao, B.Y.; Wu, Y.T.; Chen, J.; Zhang, H.B.; Wang, M. An Improved End-to-End Autoencoder Based on Reinforcement Learning by Using Decision Tree for Optical Transceivers. *Micromachines* **2022**, *13*, 31. [CrossRef]

47. Yeom, S.; Giacomelli, I.; Menaged, A.; Fredrikson, M.; Jha, S. Overfitting, robustness, and malicious algorithms: A study of potential causes of privacy risk in machine learning. *J. Comput. Secur.* **2020**, *28*, 35–70. [CrossRef]

48. Haboudane, D.; Miller, J.R.; Pattey, E.; Zarco-Tejada, P.J.; Strachan, I.B. Hyperspectral vegetation indices and novel algorithms for predicting green LAI of crop canopies: Modeling and validation in the context of precision agriculture. *Remote Sens. Environ.* **2004**, *90*, 337–352. [CrossRef]

49. Roujean, J.-L.; Breon, F.-M. Estimating PAR absorbed by vegetation from bidirectional reflectance measurements. *Remote Sens. Environ.* **1995**, *51*, 375–384. [CrossRef]

50. Liu, H.; Huete, A. A feedback based modification of the NDVI to minimize canopy background and atmospheric noise. *IEEE Trans. Geosci. Remote Sens.* **1995**, *33*, 457–465. [CrossRef]

51. Haboudane, D.; Miller, J.R.; Tremblay, N.; Zarco-Tejada, P.J.; Dextraze, L. Integrated narrow-band vegetation indices for prediction of crop chlorophyll content for application to precision agriculture. *Remote Sens. Environ.* **2002**, *81*, 416–426. [CrossRef]

52. Penuelas, J.; Frederic, B.; Filella, I. Semi-empirical indices to assess carotenoids/chlorophyll A ratio from leaf spectral reflectances. *Photosynthetica* **1995**, *31*, 221–230.

53. Chappelle, E.; Kim, M.; McMurtrey, J. Ratio Analysis of Reflectance Spectra (RARS): An Algorithm for the Remote Estimation of the Concentrations of Chlorophyll A, Chlorophyll B, and Carotenoids in Soybean Leaves. *Remote Sens. Environ.* **1992**, *39*, 239–247. [CrossRef]

54. Merzlyak, M.; Gitelson, A.; Chivkunova, O.; Rakitin, V. Non-destructive optical detection of pigment changes during leaf senescence and fruit ripening. *Physiol. Plant.* **1999**, *106*, 135–141. [CrossRef]

55. Garrity, S.; Bohrer, G.; Maurer, K.; Mueller, K.; Vogel, C.; Curtis, P. A comparison of multiple phenology data sources for estimating seasonal transitions in deciduous forest carbon exchange. *Agric. For. Meteorol.* **2011**, *151*, 1741–1752. [CrossRef]

56. Calderón Madrid, R.; Navas Cortés, J.; Lucena, C.; Zarco-Tejada, P. High-resolution airborne hyperspectral and thermal imagery for early detection of Verticillium wilt of olive using fluorescence, temperature and narrow-band spectral indices. *Remote Sens. Environ.* **2013**, *139*, 231–245. [CrossRef]

57. Mahlein, A.K.; Rumpf, T.; Welke, P.; Dehne, H.W.; Plumer, L.; Steiner, U.; Oerke, E.C. Development of spectral indices for detecting and identifying plant diseases. *Remote Sens. Environ.* **2013**, *128*, 21–30. [CrossRef]

58. Zarco-Tejada, P.; Miller, J.; Mohammed, G.; Noland, T.; Sampson, P. Chlorophyll Fluorescence Effects on Vegetation Apparent Reflectance: II. Laboratory and Airborne Canopy-Level Measurements with Hyperspectral Data. *Remote Sens. Environ.* **2000**, *74*, 596–608. [CrossRef]

59. Merton, R.; Huntington, J. Early simulation results of the ARIES-1 satellite sensor for multi-temporal vegetation research derived from AVIRIS. In Proceedings of the Eighth Annual JPL Airborne Earth Science Workshop, Pasadena, CA, USA, 8–14 February 1999.

 **agronomy**

Article

Rapeseed Seed Coat Color Classification Based on the Visibility Graph Algorithm and Hyperspectral Technique

Chaojun Zou [1,2], Xinghui Zhu [1,*], Fang Wang [3,*], Jinran Wu [4] and You-Gan Wang [5]

1 College of Information and Intelligent Science and Technology, Hunan Agricultural University, Changsha 410125, China; zouchaojun@stu.hunau.edu.cn
2 School of Economics and Management, Changsha Normal University, Changsha 410125, China
3 Key Laboratory of Intelligent Computing and Information Processing of Ministry of Education and Hunan Key Laboratory for Computation and Simulation in Science and Engineering, Xiangtan University, Xiangtan 411105, China
4 Institute for Positive Psychology and Education, Australian Catholic University, Banyo, Brisbane 4014, Australia; ryan.wu@acu.edu.au
5 School of Mathematics and Physics, The University of Queensland, St Lucia, Brisbane 4067, Australia; you-gan.wang@uq.edu.au
* Correspondence: zhuxh@hunau.edu.cn (X.Z.); fwang4@xtu.edu.cn (F.W.)

Abstract: Information technology and statistical modeling have made significant contributions to smart agriculture. Machine vision and hyperspectral technologies, with their non-destructive and real-time capabilities, have been extensively utilized in the non-destructive diagnosis and quality monitoring of crops and seeds, becoming essential tools in traditional agriculture. This work applies these techniques to address the color classification of rapeseed, which is of great significance in the field of rapeseed growth diagnosis research. To bridge the gap between machine vision and hyperspectral technology, a framework is developed that includes seed color calibration, spectral feature extraction and fusion, and the recognition modeling of three seed colors using four machine learning methods. Three categories of rapeseed coat colors are calibrated based on visual perception and vector-square distance methods. A fast-weighted visibility graph method is employed to map the spectral reflectance sequences to complex networks, and five global network attributes are extracted to fuse the full-band reflectance as model input. The experimental results demonstrate that the classification recognition rate of the fused feature reaches 0.943 under the XGBoost model, confirming the effectiveness of the network features as a complement to the spectral reflectance. The high recognition accuracy and simple operation process of the framework support the further application of hyperspectral technology to analyze the quality of rapeseed.

Keywords: hyperspectral reflectance; visibility graph algorithm; vector-square distance; color classification; machine learning

Citation: Zou, C.; Zhu, X.; Wang, F.; Wu, J.; Wang, Y.-G. Rapeseed Seed Coat Color Classification Based on the Visibility Graph Algorithm and Hyperspectral Technique. *Agronomy* 2024, 14, 941. https://doi.org/10.3390/agronomy14050941

Academic Editor: Giovanni Cabassi

Received: 27 March 2024
Revised: 22 April 2024
Accepted: 26 April 2024
Published: 30 April 2024

1. Introduction

Rape (*Brassica napus* L.) is a significant and widely grown oilseed crop whose seeds, known as rapeseed, are rich in fatty acids and proteins and serve as the raw material not only of edible vegetable oils but also of premium-grade proteins for animal feed. In general, yellow rapeseed has a thinner seed coat, a lower percentage of hulls, and a larger embryo compared to other colors. Research indicates that rapeseeds with a more yellowish and lighter color tend to have higher oil content, as well as higher levels of oleic and linoleic acids, while also exhibiting higher protein content and lower fiber content [1,2]. Since the color of the rapeseed coat is controlled by its corresponding gene, it has become an important trait for breeding. Yellow-seeded *Brassica napus* is a hybridization of other varieties [3,4]. Therefore, it is of great value economically and very meaningful for breeding to identify rapeseed coat color and screen lighter yellow seeds.

Some researchers have explored various methods for the color recognition of rapeseed seeds. Usually, there are several types: (1) Artificial visual recognition. The classification of rapeseed seed coat color through manual observation or a magnifying glass. (2) The use of instruments, including colorimeters, color analyzers, and other instruments to measure and analyze colors [1]. (3) Spectral analysis methods. Using equipment such as spectrophotometers, near-infrared spectroscopy, and Fourier transform infrared photoacoustic spectroscopy to obtain spectral reflectance information, based on the principle of color formation and the relationship between spectra, spectral data are used to analyze or identify colors [5–7]. (4) Digital image analysis methods. Obtain images using a digital camera or scanner, and use computer image processing software or other specialized software for recognition [8,9]. In addition, yellow seeds are identified with the help of chemical solutions [10]. Among them, visual methods are simple and convenient but rely on experience, lack unified standards, and have significant differences. The objective results with a certain degree of accuracy using optical and spectral analysis techniques have low accuracy in color segmentation due to resolution reasons. The popularity of digital devices has made the digital image analysis of rapeseed color an efficient and non-destructive method. Hyperspectral imaging technology can simultaneously obtain spectral information from multiple bands, providing richer spectral information (including images and data) and characterizing the inherent features of substances. Due to its high resolution and comprehensive spectrum, it has achieved successful applications in many fields such as agriculture, geology, and environment [11,12], for example, the identification of rapeseed varieties and the prediction of oleic acid, protein, and other components by hyperspectral [13,14].

Different species have specific reflectance absorption characteristics at different wavelengths. In traditional hyperspectral analysis techniques, spectral indices derived from the combination of spectral values in one or more bands are often used as research variables. In studying plant traits, some spectral vegetation indices have been proposed, such as NDVI, RVI, and DVI, and have been widely applied in remote sensing fields such as soil and water [15]. At the same time, some scholars have constructed spectral indices related to color, including the anthocyanin index (ARI) [16], carotenoid reflectance index (CRI) [17,18], red to green ratio index (RGRI) [19], chlorophyll absorption index (CARI) [20], photochemical vegetation index (PRI) [21,22], and improved chlorophyll absorption ratio index (MCARI) [23,24]. Although these studies have achieved certain results, extracting single or multiple spectral indices as parameter variables can result in the loss of some information from hyperspectral data, thereby limiting their effectiveness in identifying or inverting species varieties or traits. High-dimensional spectra can be viewed as sequences or curves, whose features will be hidden in their topological structures [25]. A complex network is a powerful tool for studying sequence topology relationships, by mining the inherent properties between objects and the entire system. Given the advantages of this technology in mining spectral reflectance sequences, in this work, we attempt to explore the topological feature of hyperspectral reflectance from the network angle and build a bridge between RGB-based machine vision systems and hyperspectral signals to challenge the rapeseed coat classification, which is a difficult task in traditional rapeseed growth diagnose, as mentioned above.

In practice, our goal is to achieve the following objectives:

- Using vector and angular distance formulations for color difference calculations in the RGB color space to achieve color calibration.
- Constructing complex networks of spectral reflectance by fast-weighted visibility graph algorithm.
- Establishing an intelligent model for rapeseed seed coat color classification by combining hyperspectral technology and machine vision systems.

By utilizing complex networks and machine learning methods, the precise automatic recognition of rapeseed seed coat color can be achieved, which provides a scientific basis for further exploring the close relationship between rapeseed color and other physical properties, as well as oil content, fatty acid composition, breeding research, and industrial quality optimization.

The rest of the paper is structured as follows: in Section 2, we describe the material preparation, data acquisition, and methodology. Section 3 presents the experimental results. We give discussions and conclusions in Sections 4 and 5, respectively.

2. Materials and Methods

2.1. Data Acquisition and Preprocessing

Two original varieties of *Brassica napus* seeds are employed in our study, namely, Xiangyou 708 and Xiangyou 710, of high oleic acid rapeseed. They were cultivated in the paddy fields of Yunyuan Experimental Base (28°23′ N, 112°93′ E) at Hunan Agricultural University, Changsha City, Hunan Province in September 2020 and harvested in April 2021. The two varieties, having been cultivated for numerous years, which are widely promoted in the Hunan province of China, serve as representative examples in our study. The SOC710 portable hyperspectral imager (Spectral range: 380~1091 nm; Resolution: 4.9458 nm; Spectral channels: 128; Sensor material: CCD; Spatial pixels: 520; Spatial resolution (Avg. RMS spot radius): <40 microns; Manufacturer: Surface Optics Corporation of the United States) and its darkroom system are used to collect the rapeseed hyperspectral data. The sample collection process strictly follows the spectral collection specifications. Rapeseeds are placed flat (see Figure 1) in a transparent petri dish in a dark room with a light source. The hyperspectral imager is placed in a small task box 370 mm vertically above the rapeseed samples. On each seed's surface, the SOC710 hyperspectral imager is used to measure the spectral reflectance five times randomly in some regions of interest (ROIs) and followed by an arithmetic averaging operation, which is as the reflectance of the sample. Meanwhile, the image color information of the three primary colors (red, green, and blue) of the samples is collected in the same ROIs. We finally obtained 282 samples in all, including 282 hyperspectral reflectance data and the corresponding RGB values.

Since the noise is always hidden at the end of bands, we remove the two ends of bands to obtain the reflectance from 400 nm to 1000 nm. Then, we perform linear interpolation with a resampling interval of 1 nm, and thus 601 reflectance values are obtained for each sample. The reflectance values characterized the physical (e.g., seed coat color) and chemical (e.g., content of components such as fatty acids, proteins) traits of the different samples of rapeseeds. The spectral curves are smoothed and filtered using the convolutional (Savitzky–Golay) smoothing method with noise reduction. Finally, the spectral data are normalized.

Figure 1. Rapeseed samples.

2.2. Framework of the Classification Model

The traits of rapeseed offspring are often reflected in their seed coat color; therefore, identifying and classifying seed coat color will help screen high-quality offspring for breeding and cultivation. For example, the genetic improvement of yellow-seeded rapeseed has garnered more attention than other color variations within the spectrum of rapeseed due to its favorable characteristics such as reduced lignin content, elevated oil content, and increased protein content.

As stated above, in traditional agriculture, relying on human eye recognition is inefficient and has a high error rate, which motivates us to develop an intelligent approach to fulfill this task. In this work, modern tools, machine vision technology, and hyperspectral technology are employed for our consideration. We aim to establish a bridge between them. Our model framework is shown in Figure 2. In the following, we present the color calibration process by machine vision system (Section 2.3). Then, we introduce a tool that prevails in the field of time series analysis, the visibility graph algorithm, which is used to extract hyperspectral intrinsic features (Section 2.4). Next, the four classifiers are introduced briefly in Section 2.5, which will be used to execute the classification task. Finally, four indicators are given in the Section 2.6 to evaluate the performance of the models.

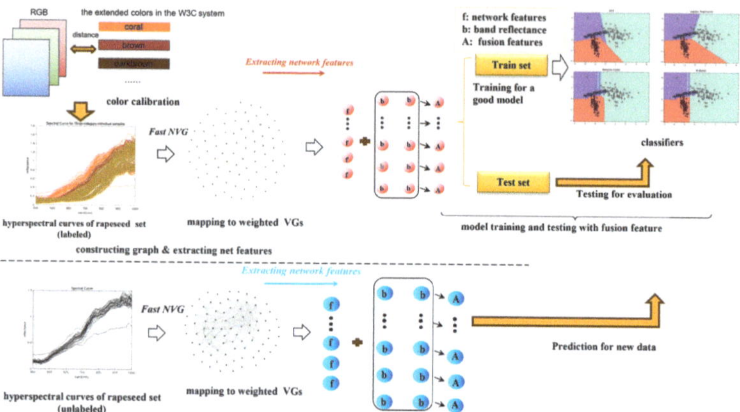

Figure 2. Framework of recognition model based on the hyperspectral technology and complex network analysis. The abbreviation "VG" means visible graph algorithm, and "W3C" refers to World Wide Web Consortium.

2.3. Color Calibration

Color calibration is necessary for the task of seed color separation. It is often performed by colorimeters, color cards, artificial vision, and machine vision systems. Ensuring definite standards and processes is key to accurately comparing the color of rapeseed from different batches. The RGB color space (which refers to red, green, and blue, respectively) is the most commonly used color system in machine vision. In what follows, we introduce vector-angular distance to fulfill the task of color calibration.

2.3.1. Vector-Angular Distance (VAD)

The Vector-Angular Distance (VAD) color difference equation is a method used to quantify the difference between two colors in the RGB color space [26–28]. The VAD equation [29] takes into account both the magnitude and the angular information of the color vectors, as defined by

$$dist = \sqrt{\frac{S_r^2 \cdot w_r \cdot (r_1 - r_2)^2 + S_g^2 \cdot w_g \cdot (g_1 - g_2)^2 + S_b^2 \cdot w_b \cdot (b_1 - b_2)^2}{(w_r + w_g + w_b) \cdot 255^2} + \theta^2 \cdot S_\theta \cdot S_{ratio}}, \tag{1}$$

where w_r, w_g, and w_b are the weights of the human eye's sensitivity to changes in the red, green, and blue components, respectively, with the range [0,1] (in this work, we took the equality weight for them). S_i, $i = r, g, b$ denotes the importance of the R, G, and B color channels, respectively, determined by

$$S_i = \min\left(\frac{3(i_1 + i_2)}{r_1 + r_2 + g_1 + g_2 + b_1 + b_2}, 1\right), \quad i = r, g, b. \tag{2}$$

Subscripts 1 and 2 denote two different colors. θ is the normalized angle between two color vectors in RGB space, given by

$$\theta = \frac{2}{\pi} \cdot \arccos\left(\frac{r_1 \cdot r_2 + g_1 \cdot g_2 + b_1 \cdot b_2}{\sqrt{(r_1^2 + g_1^2 + b_1^2) \cdot (r_2^2 + g_2^2 + b_2^2)}}\right). \tag{3}$$

S_θ is the contribution of the three colors' difference to the vector angle of two colors to be compared, which is adjusted by θ under the importance of each channel

$$S_\theta = S_{\theta_r} + S_{\theta_g} + S_{\theta_b}, \tag{4}$$

and

$$S_{\theta_i} = \frac{\frac{|i_1-i_2|}{i_1+i_2}}{\frac{|r_1-r_2|}{r_1+r_2} + \frac{|g_1-g_2|}{g_1+g_2} + \frac{|b_1-b_2|}{b_1+b_2}} \cdot S_i^2, \quad i = r, g, b. \tag{5}$$

and

$$S_{ratio} = \frac{\max(r_1, r_2, g_1, g_2, b_1, b_2)}{255} \tag{6}$$

According to Equations (1)–(6), the RGB values are dynamically adjusted to compensate for the inhomogeneity of the RGB system. Thus, the distance between each pair of colors composed of RGB is obtained.

2.3.2. Rapeseed Seed Color Calibration

Comparing the rapeseed seed colors involved in our experimental as shown in Figure 1 with the standard and extended colors in the W3C (https://www.w3.org/wiki/CSS3/Color/Extended_color_keywords, accessed on 19 October 2023), referring to the world wide web consortium, an international organization tasked with developing web technology standards, 14 colors in the families of yellow, orange, and brown, namely, dark gray, red, brown, dark brown, tan, maroon, orange, dark orange, coral orange-red, green yellow, yellow, light yellow, and golden, are selected as benchmarks for the calibration of seed color, as shown in Figure 3. According to the VAD method presented in Section 2.3.1, the distance between the RGB values of the 14 colors and those of our 282 samples is calculated. The category with the closest distance is taken as the color category of the samples. It shows that the 282 samples are categorized into three color categories of coral (75 samples, denoted as Class 0), brown (20 samples, denoted as Class 1), and dark brown (187 samples, denoted as Class 2), whose corresponding HSV systems are (0.0448, 0.6863, 1), (0.0000, 0.7455, 0.6471), (0.0833, 0.6733, 0.3961), respectively. The HSV color model is a color space, also known as the hexane model, where H stands for Hue, which describes the position of color in the color spectrum; S stands for saturation, which describes the vividness of color; and V stands for Value, which is the luminance value of the color, with higher values indicating lighter colors and lower values referring to darker colors. From the corresponding HSV values, it can be concluded that the coral category of rapeseed is the lightest in color, followed by the brown category, and the darkest in color is the dark brown category of rapeseed; thus, the coral category of rapeseed owns a higher oil content and is a target for breeding and industrial screening. In this respect, the identification of seed color plays an important role in predicting oil content and breeding. Because the different methods of light absorption and scattering will lead to different colors on the seed's surface, spectral reflectance is an ideal measure used to characterize the color of the seed. To verify this visually, we lay the above three categories of samples into a three-dimensional space constructed by the top three principal components with the highest contribution of spectral reflectance, as shown in Figure 4. The samples of the three categories are separated basically, which prompts us to further study the identification model of the seed color separation by using the hyperspectral feature. We also extract nine color indices, as listed in Table 1, which are highly related to the visual inspection of the seed coat color.

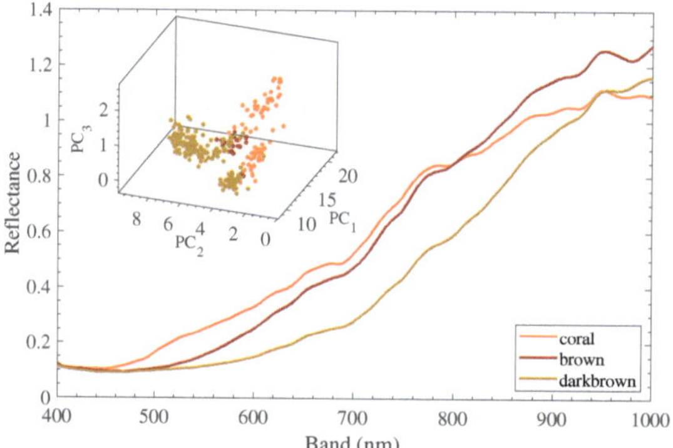

Figure 3. Color card with RGB values. Tips: * according to the rule of W3C css-system (https://www.w3.org/TR/css-color-3/#css-system, accessed on 19 October 2023), authors can customize the color values according to the environmental needs, we add a new color card, namely, darkbrown, RGB (101, 67, 33), with reference to color-hex (https://www.color-hex.com/color/654321, accessed on 19 October 2023), that is closer to the real rapeseed seed coat color to the color card family.

Figure 4. Main panel is the average spectral reflectance of rapeseed samples of the three categories; insert is the scatter plot in a 3-dimensional space constructed by the three principal components of the spectral reflectance.

Table 1. Nine selected color indices.

	Name	Abbreviation	Formula	Source
1	chlorophyll absorption ratio index	CARI	$(R_{750} - R_{705})/(R_{750} + R_{705})$	Ref. [17]
2	photochemical reflectance index	PRI	$(R_{531} - R_{570})/(R_{531} + R_{570})$	Ref. [22]
3	red ratio spectral index	RRSI	R_{705}/R_{910}	Ref. [30]
4	green ratio spectral index	GRSI	R_{582}/R_{910}	Ref. [30]
5	red difference spectral index	RDSI	$R_{418} - R_{610}$	Ref. [30]
6	green difference spectral index	GDSI	$R_{421} - R_{536}$	Ref. [30]
7	red normalized difference spectral index	RNDSI	$(R_{646} - R_{995})/(R_{646} + R_{995})$	Ref. [30]
8	green normalized difference spectral index	GNDSI	$(R_{568} - R_{988})/(R_{568} + R_{988})$	Ref. [30]
9	blue normalized difference spectral index	RNDSI	$(R_{531} - R_{571})/(R_{531} + R_{571})$	Ref. [30]

2.4. Visibility Graph Algorithm

Spectral reflectance is the ratio of the reflected flux of a ground object in a certain wavelength band to the incident flux in that band, which contains two aspects of information, namely, the data value and the order of bands. The spectral data value is of importance for characterizing the characteristics of ground objects. However, it only represents the absolute reflection of light from different wavebands on an object, but cannot reflect the difference in their reflectance at different bands. We need a tool to effectively capture the difference in the reflectance among every band associated with their order. Here, we use an idea of the visibility graph (VG) algorithm [31] to transform each spectral reflectance series into a network firstly, where the network nodes are the bands and the link among them refers to the relationship of the reflectance according to visibility rules. It has the advantage of retaining the information on the reflectance values and the order of bands. Then, five network global attributes are extracted to be used for the complements to the full-band reflectance data value. Finally, the full-band reflectance and the net attributes are fused to be used as features and input into the intelligent models for seed color identification.

2.4.1. Introduction of Visibility Graph Algorithm

The VG algorithm was proposed to analyze time series from a higher dimensional view. With the merit of operability without any assumptions, the algorithm was widely applied in various fields [32–36]. The process of VG consists of two parts: firstly, the reflectance values are arranged on the axes as a bar chart in sequential order; then, a line is drawn at the top of the bar chart according to the visibility criterion. The visibility criterion is defined below:

$$x_k < x_j + (x_i - x_j) \cdot \frac{t_k - t_j}{t_i - t_j}, \quad \forall k \in (i, j), \tag{7}$$

where t and x denote the horizontal and vertical coordinates, respectively. (t_i, x_i) and (t_j, x_j) are two points, and (t_k, x_k) is a arbitrary point between them. According to this criterion, it determines whether there is a link between two given bands and thus yields a so-called VG network. Using this criterion on the hyperspectral data can reflect the relative values of the spectral reflectance between any two bands. To capture more accurate information on the relative changes in reflectance between two bands, in this work, we consider a weight edge connection based on the original criterion. The edge weight is defined as the angle between the line connecting the reflectance of two visible bands and the horizontal direction,

$$w_{ij} = \frac{1}{2}\arctan(\frac{x_i - x_j}{t_i - t_j}), \tag{8}$$

where x_i and x_j are the two points that satisfy Equation (7). Accordingly, the weighted VG networks can be constructed by using the weighted links.

The idea of VG is simple, and the corresponding VG network possesses the proprieties of connectivity, anisotropy, and affine invariance, which makes it inherit the intrinsic structural characteristics of the original series. However, when examining the visibility between every pair of points, it is necessary to judge the value of each point between them by Equation (7), which leads to high time complexity. Therefore, in this work, we use a fast conversion algorithm based on the partition strategy introduced by [37]. The main idea is: firstly, divide the original bands into left and right segments at maximum value; then, starting from the maximum point, examine the visibility by comparing the angle between adjacent two bands (given by Equation (7)) and traverse each point of the left and right bands, respectively, meanwhile, the weight is calculated by Equation (8); and lastly, continue the above connection process recursively until the band on both sides is no longer divisible. Accordingly, by mapping every band into network nodes and using the fast-weighted VG algorithm, 282 spectral-weighted VG networks can be obtained.

2.4.2. Network Global Attributes

Complex networks are characterized by simplicity intuition, and structural robustness [38,39]. For each hyperspectral reflectance signal, its VG network $G = (N, L, W)$. N is the node set of all nodes with the number n; L is the set of all links with the number l, and W is the set of weight values of all links, with the element w_{ij} given by Equation (8). (i, j) is a link between nodes i and j with the weight w_{ij}, $(i, j \in N)$. The interpretation of the spectral data can be achieved by extracting its corresponding VG network topological properties. To this end, five global network attributes are employed for our consideration.

(1) Clustering coefficient

The clustering coefficient characterizes the connectivity of the neighboring nodes of a node. The global clustering coefficient can be defined as the average of the clustering coefficients of all nodes, or it can be expressed as the ratio of three times the number of all triangles in the network to the ratio of all connected triples in the network. The two definitions are formally different and not exactly equivalent, and the second definition is used in this study, given by

$$\langle CC \rangle = \frac{1}{n} \sum_{i=1}^{n} \frac{\sum_{j \neq k} w_{ij} \cdot w_{ik} \cdot w_{jk}}{\max_j(w_{ij}) \sum_{j \neq k} w_{ij} \cdot w_{ik}}. \tag{9}$$

(2) Assortativity coefficient

The assortativity coefficient measures the degree of similarity or interconnectedness of nodes in a network and provides a quantitative way to analyze the network's similarity and connectivity patterns [38]. Positive assortativity coefficients indicate that nodes tend to interconnect with other nodes with similar or the same strength, which means that nodes with similar properties are more likely to form connections. The weighted assortativity coefficient was developed by [40], calculated by

$$\gamma = \frac{l^{-1} \sum_{(i,j) \in L} w_{ij} \cdot k_i^w \cdot k_j^w - [l^{-1} \sum_{(i,j) \in L} w_{ij} \cdot (k_i^w + k_j^w)/2]^2}{l^{-1} \sum_{(i,j) \in L} w_{ij} \cdot (k_i^{w2} + k_j^{w2})/2 - [l^{-1} \sum_{(i,j) \in L} w_{ij} \cdot (k_i^w + k_j^w)/2]^2}, \tag{10}$$

where k_i^w is the weighted degree of node i, calculated as $k_i^w = \sum_{j \in N} w_{ij}$.

(3) Global efficiency

Global efficiency describes the speed and efficiency of information transfer between any two nodes in a network. Global efficiency can be calculated by the following steps: for each pair of nodes in the network, calculate its shortest path length (shortest path length). The shortest path length is the sum of the minimum number of edges or weights required for a connection between two nodes. The reciprocal of all the shortest path lengths is added and divided by the total number of nodes in the network minus 1. Here, the reciprocal indicates that the shorter the distance of the path, the more efficient it is. The final result obtained is the global efficiency. The weighted efficiency (denoted by Eff) is computed using an auxiliary connection-length matrix related to the edge weight [38], determined by

$$Eff = \frac{1}{n(n-1)} \sum_{i \in N} \sum_{j \in N, j \neq i} (d_{ij}^w)^{-1}, \tag{11}$$

where d_{ij}^w is the shortest weighted path length between nodes i and j computed by Dijkstra's algorithm.

(4) Structural entropy

In information theory, entropy is used to measure the uncertainty of information. Structural entropy, on the other hand, is used to measure the complexity of the structure of a system. Network structural entropy is a measure of the structural complexity of a network [41,42]. A higher structural entropy of a network indicates a more complex structure, implying a more even or diverse distribution of node degree values within the network. Conversely, a lower structural entropy suggests a simpler structure, with a more concentrated or uneven distribution of node degrees. Calculating network structural entropy is a common method used to assess network complexity; it is based on the entropy of node degree distribution [43,44], which is determined by

$$Ent = -\sum_{i=1}^{n} I_i \cdot \ln I_i,$$

where I_i is the importance of node i, defined by $I_i = k_i / \sum_{i \in N} k_i$, and k_i is the degree of node i. In calculating the structural entropy of the network, we excluded nodes with degree 0.

The two extreme values are $Ent_{max} = \ln n$ when all nodes in the network have the same degree, where $I_i = 1/n$ while $Ent_{min} = [\ln 4(n-1)]/2$, when the network has a highly skewed distribution of node degrees, with some nodes having very high degrees and others having very low degrees, where $I_1 = 1/2$ and $I_j = 1/[2(n-1)]$ for $j > 1$. To mitigate the influence of the network size n on the Ent, we can normalize Ent using the following approach to ensure that $0 \leq NEnt \leq 1$:

$$NEnt = \frac{Ent - Ent_{min}}{Ent_{max} - Ent_{min}} = \frac{-2\sum_{i=1}^{n} I_i \cdot \ln I_i - \ln 4(n-1)}{2\ln n - \ln 4(n-1)} \tag{12}$$

(5) Graph density

Graph density [38] is the ratio between the number of actual edges and the number of possible edges in a network and measures how closely the nodes in the network interact with each other and the connectivity of the graph. The graph density ranges from 0 to 1. When the graph density is close to 1, it means that the nodes in the network are very tightly connected and are close to being fully connected. When the graph density is close to 0, it means that the nodes in the network are very sparsely connected, and the correlation between the nodes is weak. The graph density only considers the number of edges and does not consider the edge weights, which are calculated as follows:

$$\rho = \frac{2l}{n(n-1)}. \tag{13}$$

We briefly conclude how the VG network works for the hyperspectral feature extraction as follows: Firstly, each hyperspectral reflectance curve is mapped into the weighted network according to the fast-weighted VG algorithm. Each network can be represented by a weighted adjacency matrix. Then, the five network attributes are calculated by using Equations (9)–(13) based on the weighted adjacency matrix, which is used to characterize the topological structure of each reflectance curve. As mentioned above, the VG network inherits the intrinsic structural features of the original sequence, reflecting the relationship between the reflectance of different bands. It can be considered complementary to the full-band reflectance values. It is important to note that the network attributes cannot replace the reflectance in our model since the VG network properties reflect the relationship between the reflectances (relative difference) among every band rather than the reflectance information itself. Next, the five attributes are combined as the network feature. The fusion feature of the full-band reflectance values and the network feature is input into the intelligent models (see the next subsection) to implement the rapeseed color classification task, which is expected to achieve higher accuracy.

2.5. Classifiers

As a central task in supervised learning, classification has received widespread attention in various scientific fields. Up to now, lots of excellent classification models have been proposed. In this work, four classifiers, namely, Logistic Regression (Logit-R), Support Vector Machines (SVM), Random Forests (RF), and Extreme Gradient Boosting (XGBoost, we use xgboost with version number 1.7.6 in Python with version number 3.8.13), are employed for rapeseed color classification. Logit-R is a classical generalized linear regression analysis model, which was introduced in the early 20th century. SVM was generated in the 1960s, which performs well in high-dimensional spaces and is especially suitable for the case of a few samples. RF was developed in 1995, contains multiple decision trees, and has good immunity to outliers. As one of the most outstanding models in machine learning in the past decade, XGBoost is a significant promotion of the gradient-boosting decision tree. Next, we introduce them briefly.

2.5.1. Logit-R

Logit-R is a generalized linear regression model. It first estimates the probability of an event occurring based on a given dataset of independent variables and then uses a step function or sigmoid function to achieve the purpose of sample classification. With this simple and easy-to-operate idea, Logit-R has become one of the frequently employed techniques for classification in the field of machine learning. It was originally used for binary classification tasks. Some meaningful extensions, such as multinomial Logit-R, are used to handle multi-classification tasks. Meanwhile, regularized versions, like L_1 or L_2 regularization, help to prevent overfitting by penalizing large coefficients. However, Logit-R cannot solve linear inseparable problems and nonlinear problems since the essence of this method is to classify by generating a straight line or a hyperplane.

2.5.2. SVM

SVM is a popular statistical theory-based learning method. The core idea is to find a segmentation hyperplane so that the points on both sides are as far away as possible from the hyperplane, i.e., the maximum spacing hyperplane. Mapping data to higher dimensional spaces using kernel functions makes the samples easier to separate in a new space; in the meantime, it avoids the computation of nonlinear surface segmentation in the original input space. A linearly separable binary classification hyperplane can be expressed as $\omega \cdot x + b = 0$ with a constraint of $y(\omega \cdot x + b) \geq 1$, whose optimal solution can be obtained via the Lagrange function. Additionally, by introducing a regularization parameter, it is easy to control the penalty for classification errors. In the method of SVM, the penalty coefficient C is a critical parameter. The smaller values of C will allow some sample classification errors but will produce larger intervals. Conversely, the larger C will strictly penalize classification errors and may lead to more compact but complex decision boundaries. SVM is an efficient method for representing complex functional dependencies in high-dimensional spaces [45–47]. SVM can be used to process the data with nonlinear features; however, its classification results mainly depend on the selection of kernel functions. Therefore, parameter tuning is important when applying support vector machine modeling for classification.

2.5.3. RF

RF is a parallel ensemble learning algorithm through random sampling and selecting features, which is based on bootstrap aggregation. It constructs multiple decision trees and integrates them to improve the model performance and generalization ability. With these advantages, it has a good ability to process high-dimensional and large-scale datasets with good immunity to noisy data. In addition, overfitting will not occur even without the need for complex parameter tuning in real data modeling.

2.5.4. XGBoost

XGBoost is a class of integrated algorithms based on gradient-boosted trees proposed by Chen [48]. Gradient boosting is an integrated learning method that iteratively adds multiple base learners (decision trees) by learning on the residuals of the previous round of base learners, thereby reducing the prediction error and improving the overall performance. With the base learner of classification and regression tree (CART), XGBoost is a scalable machine learning system that can be used to solve classification and regression problems. It predicts the output by using the K additivity function consisting of multiple weak classifiers:

$\hat{y} = \sum_{k=1}^{K} f_k(x_i), f_k \in F$, where F is the ensemble space of CART. To learn the set of functions, a regularization objective is typically included to minimize the objective function [49]: $L(\phi) = \sum_i \tau(\hat{y}_i, y_i) + \sum_k \Omega(f_k)$, where $\Omega(f) = \gamma T + \frac{1}{2}\lambda|\omega|^2$. By using gradient boosting methods and iterative strategy, XGBoost adopts parallel and distributed computation.

Moreover, the XGBoost makes full use of hardware acceleration and multi-threaded processing to accelerate the training process, which greatly improves computational efficiency; therefore, it can be used to solve large-scale problems. Moreover, XGBoost can handle advanced, complex feature spaces and, through techniques such as weight adjustment and sampling strategies, can handle data that are highly unbalanced in class distribution to make accurate predictions [48,50–54].

The XGBClassifier function of the XGBoost module in the scikit-learn extension library has three types of arguments: regular parameters, model parameters, and learning parameters. *n_estimators* and *learning_rate* are some of the most important parameters in the model parameters and learning parameters, respectively. Among them, *n_estimators* is the number of decision trees used in constructing the gradient boosting model. A larger value of *n_estimators* means that the model has more decision trees and therefore higher complexity and learning capacity. Increasing *n_estimators* can improve model performance, especially when working with complex problems or large datasets. However, if *n_estimators* is set to be too large, the model may overfit the training data, resulting in a decrease in generalization performance on new data and an increase in training time. *learning_rate* is the step size that controls how much each base learner (decision tree) learns about the residuals during an iteration. A smaller value of *learning_rate* means that the model weights are updated less at each iteration, so training will be slower. The value is a floating point number between 0 and 1, with a default value of 0.3. A smaller *learning_rate* improves the stability and generalization of the model and reduces the risk of overfitting. However, training time may increase with a smaller *learning_rate*. Optimization techniques can be used to find the best combination of hyperparameters to obtain the best model performance.

2.6. Performance Evaluation

Accuracy indicates the proportion of samples correctly classified by the model for the entire test dataset. The average accuracy (*Acc*) is the most widely employed performance measure for classification models. For the unbalanced classification data, however, there is a flaw in this metric that cannot truly reflect the classification level of the model. Naturally, several metrics such as precision (*Pre*), recall (*Rec*), and F_1 score are designed to supplement the average accuracy rate. In this work, we focus on the 3-categorization rapeseed seed color classification task. The above-evaluating indicators are calculated separately for each category. The category of interest is regarded as the positive class, and all others are negative classes. Four kinds of predictions are defined for a certain class on the test set: (1) TP: refers to the number of samples in the positive class that the model accurately predicts; (2) FN: refers to cases where the model misclassifies a sample that is a positive class as a negative class; (3) FP: refers to the number of positive classes is mistakenly predicted as the negative class; and (4) TN: represents the number of samples in the negative class that the model correctly predicts. The confusion matrix is shown in Figure 5. Accordingly, Acc, Pre, Rec, and F_1 are given by Equation (14). The K-fold cross-validation is used to evaluate the performance of models. The calculation process is then repeated 100 times to mitigate the influence of randomness.

$$\begin{cases} Acc = \dfrac{TP + TN}{TP + FN + FP + TN}, \\ Pre = \dfrac{TP}{TP + FP}, \\ Rec = \dfrac{TP}{TP + FN}, \\ F_1 = \dfrac{2TP}{TP + FP + FN}. \end{cases} \tag{14}$$

<table>
<tr><td></td><td colspan="3" align="center">predicted</td></tr>
<tr><td rowspan="4" style="writing-mode:vertical-rl">observed</td><td></td><td>class 1</td><td>class 2</td><td>class 3</td></tr>
<tr><td>class 1</td><td>a11</td><td>a12</td><td>a13</td></tr>
<tr><td>class 2</td><td>a21</td><td>a22</td><td>a23</td></tr>
<tr><td>class 3</td><td>a31</td><td>a32</td><td>a33</td></tr>
</table>

<table>
<tr><td></td><td colspan="3" align="center">predicted</td></tr>
<tr><td rowspan="4" style="writing-mode:vertical-rl">observed</td><td></td><td>class 1</td><td>class 2</td><td>class 3</td></tr>
<tr><td>class 1</td><td>TP</td><td colspan="2">FN</td></tr>
<tr><td>class 2</td><td rowspan="2">FP</td><td colspan="2" rowspan="2">TN</td></tr>
<tr><td>class 3</td></tr>
</table>

Figure 5. Confusion matrix of 3-categorization (**left**), where a11, a22, and a33 denote the correct categorization for each class, respectively. Taking class 1 as an example of interest, TP, TN, FP, and FN are shown on the (**right**) panel.

3. Results

According to the framework shown in Figure 2, each sample in the hyperspectral dataset is transferred into a network (graph) structure by using the VG algorithm. For each network, the five network attributes are extracted, whose basic statistics are shown in Table 2. The numerical differences in different categories help to classify them. Following this, we enter the experimental process of the classification task. Three aspects of comparison experiments, namely, comparison among classifiers, comparison among features, and impact of model parameters, are considered in this process. The results are reported in each subsection.

Table 2. Statistical description of the global network attributes of the three categories samples.

Attribute	Categories	Mean	Std	Min	25%	50%	75%	Max
$\langle CC \rangle$	coral	3.60×10^{-3}	8.18×10^{-4}	2.11×10^{-3}	3.08×10^{-3}	3.50×10^{-3}	4.19×10^{-3}	5.66×10^{-3}
	brown	3.78×10^{-3}	6.11×10^{-4}	2.58×10^{-3}	3.30×10^{-3}	3.86×10^{-3}	4.17×10^{-3}	5.21×10^{-3}
	dark brown	3.71×10^{-3}	8.35×10^{-4}	2.12×10^{-3}	3.18×10^{-3}	3.54×10^{-3}	4.06×10^{-3}	6.71×10^{-3}
$NEnt$	coral	0.89	0.01	0.86	0.88	0.89	0.89	0.93
	brown	0.9	0.01	0.88	0.9	0.9	0.91	0.92
	dark brown	0.93	0.03	0.87	0.91	0.93	0.95	0.98
γ	coral	0.19	0.11	-0.03	0.1	0.19	0.27	0.46
	brown	0.19	0.13	-0.16	0.14	0.19	0.27	0.45
	dark brown	0.09	0.15	-0.36	0.02	0.12	0.19	0.38
Eff	coral	2.71×10^{-3}	7.75×10^{-4}	1.40×10^{-3}	2.25×10^{-3}	2.61×10^{-3}	3.02×10^{-3}	4.75×10^{-3}
	brown	3.54×10^{-3}	6.54×10^{-4}	2.49×10^{-3}	3.18×10^{-3}	3.47×10^{-3}	3.83×10^{-3}	5.44×10^{-3}
	dark brown	4.32×10^{-3}	1.27×10^{-3}	2.00×10^{-3}	3.43×10^{-3}	4.31×10^{-3}	5.00×10^{-3}	9.01×10^{-3}
ρ	coral	0.16	0.04	0.09	0.13	0.15	0.18	0.27
	brown	0.25	0.03	0.19	0.24	0.25	0.27	0.31
	dark brown	0.36	0.10	0.14	0.28	0.37	0.44	0.55

3.1. Performance Comparison among Four Classifiers

With the different working principles, the performance of the classifiers is influenced by the features as well as the datasets. To comprehensively examine the validity of the proposed framework of the rapeseed color classification, the four classifiers mentioned in Section 2.5 are employed in our dataset. We input the fusion features composed of the full-band reflectance and the five network attributes into the classifier of RF, SVM, Logit-R, and XGBoost, respectively. The K-fold cross-validation with $K = 2 \sim 10$ is executed on the four models. The model performance presented by Acc is shown in Figure 6. The standard deviation of the Acc calculated from 10 repetitions is also shown on each bar. Clearly, although there is different model performance obtained from the four considered classifiers, the fact of the Acc with more than 0.9 shows our model is workable. The best performance comes from the XGBoost model, with its highest Acc reaching at 0.9315 under 9-fold cross-validation. This is not surprising since this popular method can adjust weights on imbalanced samples adaptively, which applies to our data exactly (in our dataset, the numbers of the three color categories are 75, 20, and 187). By contrast, the performance of the SVM classifier is the worst. One of the reasons is just the impact of imbalanced samples in our dataset.

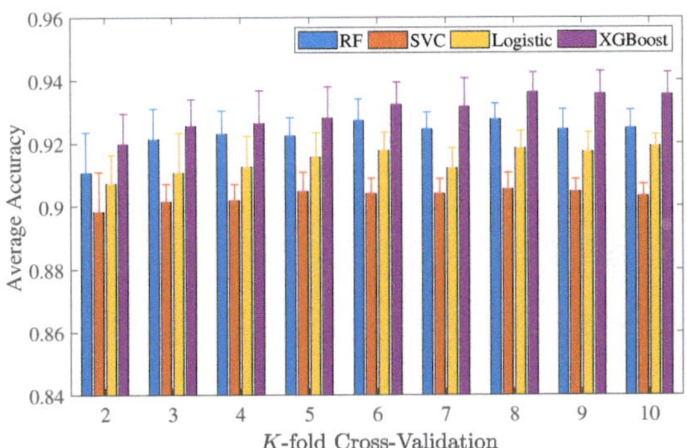

Figure 6. Comparison of the average *Acc* among four classifiers of RF, SVM, Logit-R, and XGBoost concerning *K*-fold cross-validation. The error bar represents the standard deviation calculated from 10 repetitions.

3.2. Performance Comparison among Features

In this subsection, we focus on the comparison of the model performance among different types of hyperspectral features. Besides the network attributes and full-band reflectance, we also investigate the model result from the fusion feature of nine color indices, as listed in Table 1, which is the full-band reflectance. Therefore, we have six types of features, namely, spectral reflectance, color index, network attributes, the fusion feature of color index and reflectance, the fusion feature of network attributes and reflectance, and the fusion feature of color index and network attributes. We perform the same experimental process on them. At this time, only the XGBoost classifier is employed for this purpose due to its best performance. The result is presented in Figure 7. As expected, the fusion feature displays better model performance than the full-band reflectance with every *K*-fold cross-validation. Although the only network attributes produce unsatisfactory results, they help to upgrade the model performance via the feature fusion with the spectral reflectance, demonstrating that the network feature is an important supplement for the original spectral reflectance. We will discuss this detail in Section 4. After fusing the two types of features, the *Acc* is improved obviously under all folds of cross-validation, with the highest upgrade rate being 1.33%.

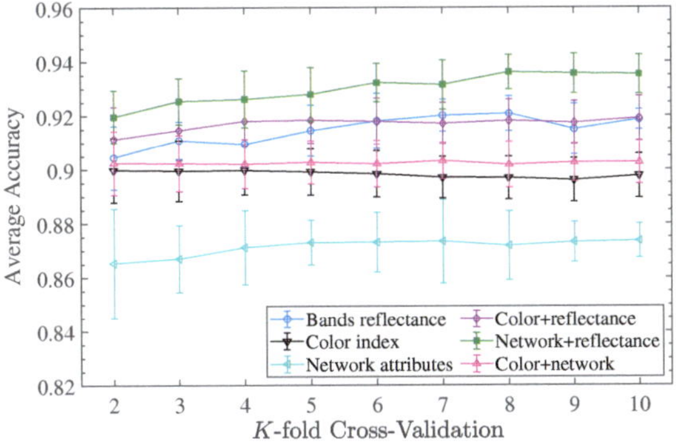

Figure 7. Comparison of the average *Acc* between six types of features with *K*-fold cross-validation by using XGBoost. The error bar represents the standard deviation calculated from 10 repetitions.

3.3. Performance of Parameter Tuning

As is well as known, hyper-parameter selection is critical in machine learning, which affects the model performance greatly. In this subsection, we try to achieve optimal model performance by parameter tuning. Generally, grid search is a universal method for exploring model hyper-parameters; however, it has huge time complexity. To efficiently address the hyperparameter tuning, the sparrow algorithm [55] was proposed, which is a metaheuristic bio-inspired optimization algorithm inspired by the foraging and evasion behaviors of sparrows. In the sparrow algorithm, individuals are divided into two categories: producers and foragers. Producers behave like resource seekers, actively searching for potential solutions in the solution space, akin to foraging. Foragers are responsible for collecting these solutions through interactions with producers, similar to how sparrows obtain food from their peers. The sparrow algorithm utilizes random optimization techniques to perform global and local searches in the solution space for model parameter optimization. In what follows, we apply the sparrow algorithm to find the optimal parameter in the XGBoost model. The two optimal parameters of *n_estimators* and *learning_rate* of the XGBoost model are autonomously searched, as listed in Table 3 for K-fold cross-validation ($K = 2{\sim}10$), which is expected to bring better model performance. Figure 8 illustrates the results obtained from the fusion feature of network and bands before and after parameter tuning under 2~10-fold cross-validation. It is seen that the *Acc* has improved slightly under all folds cross-validation after tuning with the maximum upgrade rate of 1%, as shown in the insert plot. Quantitatively, the average classification result under five-fold cross-validation is listed in Table 4, where the three evaluative indicators defined by Equation (14) are shown on the left side and the confusion matrix is shown on the right side.

Table 3. Optimal parameters of the XGBoost model for each fold of K-fold Cross-validation ($K = 2$ to 10) Using the Sparrow Algorithm.

K	2	3	4	5	6	7	8	9	10
n_estimators	1500	1500	10	1500	620	77	1500	1500	1443
learning_rate	0.3	0.3	0.3	0.3	0.3	0.01	0.3	0.3	0.3
fitness value	0.397	0.159	0.162	0.144	0.106	0.102	0.095	0.063	0.073

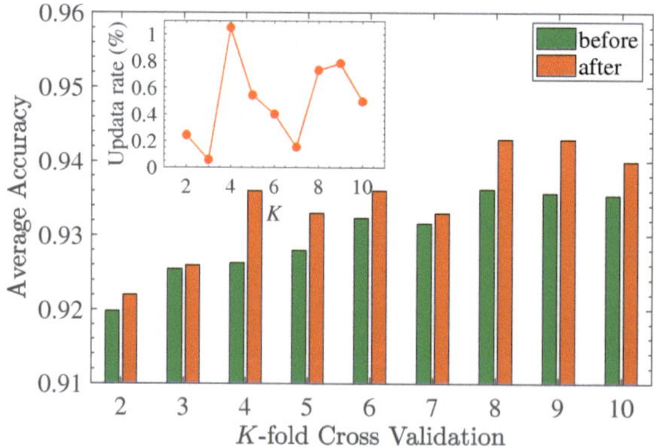

Figure 8. Comparison of the average *Acc* calculated from the fusion feature of network and spectral reflectance before and after parameter tuning.

Table 4. The average classification result under five-fold cross-validation. The left one shows the results of precision, recall, and F_1-score. The right one shows the corresponding confusion matrix.

	Pre	*Rec*	F_1-Score	Support		0	1	2
0	0.92	0.95	0.93	75	0	71	2	2
1	0.75	0.70	0.67	20	1	2	12	6
2	0.96	0.97	0.96	187	2	4	2	181
Acc			0.94	282				

3.4. Misclassification Sample Analysis

In this subsection, we focus on the misclassified samples, which may help us better understand the rapeseed classification mechanism. According to the confusion matrix shown on the right side of Table 4, we record the 18 misclassified samples in Table 5, where 'Center label' refers to the class label that has the shortest average Euclidean distance between the misclassified sample under consideration and the classes in the training set, and 'Closest label' refers to the class label of the nearest neighbor (denoted as Closest ID, see the second column on the right of the table) of the misclassified sample being considered. Three situations can be concluded: namely, the sample is misclassified to the class with the center label (Case I), the sample is misclassified to the class with the closest label (Case II), and the other case (case III).

Table 5. The three situations of the 18 misclassified samples.

	Misclassified ID	True Label	Pred Label	Center Label	Closest ID	Closest Label
	30	0	1	1	37	2
	41	0	1	1	121	2
	66	0	2	2	136	2
	73	0	2	2	81	2
Case I	71	1	2	2	77	1
	77	1	2	2	179	2
	153	1	2	2	168	2
	121	2	1	1	41	0
	122	2	1	1	109	2
	84	1	2	1	124	2
	93	1	2	1	77	2
	98	1	2	1	102	2
Case II	37	2	0	1	30	0
	54	2	0	2	47	0
	131	2	0	1	94	0
	134	2	0	2	55	0
Case III	86	1	0	1	99	1
	100	1	0	1	36	1

For case I, there are nine misclassified samples (#66, #73, ..., #121), whose predicted label is the class with the center label, suggesting that they are closer to these classes. We show the samples with index #73 and #77 as an example in Figure 9a. Both of them are closest to class 2 from the spectral reflectance curve; thus they were misclassified to class 2. For case II, there are seven misclassified samples (#84, #93, ..., #134), whose predicted label is the class with the closest label according to their nearest neighbor's label. The sample with index #84 is exactly this case, as shown in Figure 9b, whose nearest neighbor is #124 with label 2; hence, it was misclassified to class 2. In both figures, *Dis* means Euclidean distance. Now, let us pay attention to samples #100 and #86, which do not belong to the above two cases. They are misclassified to Class 0; however, the closest sample to them is not in Class 0. This forces us to search for new potential evidence. According to the decision-making process of the XGBoost model, which makes bifurcation decisions based on the importance of features, say, the distribution of the reflectance may affect the classification. To this end,

Jensen − Shannon divergence (denoted as D_{J-S}) is employed for our consideration, which is always used to appraise the difference between two distributions. Figure 9c and 9d illustrate the probability density function (PDF) of the reflectance of the samples #100 and #86, respectively. We calculate the D_{J-S} between one of them and the average reflectance distribution of training samples for each class, as also shown in the legend of the figures. For sample #100, the value of D_{J-S} between its PDF and the PDF of the averaged reflectance of class 0 is the smallest, indicating that the sample #100 may be classified as class 0. For the sample #86, although the class with the smallest D_{J-S} still comes from its true label class (Class 1), its nearest neighbor (in the sense of minimum divergence) is the training sample #35, which belongs to Class 0. Therefore, sample #86 may be misclassified to Class 0.

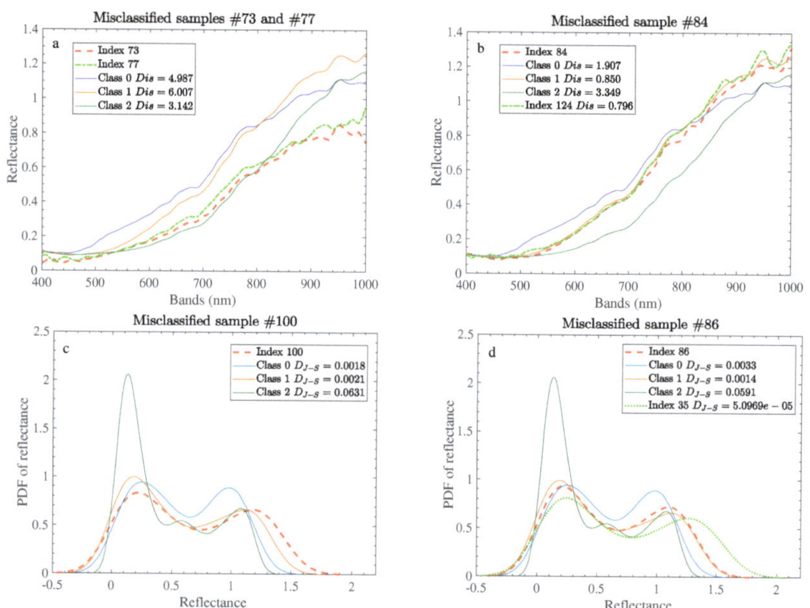

Figure 9. Three cases of misclassification. (**a**) is an example of misclassified samples #73 and #77, which belongs to case I; (**b**) is an example of a misclassified sample #84, which belongs to case II; (**c**) and (**d**) are examples of misclassified samples #100 and #86, respectively, which belong to the case III. *Dis* in upper panels means Euclidean distance, and D_{J-S} in bottom panels indicates *Jensen − Shannon* divergence.

4. Discussions

Hyperspectral technology, as a crucial near-earth remote sensing technique, boasts advantages such as non-destructiveness and efficiency, making it suitable for various crop growth monitoring tasks. Utilizing hyperspectral sensors enables the acquisition of spectral images and data across hundreds of bands. Different objects or surface materials exhibit distinct spectral reflectance in different bands, forming a unique spectral fingerprint. Leveraging spectral fingerprints allows for the effective differentiation of species types or prediction of surface properties and chemical compositions [13,56–58]. The complex network is a powerful tool used to understand the topological and dynamic characteristics of complex systems [34,38,41,59]. By applying complex network analysis methods, we can delve into the structural characteristics of spectral fingerprints and unearth special information embedded within them.

In this work, a visibility graph algorithm was adopted to extract global network features of spectral reflectance, revealing non-linear structural relationships within spectral fingerprints. Through integrating original spectral reflectance with the network features and employing a machine learning classification model for rapeseed seed coat color identification, we observed a significant performance improvement compared to methods using only original reflectance. We also put the fusion features consisting of the original reflectance and spectral color index in the same model. The result shows that the recognition accuracy is not better than using only the original spectral reflectance, which indicates that the spectral color index cannot form a benign complementarity for the original spectral reflectance, while the network features can do so [60]. One potential reason may lie in the design of VG algorithms, i.e., the VG can capture the difference in the reflectance among every band associated with their order, which compensates for the differential information in variable bands of the original spectral reflectance. In addition, the fast-weighted VG algorithm enhances this difference numerically and improves the efficiency of network construction.

Machine learning, which has played an important role in the growth diagnosis of modern smart agriculture, helps train intelligent models for our rapeseed color classification. As a popular machine learning technology, XGBoost, renowned for its gradient boosting framework, weighted learning strategies, and parallel processing, typically outperforms traditional machine learning methods in terms of both performance and computational efficiency [48,50]. In this work, the XGBoost model is also confirmed to bring the best performance to the rapeseed color classification task.

Since this work involves two key issues in machine learning (feature extraction and model parameter selection), we will now discuss the sensitivity of these two issues to the results. On the one hand, the parameter tuning, as mentioned above, is critical to the machine learning model. In Section 3.3, we utilize the sparrow algorithm to find the optimal parameter combination of the XGBoost model for K-fold cross-validation (K = 2 to 10). The optimal parameter combinations of {*n_estimators, learning_rate*} are quite different from those obtained by the sparrow algorithm as listed in Table 3. Here, we reconsider the parameter selection for the XGBoost model. At this time, we use a global search strategy to search for the optimal parameters of *n_estimators* and *learning_rate* under five-fold cross-validation. The *n_estimators* and *learning_rate* are set from 5 to 1630 and from 0.005 to 0.4, respectively, with equidistant 26 and 25 values. The *Acc* concerning the two parameters is shown in Figure 10. It is explained that the sparrow algorithm is a fast optimization method that may fall into local optima. As seen from Figure 10, there are five sets of parameter combinations (see the red triangles) that enable the *Acc* to reach its maximum value of 0.9397; interestingly, the corresponding optimal *learning_rate* is at a very small level, i.e., 0.005~0.01, indicating that the trained model is relatively complex. By contrast, the optimal *n_estimators* lacks an obvious pattern (the five optimal selections are 70, 265, 720,785, and 1110), suggesting that the contribution of *n_estimators* to the model is not as significant as that of *learning_rate*

On the other hand, to investigate the performance of the five network attributes, a sensitive analysis is conducted for them. We employ the concept of information gain [61] to measure the importance of features. In practice, for every feature, calculate the information gain (denoted by *Gain*), defined by

$$Gain = H(D) - \sum_{v=1}^{V} \frac{N_{D_v}}{N_D} H(D_v), \tag{15}$$

where *Gain* is the information gain of each feature, referring to the feature importance, N_D and N_{D_V} denote the number of samples in the dataset and feature subset, and $H(D)$ is the Shannon entropy, defined as

$$H(D) = - \sum_{i=1}^{c} p_i \log p_i, $$

where p_i is the ratio of the class i in the dataset D produced by dividing the dataset D into different subsets, and then we summarize these information gains. In general, the greater the *Gain*, the greater the contribution of the studied feature to classification, and the higher the importance of the feature. We use this notion to calculate the importance of the five network attributes, which are illustrated in Figure 11a according to the model performance based on the fusion feature of the full-band reflectance and the network features. It illustrates that the most important network attribute is the graph density ρ whose importance is higher than other attributes significantly. In addition, we calculate the *Acc* of the model under the fusion feature of full-band reflectance and the different number of the optimal network attributes (according to the importance sort shown in Figure 11a) as shown in Figure 11b. The result is obtained under a five-fold cross-validation. It shows that there are no significant differences between the five *Acc*, say, when we use only one network attribute (i.e., the graph density ρ), the *Acc* reaches 0.9325, which is slightly lower than 0.9397 in the case that we use all the five network attributes. The result is calculated under the optimal parameter combination discussed above. In this regard, the graph density is the only key feature, which can be suggested to execute the rapeseed coat color classification task.

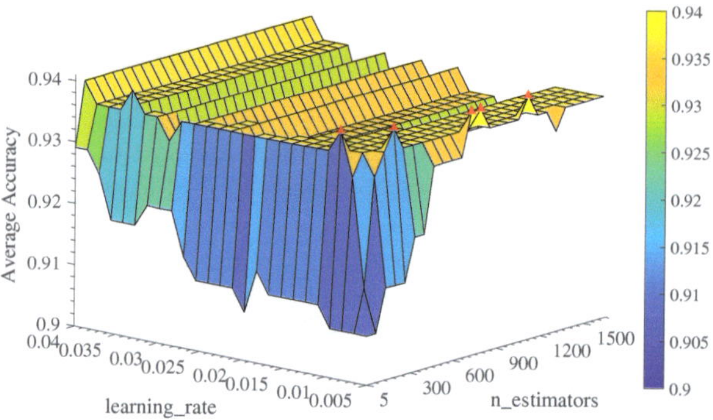

Figure 10. Parameter sensitivity analysis of XGBoost model. The best *Acc* is 0.9397; there are five corresponding optimal parameter combinations of {*n_estimators, learning_rate*}, as presented by the red triangle.

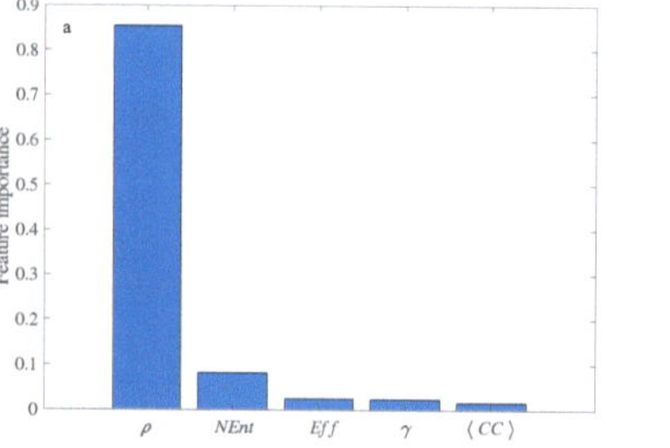
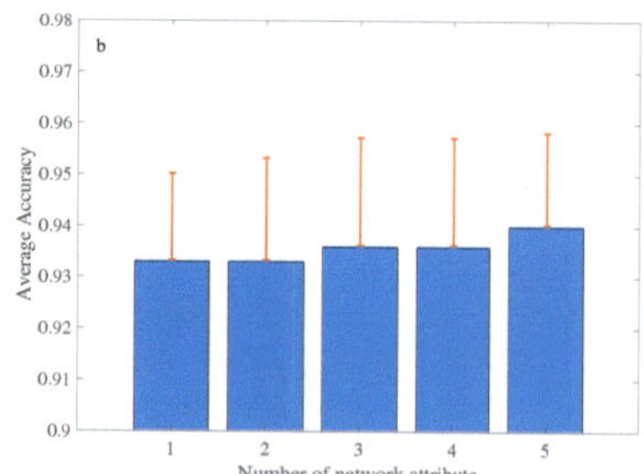

Figure 11. Network feature sensitivity analysis. (**a**) is the importance rate of the five network attributes, which is according to the model performance based on the fusion feature of the full-band reflectance and the network features. (**b**) is the *Acc* under the fusion feature of full-band reflectance and the different number of optimal network attributes with five-fold cross-validation.

Misclassification sample analysis is necessary for the classification task, especially when using machine learning models. By carefully analyzing the misclassification samples based on the confusion matrix, we can conclude there are three cases of misclassification. The mainstream cases are that the samples are misclassified to the class with the center label and the closest label, both in the sense of minimum Euclidean distance. The other two samples are exceptions, which are misclassified to the class with their closest class in the sense of minimum divergence.

One limitation of this study is the restricted sample size and color range analyzed. The experimental data are derived from two *Brassica napus* varieties grown in a single field trial season (2020–2021) in the Changsha City region. Due to the relatively small sample size, only 14 main colors representing each color family from the W3C system were selected as baseline color values. It is important to note that our integration of hyperspectral technology and machine vision for rapeseed seed coat color identification represents an initial effort in this field. This study contributes to a better understanding of the relationship between hyperspectral signals and image color information of crops, which are two critical technologies in smart agriculture. Furthermore, this work offers innovative insights into hyperspectral data mining from the perspective of complex networks, providing a necessary complement to the original reflectance data.

5. Conclusions

Machine vision technology has been widely applied in the growth monitoring and diagnosis of agricultural informatization, serving as a crucial auxiliary tool in traditional agriculture. Over the past decade, hyperspectral technology, as a novel information technology, has found extensive applications in both military and civilian sectors. In this work, we have proposed a framework that combines machine vision and hyperspectral technology with machine learning methods to achieve intelligent seed classification tasks. This approach provides a non-destructive and highly effective alternative to traditional labor-intensive and laboratory techniques for identifying the color of rapeseed seeds. With the utilization of inherent rich spectral and image data from rapeseed samples, this technology holds significant potential in the field of agriculture.

The proposed framework establishes a bridge between machine vision and hyperspectral technology. Specifically, our focus is on the color classification task of rapeseed seeds. Initially, three color categories are calibrated using visual perception and the vector angle distance method: 75 samples for coral color, 20 samples for brown, and 187 samples for dark brown. Subsequently, the weighted VG algorithm is used to map each of the hyperspectral samples into a network. Five network attributes are extracted and fused with the full-band spectral reflectance, which is input into the XGBoost model to achieve rapeseed color recognition. The experimental results show that the best Acc of the fused feature reaches 0.932 (under 8-fold cross-validation), which is higher than the results from the original full-band reflectance ($Acc = 0.922$) and other combination features significantly. The average classification recognition rate can be up to 0.943 through tuning parameters by the sparrow algorithm.

Theoretically, hyperspectral technology covers a wide range of bands, including visible light and near-infrared, with the advantages of high resolution and spectral integration, offering the potential for higher accuracy. However, due to the high time and cost associated with acquiring various sample data, the experimental sample data in this study are relatively limited. When the sample size is sufficient and relatively balanced, theoretically, a more accurate identification of a greater range of colors can be achieved by selecting more calibrated colors. Hyperspectral analysis is a key technology and method for achieving high-throughput analysis of crops [1,58,62–65].

This study proposes a novel framework that employs hyperspectral reflectance data and computer vision techniques to classify the color of rapeseed coats. By establishing a correlation between spectral characteristics and visual color perception, we introduce an intelligent model that offers a remarkably efficient alternative to conventional empirical and laboratory-based methods for determining the color of rapeseed seeds. Leveraging the inherent spectral and image information contained within rapeseed samples, this technology holds promising applications across various domains within the agricultural sector. The use of hyperspectral technology for identifying rapeseed seed coat colors has become part of research efforts, including identifying rapeseed varieties and quality, predicting rapeseed yield, and analyzing oil content. We aim to extend this research to address other agricultural challenges. One key objective is to establish a connection between rapeseed of different colors and their genetic information, enabling the classification of rapeseed quality based on color. This approach would provide support for traditional laboratories in identifying rapeseed quality through biological and chemical methods, complementing their existing techniques. Ultimately, the researchers plan to leverage the established connection between rapeseed color and genetic information to grade rapeseed quality. This grading system would assist breeding scientists in cultivating high-quality rapeseed varieties, contributing to the development of superior rapeseed cultivars.

Supplementary Materials: The following supporting information can be downloaded at https://www.mdpi.com/article/10.3390/agronomy14050941/s1, Table S1: The hyperspectral reflectance and the corresponding RGB data.

Author Contributions: C.Z.: data curation, formal analysis, investigation, validation, writing—original draft, and writing—review and editing; X.Z.: conceptualization, project administration; F.W.: conceptualization, funding acquisition, methodology, and writing—review and editing; J.W.: software, validation, and writing—review and editing; Y.-G.W.: software, funding acquisition, and writing—review and editing. All authors have read and agreed to the published version of the manuscript.

Funding: This research was supported partially by the Key Research Project of the Department of Education of Hunan Province (CN) (Grant No. 22A0135), the "Chunhui" Program Collaborative Scientific Research Project (202202004), the Research Project of the Department of Education of Hunan Province (CN) (Grant No. KN2022026), the Research Project of Educational Science Planning in Hunan Province (CN) (Grant No. 19JYKX007), and the Australian Research Council project (DP160104292).

Data Availability Statement: The data used in this work are openly available in the Supplementary Material.

Acknowledgments: The authors wish to thank the anonymous reviewers and the handling editor for their constructive comments and suggestions, which led to a great improvement in the presentation of this work.

Conflicts of Interest: The authors declare that they have no competing interests.

References

1. Łopatyńska, A.; Wolko, J.; Bocianowski, J.; Cyplik, A.; Gacek, K. Statistical Multivariate Methods for the Selection of High-Yielding Rapeseed Lines with Varied Seed Coat Color. *Agriculture* **2023**, *13*, 992. [CrossRef]
2. Rahman, M.; McVetty, P. A review of Brassica seed color. *Can. J. Plant Sci.* **2011**, *91*, 437–446. [CrossRef]
3. Chen, B.; Heneen, W. Inheritance of seed colour in *Brassica campestris* L. and breeding for yellow-seeded *B. napus* L. *Euphytica* **1992**, *59*, 157–163. [CrossRef]
4. Rahman, M. Production of yellow-seeded Brassica napus through interspecific crosses. *Plant Breed.* **2001**, *120*, 463–472. [CrossRef]
5. Michalski, K. Seed color assessment in rapeseed seeds using Color and Near Infrared Reflectance Spectrometers. *Rośl. Oleist.* **2009**, *30*, 119–132.
6. Van Deynze, A.; Pauls, K. Seed colour assessment in Brassica napus using a Near Infrared Reflectance spectrometer adapted for visible light measurements. *Euphytica* **1994**, *76*, 45–51. [CrossRef]
7. Velasco, L.; Fernández-Martínez, J.M.; De Haro, A. An efficient method for screening seed colour in Ethiopian mustard using visible reflectance spectroscopy and multivariate analysis. *Euphytica* **1996**, *90*, 359–363. [CrossRef]
8. Tańska, M.; Ambrosewicz-Walacik, M.; Jankowski, K.; Rotkiewicz, D. Possibility use of digital image analysis for the estimation of the rapeseed maturity stage. *Int. J. Food Prop.* **2017**, *20*, S2379–S2394. [CrossRef]
9. Tańska, M.; Rotkiewicz, D.; Kozirok, W.; Konopka, I. Measurement of the geometrical features and surface color of rapeseeds using digital image analysis. *Food Res. Int.* **2005**, *38*, 741–750. [CrossRef]
10. Lu, Y.; Liu, X.; Liu, S.; Yue, Y.; Guan, C.; Liu, Z. A simple and rapid procedure for identification of seed coat colour at the early developmental stage of Brassica juncea and Brassica napus seeds. *Plant Breed.* **2012**, *131*, 176–179. [CrossRef]
11. Pande, C.B.; Moharir, K.N. Application of hyperspectral remote sensing role in precision farming and sustainable agriculture under climate change: A review. In *Climate Change Impacts on Natural Resources, Ecosystems and Agricultural Systems*; Springer: Berlin/Heidelberg, Germany, 2023; pp. 503–520.
12. Moharram, M.A.; Sundaram, D.M. Land Use and Land Cover Classification with Hyperspectral Data: A comprehensive review of methods, challenges and future directions. *Neurocomputing* **2023**, *536*, 90–113. [CrossRef]
13. Liu, F.; Wang, F.; Wang, X.; Liao, G.; Zhang, Z.; Yang, Y.; Jiao, Y. Rapeseed variety recognition based on hyperspectral feature fusion. *Agronomy* **2022**, *12*, 2350. [CrossRef]

14. Rybacki, P.; Niemann, J.; Bahcevandziev, K.; Durczak, K. Convolutional neural network model for variety classification and seed quality assessment of winter rapeseed. *Sensors* **2023**, *23*, 2486. [CrossRef] [PubMed]
15. Field, C.; Gamon, J.; Peñuelas, J. Remote Sensing of Terrestrial Photosynthesis. In *Ecophysiology of Photosynthesis*; Springer: Berlin/Heidelberg, Germany, 1995; pp. 511–527.
16. Gitelson, A.A.; Merzlyak, M.N.; Chivkunova, O.B. Optical properties and nondestructive estimation of anthocyanin content in plant leaves. *Photochem. Photobiol.* **2001**, *74*, 38–45. [CrossRef]
17. Gitelson, A.A.; Zur, Y.; Chivkunova, O.B.; Merzlyak, M.N. Assessing carotenoid content in plant leaves with reflectance spectroscopy. *Photochem. Photobiol.* **2002**, *75*, 272–281. [CrossRef] [PubMed]
18. Wang, J.; Xian, X.; Xu, X.; Qu, C.; Lu, K.; Li, J.; Liu, L. Genome-wide association mapping of seed coat color in Brassica napus. *J. Agric. Food Chem.* **2017**, *65*, 5229–5237. [CrossRef] [PubMed]
19. Gamon, J.; Surfus, J. Assessing leaf pigment content and activity with a reflectometer. *New Phytol.* **1999**, *143*, 105–117. [CrossRef]
20. Kim, M.S.; Daughtry, C.; Chappelle, E.; McMurtrey, J.; Walthall, C. The use of high spectral resolution bands for estimating absorbed photosynthetically active radiation (A par). In Proceedings of the 6th International Symposium on Physical Measurements and Signatures in Remote Sensing, CNES, Val d'Isère, France, 17–21 January 1994.
21. Gamon, J.; Penuelas, J.; Field, C. A narrow-waveband spectral index that tracks diurnal changes in photosynthetic efficiency. *Remote Sens. Environ.* **1992**, *41*, 35–44. [CrossRef]
22. Gamon, J.; Serrano, L.; Surfus, J. The photochemical reflectance index: An optical indicator of photosynthetic radiation use efficiency across species, functional types, and nutrient levels. *Oecologia* **1997**, *112*, 492–501. [CrossRef]
23. Daughtry, C.S.; Walthall, C.; Kim, M.; De Colstoun, E.B.; McMurtrey Iii, J. Estimating corn leaf chlorophyll concentration from leaf and canopy reflectance. *Remote Sens. Environ.* **2000**, *74*, 229–239. [CrossRef]
24. Matoušková, E.; Kovářová, K.; Cihla, M.; Hodač, J. Monitoring biological degradation of historical stone using hyperspectral imaging. *Eur. J. Remote Sens.* **2023**, 2220565. [CrossRef]
25. Deng, Y.J.; Yang, M.L.; Li, H.C.; Long, C.F.; Fang, K.; Du, Q. Feature Dimensionality Reduction with L 2, p-Norm-Based Robust Embedding Regression for Classification of Hyperspectral Images. *IEEE Trans. Geosci. Remote Sens.* **2024**, *62*, 5509314. [CrossRef]
26. Zhao, Y.; Kwon, K.C.; Piao, Y.l.; Jeon, S.H.; Kim, N. Depth-layer weighted prediction method for a full-color polygon-based holographic system with real objects. *Opt. Lett.* **2017**, *42*, 2599–2602. [CrossRef] [PubMed]
27. Sağlam, A.; Baykan, N.A. A new color distance measure formulated from the cooperation of the Euclidean and the vector angular differences for lidar point cloud segmentation. *Int. J. Eng. Geosci.* **2021**, *6*, 117–124. [CrossRef]
28. Long, X.; Sun, J. Image segmentation based on the minimum spanning tree with a novel weight. *Optik* **2020**, *221*, 165308. [CrossRef]
29. Yang, Z.; Wang, Y.; Yang, Z. Vector-angular distance color difference formula in RGB color space. *Comput. Eng. Appl.* **2010**, *46*, 154–156.
30. Liu, F. Identification of Rapeseed Variety and Modeling of Fatty Acid Content Using Hyperspectral Features Fusion. Ph.D. Thesis, Hunan Agricultural University, Changsha, China, 2021.
31. Lacasa, L.; Luque, B.; Ballesteros, F.; Luque, J.; Nuno, J.C. From time series to complex networks: The visibility graph. *Proc. Natl. Acad. Sci. USA* **2008**, *105*, 4972–4975. [CrossRef]
32. Ahmadlou, M.; Adeli, H. Visibility graph similarity: A new measure of generalized synchronization in coupled dynamic systems. *Phys. D Nonlinear Phenom.* **2012**, *241*, 326–332. [CrossRef]
33. Lacasa, L.; Luque, B.; Luque, J.; Nuno, J.C. The visibility graph: A new method for estimating the Hurst exponent of fractional Brownian motion. *Europhys. Lett.* **2009**, *86*, 30001. [CrossRef]
34. Zhang, R.; Ashuri, B.; Shyr, Y.; Deng, Y. Forecasting construction cost index based on visibility graph: A network approach. *Phys. A Stat. Mech. Appl.* **2018**, *493*, 239–252. [CrossRef]
35. Silva, V.F.; Silva, M.E.; Ribeiro, P.; Silva, F. Time series analysis via network science: Concepts and algorithms. *Wiley Interdiscip. Rev. Data Min. Knowl. Discov.* **2021**, *11*, e1404. [CrossRef]
36. Wen, T.; Chen, H.; Cheong, K.H. Visibility graph for time series prediction and image classification: A review. *Nonlinear Dyn.* **2022**, *110*, 2979–2999. [CrossRef] [PubMed]
37. Lan, X.; Mo, H.; Chen, S.; Liu, Q.; Deng, Y. Fast transformation from time series to visibility graphs. *Chaos Interdiscip. J. Nonlinear Sci.* **2015**, *25*, 083105. [CrossRef] [PubMed]
38. Silva, T.C.; Zhao, L. *Machine Learning in Complex Networks*; Springer: Berlin/Heidelberg, Germany, 2016.
39. Tian, M.; Feng, J.; Rivard, B.; Zhao, C. A method to compute the n-dimensional solid spectral angle between vectors and its use for band selection in hyperspectral data. *Int. J. Appl. Earth Obs. Geoinf.* **2016**, *50*, 141–149. [CrossRef]
40. Leung, C.; Chau, H. Weighted assortative and disassortative networks model. *Phys. A Stat. Mech. Appl.* **2007**, *378*, 591–602. [CrossRef]
41. Almog, A.; Shmueli, E. Structural entropy: Monitoring correlation-based networks over time with application to financial markets. *Sci. Rep.* **2019**, *9*, 10832. [CrossRef]
42. Girvan, M.; Newman, M.E. Community structure in social and biological networks. *Proc. Natl. Acad. Sci. USA* **2002**, *99*, 7821–7826. [CrossRef]
43. Zhang, Z.; Wang, F.; Shen, L.; Xie, Q. Multiscale time-lagged correlation networks for detecting air pollution interaction. *Phys. A Stat. Mech. Appl.* **2022**, *602*, 127627. [CrossRef]

44. Cai, M.; Du, H.; Feldman, M.W. A new network structure entropy based on maximum flow. *Acta Phys. Sin.* **2014**, *63*, 102–112.
45. Kok, Z.H.; Shariff, A.R.M.; Alfatni, M.S.M.; Khairunniza-Bejo, S. Support vector machine in precision agriculture: A review. *Comput. Electron. Agric.* **2021**, *191*, 106546. [CrossRef]
46. Vapnik, V. *The Nature of Statistical Learning Theory*; Springer Science & Business Media: Berlin/Heidelberg, Germany, 1999.
47. Nasien, D.; Yuhaniz, S.S.; Haron, H. Statistical learning theory and support vector machines. In Proceedings of the 2010 Second International Conference on Computer Research and Development, Kuala Lumpur, Malaysia, 7–10 May 2010; pp. 760–764.
48. Chen, T.; Guestrin, C. Xgboost: A scalable tree boosting system. In Proceedings of the 22nd ACM SIGKDD International Conference on Knowledge Discovery and Data Mining, San Francisco, CA, USA, 13–17 August 2016; pp. 785–794.
49. Chung, C.C.; Su, E.C.Y.; Chen, J.H.; Chen, Y.T.; Kuo, C.Y. XGBoost-based simple three-item model accurately predicts outcomes of acute ischemic stroke. *Diagnostics* **2023**, *13*, 842. [CrossRef] [PubMed]
50. Gertz, M.; Große-Butenuth, K.; Junge, W.; Maassen-Francke, B.; Renner, C.; Sparenberg, H.; Krieter, J. Using the XGBoost algorithm to classify neck and leg activity sensor data using on-farm health recordings for locomotor-associated diseases. *Comput. Electron. Agric.* **2020**, *173*, 105404. [CrossRef]
51. Lundberg, S.M.; Erion, G.; Chen, H.; De Grave, A.; Prutkin, J.M.; Nair, B.; Katz, R.; Himmelfarb, J.; Bansal, N.; Lee, S.I. From local explanations to global understanding with explainable AI for trees. *Nat. Mach. Intell.* **2020**, *2*, 56–67. [CrossRef]
52. Obsie, E.Y.; Qu, H.C.; Drummond, F. Wild blueberry yield prediction using a combination of computer simulation and machine learning algorithms. *Comput. Electron. Agric.* **2020**, *178*, 105778. [CrossRef]
53. Gomes, W.P.C.; Gonçalves, L.; da Silva, C.B.; Melchert, W.R. Application of multispectral imaging combined with machine learning models to discriminate special and traditional green coffee. *Comput. Electron. Agric.* **2022**, *198*, 107097. [CrossRef]
54. Babajide Mustapha, I.; Saeed, F. Bioactive molecule prediction using extreme gradient boosting. *Molecules* **2016**, *21*, 983. [CrossRef] [PubMed]
55. Xue, J.; Shen, B. A novel swarm intelligence optimization approach: Sparrow search algorithm. *Syst. Sci. Control. Eng.* **2020**, *8*, 22–34. [CrossRef]
56. Jiang, S.; Wang, F.; Shen, L.; Liao, G. Local detrended fluctuation analysis for spectral red-edge parameters extraction. *Nonlinear Dyn.* **2018**, *93*, 995–1008. [CrossRef]
57. Jiang, S.; Wang, F.; Shen, L.; Liao, G.; Wang, L. Extracting sensitive spectrum bands of rapeseed using multiscale multifractal detrended fluctuation analysis. *J. Appl. Phys.* **2017**, *121*, 104702. [CrossRef]
58. Liu, F.; Wang, F.; Liao, G.; Lu, X.; Yang, J. Prediction of oleic acid content of rapeseed using hyperspectral technique. *Appl. Sci.* **2021**, *11*, 5726. [CrossRef]
59. Holme, P.; Park, S.M.; Kim, B.J.; Edling, C.R. Korean university life in a network perspective: Dynamics of a large affiliation network. *Phys. A Stat. Mech. Appl.* **2007**, *373*, 821–830. [CrossRef]
60. Saire, J.C.; Zhao, L. Complex Network-Based Data Classification Using Minimum Spanning Tree Metric and Optimization. In Proceedings of the 2023 International Joint Conference on Neural Networks (IJCNN), Gold Coast, Australia, 18–23 June 2023; pp. 1–7.
61. Chen, J.; Luo, D.L.; Mu, F.X. An improved ID3 decision tree algorithm. In Proceedings of the 2009 4th International Conference on Computer Science & Education, Nanning, China, 25–28 July 2009; pp. 127–130.
62. Kumari, P.; Gangwar, H.; Kumar, V.; Jaiswal, V.; Gahlaut, V. Crop Phenomics and High-Throughput Phenotyping. In *Digital Agriculture: A Solution for Sustainable Food and Nutritional Security*; Springer: Berlin/Heidelberg, Germany, 2024; pp. 391–423.
63. Rehman, T.U.; Ma, D.; Wang, L.; Zhang, L.; Jin, J. Predictive spectral analysis using an end-to-end deep model from hyperspectral images for high-throughput plant phenotyping. *Comput. Electron. Agric.* **2020**, *177*, 105713. [CrossRef]
64. Bohorquez, J.C.S.; Reyes, J.M.R.; Alama, W.I.; Alama, C.C. New Hyperspectral Index for Determining the State of Fermentation in the Non-Destructive Analysis for Organic Cocoa Violet. *IEEE Lat. Am. Trans.* **2018**, *16*, 2435–2440. [CrossRef]
65. Dutta, A.; Tyagi, R.; Chattopadhyay, A.; Chatterjee, D.; Sarkar, A.; Lall, B.; Sharma, S. Early detection of wilt in Cajanus cajan using satellite hyperspectral images: Development and validation of disease-specific spectral index with integrated methodology. *Comput. Electron. Agric.* **2024**, *219*, 108784. [CrossRef]

 *agronomy*

Article

Improving Wheat Leaf Nitrogen Concentration (LNC) Estimation across Multiple Growth Stages Using Feature Combination Indices (FCIs) from UAV Multispectral Imagery

Xiangxiang Su [1], Ying Nian [1], Hu Yue [1], Yongji Zhu [1], Jun Li [1], Weiqiang Wang [1], Yali Sheng [1], Qiang Ma [1], Jikai Liu [1,2], Wenhui Wang [3] and Xinwei Li [1,2,4,*]

[1] College of Resource and Environment, Anhui Science and Technology University, Fengyang 233100, China; suxxahstu@163.com (X.S.); nianying1010@163.com (Y.N.); yuehahstu@163.com (H.Y.); yongjizhu_21@163.com (Y.Z.); lijahstu@163.com (J.L.); w000910@163.com (W.W.); shengyali333@163.com (Y.S.); maq@ahstu.edu.cn (Q.M.); liujk@ahstu.edu.cn (J.L.)
[2] Anhui Engineering Research Center of Smart Crop Planting and Processing Technology, Fengyang 233100, China
[3] College of Life Sciences, Langfang Normal University, Langfang 065000, China; 1172139@lfnu.edu.cn
[4] Anhui Province Agricultural Waste Fertilizer Utilization and Cultivated Land Quality Improvement Engineering Research Center, Anhui Science and Technology University, Fengyang 233100, China
* Correspondence: lixw@ahstu.edu.cn

Citation: Su, X.; Nian, Y.; Yue, H.; Zhu, Y.; Li, J.; Wang, W.; Sheng, Y.; Ma, Q.; Liu, J.; Wang, W.; et al. Improving Wheat Leaf Nitrogen Concentration (LNC) Estimation across Multiple Growth Stages Using Feature Combination Indices (FCIs) from UAV Multispectral Imagery. *Agronomy* **2024**, *14*, 1052. https://doi.org/10.3390/agronomy14051052

Academic Editor: Alberto San Bautista

Received: 28 March 2024
Revised: 2 May 2024
Accepted: 13 May 2024
Published: 15 May 2024

Abstract: Leaf nitrogen concentration (LNC) is a primary indicator of crop nitrogen status, closely related to the growth and development dynamics of crops. Accurate and efficient monitoring of LNC is significant for precision field crop management and enhancing crop productivity. However, the biochemical properties and canopy structure of wheat change across different growth stages, leading to variations in spectral responses that significantly impact the estimation of wheat LNC. This study aims to investigate the construction of feature combination indices (FCIs) sensitive to LNC across multiple wheat growth stages, using remote sensing data to develop an LNC estimation model that is suitable for multiple growth stages. The research employs UAV multispectral remote sensing technology to acquire canopy imagery of wheat during the early (Jointing stage and Booting stage) and late (Early filling and Late filling stages) in 2021 and 2022, extracting spectral band reflectance and texture metrics. Initially, twelve sensitive spectral feature combination indices (SFCIs) were constructed using spectral band information. Subsequently, sensitive texture feature combination indices (TFCIs) were created using texture metrics as an alternative to spectral bands. Machine learning algorithms, including partial least squares regression (PLSR), random forest regression (RFR), support vector regression (SVR), and Gaussian process regression (GPR), were used to integrate spectral and texture information, enhancing the estimation performance of wheat LNC across growth stages. Results show that the combination of Red, Red edge, and Near-infrared bands, along with texture metrics such as Mean, Correlation, Contrast, and Dissimilarity, has significant potential for LNC estimation. The constructed SFCIs and TFCIs both enhanced the responsiveness to LNC across multiple growth stages. Additionally, a sensitive index, the Modified Vegetation Index (MVI), demonstrated significant improvement over NDVI, correcting the over-saturation concerns of NDVI in time-series analysis and displaying outstanding potential for LNC estimation. Spectral information outperforms texture information in estimation capability, and their integration, particularly with SVR, achieves the highest precision (coefficient of determination (R^2) = 0.786, root mean square error (RMSE) = 0.589%, and relative prediction deviation (RPD) = 2.162). In conclusion, the sensitive FCIs developed in this study improve LNC estimation performance across multiple growth stages, enabling precise monitoring of wheat LNC. This research provides insights and technical support for the construction of sensitive indices and the precise management of nitrogen nutrition status in field crops.

Keywords: leaf nitrogen concentration (LNC); wheat; UAV; multiple growth stages; feature combination indices (FCIs)

1. Introduction

Wheat, one of the world's most important cereal crops, significantly impacts global grain markets. Its production levels directly affect the dynamics of food security, which are crucial for human populations to survive and grow [1,2]. Leaf nitrogen concentration (LNC) is a vital parameter for assessing the nitrogen nutritional status of crops, serving as a critical indicator for nitrogen fertilization management in wheat during the early growth stages [3]. LNC directly affects grain quality development and yield in later growth stages [4,5]. Consequently, the rapid, precise, and dynamic acquisition of LNC information is essential to making informed nitrogen fertilization decisions and enhancing crop production.

Traditional methods for measuring LNC in crops often rely on destructive field sampling and chemical analysis in the laboratory. These procedures are not only time-consuming and labor-intensive but also expensive [6], and they present certain safety risks, making them unsuitable for general use. Remote sensing technology has emerged as an effective tool for monitoring crop growth conditions in precision agriculture due to its accuracy and speed [7], providing critical technical support for the efficient and non-destructive acquisition of LNC information [8]. Nowadays, crop development is monitored using a variety of remote sensing platforms, such as satellites [9], unmanned aerial vehicles (UAVs) [8], and ground-based hyperspectral sensors [10]. Satellite platforms can undertake large-scale LNC monitoring [11,12], but their low revisit frequency and spatial-temporal resolution restrict real-time high-precision monitoring. Although ground-based hyperspectral platforms may collect high-precision remote sensing data [13], their limited sampling points and high costs make them unsuitable for large area applications. UAVs are commonly used for crop LNC monitoring due to their flexibility, convenience, and low cost compared to satellite or ground-based hyperspectral systems [6,14].

A range of sensors, including RGB [15], multispectral [16], hyperspectral [17], and LiDAR [18], can be mounted on UAV platforms to collect data from field remote sensing. While hyperspectral sensors and LiDAR are expensive and unsuitable for practical field-scale applications, RGB sensors cannot capture the Red edge and Near-infrared bands, which are sensitive to vegetation features [19]. With their superior performance and low cost, multispectral sensors may acquire high-spatial–temporal-resolution images across multiple crop growth stages, which makes them more suitable for rapid, non-destructive crop growth monitoring [7].

Vegetation indices (VIs) derived from UAV remote sensing spectral band data are closely related to agronomic parameters and are commonly used for crop growth monitoring [20] and yield estimation [16]. Crop canopy structure and biochemical properties change according to growth stage [13], with particularly significant variations between vegetative and reproductive development stages [21]. The heterogeneity of canopies across growth stages can lead to variations in spectral responses, affecting the accuracy of LNC estimates using VIs. Compared to the early growth stage, increased canopy cover in the later stage might cause spectral saturation effects [22], reducing the sensitivity of VIs to crop biochemical properties. These factors indicate that there are still limitations in using existing VIs for precise monitoring of LNC across multiple growth stages of crops.

High-resolution images acquired by UAV multispectral sensors capture the spectral characteristics of wheat canopies and the texture features that reflect crop structure changes. These texture features can reflect variations in wheat leaf color across various varieties and fertilizer treatments and respond to changes in canopy structure during different growth stages [23]. Texture metrics and crop biochemistry are closely related, but developing trustworthy crop information prediction models based only on texture metrics is difficult due to the absence of a formulaic method to enhance their response to vegetation physiology [24]. Prior studies often used texture metrics in models to complement VIs and improve model performance [25].

Although combining texture metrics with VIs improved the estimation accuracy of crop LNC [14], most studies have relied on pre-existing texture metrics and VIs constructed from specific spectral bands and combination formulas [26], ignoring the potential of feature combination applications. VIs such as Normalized Difference Vegetation Index (NDVI) and Ratio Vegetation Index (RVI), which are constructed using simple formulas, offer advantages like reducing the impact of background noise and effectively mitigating the interference from solar zenith angle, thereby enhancing their responsiveness to vegetation features [27,28]. However, the current fixed pairing of bands and formulas in the VI design limits the application potential of feature combinations in crop growth monitoring [19,29]. This study aims to improve the estimation capability of LNC across growth stages and meet the demands of modern agricultural production by further investigating the feasibility of constructing feature combination indices (FCIs) sensitive to LNC across multiple growth stages using remote sensing data. It will allow for the full exploration of feature combinations and their respective benefits. By leveraging spectral band information, the study developed spectral feature combination indices (SFCIs) sensitive to LNC across various growth stages. In addition, using the same methodology, the study has created texture feature combination indices (TFCIs).

The potential of SFCIs in the field of precision agriculture is gradually being revealed by scholars. Inoue et al. [29] constructed a series of spectral indices based on simple formulas of NDVI and RVI, and found that the spectral index composed of the first-order derivatives at 740 nm and 522 nm (RSI: D740, D522) is the most accurate and stable indicator for monitoring canopy nitrogen content. Yao et al. [19] utilized spectral information combined with simple formulas of NDVI and RVI to create a series of novel spectral indices that achieves precise monitoring of leaf nitrogen accumulation in winter wheat. TFCIs are also being progressively developed for crop growth monitoring. Fan et al. [21] demonstrated that the use of feature combination formulas integrated with texture information can significantly enhance the estimation accuracy of plant nitrogen content in potatoes. Yuan et al. [19,30] constructed four texture indices based on hyperspectral texture information to accurately estimate the leaf area index of rice. However, previous studies have rarely constructed both SFCIs and TFCIs or systematically compared them with traditional variables (VIs and texture metrics) in terms of estimation potential.

Based on the previously mentioned data, this study constructed 12 FCIs to reduce the impact of canopy heterogeneity caused by variations in growth stages. It used spectral bands and texture metrics from high-resolution UAV multispectral imagery at the Jointing, Booting, Early filling, and Late filling stages. The objectives of this study are as follows: (1) to thoroughly consider all possible combinations of the 12 developed FCIs (SFCIs, TFCIs) based on UAV multispectral data (five spectral bands, 40 texture metrics) and to assess the feasibility; (2) to compare the capabilities of the newly developed FCIs with traditional variables (VIs, Tm) in estimating wheat LNC across multiple growth stages; and (3) to evaluate the potential of using FCIs to enhance the accuracy of LNC estimation across multiple growth stages and to investigate the feasibility of integrating multi-characteristic features for wheat LNC estimation.

2. Materials and Methods

2.1. Experimental Area and Experimental Design

A two-year field trial of winter wheat was conducted in Chuzhou City, Anhui Province (32°48′52″ N, 117°46′7″ E). The experimental site is located in the middle and lower reaches of the Yangtze River, characterized by a humid climate with distinct seasons, typical of a subtropical monsoon climate. The average elevation is 31 m, with an annual average temperature of 15.4 °C, an annual average precipitation of 1000–1100 mm, approximately 144 days of rainfall per year, and a frost-free period of about 240 days throughout the year.

The winter wheat field trial primarily considered two variables: different nitrogen application rates and different varieties, with three replicates for each, as detailed in Table 1. In Exp. 1, there were a total of 36 sampling plots, each with an area of 16 m^2 (2 m × 8 m), and the wheat was planted with a row spacing of 0.3 m. Exp. 2 consisted of 36 sampling plots, each with an area of 10 m^2 (2 m × 5 m), and the wheat was planted using manual strip sowing with the same row spacing of 0.3 m (Figure 1). Both Exp. 1 and Exp. 2 employed the same sowing techniques and fertilization levels, with phosphorus and potassium fertilizers applied as basal fertilizers before sowing (P = 90 kg/ha, K = 135 kg/ha), followed by topdressing at the jointing stage with a 6:4 ratio of basal to topdressing nitrogen fertilizers, in accordance with local agricultural practices. The field trial was conducted following the local farmers' field management practices, and no pest or weed infestations occurred during the trial period.

Table 1. Detailed information regarding the winter wheat field trial and sampling times.

Growing Season	Variety	N Treatments (kg/ha)	Sampling Stage
Exp. 1 2021	V1: Huaimai 44 V2: Yannong 999 V3: Ningmai 13	N0 (0) N1 (100) N2 (200) N3 (300)	Jointing (J, 14 March) Booting (B, 8 April) Early filling (EF, 9 May) Late filling (LF, 24 May)
Exp. 1 2022	V1: Huaimai 44 V2: Yannong 999 V3: Ningmai 13	N0 (0) N1 (100) N2 (200) N3 (300)	Jointing (J, 16 March) Booting (B, 10 April) Early filling (EF, 5 May) Late filling (LF, 21 May)

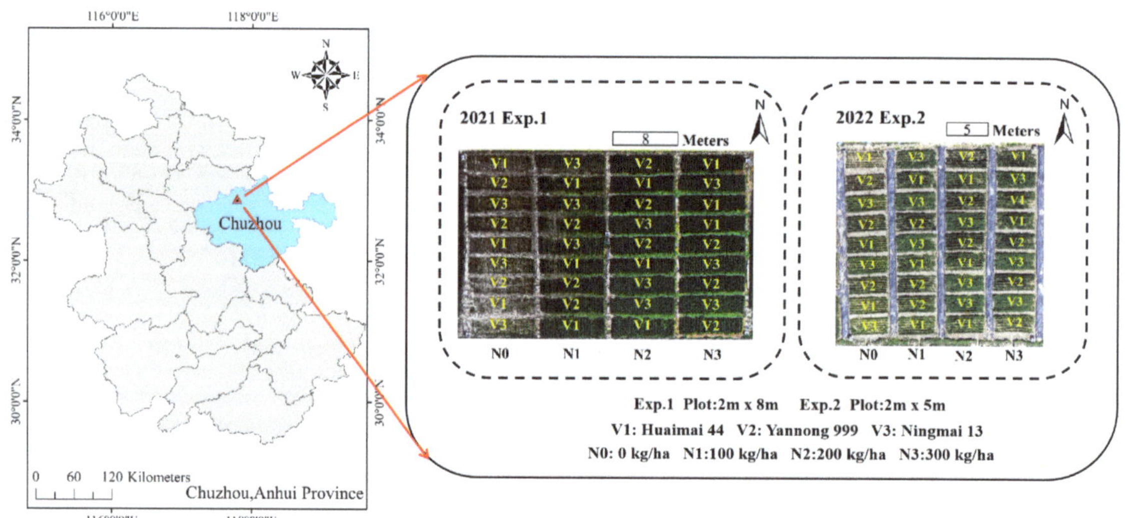

Figure 1. Study location and field experimental layout.

2.2. LNC Measurements

In the field, destructive sampling was conducted on each plot, with random samples of winter wheat representing the average growth within the area taken and placed into sealed bags for immediate transport to the laboratory. The wheat plants were then subjected to stem and leaf separation, and the leaves were dried in an oven at 105 °C for 0.5 h, followed by constant temperature drying at 75 °C for over 48 h until a stable weight was achieved. The dried leaves were ground into a powder form. Then, 2 to 4 mg of the sample was weighed and analyzed using the Euro Vector EA3000 automatic elemental analyzer to determine the nitrogen concentration (%) of the wheat leaf tissue.

2.3. UAV Image Collection and Preprocessing

During the Jointing (J), Booting (B), Early filling (EF), and Late filling (LF) stages of winter wheat, multispectral canopy imagery data were acquired using a Phantom 4 Multispectral RTK (DJI Technology Co., Shenzhen, China) unmanned aerial vehicle (UAV) (Figure 2a). The UAV is equipped with five monochrome high-resolution sensors for multispectral imaging, each with a resolution of 2.08 million effective pixels. It also features individual filters for Blue (450 ± 16 nm), Green (560 ± 16 nm), Red (650 ± 16 nm), Red edge (730 ± 16 nm), and Near-infrared (840 ± 26 nm) bands, with spectral bandwidths of 20 nm, 20 nm, 10 nm, 10 nm, and 40 nm, respectively. The UAV is equipped with an RTK positioning system that offers a vertical accuracy of ±1.5 cm and a horizontal accuracy of ±1 cm, enabling the acquisition of high-precision spectral and texture information from the wheat canopy. Data collection was conducted between 11:00 and 13:00 on clear days with stable solar radiation intensity and no wind or clouds. The UAV flight routes for the two-year winter wheat field trial were autonomously defined using DJI GS PRO software (https://www.dji.com/cn/ground-station-pro/), with the UAV's autopilot system executing the predefined flight plans. The flight had a forward and side overlap of 90% and 85%, respectively, with a flight speed of 2 m/s and an altitude of 30 m. The images were captured in TIFF format and stored on an SD card.

Figure 2. UAV (**a**), plates that correct 5%, 10%, 20%, and 40% reflectivity of calibrate differences (**b**).

The multispectral (MS) data were preprocessed using PIX4Dmapper software (version 4.4.12, Pix4D SA, Prilly, Switzerland) to obtain high-resolution wheat MS imagery. This process included image calibration, generation of high-density point clouds, mesh creation, and texture generation. Four radiance calibration panels with reflectance values of 5%, 10%, 20%, and 40% (each 0.5 m × 0.5 m in size) were placed within the UAV data collection area (Figure 2b). The empirical line method (ELM) was employed to radiometrically calibrate the five bands of the MS imagery [31], converting digital numbers into reflectance values.

2.4. Vegetation Index Calculation

Vegetation indices (VIs) are obtained by calculations or combinations of characteristic bands, providing robust vegetation information factors and enhancing, to a certain extent, the expressive ability of remote sensing data [19]. Building upon previous research, this study initially selected ten commonly used VIs that are associated with estimating chlorophyll and nitrogen content to analyze their correlation with wheat LNC across multiple growth stages and to establish estimation models (Table 2).

Table 2. Vegetation indices used in this paper.

Name	Abbreviations	Formulation	References
Normalized Difference Vegetation Index	NDVI	$(NIR - R)/(NIR + R)$	[27]
Coloration Index	CI	$(R - B)/R$	[32]
Normalized Pigment Chlorophyll ratio Index	NPCI	$(R - B)/(R + B)$	[33]
Green Chlorophyll Vegetation Index	GCVI	$NIR/G - 1$	[34]
Greenness Index	GI	G/R	[35]
Triangular Vegetation Index	TVI	$0.5 \times (120 \times (RE - G) - 200 \times (R - G))$	[36]
Plant Senescence Reflectance Index	PSRI	$(R - G)/RE$	[37]
Blue Red pigment Index	BRI	B/R	[38]
MERIS Terrestrial Chlorophyll Index	MTCI	$(NIR - RE)/(RE - R)$	[39]
Normalized Difference Red-Edge Index	NDREI	$(RE - G)/(RE + G)$	[40]

2.5. Texture Metrics Extraction

Texture is a visual feature that reflects the spatial arrangement of an image independent of its brightness, capturing information about changes in crop canopy structure by calculating the gray-level spatial correlation properties between two pixels [41]. This study employs the widely used Gray Level Co-occurrence Matrix (GLCM) method to extract texture information from winter wheat. GLCM requires user-defined parameters such as window size and orientation. To enhance the generalizability of the subsequent indices and to facilitate comparative analysis with traditional methods, the study adopts the most commonly used parameters for extracting texture metrics (Tm), specifically a window size of 3 pixels × 3 pixels, an extraction direction of 45°, and grayscale quantization levels set to 64 as default parameters. Detailed information is provided in Table 3.

Table 3. Texture metrics extracted based on the GLCM method.

Numbering	Abbreviation	Tm	Formulation	Description
1	Mea	Mean	$\sum_{i=0}^{N-1}\sum_{j=0}^{N-1} i \times p(i,j)$	The mean value in the texture
2	Var	Variance	$\sum_{i}\sum_{j}(i - u)^2 p(i,j)$	The size of the texture change
3	Hom	Homogeneity	$\sum_{i}\sum_{j}\frac{1}{1+(i-j)^2}p(i,j)$	The homogeneity of grey level in the texture
4	Con	Contrast	$\sum_{n=0}^{N_g-1} n^2 \left\{ \sum_{i=1j=1}^{N_g}\sum^{N_g} p(i,j) \right\}_{|i-j|=n}$	The clarity in the texture Same as contrast
5	Dis	Dissimilarity	$\sum_{n=1}^{N_g-1} n \left\{ \sum_{i=1j=1}^{N_g}\sum^{N_g} p(i,j) \right\}_{|i-j|=n}$	The similarity of the pixels in the texture
6	Ent	Entropy	$-\sum_{i}\sum_{j} p(i,j) \log(p(i,j))$	The diversity of the pixels in the texture
7	Sem	Second moment	$\sum_{i}\sum_{j} \{p(i,j)\}^2$	The uniformity of greyscale in the texture
8	Cor	Correlation	$\frac{\sum_i \sum_j (i,j) p(i,j) - \mu_i \mu_j}{\sigma_i \sigma_j}$	The consistency in the texture

Note: The parameters u_i, u_j, σ_i, and σ_j represent the average and standard deviation of the row and column sums of the GLCM, i and j represent the row and column indices, respectively, while p denotes a matrix. $p(i,j)$ signifies the ratio of the value at the corresponding row and column in the matrix to the sum of all values within the matrix.

2.6. Feature Combination Index Construction

To fully explore and exploit the potential of feature combination applications, this study has selected 12 different constructed formulas based on previous research to attempt the construction of FCIs that remain sensitive to LNC across multiple growth stages. These include 6 dual-index formulas and 6 triple-index formulas, with the detailed information of these formulas provided in Table 4.

Table 4. Formulation and basis used for constructing feature combination indices (FCIs) in this study.

Type	Abbreviation	Formulation	References
	$SFCI_D$ 1 $TFCI_D$ 1	$(\lambda1 - \lambda2)/(\lambda1 + \lambda2)$	[27]
	$SFCI_D$ 2 $TFCI_D$ 2	$\lambda1 - \lambda2$	[42]
	$SFCI_D$ 3 $TFCI_D$ 3	$\lambda1/\lambda2$	[28]
	$SFCI_D$ 4 $TFCI_D$ 4	$(\lambda1 \times \lambda1 - \lambda2)/(\lambda1 \times \lambda1 + \lambda2)$	[43]
	$SFCI_D$ 5 $TFCI_D$ 5	$(\lambda1 - \lambda2)/\lambda1$	[32]
	$SFCI_D$ 6 $TFCI_D$ 6	$1.5 \times (\lambda1 - \lambda2)/(\lambda1 + \lambda2 + 0.5)$	[44]
Spectral feature combination indices (SFCIs) Texture feature combination indices (TFCIs)	$SFCI_T$ 1 $TFCI_T$ 1	$(\lambda1 - \lambda2)/(\lambda2 + \lambda3)$	[45]
	$SFCI_T$ 2 $TFCI_T$ 2	$\lambda1/(\lambda2 + \lambda3)$	[46]
	$SFCI_T$ 3 $TFCI_T$ 3	$\lambda1/(\lambda2 \times \lambda3)$	[47]
	$SFCI_T$ 4 $TFCI_T$ 4	$(\lambda1 \times \lambda2)/\lambda3$	[45]
	$SFCI_T$ 5 $TFCI_T$ 5	$(\lambda1 + \lambda2)/\lambda3$	[45]
	$SFCI_T$ 6 $TFCI_T$ 6	$(\lambda1 - \lambda2)/(\lambda2 - \lambda3)$	[39]

Note: $SFCI_D$ is a SFCI composed of two spectral bands, while $SFCI_T$ is a SFCI constructed from three spectral bands. Similarly, $TFCI_D$ is a TFCI composed of two texture metrics, and $TFCI_T$ is a TFCI constructed from three texture metrics.

In the SFCIs constructed based on spectral band information, $\lambda1$, $\lambda2$, and $\lambda3$ represent the reflectance of the respective spectral bands. In the TFCIs constructed based on texture information, $\lambda1$, $\lambda2$, and $\lambda3$ represent the texture metrics (Tm) extracted using the GLCM method. For subsequent comparative analysis, this study categorizes spectral information into two groups: VIs (Table 2) and SFCIs (Table 4), and texture information into two groups: Tm (Table 3) and TFCIs (Table 4).

2.7. Model Construction and Accuracy Assessment

To develop a model for estimating LNC across multiple growth stages of wheat, this study tested four representative machine learning algorithms: partial least squares regression (PLSR), random forest regression (RFR), support vector regression (SVR), and Gaussian process regression (GPR). PLSR is a linear regression model that combines the strengths of principal component analysis, canonical correlation analysis, and linear regression analysis. It projects the predictive variables and observed variables into a new space, aiming to find the multidimensional direction that explains the maximum variance in the predictive variable space, extracting the maximum information reflecting data variability. PLSR is adept at addressing multicollinearity among independent variables and reducing the impact of random noise [48]. RFR is a non-parametric regression method based on the decision tree approach. RFR constructs multiple decision trees on different subsets of the data, training models on each subset. By aggregating the predictions from all the individual trees, RFR enhances the model's accuracy and reduces the prediction error [49]. RFR exhibits stable performance even in the presence of noise, effectively managing issues of information redundancy and overfitting. SVR is a non-linear regression model that operates on the principle of structural risk minimization. It seeks to find the hyperplane that maximizes the margin for feature variables by minimizing the cost function, resulting in a model with strong robustness and generalizability [50]. GPR is a Bayesian non-parametric method that performs regression by using a Gaussian process prior on the data. It also provides posterior predictions, offering good generalizability and interpretability, making it suitable for handling small datasets and non-linear problems [51]. The four machine learning methods mentioned above were implemented using R language version 4.3.1 (R Foundation, Vienna, Austria). In the study, the hyperparameters for each algorithm were specified as follows. For PLSR, the number of components (ncomp) was set to a range of 1–5. In RFR, the number of predictors randomly selected (mtry) was set to 3, and the number of trees (ntrees) was set to 1000. For SVR with a Gaussian kernel, the standard deviation (sigma) of the kernel function ranged from 0.01 with an increment of 0.01 up to 0.2. The regularization parameter (C) ranged from 0.01 with an increment of 0.1 up to 2. GPR utilized a linear kernel for the regression task. To ensure the reproducibility of the results, a random seed was set for the execution of the algorithms. Additionally, all the aforementioned algorithms opted for data preprocessing through centering and scaling.

This study randomly divided the wheat LNC dataset from four critical growth stages over two years (n = 288) into calibration and validation datasets. One-third of the data were used to construct the LNC estimation model (Calibration, n = 96), and two-thirds were used for model evaluation (Validation, n = 192). The study employed three statistical metrics to evaluate the model performance, including coefficient of determination (R^2), Root Mean Square Error (RMSE), and the Relative Prediction Deviation (RPD). R^2 represents the proportion of variance in the dependent variable that is explained by the model. It serves as a general indicator of model performance. The closer the R^2 value is to 1, the better the correlation between predicted and actual values. The RMSE translates the squared prediction errors into units that are consistent with the actual values, effectively reflecting the discrepancies between predictions and reality. The RPD also evaluates the stability of the model by reflecting the differences between predicted and actual values. A higher RPD indicates smaller discrepancies between predictions and actual values, suggesting better model stability. Additionally, the flowchart from data acquisition to model construction and evaluation is shown in Figure 3.

$$R^2 = 1 - \frac{\sum_{i=1}^{n}(y_i - \hat{y}_i)^2}{\sum_{i=1}^{n}(y_i - \bar{y})^2} \tag{1}$$

$$RMSE = \sqrt{\frac{\sum_{i=1}^{n}(y_i - \hat{y}_i)^2}{n}} \tag{2}$$

$$RPD = \frac{SD}{RMSE} \tag{3}$$

In the formulas, y_i and $\hat{y}_i$ represent the observed and predicted values of LNC, respectively, $\bar{y}$ is the mean of the observed LNC values, n is the sample size, and SD is the standard deviation of the reference values.

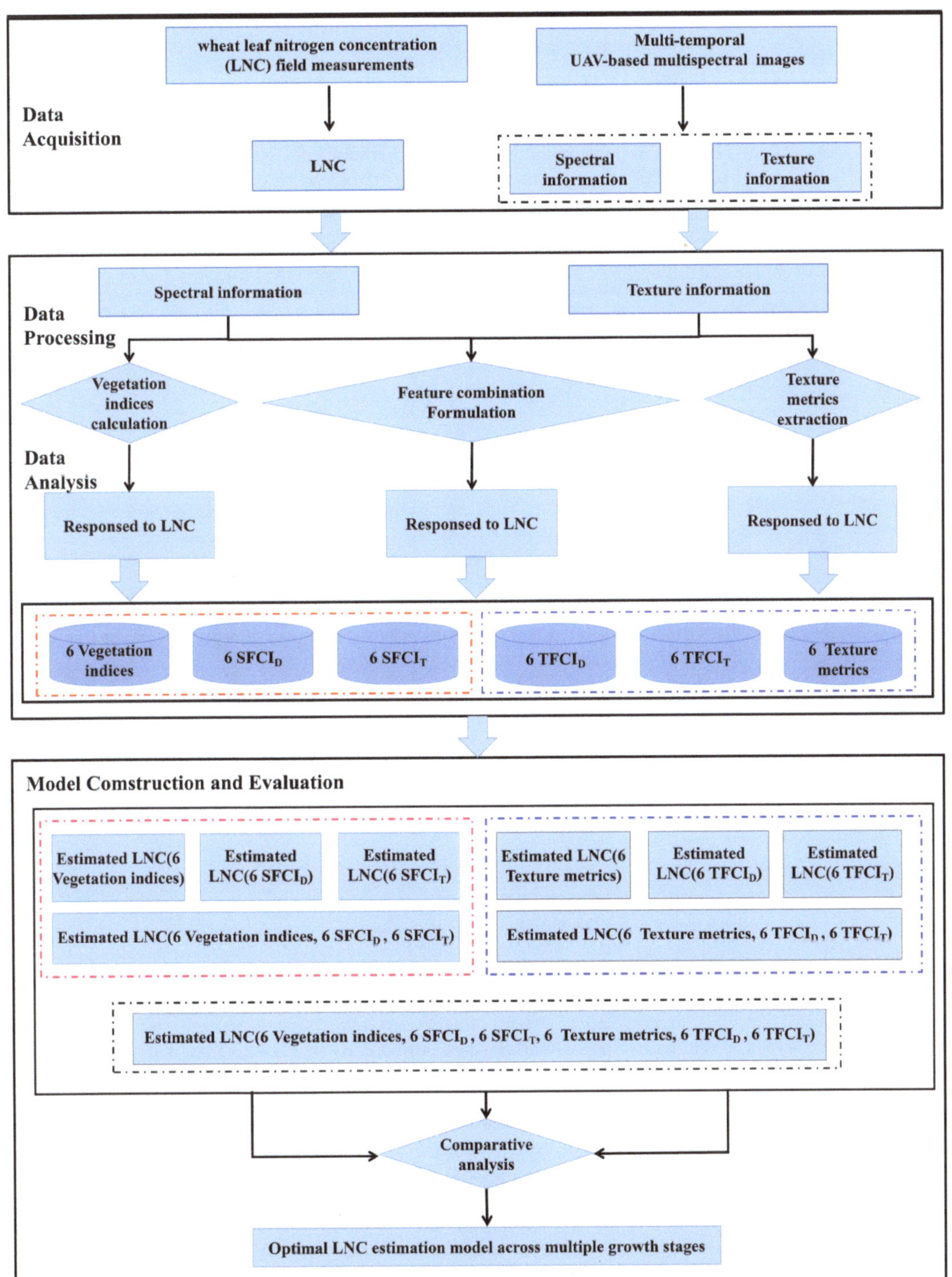

Figure 3. Flow chart of wheat LNC estimation modelling method.

3. Results

3.1. Variations in Winter Wheat LNC

Figure 4 illustrates the distribution and variation of LNC across four critical growth stages of winter wheat. From Jointing to Late filling, the mean values of each sample gradually decrease, specifically Mean J-LF: 4.141, 2.888, 2.380, and 1.253, while the standard deviation remains relatively stable, with a range of 0.668 to 0.831. The point density distribution across the four key growth stages is not uniform, indicating significant differences in LNC at each stage, and a decreasing trend in LNC values.

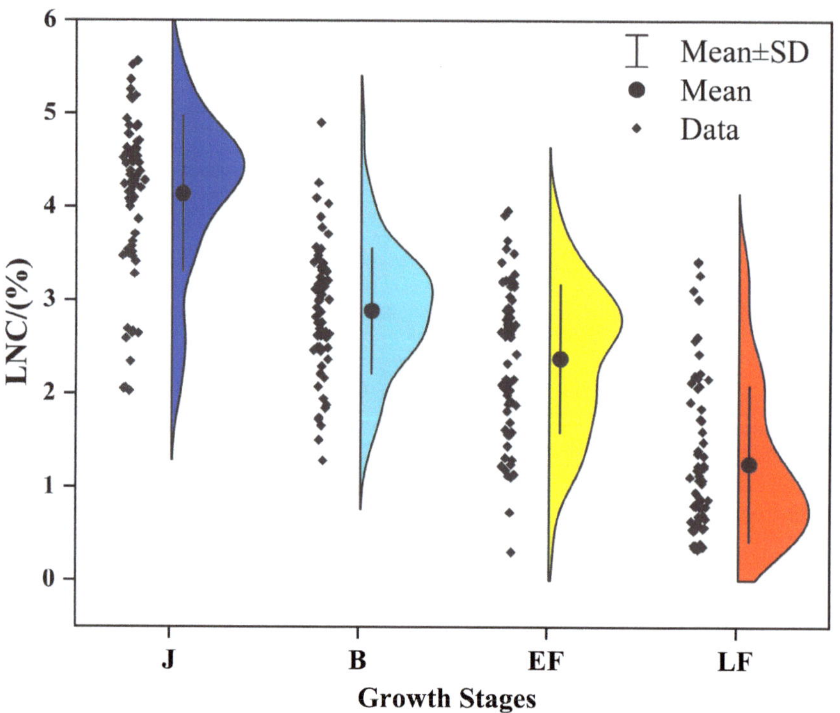

Figure 4. Description of LNC for different growth stages of winter wheat, where J–LF stands for Jointing, Booting, Early filling, and Late filling.

3.2. The Response of Spectral Information to Multiple Growth Stages LNC

3.2.1. Correlation between LNC and VIs at Multiple Growth Stages

The correlation analysis results between commonly used VIs and LNC are shown in Figure 5. The correlation coefficients between VIs and LNC vary significantly across different growth stages, with most VIs reaching their peak correlation at the Booting stage and then gradually decreasing during the Early filling and Late filling stages (Figure 5a). As indicated in Figure 5b, the VIs used in this study all show moderate correlations with LNC across multiple growth stages of winter wheat ($p < 0.05$). Among them, TVI has the strongest correlation ($|r| = 0.593$, $p < 0.001$), while GCVI has the weakest ($|r| = 0.158$, $p < 0.05$). Integrating the correlations across different growth stages (Figure 5a), the overall correlation of VIs with LNC smooths out the high and low values of correlation coefficients at each stage, reflecting the overall relationship between VIs and LNC across growth stages. Additionally, since this study constructed six $SFCI_D$ and six $SFCI_T$, to facilitate subsequent comparative analysis of the performance of commonly used VIs in LNC estimation across multiple growth stages, the top six VIs with the strongest correlations were selected for

LNC modeling (Figure 5b); these are TVI, PSRI, MTCI, NPCI, BRI, and CI, with correlation coefficients ($|r|$) of 0.593, 0.588, 0.543, 0.523, 0.517, and 0.517, respectively.

3.2.2. Correlation between LNC and SFCIs at Multiple Growth Stages

Based on the spectral reflectance of winter wheat canopies, SFCIs sensitive to LNC across multiple growth stages were constructed using the formulas from Table 4, which include six $SFCI_D$ and six $SFCI_T$. The correlation analysis results between these SFCIs and LNC during multiple growth stages of winter wheat are shown in Figure 6. To explore the potential of SFCIs for estimating LNC across wheat's multiple growth stages, the SFCIs with the strongest correlation for each FCI treatment were selected for subsequent modeling, resulting in a total of six $SFCI_D$ and six $SFCI_T$. Among the $SFCI_D$, the six with the highest correlation with LNC across winter wheat's multiple growth stages are $SFCI_D1$ (RE,R), $SFCI_D2$ (B,R), $SFCI_D3$ (R,RE), $SFCI_D4$ (R,B), $SFCI_D5$ (RE,R), and $SFCI_D6$ (B,R), with correlation coefficients (r) of 0.532, 0.652, −0.564, −0.598, 0.564, and 0.641, respectively. Among the $SFCI_T$, the six with the highest correlation with LNC are $SFCI_T1$ (RE,R,NIR), $SFCI_T2$ (R,RE,B), $SFCI_T3$ (RE,R,NIR), $SFCI_T4$ (NIR,R,RE), $SFCI_T5$(R,R,RE), and $SFCI_T6$ (R,B,RE), with correlation coefficients (r) of 0.780, −0.597, 0.713, −0.661, −0.564, and 0.663, respectively. From Figure 6, it can be observed that $SFCI_T$ has a better correlation with LNC across multiple growth stages of wheat compared to $SFCI_D$, and SFCIs constructed with multiple spectral reflectance bands show a stronger association with wheat LNC.

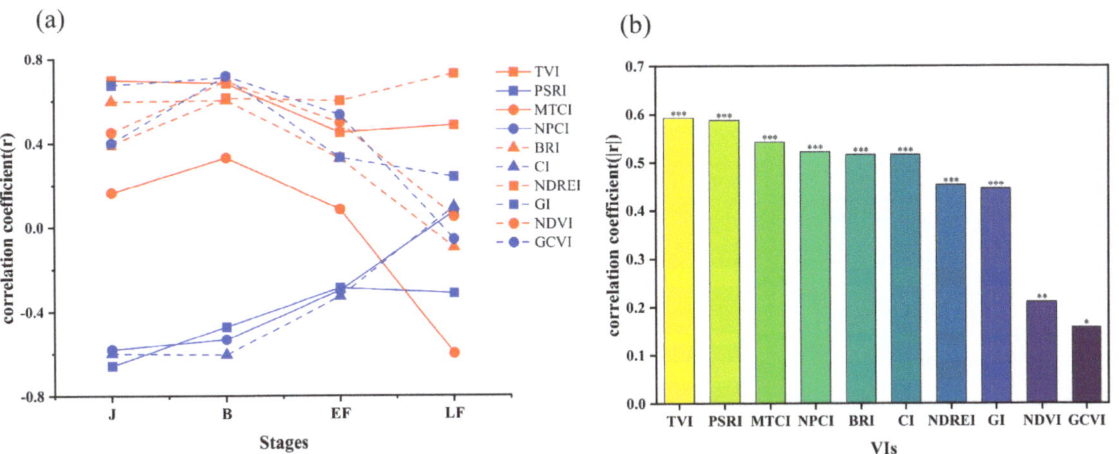

Figure 5. Correlation between LNC and VIs across multiple growth stages. (**a**) Variation in correlation at four growth stages. (**b**) Correlation analysis results across multiple growth stages. *** indicates $p < 0.001$, ** indicates $p < 0.01$, and * indicates $p < 0.05$.

3.2.3. Estimating LNC of Winter Wheat across Multiple Growth Stages Using Spectral Information

This study employed PLSR, RFR, SVR, and GPR algorithms to construct LNC estimation models based on spectral information (VIs and SFCIs) extracted from MS imagery of winter wheat canopies obtained by UAV. The validation statistics of the models are presented in Table 5. According to Table 5, VIs provide better estimation accuracy for LNC than $SFCI_D$, but overall, they perform lower than $SFCI_T$. Combining VIs with $SFCI_D$ and $SFCI_T$ to construct LNC estimation models yields higher accuracy, with the RFR model showing the best performance among the four machine learning algorithms, specifically with an $R^2 = 0.738$, RMSE = 0.653%, RPD = 1.952.

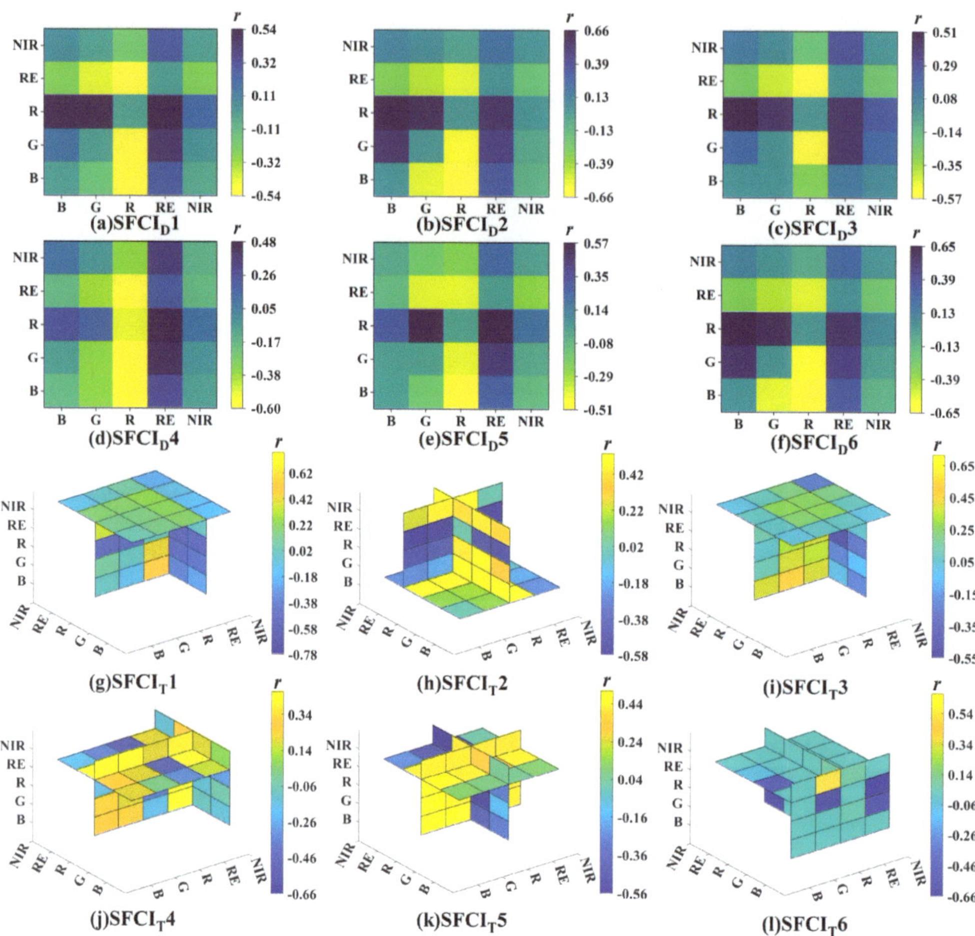

Figure 6. Correlation analysis results between SFCIs and LNC. $SFCI_D$ (**a–f**), $SFCI_T$ (**g–l**).

Table 5. Model validation statistics for estimating LNC across multiple growth stages of winter wheat using spectral information.

Data Type	Number	Metrics	PLSR	RFR	SVR	GPR
VIs	6	R^2	0.530	0.699	0.641	0.528
		RMSE (%)	0.873	0.706	0.762	0.875
		RPD	1.459	1.803	1.670	1.455
$SFCI_D$	6	R^2	0.446	0.536	0.538	0.461
		RMSE (%)	0.962	0.870	0.869	0.938
		RPD	1.324	1.463	1.465	1.357
$SFCI_T$	6	R^2	0.661	0.694	0.668	0.655
		RMSE (%)	0.744	0.704	0.737	0.748
		RPD	1.711	1.808	1.727	1.702
VIS $SFCI_D$ $SFCI_T$	18	R^2	0.671	0.738	0.696	0.682
		RMSE (%)	0.733	0.653	0.703	0.721
		RPD	1.737	1.952	1.813	1.768

Note: VIs represent Vegetation Indices, $SFCI_D$ are SFCIs composed of two spectral bands, and $SFCI_T$ are SFCIs composed of three spectral bands.

3.3. Multiple Growth Stage LNC Estimation Based on Texture Information

3.3.1. Correlation between LNC and Texture Metrics at Multiple Growth Stages

The study extracted Tm from the five bands of the multispectral (MS) imagery using the Gray Level Co-occurrence Matrix (GLCM) and analyzed their correlation with LNC across multiple growth stages of winter wheat, as shown in Figure 7. The correlation coefficients for the texture metrics across all bands are generally low, with none exceeding 0.5. Among them, the texture metric at the Near-infrared band (NIR.Con) shows the highest correlation with LNC (r = 0.499). To enhance the precision of Tm in estimating LNC across multiple growth stages and for comparative analysis with TFCIs (TFCI$_D$, TFCI$_T$), the study selected the six Tm with the highest correlation coefficients for subsequent model construction. These are NIR.Con, NIR.Dis, NIR.Var, RE.Hom, G.Cor, and RE.Dis, with correlation coefficients of 0.499, 0.487, 0.478, −0.464, −0.471, and 0.461, respectively.

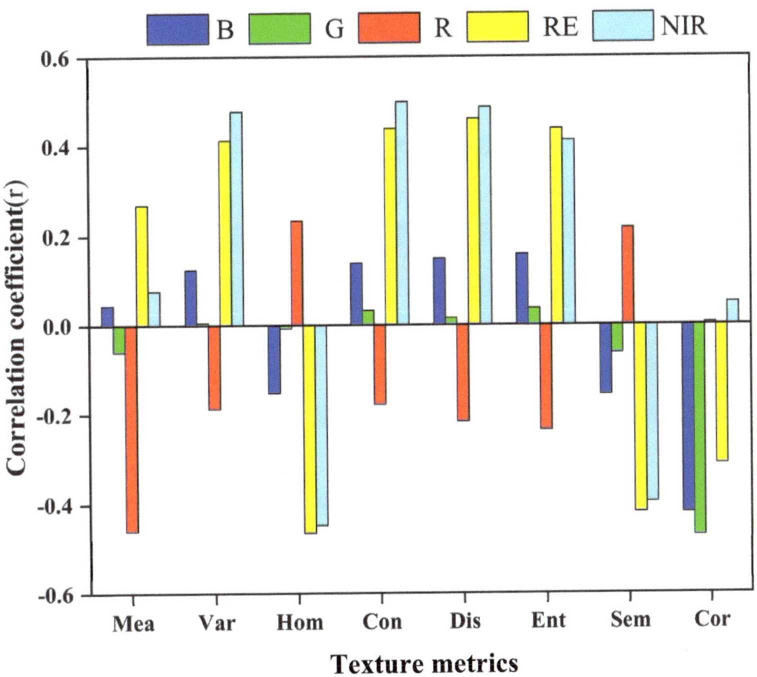

Figure 7. Correlation analysis of Tm with LNC across multiple growth stages.

3.3.2. Correlation between LNC and TFCIs at Multiple Growth Stages

Based on MS imagery, Tm were extracted using the GLCM, and 12 TFCIs were constructed using the formulas from Table 4. The optimal TFCIs were selected for subsequent modeling for each FCI treatment. Figure 8 presents the correlation results between TFCI$_D$, TFCI$_T$, and LNC across multiple growth stages of winter wheat. The numbers on each axis correspond to the Tm for each band, specifically B: 1–8, G: 9–16, R: 17–24, RE: 25–32, and NIR: 33–40, with the order of Tm for each band following the arrangement in Table 3.

From Figure 8, it can be observed that the TFCI$_D$ with the highest correlation with LNC across multiple growth stages of winter wheat are TFCI$_D$1 (NIR.Cor, RE.Cor), TFCI$_D$2 (R.Mea, B.Mea), TFCI$_D$3 (RE.Cor, NIR.Cor), TFCI$_D$4 (G.Cor, NIR.Dis), TFCI$_D$5 (NIR.Cor, RE.Cor) and TFCI$_D$6 (NIR.Cor, RE.Cor), with correlation coefficients of 0.616, −0.634, −0.623, −0.586, −0.607, and 0.609, respectively. The six TFCI$_T$ with the highest correlation are TFCI$_T$1 (RE.Cor, NIR.Cor, B.Hom), TFCI$_T$2 (RE.Mea, R.Mea, NIR.Mea), TFCI$_T$3 (NIR.Var, RE.Cor, NIR.Mea), TFCI$_T$4 (B.Dis, RE.Con, G.Var), TFCI$_T$5 (NIR.Mea, R.Mea, RE.Mea), and TFCI$_T$6 (G.Dis, B.Dis, RE.Con), with correlation coefficients of −0.656, 0.688, 0.685,

0.668, −0.662, and 0.627, respectively. The TFCIs constructed in this study show highe
correlations with LNC across multiple growth stages compared to the individual Tm, and
the TFCI$_T$ generally have better correlations with LNC than the TFCI$_D$ (Figures 7 and 8).

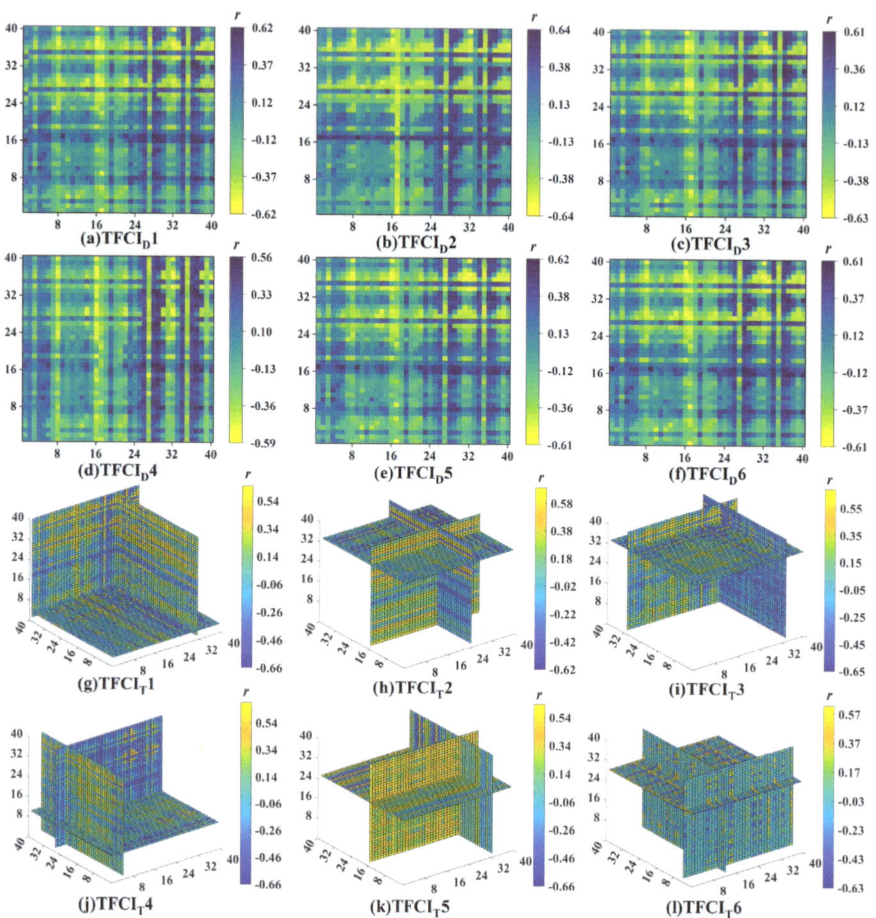

Figure 8. Correlation analysis results between TFCIs and LNC. TFCI$_D$ (**a–f**), TFCI$_T$ (**g–l**).

3.3.3. Estimating LNC of Winter Wheat across Multiple Growth Stages Based on Texture Information

This study constructs LNC estimation models across multiple growth stages of winter
wheat using texture information (Tm and TFCI) extracted from MS imagery of wheat
canopies obtained by UAV, employing PLSR, RFR, SVR, and GPR algorithms. Table 6
presents the statistical results of model validation. According to Table 6, compared to
traditional Tm, the TFCIs constructed in this study perform better in estimating LNC of
winter wheat, with the performance ranking as TFCI$_T$ > TFCI$_D$ > Tm. The SVR model with
TFCI$_T$ as the input variable achieves the highest accuracy, with R^2 = 0.626, RMSE = 0.785%,
and RPD = 1.621. When Tm and TFCIs are combined as input variables for LNC estimation
models, the model performance is significantly enhanced, clearly surpassing models using
Tm and TFCIs individually. Furthermore, among the four machine learning algorithms,
the SVR model with texture information as the input variable consistently outperforms the
other three, with the highest accuracy achieved by the SVR model integrating Tm, TFCI$_D$,
and TFCI$_T$, with specific metrics of R^2 = 0.688, RMSE = 0.714%, and RPD = 1.783.

Table 6. Model validation statistics for estimating LNC across multiple growth stages of winter wheat using texture information.

Data Type	Number	Metrics	PLSR	RFR	SVR	GPR
Tm	6	R^2	0.350	0.319	0.391	0.355
		RMSE (%)	1.035	1.054	1.006	1.030
		RPD	1.230	1.208	1.266	1.237
$TFCI_D$	6	R^2	0.556	0.542	0.560	0.556
		RMSE (%)	0.854	0.874	0.864	0.852
		RPD	1.492	1.457	1.474	1.495
$TFCI_T$	6	R^2	0.579	0.622	0.626	0.591
		RMSE (%)	0.828	0.783	0.785	0.816
		RPD	1.537	1.627	1.621	1.561
Tm $TFCI_D$ $TFCI_T$	18	R^2	0.645	0.659	0.688	0.679
		RMSE (%)	0.761	0.744	0.714	0.722
		RPD	1.675	1.711	1.783	1.763

Note: Tm represents texture metrics, $TFCI_D$ are TFCIs composed of two texture metrics, and $TFCI_T$ are TFCIs composed of three texture metrics.

3.4. Combining the UAV-Based Spectral and Texture Information for Estimating LNC across Multiple Growth Stages of Winter Wheat

Based on four different machine learning algorithms, LNC estimation models were constructed (Table 7, Figure 9). The results indicate that the optimal model for estimating LNC using spectral information is RFR, which outperforms the other three algorithms with higher accuracy: R^2 = 0.738, RMSE = 0.653%, RPD = 1.952. The best models for estimating LNC using texture information and a combination of spectral and texture information are both SVR, which shows superior performance compared to PLSR, RFR, and GPR, with specific metrics of R^2 = 0.786, RMSE = 0.589%, RPD = 2.162. From the perspective of data types, it was found that the capability of estimating LNC based on spectral information across all four machine learning algorithms is superior to that of texture information. Moreover, combining spectral information with texture information maximizes the estimation accuracy of LNC across multiple growth stages compared to using either type of information alone.

Table 7. Performance evaluation of winter wheat estimation models based on combination of spectral information and texture information.

Data Type	Number	Metrics	PLSR	RFR	SVR	GPR
Spectral information (VIs $SFCI_D$ $SFCI_T$)	18	R^2	0.671	0.738	0.696	0.682
		RMSE (%)	0.733	0.653	0.703	0.721
		RPD	1.737	1.952	1.813	1.768
Texture information (Tm $TFCI_D$ $TFCI_T$)	18	R^2	0.645	0.659	0.688	0.679
		RMSE (%)	0.761	0.744	0.714	0.722
		RPD	1.675	1.711	1.783	1.763
Spectral and texture information	36	R^2	0.747	0.783	0.786	0.775
		RMSE (%)	0.638	0.596	0.589	0.604
		RPD	1.995	2.139	2.162	2.108

This study, based on the optimal estimation model for LNC across multiple growth stages (SVRval: R^2 = 0.786, RMSE = 0.589%, and RPD = 2.162), outlined the feature importance of the model (Figure 10). The feature importance scale runs from 0 to 100, where 100 represents the most contributing feature, and 0 represents the least contributing feature. $SFCI_T1$ had the highest feature relevance, while RE.Dis had the least. The top ten most important features are FCIs. We created a ranking chart of feature type relevance by further integrating the significance of various features by type and calculating the average

(Figure 11). Figure 11 illustrates that developed FCIs are more significant than VIs and Tm. The overall importance of feature types is as follows: SFCI$_T$ > TFCI$_T$ > SFCI$_D$ > TFCI$_D$ > VIs > Tm, with SFCI$_T$ outperforming the rest.

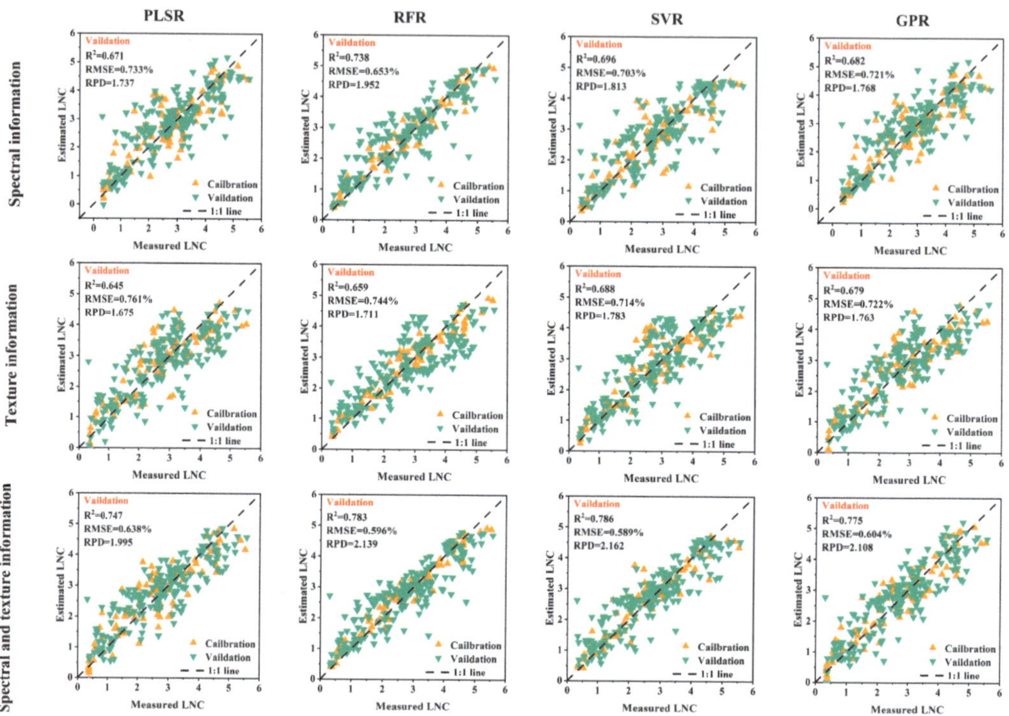

Figure 9. Scatter plot of fit for winter wheat LNC estimation model based on spectral and textural information.

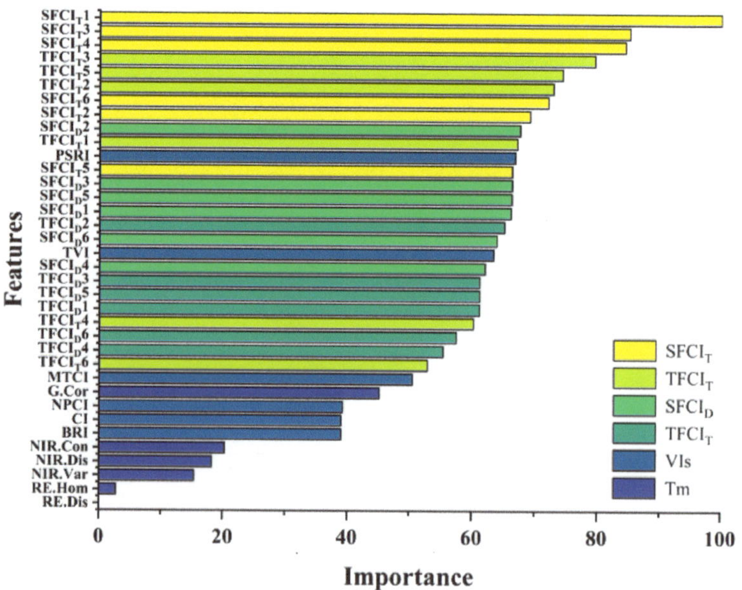

Figure 10. Feature importance of the optimal estimation model for LNC across multiple growth stages.

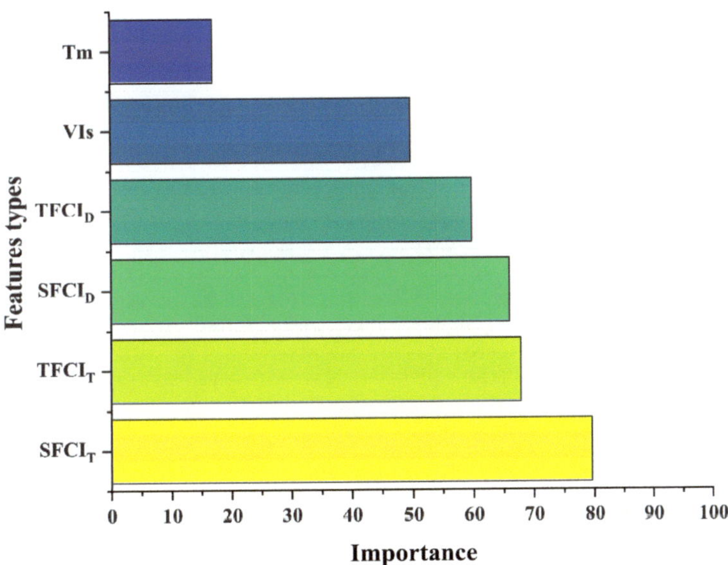

Figure 11. Importance of feature types in the optimal estimation model for LNC across multiple growth stages.

4. Discussion

4.1. Response of Spectral Information to LNC across Multiple Growth Stages

Spectral information, a type of remote sensing data that reflect the growth status of crops, and crop phenotypic information are closely associated [52]. VIs derived from band operations moderately correlate with LNC because the spectral bands extracted from multispectral images responded differently to LNC (Figure 5). In contrast to previous research [14,53], our study's correlation between NDVI and LNC is not optimal. In addition, the results indicate that, like NDVI, GCVI, which represents the amount of chlorophyll and nitrogen in crops, has the lowest correlation with LNC across multiple growth stages. This may be related to the growth stages of winter wheat used in this study, where canopy heterogeneity across multiple growth stages affects the comprehensive performance of NDVI and GCVI. NDVI and GCVI gradually increased during the Jointing and Booting stages followed by a sharp decrease during the Early and Late filling stages. The UAV-obtained canopy images of winter wheat at the Jointing stage are susceptible to soil background effects, and the complex spectral mixing effect [8] may affect the NDVI and GCVI information. The canopy growth variations during the Booting stage mask the background effects, enhancing NDVI and GCVI accuracy. Previously, Su et al. [16] also demonstrated that the spectral accuracy during the Booting stage is superior to other periods. However, NDVI and GCVI information tends to become saturated in the Early and Late filling stages due to the wheat canopy closure [54], decreasing their sensitivity to LNC.

Previous research has shown that specific feature combination algorithms can provide spectral information reflecting crop growth conditions [55]. This study used feature combination formulas to construct indices sensitive to LNC across multiple growth stages. The findings indicate that, under each SFCID treatment, the spectral indices with the best correlation with LNC in various growth stages are primarily composed of the R and RE bands and that, under each SFCI$_T$ treatment, the spectral features with the best correlation are mainly composed of the R, RE, and NIR bands (Figure 6). Thus, information from the R, RE, and NIR bands has the potential for estimating LNC across multiple growth stages. Nitrogen is a constituent of the chlorophyll molecule, a vital pigment in plant photosynthesis, and variations in the nitrogen content in wheat leaves can affect the amount of chlorophyll present. Chlorophyll is particularly sensitive to red light, as the pigment predominantly absorbs it [13]. Thus, an increase in LNC may result in a lower reflectance in the red spectral band, reducing the amount of red light reflected to the sensor from the plant surface. The RE band represents a transitory zone in the plant's spectral reflectance, characterized by a sharp increase. The position of this "red edge" correlates with the chlorophyll content [19]. An increase in LNC usually causes the RE to shift towards longer wavelengths, as more chlorophyll absorbs light in the red band and extends the decline in reflectance to longer wavelengths. While chlorophyll hardly absorbs any light in the NIR spectrum, variations in this band significantly correlate with the internal structure and biochemical composition of the plant leaves. The NIR band effectively represents crops' health status since plant cellular structure reflects a high percentage of the NIR spectrum [55].

In line with Fan et al.'s [56] findings, our investigation showed that $SFCI_T$ outperforms $SFCI_D$ in estimating LNC across multiple growth stages, providing a more accurate reflection of wheat LNC information (Figure 6). As wheat biochemical properties change over time, complex canopy heterogeneity causes variations in spectral responses. Three-band combinations respond better to these changes than two-band combinations, improving correlation with LNC and the ability to estimate LNC. This also clarifies why LNC estimation models constructed using VIs have better precision than $SFCI_D$ but worse precision than $SFCI_T$ (Table 5). Another important factor contributing to this diversity is the variations in feature combination formulas. The study found that using the same band information but different feature combination formulas has varying effects on mitigating the growth stage effect. Zheng et al. [55] reported similar results when estimating LNC. The results indicate that band information and feature combination formulas concurrently influence the ability of SFCIs to estimate LNC across multiple growth stages.

Furthermore, with a correlation coefficient of r = 0.780, the study preliminarily determined that $SFCI_T1$ (RE, R, NIR) is the most promising feature combination index for estimating LNC across multiple growth stages. The formulas in Table 4 indicate that $SFCI_T1$ (RE, R, NIR) is equivalent to an improved NDVI index (called the Modified Vegetation Index, MVI), where the numerator changes from (NIR-R) in NDVI to (RE-R), while the denominator stays the same. The correlation coefficient (r) between $SFCI_T1$ and wheat LNC across multiple growth stages increased by 268% compared to NDVI.

$$MVI = (RE - R)/(NIR + R) \tag{4}$$

Using the RE band, which is sensitive to leaf chlorophyll [57], rather than the NIR band, which is strongly reflected by leaves [13], capitalizes on the physiological fact that nitrogen is a key component of chlorophyll and is closely linked to it. Additionally, the RE band has a lower reflectance rate than the NIR band regarding vegetation spectral characteristics. MVI showed a generally acceptable match and a favorable response to LNC across wheat growth stages (Figure 12a). The MVI and LNC density curves exhibited a similar pattern, gradually increasing and peaking when LNC was around 3 and MVI was around 0.3, before declining. Figure 12b shows a poor match with scattered data points between the NDVI and LNC across multiple growth stages. Moreover, the NDVI density curve peaked between 0.8 and 1.0, exhibiting a state of hyper-aggregation in this range. The canopy heterogeneity during the different growth stages of wheat might be responsible for this. NDVI may reach saturation during the reproductive growth stage, when the wheat canopy is closed, with minimal overall variation and diminished spectral sensitivity. However, the FCIs in this study were developed based on a two-year dataset of wheat LNC during the Jointing, Booting, Early filling and Late filling stages. Future work will include validating their performance on other crops and agronomic indicators at different experimental sites and attempting to refine them into remote sensing indices with potential for spatiotemporal transferability.

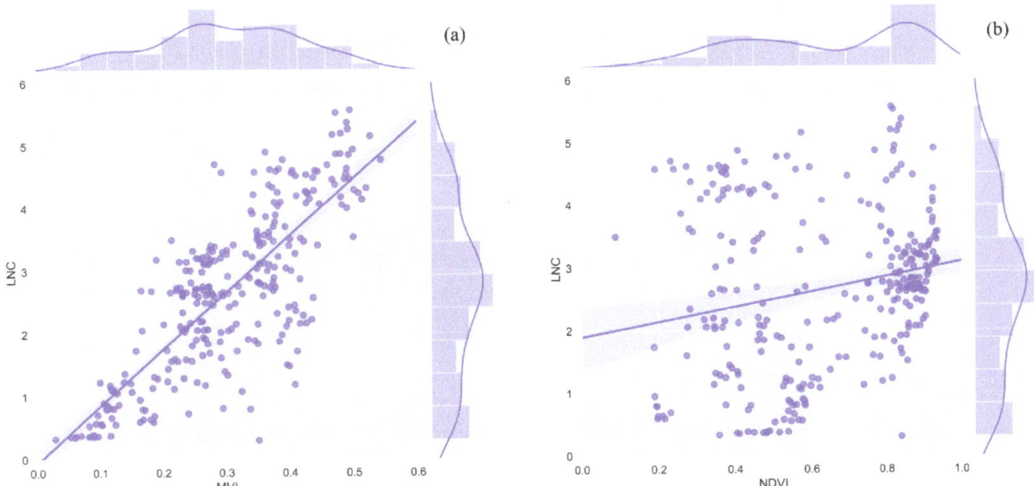

Figure 12. Scatter plot fit of MVI and NDVI against wheat LNC across multiple growth stages: (**a**) MVI-LNC, (**b**) NDVI-LNC.

In summary, the FCIs developed in this study consider both the vegetative and reproductive growth stages, efficiently adapting to variations in LNC throughout wheat growth. The Jointing, Booting, Early and Late filling stages of a two-year wheat LNC dataset were the basis for developing these FCIs. Future work will verify their efficacy on other crops and agronomic indicators across diverse experimental sites and attempt to enhance them into remote sensing indices capable of spatiotemporal transferability.

4.2. Contribution of Texture Information to LNC Estimation across Multiple Growth Stages

Texture, which reflects the spatial arrangement of crop canopy structure without relying on brightness, is frequently combined with spectral information to construct crop information estimation models, improving model performance [58,59]. Our study (Figure 7) supports the widely accepted notion that a weak correlation exists between Tm and agronomic parameters [58]. Thus, developing reliable models for crop phenotypic estimation using just Tm is challenging. This study used Tm to construct 12 sensitive TFCIs across four growth stages of wheat. The findings indicate that NIR information is superior to other bands in estimating LNC, as the six Tm with the highest correlation with LNC were primarily found in the NIR band [13]. Guo et al. [60] discovered that Con had the highest accuracy in extracting maize heading date. Liu et al. [23] demonstrated that Con outperformed other Tm in estimating rice above-ground biomass (AGB). The study's optimal TFCI$_D$ and TFCI$_T$ treatments primarily consisted of Mea, Cor, Con, and Dis. This is consistent with Liu et al. [61], who identified Mea, Con, and Dis as essential parameters for estimating winter wheat AGB. Mea and Cor are low-frequency information reflectors within the window that depict internal information about the growth and development of winter wheat. They show the parts of the plant growing quickly and consistently, documenting the developmental variations in wheat at various growth stages. Con and Dis focus on high-frequency information, demonstrated by Liu et al. [23] to help estimate crop information, such as the degree of gray-level fluctuations and the relationship between pixel distances and diagonal. Sensitive Tm enhanced the response to LNC across multiple growth stages by employing feature combination formulas, which reflected the changes in LNC over these times. In line with Yang et al.'s [45] findings, our investigation discovered that all 12 TFCIs correlated more with LNC than Tm. It could be attributed to the feature combination formulae highlighting crop canopy information by minimizing interference from soil background, solar angle, terrain, and shadows after optimized band information [62,63]. Our study confirms the feasibility of optimizing Tm using feature combination formulas, providing a practical approach to enhance Tm's responsiveness to LNC. Sarker et al. [64] also suggested that texture information processed by formulas could improve the estimation accuracy of forest biomass. Furthermore, the study indicated that $TFCI_T$ had a higher correlation with LNC than both Tm and $TFCI_D$. In contrast to $TFCI_D$ and Tm, $TFCI_T$ provides an extra dimension of texture information [45], capturing the changes in wheat LNC over time in more detail and further boosting the responsiveness to LNC.

4.3. The Significance of Combining Spectral and Texture Information

We found that the precision of LNC models constructed by combining spectral and texture information was higher than when using either spectral or texture information alone (Table 7), which is similar to the findings of Freitas et al. [25] and Zhang et al. [65] regarding other physiological and biochemical parameters. Spectral and texture information reflect different aspects of LNC changes, and integrating data from multiple feature types into a more comprehensive feature set can enhance the performance of predictive models and increase their interpretability [26]. However, spectral and texture information integration does not always result in improved model performance. Using five machine learning algorithms and two ensemble learning algorithms, Shu et al. [66] estimated corn leaf area index, fresh weight, and dry weight. They discovered that the combined spectral and texture information decreased the estimation accuracy, with texture information demonstrating higher estimation capability than spectral information. Liu et al. [23] found that integrating spectral and texture information for estimating rice AGB did not improve model precision. Multiple factors influence this phenomenon. (1) Different experimental settings directly affect the precision of spectral or texture information acquisition, and incorporating less precise data into the model may diminish model accuracy [67]. In particular, distinct crop types and varieties may result in variations in canopy structure [68], which can alter the correlation between spectral and texture information and crop physiological and biochemical parameters. All of this could have an impact on the accuracy of estimation models. Different crop planting directions [69] and different UAV data acquisition times [70] can cause different canopy shadows in remote sensing images, while different crop planting densities [71] directly affect canopy structure arrangement. Climate conditions [72], sensor types [66], image resolution [73], crop growth stages [74], and different UAV flight altitudes [75] also affect the precision of spectral and texture information. (2) Machine learning algorithms, also referred to as "black box" models, differ in the precision of the models they construct [76]. By changing the model input variables, feature selection, a critical step in machine learning algorithms, can directly affect the precision of estimation models. Multiple-feature models are not always stable, and Liu et al. [77] found that the number of variables in a model does not always equate to its performance. Overlapping features can cause data redundancy, reducing model performance. Zhou et al. [67] discovered that combining agronomic practice information (API) with spectral and texture features to estimate rice yield decreased model precision. The contribution and mutual influence of multiple feature characteristics affect the performance of models that incorporate spectral and texture information. However, this study aimed to explore the potential of feature combination formulas in estimating LNC across multiple growth stages, and thus, feature selection was not a priority. The study selected features using the standard Pearson correlation analysis method, focusing primarily on the response capability of spectral and texture information to LNC while neglecting the mutual influence of features within the estimation model. Future work will explore the relationship between feature interactions and model performance to improve accuracy.

In summary, the significance of combining spectral and texture information in this work lies in the ability to estimate wheat LNC more accurately and reliably. Spectral information reveals the biochemical and physiological state of crops [19], while texture information reflects the spatial arrangement and structural characteristics of the crop canopy [23]. Integrating these two data types allows the model to better account for the complex relationships between crop growth, canopy structure, and environmental factors [78], leading to more precise estimation of crop traits. This integrated approach leverages the strengths of both spectral and texture analyses, resulting in a more detailed understanding of crop health and nutritional status [71]. Spectral indices may be affected by factors such as soil background and crop canopy, and including texture information can mitigate these effects. On the other hand, texture information can reveal additional insights about the canopy's microstructure that spectral information alone may not capture [23]. This integrated method can lead to more effective crop management strategies, optimized fertilization, and higher crop yields and quality. It also encourages the development of more powerful remote sensing tools for precision agriculture, enabling farmers to make informed decisions based on comprehensive and accurate data.

4.4. The Comparability of Various Machine Learning Algorithms in Estimating LNC

Different machine learning algorithms are suited to different environments due to their inherent limitations. In particular, the PLSR algorithm is sensitive to outliers, affecting the model's stability and predictive power [79]. In addition, inappropriate selection of model parameters (number of components, regularization parameters) might reduce estimation performance. While the RFR algorithm reduces the risk of overfitting based on individual trees, it can still occur with high-dimensional datasets [79,80]. The SVR algorithm requires the storage of support vectors, which can be memory-intensive for large datasets. Moreover, SVR operations require considerable time and computational resources [71]. Selecting an appropriate kernel function for GPR can be challenging since different kernel functions may be suitable for various data types. However, there are no universal guidelines for determining the optimal kernel function. Furthermore, GPR's performance on large-scale datasets is relatively average, since its computational complexity rises with the number of data points [81]. This study used four machine learning algorithms, PLSR, RFR, SVR, and GPR, to construct LNC estimation models for multiple growth stages of wheat. We discovered that the optimal algorithm for estimating LNC across multiple wheat growth stages seems to depend on the data type, which is consistent with the conclusions of Yu et al. [82] that the effectiveness of RFR and SVR models in estimating wheat N content is related to the type of imagery. Despite diverse feature input conditions, the RFR algorithm demonstrated high robustness in LNC estimation models using spectral information, with high R^2, RPD, and low RMSE. The model utilizing VIs + $SFCI_D$ + $SFCI_T$ as input variables had the highest precision, with R^2 = 0.738, RMSE = 0.653%, and RPD = 1.952. Li et al. [71] confirmed the stability of the RFR algorithm in monitoring the crop nitrogen nutritional status, using the RFR algorithm combined with spectral information to construct a nitrogen monitoring model for wheat, with specific performance indicators of R^2 = 0.74, RMSE = 1.59 mg g^{-1}, and RPD = 1.25. Compared to this, the RFR model in this study showed a similar R^2 but a significantly higher RPD, which may be related to the features included in the model.

The accuracy of LNC estimation may be improved by integrating $SFCI_D$ and $SFCI_T$, which can reflect crop LNC information [71]. The results of this study indicate that the RFR algorithm is better suited for estimating LNC using spectral information, owing to improved model performance and stability. The SVR algorithm outperformed PLSR, RFR, and GPR in constructing wheat LNC estimation models based on texture information. Among the four feature input sets of texture information, SVR consistently had the highest R^2, RPD, and the smallest RMSE, (R^2 = 0.688, RMSE = 0.714%, and RPD = 1.783). This precision is lower than that of the SPAD estimation model constructed by Xie et al. [83] using the SVR algorithm combined with texture features based on the Litchi Fruit Growth Period but higher than the model built using the combination of the Litchi Fruit Growth Period and the Autumn Shoot Period. Models based on a single growth period are more robust, as they do not have to account for the impact of canopy heterogeneity across multiple growth stages on model estimation, further demonstrating the potential of the TFCIs developed in this study for LNC estimation. When integrated with texture information, the SVR model constructed in this study outperformed the model that only used Tm as an input variable alone (Table 6).

Interestingly, with the combined spectral and texture information dataset, the SVR algorithm performed well in constructing LNC estimation models utilizing texture information, presumably because half of the feature information was present. These results are consistent with the findings of Zhu et al. [84], who used vegetation indices and texture features in conjunction with machine learning algorithms to construct a wheat scab-monitoring model and found that the SVR algorithm had the highest model accuracy. This could be because the SVR algorithm, which applies the structural risk minimization principle [85], performs better at predicting LNC across various growth stages than RFR since it can manage the non-linear mapping relationship between texture information and LNC.

4.5. Limitations and Future Research Perspectives

The study entailed a two-year field trial of winter wheat with three high-yielding varieties and four nitrogen gradients, using feature combination formulas to construct SFCIs and TFCIs sensitive to LNC at several growth stages. Integrating these indices with four machine learning algorithms yielded better LNC estimation across growth stages, which can be improved further. With the rapid development of sensor technology, the fusion of multi-source remote sensing data can enhance the accuracy of crop nitrogen status estimation [71]. Future research will investigate using several sensors (such as RGB, hyperspectral, and LiDAR) for collaborative observation to improve LNC estimation performance in winter wheat. The remote sensing platform used in this study is relatively unique, and the limits of UAV working hours and flight altitude restrict its application potential on a large regional scale. The consistent development of satellite remote sensing technology provides a solid foundation for large-scale agricultural surveillance. Future studies will incorporate satellite remote sensing platforms and determine how to integrate the advantages of various remote sensing platforms. In addition, our research identified a vegetation index, MVI, with the potential to estimate LNC across multiple growth stages. It would be interesting to test its performance in the future for other agronomic parameters, such as LAI and AGB. Although this study achieved adequate precision in estimating LNC across multiple growth stages using two years of trial data, the single trial location limits its generalizability, and future work will test the model's transferability across different trial sites.

The integration of spectral and texture information to enhance LNC estimation accuracy in winter wheat across several growth stages still has limitations, primarily due to two factors. First, soil noise significantly impacts the early growth stages of wheat. When UAVs acquire remote sensing data during the early growth stages, the exposed soil background increases sensor noise, lowering sensor precision [86]. Second, after wheat heading, UAV imagery consists predominantly of leaves, spikes, and a small amount of soil, and the complex spectral mixing reduces spectral sensitivity [16]. Therefore, future research will focus on using various sensor signal-processing algorithms or removing soil background pixels to minimize the impact of soil background. We would also employ mixed spectral decomposition techniques to eliminate the effects of spectral mixing in the later growth stages, thereby enhancing model accuracy.

5. Conclusions

This study used multispectral remote sensing data to constructed feature combination indices (FCIs) sensitive to LNC across multiple growth stages based on multispectral remote sensing data and compared the performance of spectral and texture information in building LNC estimation models using four machine learning algorithms: PLSR, RFR, SVR, and GPR. It also examined the potential of estimating LNC by combining spectral and texture information. The results indicate that the combination of Red, Red edge, and Near-infrared bands has a high potential for estimating LNC across multiple growth stages, effectively capturing wheat's developmental changes over time. Texture metrics such as

Mea, Cor, Con, and Dis are highly sensitive to wheat LNC while showing modest resilience to canopy heterogeneity across various growth stages. The SFCIs and TFCIs constructed using feature combination formulas significantly improved the response to LNC across growth stages. With a correlation coefficient increase of 268%, a preliminary finding of a vegetation index, MVI, demonstrated significant improvement over NDVI, correcting the over-saturation concerns of NDVI in time-series analysis and displaying outstanding potential for LNC estimation. In addition, spectral information performed better than textural information in estimating LNC across multiple growth stages. Integrating spectral and texture information increased LNC estimation performance across growth stages, with the SVR algorithm achieving the highest precision (R^2 = 0.786, RMSE = 0.589%, and RPD = 2.162). Our study has made it possible to precisely monitor LNC over multiple crop growth stages, providing scientific guidance for more accurate field nitrogen fertilization management and refined crop nutrition management.

Author Contributions: Conceptualization, X.S. and X.L.; Methodology, X.S., J.L. (Jun Li), W.W. (Weiqiang Wang) and Y.S.; Software, X.S., H.Y. and Y.Z.; Validation, Y.Z.; Formal analysis, Y.N. and W.W. (Wenhui Wang); Investigation, Q.M.; Resources, X.S.; Data curation, Y.N., H.Y., J.L. (Jun Li), W.W. (Weiqiang Wang) and Y.S.; Writing—original draft, X.S.; Writing—review & editing, X.S., Y.Z., Q.M., J.L. (Jikai Liu), W.W. (Wenhui Wang) and X.L.; Visualization, Y.Z.; Supervision, Q.M., J.L. (Jikai Liu) and X.L.; Project administration, J.L. (Jikai Liu) and X.L.; Funding acquisition, J.L. (Jikai Liu), W.W. (Wenhui Wang) and X.L. All authors have read and agreed to the published version of the manuscript.

Funding: This research was funded by Scientific research projects in higher education institutions of Anhui Province (no. 2023AH051855; 2022AH051623); Anhui Province Crop Intelligent Planting and Processing Technology Engineering Research Center Open Research Project (no. ZHKF03); Natural Science Foundation of Hebei Province (no. C2023408010); Scientific research projects in higher education institutions of Hebei Province (no. QN2024158).

Data Availability Statement: The data presented in this study are available on request from the corresponding author due to the need for follow-up studies.

Conflicts of Interest: The authors declare no conflicts of interest.

References

1. Edae, E.A.; Byrne, P.F.; Haley, S.D.; Lopes, M.S.; Reynolds, M.P. Genome-wide association mapping of yield and yield components of spring wheat under contrasting moisture regimes. *Theor. Appl. Genet.* **2014**, *127*, 791–807. [CrossRef] [PubMed]
2. Ma, F.; Xu, Y.; Wang, R.; Tong, Y.; Zhang, A.; Liu, D.; An, D. Identification of major QTLs for yield-related traits with improved genetic map in wheat. *Front. Plant Sci.* **2023**, *14*, 1138696. [CrossRef] [PubMed]
3. Hansen, P.M.; Schjoerring, J.K. Reflectance measurement of canopy biomass and nitrogen status in wheat crops using normalized difference vegetation indices and partial least squares regression. *Remote Sens. Environ.* **2003**, *86*, 542–553. [CrossRef]
4. Lu, N.; Wu, Y.; Zheng, H.; Yao, X.; Zhu, Y.; Cao, W.; Cheng, T. An assessment of multi-view spectral information from UAV-based color-infrared images for improved estimation of nitrogen nutrition status in winter wheat. *Precis. Agric.* **2022**, *23*, 1653–1674. [CrossRef]
5. Tan, C.; Guo, W.; Wang, J. Predicting grain protein content of winter wheat based on landsat TM images and leaf nitrogen Content. In Proceedings of the 2011 International Conference on Remote Sensing, Environment and Transportation Engineering, Nanjing, China, 24–26 June 2011; IEEE: Piscataway, NJ, USA, 2011; pp. 5165–5168. [CrossRef]
6. Ma, X.; Chen, P.; Jin, X. Predicting wheat leaf nitrogen content by combining deep multitask learning and a mechanistic model using UAV hyperspectral images. *Remote Sens.* **2022**, *14*, 6334. [CrossRef]
7. Zhu, Y.; Liu, J.; Tao, X.; Su, X.; Li, W.; Zha, H.; Wu, W.; Li, X. A Three-Dimensional Conceptual Model for Estimating the Above-Ground Biomass of Winter Wheat Using Digital and Multispectral Unmanned Aerial Vehicle Images at Various Growth Stages. *Remote Sens.* **2023**, *15*, 3332. [CrossRef]
8. Wang, W.; Zheng, H.; Wu, Y.; Yao, X.; Zhu, Y.; Cao, W.; Cheng, T. An assessment of background removal approaches for improved estimation of rice leaf nitrogen concentration with unmanned aerial vehicle multispectral imagery at various observation times. *Field Crop. Res.* **2022**, *283*, 108543. [CrossRef]
9. Mutanga, O.; Adam, E.; Adjorlolo, C.; Abdel-Rahman, E.M. Evaluating the robustness of models developed from field spectral data in predicting African grass foliar nitrogen concentration using WorldView-2 image as an independent test dataset. *Int. J. Appl. Earth Obs. Geoinf.* **2015**, *34*, 178–187. [CrossRef]

10. Song, D.; Gao, D.; Sun, H.; Qiao, L.; Zhao, R.; Tang, W.; Li, M. Chlorophyll content estimation based on cascade spectral optimizations of interval and wavelength characteristics. *Comput. Electron. Agric.* **2021**, *189*, 106413. [CrossRef]

11. Eitel, J.; Long, D.S.; Gessler, P.E.; Smith, A. Using in-situ measurements to evaluate the new RapidEye™ satellite series for prediction of wheat nitrogen status. *Int. J. Remote Sens.* **2007**, *28*, 4183–4190. [CrossRef]

12. Huang, S.; Miao, Y.; Yuan, F.; Gnyp, M.L.; Yao, Y.; Cao, Q.; Wang, H.; Lenz-Wiedemann, V.I.; Bareth, G. Potential of RapidEye and WorldView-2 satellite data for improving rice nitrogen status monitoring at different growth stages. *Remote Sens.* **2017**, *9*, 227 [CrossRef]

13. Gao, D.; Qiao, L.; An, L.; Zhao, R.; Sun, H.; Li, M.; Tang, W.; Wang, N. Estimation of spectral responses and chlorophyll based on growth stage effects explored by machine learning methods. *Crop J.* **2022**, *10*, 1292–1302. [CrossRef]

14. Xu, S.; Xu, X.; Zhu, Q.; Meng, Y.; Yang, G.; Feng, H.; Yang, M.; Zhu, Q.; Xue, H.; Wang, B. Monitoring leaf nitrogen content in rice based on information fusion of multi-sensor imagery from UAV. *Precis. Agric.* **2023**, 1–23. [CrossRef]

15. Walter, J.; Edwards, J.; McDonald, G.; Kuchel, H. Photogrammetry for the estimation of wheat biomass and harvest index. *Field Crop. Res.* **2018**, *216*, 165–174. [CrossRef]

16. Su, X.; Wang, J.; Ding, L.; Lu, J.; Zhang, J.; Yao, X.; Cheng, T.; Zhu, Y.; Cao, W.; Tian, Y. Grain yield prediction using multi-temporal UAV-based multispectral vegetation indices and endmember abundance in rice. *Field Crop. Res.* **2023**, *299*, 108992. [CrossRef]

17. Yin, C.; Lv, X.; Zhang, L.; Ma, L.; Wang, H.; Zhang, L.; Zhang, Z. Hyperspectral UAV images at different altitudes for monitoring the leaf nitrogen content in cotton crops. *Remote Sens.* **2022**, *14*, 2576. [CrossRef]

18. Ten Harkel, J.; Bartholomeus, H.; Kooistra, L. Biomass and crop height estimation of different crops using UAV-based LiDAR. *Remote Sens.* **2019**, *12*, 17. [CrossRef]

19. Yao, X.; Zhu, Y.; Tian, Y.; Feng, W.; Cao, W. Exploring hyperspectral bands and estimation indices for leaf nitrogen accumulation in wheat. *Int. J. Appl. Earth Obs. Geoinf.* **2010**, *12*, 89–100. [CrossRef]

20. Yang, H.; Hu, Y.; Zheng, Z.; Qiao, Y.; Zhang, K.; Guo, T.; Chen, J. Estimation of Potato Chlorophyll Content from UAV Multispectral Images with Stacking Ensemble Algorithm. *Agronomy* **2022**, *12*, 2318. [CrossRef]

21. Fan, Y.; Feng, H.; Yue, J.; Jin, X.; Liu, Y.; Chen, R.; Bian, M.; Ma, Y.; Song, X.; Yang, G. Using an optimized texture index to monitor the nitrogen content of potato plants over multiple growth stages. *Comput. Electron. Agric.* **2023**, *212*, 108147. [CrossRef]

22. Jay, S.; Gorretta, N.; Morel, J.; Maupas, F.; Bendoula, R.; Rabatel, G.; Dutartre, D.; Comar, A.; Baret, F. Estimating leaf chlorophyll content in sugar beet canopies using millimeter-to centimeter-scale reflectance imagery. *Remote Sens. Environ.* **2017**, *198*, 173–186. [CrossRef]

23. Liu, J.; Zhu, Y.; Song, L.; Su, X.; Li, J.; Zheng, J.; Zhu, X.; Ren, L.; Wang, W.; Li, X. Optimizing window size and directional parameters of GLCM texture features for estimating rice AGB based on UAVs multispectral imagery. *Front. Plant Sci.* **2023**, *14*, 1284235. [CrossRef]

24. Fu, Y.; Yang, G.; Song, X.; Li, Z.; Xu, X.; Feng, H.; Zhao, C. Improved estimation of winter wheat aboveground biomass using multiscale textures extracted from UAV-based digital images and hyperspectral feature analysis. *Remote Sens.* **2021**, *13*, 581. [CrossRef]

25. Freitas, R.G.; Pereira, F.R.; Dos Reis, A.A.; Magalhães, P.S.; Figueiredo, G.K.; Do Amaral, L.R. Estimating pasture aboveground biomass under an integrated crop-livestock system based on spectral and texture measures derived from UAV images. *Comput. Electron. Agric.* **2022**, *198*, 107122. [CrossRef]

26. Li, Z.; Zhou, X.; Cheng, Q.; Fei, S.; Chen, Z. A Machine-Learning Model Based on the Fusion of Spectral and Textural Features from UAV Multi-Sensors to Analyse the Total Nitrogen Content in Winter Wheat. *Remote Sens.* **2023**, *15*, 2152. [CrossRef]

27. Rouse, J.W.; Haas, R.H.; Schell, J.A.; Deering, D.W. Monitoring vegetation systems in the Great Plains with ERTS. *Nasa Spec. Publ.* **1974**, *351*, 309.

28. Birth, G.S.; McVey, G.R. Measuring the color of growing turf with a reflectance spectrophotometer 1. *Agron. J.* **1968**, *60*, 640–643. [CrossRef]

29. Inoue, Y.; Sakaiya, E.; Zhu, Y.; Takahashi, W. Diagnostic mapping of canopy nitrogen content in rice based on hyperspectral measurements. *Remote Sens. Environ.* **2012**, *126*, 210–221. [CrossRef]

30. Yuan, W.; Meng, Y.; Li, Y.; Ji, Z.; Kong, Q.; Gao, R.; Su, Z. Research on rice leaf area index estimation based on fusion of texture and spectral information. *Comput. Electron. Agric.* **2023**, *211*, 108016. [CrossRef]

31. Di Gennaro, S.F.; Toscano, P.; Gatti, M.; Poni, S.; Berton, A.; Matese, A. Spectral comparison of UAV-Based hyper and multispectral cameras for precision viticulture. *Remote Sens.* **2022**, *14*, 449. [CrossRef]

32. Kumar, S.; Gautam, G.; Saha, S.K. Hyperspectral remote sensing data derived spectral indices in characterizing salt-affected soils: A case study of Indo-Gangetic plains of India. *Environ. Earth Sci.* **2015**, *73*, 3299–3308. [CrossRef]

33. Peñuelas, J.; Gamon, J.A.; Fredeen, A.L.; Merino, J.; Field, C.B. Reflectance indices associated with physiological changes in nitrogen-and water-limited sunflower leaves. *Remote Sens. Environ.* **1994**, *48*, 135–146. [CrossRef]

34. Gitelson, A.A.; Viña, A.; Arkebauer, T.J.; Rundquist, D.C.; Keydan, G.; Leavitt, B. Remote estimation of leaf area index and green leaf biomass in maize canopies. *Geophys. Res. Lett.* **2003**, *30*. [CrossRef]

35. Zhou, L.; He, H.; Sun, X.; Zhang, L.; Yu, G.; Ren, X.; Wang, J.; Zhao, F. Modeling winter wheat phenology and carbon dioxide fluxes at the ecosystem scale based on digital photography and eddy covariance data. *Ecol. Inform.* **2013**, *18*, 69–78. [CrossRef]

36. Broge, N.H.; Leblanc, E. Comparing prediction power and stability of broadband and hyperspectral vegetation indices for estimation of green leaf area index and canopy chlorophyll density. *Remote Sens. Environ.* **2001**, *76*, 156–172. [CrossRef]

37. Blackburn, G.A. Spectral indices for estimating photosynthetic pigment concentrations: A test using senescent tree leaves. *Int. J. Remote Sens.* **1998**, *19*, 657–675. [CrossRef]

38. Hunt, E.R.; Cavigelli, M.; Daughtry, C.S.; Mcmurtrey, J.E.; Walthall, C.L. Evaluation of digital photography from model aircraft for remote sensing of crop biomass and nitrogen status. *Precis. Agric.* **2005**, *6*, 359–378. [CrossRef]

39. Dash, J.; Curran, P.J.; Tallis, M.J.; Llewellyn, G.M.; Taylor, G.; Snoeij, P. Validating the MERIS Terrestrial Chlorophyll Index (MTCI) with ground chlorophyll content data at MERIS spatial resolution. *Int. J. Remote Sens.* **2010**, *31*, 5513–5532. [CrossRef]

40. Hassan, M.A.; Yang, M.; Rasheed, A.; Jin, X.; Xia, X.; Xiao, Y.; He, Z. Time-series multispectral indices from unmanned aerial vehicle imagery reveal senescence rate in bread wheat. *Remote Sens.* **2018**, *10*, 809. [CrossRef]

41. Zhou, M.; Zheng, H.; He, C.; Liu, P.; Awan, G.M.; Wang, X.; Cheng, T.; Zhu, Y.; Cao, W.; Yao, X. Wheat phenology detection with the methodology of classification based on the time-series UAV images. *Field Crop. Res.* **2023**, *292*, 108798. [CrossRef]

42. Richardson, A.J.; Everitt, J.H. Using spectral vegetation indices to estimate rangeland productivity. *Geocarto Int.* **1992**, *7*, 63–69. [CrossRef]

43. Chen, J.M. Evaluation of vegetation indices and a modified simple ratio for boreal applications. *Can. J. Remote Sens.* **1996**, *22*, 229–242. [CrossRef]

44. Huete, A.R. A soil-adjusted vegetation index (SAVI). *Remote Sens. Environ.* **1988**, *25*, 295–309. [CrossRef]

45. Yang, N.; Zhang, Z.; Zhang, J.; Guo, Y.; Yang, X.; Yu, G.; Bai, X.; Chen, J.; Chen, Y.; Shi, L. Improving estimation of maize leaf area index by combining of UAV-based multispectral and thermal infrared data: The potential of new texture index. *Comput. Electron. Agric.* **2023**, *214*, 108294. [CrossRef]

46. Tian, Y.C.; Yao, X.; Yang, J.; Cao, W.X.; Hannaway, D.B.; Zhu, Y. Assessing newly developed and published vegetation indices for estimating rice leaf nitrogen concentration with ground-and space-based hyperspectral reflectance. *Field Crop. Res.* **2011**, *120*, 299–310. [CrossRef]

47. Chappelle, E.W.; Kim, M.S.; McMurtrey III, J.E. Ratio analysis of reflectance spectra (RARS): An algorithm for the remote estimation of the concentrations of chlorophyll a, chlorophyll b, and carotenoids in soybean leaves. *Remote Sens. Environ.* **1992**, *39*, 239–247. [CrossRef]

48. Herrmann, I.; Pimstein, A.; Karnieli, A.; Cohen, Y.; Alchanatis, V.; Bonfil, D.J. LAI assessment of wheat and potato crops by VENμS and Sentinel-2 bands. *Remote Sens. Environ.* **2011**, *115*, 2141–2151. [CrossRef]

49. Yu, N.; Li, L.; Schmitz, N.; Tian, L.F.; Greenberg, J.A.; Diers, B.W. Development of methods to improve soybean yield estimation and predict plant maturity with an unmanned aerial vehicle based platform. *Remote Sens. Environ.* **2016**, *187*, 91–101. [CrossRef]

50. Wu, Z.; Luo, J.; Rao, K.; Lin, H.; Song, X. Estimation of wheat kernel moisture content based on hyperspectral reflectance and satellite multispectral imagery. *Int. J. Appl. Earth Obs. Geoinf.* **2024**, *126*, 103597. [CrossRef]

51. Sun, A.Y.; Wang, D.; Xu, X. Monthly streamflow forecasting using Gaussian process regression. *J. Hydrol.* **2014**, *511*, 72–81. [CrossRef]

52. Duan, B.; Liu, Y.; Gong, Y.; Peng, Y.; Wu, X.; Zhu, R.; Fang, S. Remote estimation of rice LAI based on Fourier spectrum texture from UAV image. *Plant Methods* **2019**, *15*, 1–12. [CrossRef]

53. Chen, X.; Li, F.; Shi, B.; Chang, Q. Estimation of Winter Wheat Plant Nitrogen Concentration from UAV Hyperspectral Remote Sensing Combined with Machine Learning Methods. *Remote Sens.* **2023**, *15*, 2831. [CrossRef]

54. Carlson, T.N.; Ripley, D.A. On the relation between NDVI, fractional vegetation cover, and leaf area index. *Remote Sens. Environ.* **1997**, *62*, 241–252. [CrossRef]

55. Zheng, H.; Li, W.; Jiang, J.; Liu, Y.; Cheng, T.; Tian, Y.; Zhu, Y.; Cao, W.; Zhang, Y.; Yao, X. A comparative assessment of different modeling algorithms for estimating leaf nitrogen content in winter wheat using multispectral images from an unmanned aerial vehicle. *Remote Sens.* **2018**, *10*, 2026. [CrossRef]

56. Fan, Y.; Feng, H.; Yue, J.; Liu, Y.; Jin, X.; Xu, X.; Song, X.; Ma, Y.; Yang, G. Comparison of Different Dimensional Spectral Indices for Estimating Nitrogen Content of Potato Plants over Multiple Growth Periods. *Remote Sens.* **2023**, *15*, 602. [CrossRef]

57. Jiang, J.; Johansen, K.; Stanschewski, C.S.; Wellman, G.; Mousa, M.A.; Fiene, G.M.; Asiry, K.A.; Tester, M.; McCabe, M.F. Phenotyping a diversity panel of quinoa using UAV-retrieved leaf area index, SPAD-based chlorophyll and a random forest approach. *Precis. Agric.* **2022**, *23*, 961–983. [CrossRef]

58. Zhang, J.; Qiu, X.; Wu, Y.; Zhu, Y.; Cao, Q.; Liu, X.; Cao, W. Combining texture, color, and vegetation indices from fixed-wing UAS imagery to estimate wheat growth parameters using multivariate regression methods. *Comput. Electron. Agric.* **2021**, *185*, 106138. [CrossRef]

59. Liu, Y.; Feng, H.; Yue, J.; Fan, Y.; Bian, M.; Ma, Y.; Jin, X.; Song, X.; Yang, G. Estimating potato above-ground biomass by using integrated unmanned aerial system-based optical, structural, and textural canopy measurements. *Comput. Electron. Agric.* **2023**, *213*, 108229. [CrossRef]

60. Guo, Y.; Fu, Y.H.; Chen, S.; Bryant, C.R.; Li, X.; Senthilnath, J.; Sun, H.; Wang, S.; Wu, Z.; de Beurs, K. Integrating spectral and textural information for identifying the tasseling date of summer maize using UAV based RGB images. *Int. J. Appl. Earth Obs. Geoinf.* **2021**, *102*, 102435. [CrossRef]

61. Liu, Y.; Liu, S.; Li, J.; Guo, X.; Wang, S.; Lu, J. Estimating biomass of winter oilseed rape using vegetation indices and texture metrics derived from UAV multispectral images. *Comput. Electron. Agric.* **2019**, *166*, 105026. [CrossRef]

62. Tucker, C.J. Red and photographic infrared linear combinations for monitoring vegetation. *Remote Sens. Environ.* **1979**, *8*, 127–150. [CrossRef]

63. Huete, A.R.; Jackson, R.D.; Post, D.F. Spectral response of a plant canopy with different soil backgrounds. *Remote Sens. Environ.* **1985**, *17*, 37–53. [CrossRef]

64. Sarker, L.R.; Nichol, J.E. Improved forest biomass estimates using ALOS AVNIR-2 texture indices. *Remote Sens. Environ.* **2011**, *115*, 968–977. [CrossRef]

65. Zhang, X.; Zhang, K.; Sun, Y.; Zhao, Y.; Zhuang, H.; Ban, W.; Chen, Y.; Fu, E.; Chen, S.; Liu, J. Combining spectral and texture features of UAS-based multispectral images for maize leaf area index estimation. *Remote Sens.* **2022**, *14*, 331. [CrossRef]

66. Shu, M.; Fei, S.; Zhang, B.; Yang, X.; Guo, Y.; Li, B.; Ma, Y. Application of UAV multisensor data and ensemble approach for high-throughput estimation of maize phenotyping traits. *Plant Phenomics* **2022**. [CrossRef]

67. Zhou, L.; Meng, R.; Yu, X.; Liao, Y.; Huang, Z.; Lü, Z.; Xu, B.; Yang, G.; Peng, S.; Xu, L. Improved Yield Prediction of Ratoon Rice Using Unmanned Aerial Vehicle-Based Multi-Temporal Feature Method. *Rice Sci.* **2023**, *30*, 247–256. [CrossRef]

68. Guan, S.; Fukami, K.; Matsunaka, H.; Okami, M.; Tanaka, R.; Nakano, H.; Sakai, T.; Nakano, K.; Ohdan, H.; Takahashi, K. Assessing correlation of high-resolution NDVI with fertilizer application level and yield of rice and wheat crops using small UAVs. *Remote Sens.* **2019**, *11*, 112. [CrossRef]

69. Meggio, F.; Zarco-Tejada, P.J.; Miller, J.R.; Martín, P.; González, M.R.; Berjón, A. Row orientation and viewing geometry effects on row-structured vine crops for chlorophyll content estimation. *Can. J. Remote Sens.* **2008**, *34*, 220–234. [CrossRef]

70. Li, D.; Chen, J.M.; Zhang, X.; Yan, Y.; Zhu, J.; Zheng, H.; Zhou, K.; Yao, X.; Tian, Y.; Zhu, Y. Improved estimation of leaf chlorophyll content of row crops from canopy reflectance spectra through minimizing canopy structural effects and optimizing off-noon observation time. *Remote Sens. Environ.* **2020**, *248*, 111985. [CrossRef]

71. Li, R.; Wang, D.; Zhu, B.; Liu, T.; Sun, C.; Zhang, Z. Estimation of nitrogen content in wheat using indices derived from RGB and thermal infrared imaging. *Field Crop. Res.* **2022**, *289*, 108735. [CrossRef]

72. Duan, B.; Fang, S.; Gong, Y.; Peng, Y.; Wu, X.; Zhu, R. Remote estimation of grain yield based on UAV data in different rice cultivars under contrasting climatic zone. *Field Crop. Res.* **2021**, *267*, 108148. [CrossRef]

73. Liu, Y.; Feng, H.; Yue, J.; Jin, X.; Li, Z.; Yang, G. Estimation of potato above-ground biomass based on unmanned aerial vehicle red-green-blue images with different texture features and crop height. *Front. Plant Sci.* **2022**, *13*, 938216. [CrossRef] [PubMed]

74. Wu, Q.; Zhang, Y.; Zhao, Z.; Xie, M.; Hou, D. Estimation of Relative Chlorophyll Content in Spring Wheat Based on Multi-Temporal UAV Remote Sensing. *Agronomy* **2023**, *13*, 211. [CrossRef]

75. Yin, Q.; Zhang, Y.; Li, W.; Wang, J.; Wang, W.; Ahmad, I.; Zhou, G.; Huo, Z. Estimation of Winter Wheat SPAD Values Based on UAV Multispectral Remote Sensing. *Remote Sens.* **2023**, *15*, 3595. [CrossRef]

76. Xiong, J.; Lin, C.; Cao, Z.; Hu, M.; Xue, K.; Chen, X.; Ma, R. Development of remote sensing algorithm for total phosphorus concentration in eutrophic lakes: Conventional or machine learning? *Water Res.* **2022**, *215*, 118213. [CrossRef] [PubMed]

77. Liu, J.; Zhu, Y.; Tao, X.; Chen, X.; Li, X. Rapid prediction of winter wheat yield and nitrogen use efficiency using consumer-grade unmanned aerial vehicles multispectral imagery. *Front. Plant Sci.* **2022**, *13*, 1032170. [CrossRef] [PubMed]

78. Xu, T.; Wang, F.; Shi, Z.; Xie, L.; Yao, X. Dynamic estimation of rice aboveground biomass based on spectral and spatial information extracted from hyperspectral remote sensing images at different combinations of growth stages. *ISPRS-J. Photogramm. Remote Sens.* **2023**, *202*, 169–183. [CrossRef]

79. Metz, M.; Abdelghafour, F.; Roger, J.; Lesnoff, M. A novel robust PLS regression method inspired from boosting principles: RoBoost-PLSR. *Anal. Chim. Acta.* **2021**, *1179*, 338823. [CrossRef] [PubMed]

80. Li, Z.; Chen, Z.; Cheng, Q.; Duan, F.; Sui, R.; Huang, X.; Xu, H. UAV-based hyperspectral and ensemble machine learning for predicting yield in winter wheat. *Agronomy* **2022**, *12*, 202. [CrossRef]

81. Pan, Y.; Zeng, X.; Xu, H.; Sun, Y.; Wang, D.; Wu, J. Evaluation of Gaussian process regression kernel functions for improving groundwater prediction. *J. Hydrol.* **2021**, *603*, 126960. [CrossRef]

82. Yu, J.; Wang, J.; Leblon, B.; Song, Y. Nitrogen estimation for wheat using UAV-based and satellite multispectral imagery, topographic metrics, leaf area index, plant height, soil moisture, and machine learning methods. *Nitrogen* **2021**, *3*, 1–25. [CrossRef]

83. Xie, J.; Wang, J.; Chen, Y.; Gao, P.; Yin, H.; Chen, S.; Sun, D.; Wang, W.; Mo, H.; Shen, J. Estimating the SPAD of Litchi in the Growth Period and Autumn Shoot Period Based on UAV Multi-Spectrum. *Remote Sens.* **2023**, *15*, 5767. [CrossRef]

84. Zhu, W.; Feng, Z.; Dai, S.; Zhang, P.; Wei, X. Using UAV multispectral remote sensing with appropriate spatial resolution and machine learning to monitor wheat scab. *Agriculture* **2022**, *12*, 1785. [CrossRef]

85. Camps-Valls, G.; Bruzzone, L.; Rojo-Álvarez, J.L.; Melgani, F. Robust support vector regression for biophysical variable estimation from remotely sensed images. *IEEE Geosci. Remote Sens. Lett.* **2006**, *3*, 339–343. [CrossRef]

86. Xu, X.; Fan, L.; Li, Z.; Meng, Y.; Feng, H.; Yang, H.; Xu, B. Estimating leaf nitrogen content in corn based on information fusion of multiple-sensor imagery from UAV. *Remote Sens.* **2021**, *13*, 340. [CrossRef]

Article

Estimation of Winter Wheat Chlorophyll Content Based on Wavelet Transform and the Optimal Spectral Index

Xiaochi Liu [1,2], Zhijun Li [1,2,*], Youzhen Xiang [1,2,*], Zijun Tang [1,2], Xiangyang Huang [1,2], Hongzhao Shi [1,2], Tao Sun [1,2], Wanli Yang [1,2], Shihao Cui [1,2], Guofu Chen [1,2] and Fucang Zhang [1,2]

[1] Key Laboratory of Agricultural Soil and Water Engineering in Arid and Semiarid Areas of Ministry of Education, Northwest A&F University, Yangling, Xianyang 712100, China; 2023055903@nwafu.edu.cn (X.L.); tangzijun@nwsuaf.edu.cn (Z.T.); 2023055900@nwsuaf.edu.cn (X.H.); shihongzhao7@nwsuaf.edu.cn (H.S.); 2021050986@nwsuaf.edu.cn (T.S.); 2022012387@nwsuaf.edu.cn (W.Y.); 2022012353@nwsuaf.edu.cn (S.C.); 2022012367@nwsuaf.edu.cn (G.C.); zhangfc@nwsuaf.edu.cn (F.Z.)

[2] Institute of Water–Saving Agriculture in Arid Areas of China, Northwest A&F University, Yangling, Xianyang 712100, China

[*] Correspondence: lizhij@nwsuaf.edu.cn (Z.L.); youzhenxiang@nwsuaf.edu.cn (Y.X.)

Citation: Liu, X.; Li, Z.; Xiang, Y.; Tang, Z.; Huang, X.; Shi, H.; Sun, T.; Yang, W.; Cui, S.; Chen, G.; et al. Estimation of Winter Wheat Chlorophyll Content Based on Wavelet Transform and the Optimal Spectral Index. *Agronomy* **2024**, *14*, 1309. https://doi.org/10.3390/agronomy14061309

Academic Editor: Anatoly Gitelson

Received: 28 May 2024
Revised: 12 June 2024
Accepted: 15 June 2024
Published: 17 June 2024

Abstract: Hyperspectral remote sensing technology plays a vital role in advancing modern precision agriculture due to its non-destructive and efficient nature. To achieve accurate monitoring of winter wheat chlorophyll content, this study utilized 68 sets of chlorophyll content data and hyperspectral measurements collected during the jointing stage of winter wheat over two consecutive years (2019–2020), under various fertilization types and nitrogen application levels. Continuous wavelet transform was applied to transform the original reflectance, ranging from 2^1 to 2^{10}, and the correlation matrix method was utilized to identify the spectral index at each scale, with the highest correlation to winter wheat chlorophyll content as the optimal spectral index combination input. Subsequently, winter wheat chlorophyll content prediction models were developed using three machine learning methods: random forest (RF), support vector machine (SVM), and a genetic algorithm-optimized backpropagation neural network (GA-BP). The results indicate that the spectral data processed through continuous wavelet transform at seven scales, from 2^1 to 2^7, show the highest correlation with winter wheat chlorophyll content at a scale of 2^6, with a correlation coefficient of 0.738, compared with the correlation of 0.611 of the original reflectance, and the accuracy is improved by 20.7%. The average highest correlation value between the spectral index at scale 2^6 and winter wheat chlorophyll content is 0.752. As the scale of wavelet transform increases, the correlation between the spectral index and winter wheat chlorophyll content and the accuracy of the predictive model show a trend of first increasing and then decreasing. The optimal input variables for predicting winter wheat chlorophyll content and the best machine learning method are the spectral data at a scale of 2^6 processing combined with the GA-BP model. The optimal predictive model has a validation set coefficient of determination (R^2) of 0.859, root mean square error (RMSE) of 1.366, and mean relative error (MRE) of 2.920%. The results show that the prediction model can provide a technical basis for improving the hyperspectral inversion accuracy of winter wheat chlorophyll and modern precision agriculture.

Keywords: winter wheat; chlorophyll content; hyperspectral; wavelet transform; optimal spectral index

1. Introduction

As an important food crop in China, the planting area and yield of winter wheat play a central role China [1]. However, the planting technology of winter wheat in some areas is relatively lagging behind, and this lack of modern planting management technology and equipment affects the yield and quality [2]. Therefore, it is of great strategic significance to strengthen the quality of domestic wheat planting and increase the yield to ensure national food security [3].

Chlorophyll is one of the most important pigments in plants; in the process of photosynthesis, chlorophyll plays a crucial role in facilitating energy conversion and sustaining the growth and development of plants. Furthermore, it acts as a protective and regulatory agent in maintaining plant health, thereby aiding in plant adaptation to external environmental conditions [4]. Romina et al. [5] used different kinds of chlorophyll instruments to determine the nitrogen content in crops by measuring the chlorophyll content of sweet peppers. Warlles et al. [6] utilized a portable chlorophyll meter for measuring the chlorophyll content in sorghum leaves; a significant correlation between photosynthetic pigments extracted in DMSO and leaf nitrogen content was determined. The above studies have indicated that chlorophyll content can be used to predict crop nitrogen requirements, assess crop photosynthetic capacity, and determine nutritional levels [7]. Accurate monitoring of the chlorophyll content in winter wheat can help prevent soil and water pollution caused by excessive fertilization [8]. The traditional method of measuring chlorophyll content involves sampling in the field and conducting measurements in indoor laboratories [9]. This method is not only time-consuming and labor-intensive, but also yields delayed and unstable results, greatly limiting the accuracy and timeliness of agricultural decision making [10]. In modern precision agriculture, efficient and non-destructive monitoring of crop growth status is vital [11]. Remote sensing technology, as a novel detection approach, enables the real-time acquisition of various spectral information of land cover on a large scale and has been widely applied in modern agricultural management [12].

With the advancement of hyperspectral remote sensing technology, leaf reflectance-based hyperspectral remote sensing is increasingly applied in the estimation of the chlorophyll content [13]. Currently, feature parameters selected based on maximum or local extreme values have been widely used as sensitive spectral variables for detecting the chlorophyll content [14]. Yoder et al. [15] found that the absorption, scattering, and transmission properties of chlorophyll in the visible light spectrum were more easily detectable and quantifiable. Saberioon et al. [16] estimated the chlorophyll content in the canopy layer at different growth stages using vegetation indices, achieving an estimation accuracy of R^2 up to 0.78. Yang et al. [17] separately utilized the Modified Soil Adjusted Vegetation Index 2 (MSAVI2) and spectral reflectance at 800 nm to establish chlorophyll content estimation models, with a modeling accuracy of R^2 reaching 0.88. However, during the dynamic growth period, effectively removing the interference signals, especially random and low-frequency signals, poses challenges and issues [18]. Previous studies have found that wavelet transform can effectively utilize hyperspectral reflectance data by preserving both global and local spectral information, facilitating a better analysis and interpretation of the data. Additionally, wavelet transform exhibits remarkable denoising capabilities as it can separate noise from signals at different scales or frequencies, effectively reducing the distortions caused by noise to enhance the accuracy of hyperspectral reflectance [19]. Liu et al. [20] used hyperspectral data, coupled continuous wavelet transform with the RF method, to construct a model for estimating the nitrogen content in summer corn, achieving remote sensing estimation of the nitrogen content and improving the modeling accuracy. Cheng et al. [21] studied continuous wavelet transform, utilizing spectral data from 265 leaf samples of 47 plants to effectively estimate the water content in the samples, with an accuracy as high as 75%. The aforementioned studies indicate that continuous wavelet transform can be utilized to enhance the modeling results for chlorophyll content detection.

Spectral indices are linear or nonlinear combinations of different sensitive bands, and compared with a single band, the hyperspectral vegetation indices constructed by screening bands based on the unique spectral characteristics of green vegetation contain more adequate crop growth information [22]. The correlation matrix method was used to select the optimal spectral bands with a high correlation to winter wheat chlorophyll content among all of the available bands [23]. Hyperspectral remote sensing technology has a finer division of spectra, which provides the possibility of selecting bands with a strong correlation with chlorophyll content in all of the available spectral bands for the construction of the optimal spectral index. The aim of this study is to investigate the correlation changes of winter wheat chlorophyll content through the wavelet transformation of hyperspectral reflectance and spectral index selection. Subsequently, scales with poor correlation will be eliminated, and machine learning models will be employed to construct prediction models for the winter wheat chlorophyll content. The study will explore the impact of different scales and machine learning model combinations on the accuracy of chlorophyll content prediction, aiming to identify the optimal prediction model. Our goal is to propose a prediction model for monitoring the chlorophyll content during the growth process of winter wheat, providing a more precise and rapid scientific basis for the development of modern agriculture.

2. Materials and Methods

2.1. Overview of the Experimental Area and Experimental Design

This experiment was conducted in 2019 at the China Institute of Water-Saving Agriculture in Arid Areas, Northwest A&F University ($34°17'44''$ N, $108°4'25''$ E). The study area is a typical dryland agricultural region in northwest China, characterized by a warm temperate monsoon semi-humid climate, with rainfall concentrated from July to September. The average annual precipitation is approximately 632 mm, with an evaporation of 1500 mm, and an average temperature of around 12.9 °C. The soil texture in the test area is heavy loam, the field water holding of the 0–100 cm section is 24%, and the withering water content is 8.5%. The pH of 0–20 cm sampled soil was 8.14, organic matter 12.0 g/kg, total nitrogen content 0.89 g/kg, total phosphorus 0.60 g/kg, total potassium 14.10 g/kg, alkaline nitrogen 55.3 mg/kg, effective phosphorus 8.21 mg/kg, and rapid potassium 132 mg/kg. Four N application levels were set up: N1: 100 kg/hm^2, N2: 160 kg/hm^2, N3: 220 kg/hm^2, and N4: 280 kg/hm^2. Four fertilizer types were set up: urea (U), slow-release fertilizers (SRF), UNS1 (ratio of U to SRF is 3/7), and UNS2 (ratio of U to SRF is 1/4). The experiments were conducted with no application of nitrogen fertilizer, and the control (CK) was used as a baseline. The experiment consisted of 17 treatments, each treatment was repeated three times, and the randomized block arrangement (positioning design) was adopted. The area of each plot was 7 m × 3 m = 21 m^2 and the sowing density was 180 kg/hm^2. There was a 2 m wide protected area around the experimental area and a 0.5 m wide isolation zone was set up between two adjacent plots. The experimental design was based on that of Tang et al. [24].

2.2. Data Acquisition

2.2.1. Remote Sensing Data Acquisition of Winter Wheat Canopy

This experiment was conducted on 31 March 2019 and 3 April 2020, respectively, using an ASD Field-Spec 3 back-mounted field spectrometer (Analytical Spectral Devices, Inc., Boulder, CO, USA). The weather was clear and windless with sufficient sunshine on the day of data measurement, and the measurement time was from 11:00 a.m. to 13:00 p.m. Three representative samples were selected for each plot, and nine spectral curves were collected from each sample at a time, and the average value was used as the spectral reflectance of the sample, with a total of 68 sets of data collected.

2.2.2. Winter Wheat Chlorophyll Content Acquisition

On the day of hyperspectral data collection, the winter wheat chlorophyll content data were acquired using the Soil and Plant Analyzer Development (SPAD-502, Inc., San Francisco, CA, USA) method. Prior to the measurements, the SPAD instrument was calibrated using leaf samples with a known chlorophyll content. The measurements were conducted on a clear, sunny day with ample sunlight. In order to minimize the sampling contingency, three random leaves were selected within each experimental plot for data collection, with their mean value serving as the measured value for each experimental plot.

2.3. Hyperspectral Data Preprocessing

2.3.1. SG Smoothing Processing

In order to reduce (eliminate) the impact of background noise, baseline drift, and stray light on the spectral reflectance curve, this study applied Savitzky–Golay convolution smoothing to the original spectral data in Origin 2023 (Origin Lab, Northampton, MA, USA). In order to effectively remove noise while maintaining the trend information in the data, a second-degree polynomial with nine smoothing points was used for function fitting and noise filtering [25].

2.3.2. Continuous Wavelet Transform

Wavelet transform is a mathematical tool used to analyze the local features of the signals, images, or data [26]. By convolving the signal with different scales and wavelet basis functions, we revealed the local features of the signal in time and frequency. Wavelet transform included Continuous Wavelet Transform (CWT) and Discrete Wavelet Transform (DWT), with CWT providing more shape and position information on the absorption features in vegetation spectra [27]. Meanwhile, within the scale range of 2^1–2^{10}, a sufficient level of resolution could be guaranteed to capture subtle variations in the signal, while also avoiding excessive refinement that may lead to noise amplification [28]. In this study, CWT was chosen, and the code was executed in MATLAB R2023a (Math Works Inc., Natick, MA, USA) to perform the wavelet transform and to obtain data at different transformation scales. The calculation formula is as follows:

$$\psi_{a,b}(\lambda) = \frac{1}{\sqrt{a}}\psi\frac{(\lambda - b)}{a} \tag{1}$$

$$W_f(a,b) = \int_{-oo}^{+\infty} f(\lambda)\psi_{a,b}(\lambda)d\lambda \tag{2}$$

where $f(\lambda)$ is the crown height spectral reflectance, λ is the spectral band in the range of 350~1830 nm, $\psi_{a,b}(\lambda)$ is the wavelet basis function, a is scale factor, and b is the translation factor.

2.4. Selection and Construction of Spectral Index

To reduce noise and atmospheric interference and highlight specific spectral features, seven spectral indices, as shown in (Table 1), were selected. Among them, the relationships between RI and TVI with plant chlorophyll content and leaf area index were strong, but sensitivity decreased in the dense vegetation, mSR and mNDI optimized the specular reflectance effect of leaves and were more sensitive to leaf changes, DI, NDVI, and SAVI could reflect the background influences of the plant canopy and eliminate some errors [29]

Table 1. Selection of spectral indices.

Select Index	Computing Formula	Literature Number
Ratio vegetation index (RI)	R_i/R_j	[30]
Triangular vegetation index (TVI)	$0.5[120(R_i - R_{550}) \quad 200(R_j - R_{550})]$	[30]
Modified red edge simple ratio (mSR)	$(R_i - R_{455})/(R_j - R_{455})$	[30]
Modified normalized difference index (mNDI)	$(R_i + R_j - 2R_{455})$	[30]
Difference index (DI)	$(R_i - R_j)$	[30]
Soil-adjusted vegetation index (SAVI)	$(1 + 0.16)\frac{(R_i - R_j)}{(R_i + R_j + 0.16)}$	[30]
Normalized difference vegetation index (NDVI)	$(R_i - R_j)/(R_i + R_j)$	[30]

2.5. Model Construction

Based on the field experiments, a total of 68 samples of chlorophyll content and hyperspectral data of winter wheat canopy were collected. In this study, various optimal spectral index combinations were constructed as the input variables. Three machine learning methods, SVM, RF, and GA-BP, were employed for modeling the regression prediction of the chlorophyll content in winter wheat. After removing the outliers, two-thirds of the data were randomly selected as the training dataset and one-third as the validation dataset. The average of multiple predictions by the machine learning models was taken as the final model fitting result in this experiment.

2.5.1. RF

RF is an ensemble learning method that utilizes multiple decision trees to perform classification and regression tasks. In RF, each decision tree is trained on a random subset of the training dataset [31]. Moreover, for each decision tree, node splitting is also based on a randomly selected subset of features, thereby enhancing the model's generalization ability and reducing the risk of overfitting. During the training–testing process within the decision tree model, it is necessary to traverse each feature and method to effectively determine the optimal number of decision trees. After multiple rounds of training and error analysis, the number of decision trees in the RF model was determined to be 100.

RF derives the final prediction by averaging or voting on the predictions from each tree, thus effectively improving the accuracy and robustness of the model [32]. This model typically excels in handling large-scale datasets, high-dimensional features, and addressing missing data issues. RF models also offer good interpretability, providing feature importance rankings that aid in understanding the impact of different features on prediction outcomes.

2.5.2. SVM

SVM is a binary machine learning algorithm that utilizes Gaussian and polynomial kernels as the base kernel functions, and it optimizes the weight coefficients using the gradient descent algorithm [33]. It exhibits excellent generalization ability and robustness, without overfitting issues, and has been widely applied in pattern recognition, classification, and small-sample regression analysis [34]. Following the principle of minimizing the cross-validation error, the penalty coefficients C and γ of the SVM model in this study are set to 20 and 0.02, respectively.

2.5.3. GA-BP

GA-BP is a machine learning method that combines the genetic algorithm with the neural network. In traditional BP neural networks, the weights and bias parameters are updated by a gradient descent algorithm, but this leads to a weak model generalization ability due to the possibility of falling into problems such as local minima or overfitting.

To solve these problems, genetic algorithms can be introduced into BP neural networks to optimize the parameters of the neural networks for a better performance and generalization ability. The genetic algorithm is an optimization algorithm that simulates the process of biological evolution and searches for optimal solutions through operations such as selection, crossover, and mutation [35]. Combining the genetic algorithm with the BP neural network encodes the weights and bias parameters of the neural network as chromosomes, and these parameters are optimized by operations such as crossover and mutation of the genetic algorithm to obtain a better neural network model. By using BP neural networks optimized based on genetic algorithms, the performance and generalization ability of the model can be effectively improved, and the problem of falling into local optimal solutions can also be avoided.

2.6. Model Accuracy Validation

(1)   Evaluation indicators

The model fitting results were evaluated using the root mean square error (RMSE), coefficient of determination (R^2), and the mean relative error (MRE)—the higher the R^2, the higher the prediction accuracy of the model, and the smaller the RMSE and MRE, the more stable the prediction performance of the model and the more concentrated the prediction results [36].

(2)   Significance analysis

By calculating the Pearson correlation coefficient and referring to the significance test table for correlation coefficients, for a sample size of 66, to achieve a highly significant correlation level ($p < 0.01$) and ensure the accuracy of estimating chlorophyll content, the correlation coefficient should exceed 0.310 [37].

3. Results

3.1. Correlation Analysis of Raw Hyperspectral Reflectance, Wavelet Transformed Hyperspectral Reflectance, and Chlorophyll Content

Using the Mexican hat wavelet function as the basis function for continuous wavelet transform (CWT), winter wheat hyperspectral data were decomposed to obtain the wavelet energy coefficients at different scales. To eliminate the influence of positive and negative correlations, the correlation coefficient was squared to calculate the determination coefficient (R^2). A noticeable improvement in the correlation between spectral data after partial-scale CWT transformation and chlorophyll content in winter wheat was observed compared with the original spectral reflectance data, as shown in Figure 1. Within small scales, a peak in correlation between the wavelet-transformed spectral reflectance and winter wheat chlorophyll content was observed near 750 nm. As the decomposition scale increased, the correlation between wavelet-transformed spectral reflectance and winter wheat chlorophyll content exhibited a trend of initially increasing and then decreasing. At a decomposition scale of 2^6, the maximum correlation with chlorophyll content reached 0.738, representing an increase of 0.127 compared with the original spectral reflectance correlation of 0.611.

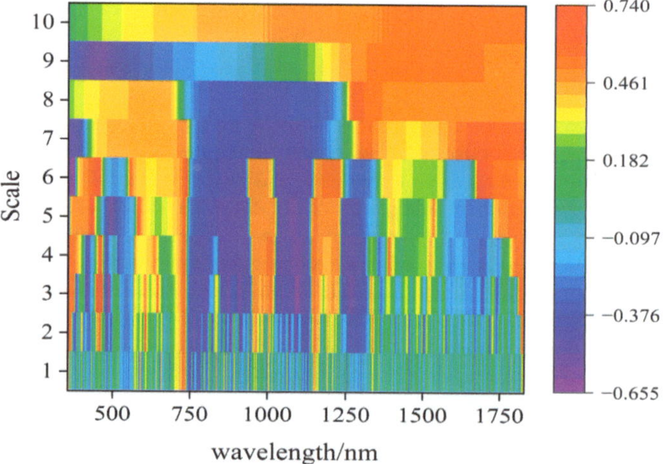

Figure 1. Correlation between wavelet transformed hyperspectral reflectance and chlorophyll content of winter wheat.

3.2. Construction of Spectral Indices and Extraction of Optimal Spectral Index Band Combinations

In order to more effectively utilize the information contained in the hyperspectral reflectance data, spectral indices were first calculated for each individual band of the hyperspectral reflectance data that had undergone CWT processing. Subsequently, a correlation analysis was conducted between these spectral indices and winter wheat chlorophyll content using the correlation matrix method. The bands with the highest correlation coefficients, i and j, were selected to construct different spectral indices. Seven spectral indices were then built, with the top five spectral indices showing the highest correlation with winter wheat chlorophyll content selected as the optimal spectral index combination. A correlation matrix plot was generated, as shown in Figures 2–4, where the color gradient from blue to red indicates the correlation between spectral indices and winter wheat chlorophyll content, ranging from negative to positive correlations. Figures 2–4 depict the correlation matrices between winter wheat chlorophyll content and hyperspectral vegetation indices obtained from wavelet transform at scales of 2^1 to 2^{10}. After undergoing continuous wavelet transform processing, the calculated optimal spectral indices showed a significantly higher correlation with winter wheat chlorophyll content compared with the original spectral data at certain scales. At the 26th scale, the average correlation coefficient between each optimal spectral index and winter wheat chlorophyll content was the highest at 0.752. The optimal spectral index combination at this scale included RI, DI, SAVI, NDVI, and mNDI. The optimal spectral index combinations for the rest of the transformed scales and the corresponding bands are shown in Table 2.

Table 2. Optimal spectral index band combinations at different transformation scales.

Transformation Scale	Spectral Index	Correlation Coefficient	Optimal Spectral Index Combination
0	RI	0.740	
	DI	0.865	
	SAVI	0.829	
	NDVI	0.731	RI, DI, SAVI, NDVI, mNDI
	TVI	0.134	
	mSR	0.368	
	mNDI	0.736	
2	RI	0.745	
	DI	0.686	
	SAVI	0.684	
	NDVI	0.719	RI, DI, SAVI, NDVI, mNDI
	TVI	0.587	
	mSR	0.490	
	mNDI	0.718	
4	RI	0.775	
	DI	0.765	
	SAVI	0.779	
	NDVI	0.738	RI, DI, SAVI, NDVI, mNDI
	TVI	0.587	
	mSR	0.458	
	mNDI	0.713	
8	RI	0.805	
	DI	0.817	
	SAVI	0.815	
	NDVI	0.774	RI, DI, SAVI, NDVI, mNDI
	TVI	0.543	
	mSR	0.299	
	mNDI	0.794	

Table 2. *Cont.*

Transformation Scale	Spectral Index	Correlation Coefficient	Optimal Spectral Index Combination
16	RI	0.744	RI, DI, SAVI, NDVI, mNDI
	DI	0.816	
	SAVI	0.841	
	NDVI	0.752	
	TVI	0.448	
	mSR	0.745	
	mNDI	0.793	
32	RI	0.764	RI, DI, SAVI, NDVI, mNDI
	DI	0.814	
	SAVI	0.845	
	NDVI	0.756	
	TVI	0.761	
	mSR	0.564	
	mNDI	0.759	
64	RI	0.750	RI, DI, SAVI, NDVI, mNDI
	DI	0.816	
	SAVI	0.842	
	NDVI	0.749	
	TVI	0.668	
	mSR	0.690	
	mNDI	0.749	
128	RI	0.715	RI, DI, SAVI, NDVI, mNDI
	DI	0.741	
	SAVI	0.755	
	NDVI	0.724	
	TVI	0.437	
	mSR	0.243	
	mNDI	0.703	
256	RI	0.716	RI, DI, SAVI, NDVI, mNDI
	DI	0.563	
	SAVI	0.731	
	NDVI	0.760	
	TVI	0.400	
	mSR	0.022	
	mNDI	0.744	
512	RI	0.758	RI, SAVI, NDVI, mSR, mNDI
	DI	0.571	
	SAVI	0.699	
	NDVI	0.759	
	TVI	0.124	
	mSR	0.674	
	mNDI	0.675	
1024	RI	0.659	RI, SAVI, NDVI, mSR, mNDI
	DI	0.607	
	SAVI	0.655	
	NDVI	0.658	
	TVI	0.526	
	mSR	0.690	
	mNDI	0.692	

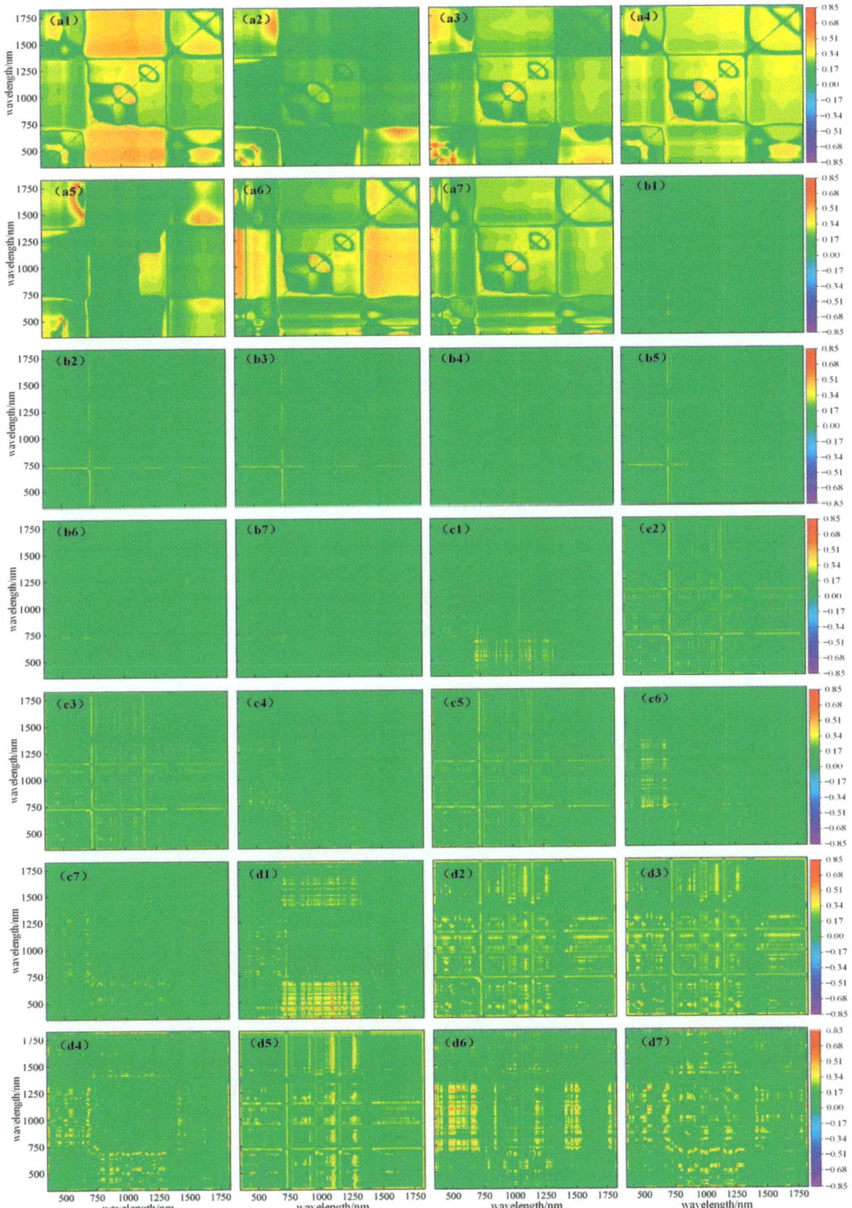

Figure 2. Correlation matrix of spectral indices at different scales: (**a1–a7**), (**b1–b7**), (**c1–c7**), (**d1–d7**) are the correlation of chlorophyll content screened by RI, DI, SAVI, NDVI, TVI, mSR, mNDI index of original reflectance at original scale, 2^1 transform scale, 2^2 transform scale and 2^3 transform scale.

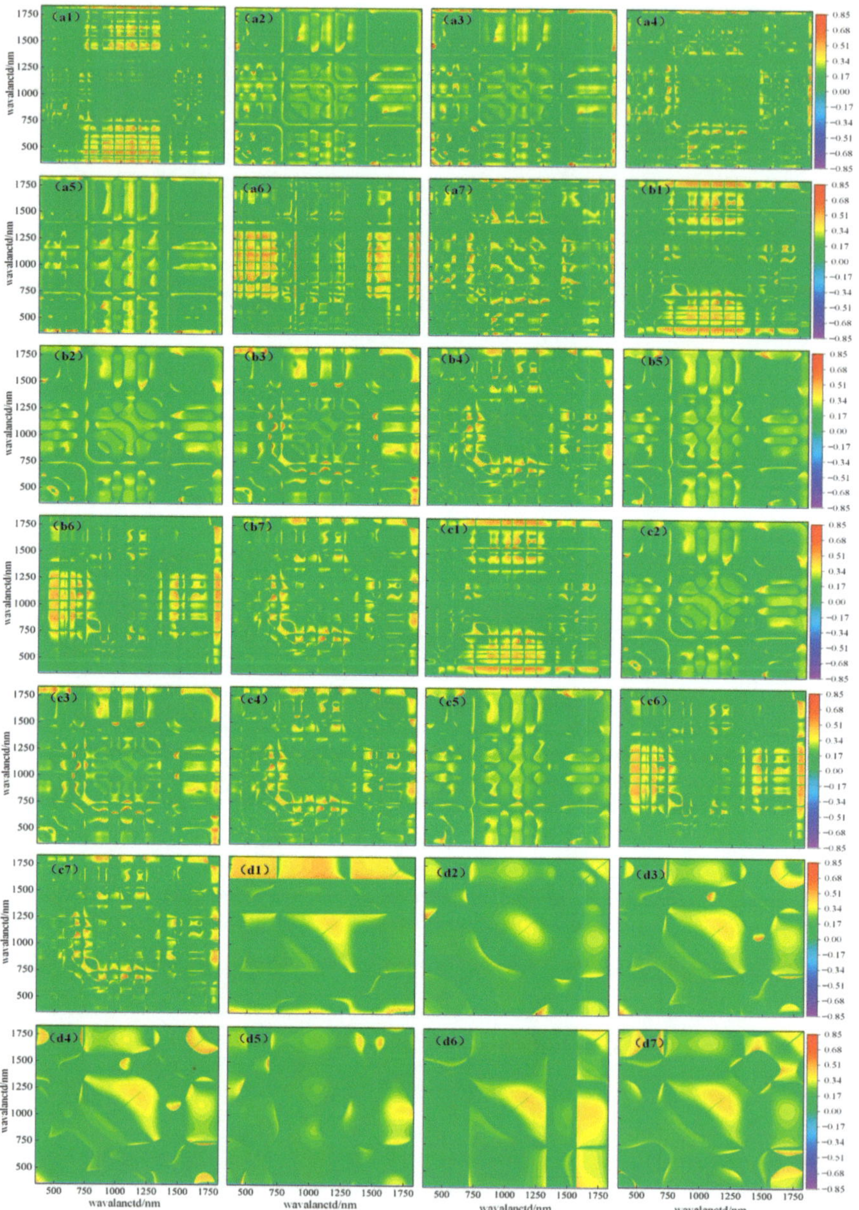

Figure 3. Correlation matrix of spectral indices at different scales: (**a1–a7**), (**b1–b7**), (**c1–c7**), (**d1–d7**) are the correlation of chlorophyll content screened by RI, DI, SAVI, NDVI, TVI, mSR, mNDI index of original reflectance at 2^4 transform scale, 2^5 transform scale, 2^6 transform scale and 2^7 transform scale.

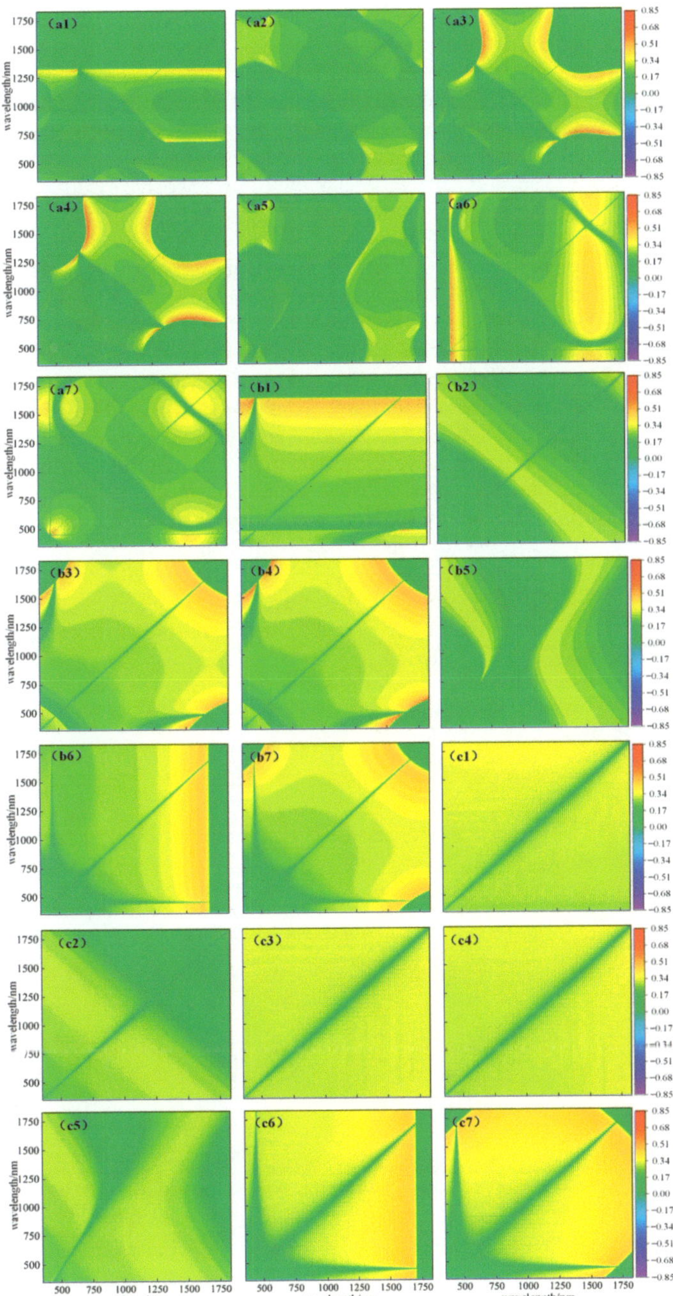

Figure 4. Correlation matrix of spectral indices at different scales: (**a1–a7**), (**b1–b7**), (**c1–c7**) are the correlation of chlorophyll content screened by RI, DI, SAVI, NDVI, TVI, mSR, mNDI index of original reflectance at 2^8 transform scale, 2^9 transform scale and 2^{10} transform scale.

3.3. Winter Wheat Chlorophyll Content Prediction Model Construction

With the increase in transformation scale, the correlation of spectral indices selected after 2^7 transformations continued to decrease. The correlation between spectral reflectance transformed from 2^8 to 2^{10} and winter wheat chlorophyll content was lower than that of the original spectral reflectance. Therefore, high-scale transformations were excluded to select the optimal spectral index combination from scales 2^1 to 2^7 and winter wheat chlorophyll content as the model input variables. SVM, GA-BP, and RF models were used to construct the model estimating the winter wheat chlorophyll content, to comprehensively evaluate the accuracy of the model from the aspects of R^2, RMSE, and MRE, respectively. R^2, RMSE, and MRE aspects of the comprehensive evaluation of the model accuracy, different modeling combinations for winter wheat chlorophyll content prediction results are shown in Table 3 and Figures 5 and 6. The results showed that R^2 of the winter wheat chlorophyll content estimation model under different scales of continuous wavelet transform were 2^6, 2^3, 2^5, 2^2, 2^0, 2^4, and 2^7 in descending order; RMSE and MRE were 2^6, 2^3, 2^5, 2^2, 2^0, 2^4, and 2^7 in descending order; and RF, GA-BP, and SVM constructed by 2^6-scale continuous wavelet transform spectral indices of winter wheat were used to predict the chlorophyll content of the model in the validation set. The validation set R^2 values of the chlorophyll content prediction model for winter wheat were 0.858 and 0.859, respectively, both higher than 0.310 ($p \leq 0.01$), indicating a highly significant correlation level with better linear fitting results. Under the same scale transform processing, the training set and validation set accuracies of the chlorophyll content prediction model of winter wheat constructed by the three modeling methods, from largest to the smallest, were GA-BP, RF and SVM. In summary, the 2^6-scale continuous wavelet transform and GA-BP model were the optimal scale and the optimal model construction method in this study, and the modeling and validation sets of the optimal winter wheat chlorophyll content prediction model constructed from these models had R^2 of 0.858 and 0.859, RMSE of 1.281 and 1.366, and MRE of 2.621% and 2.920%, respectively.

Table 3. Model accuracy analysis.

Transformation Scale	Evaluation Indicators	GA-BP		RF		SVM	
		Training Set	Validation Set	Training Set	Validation Set	Training Set	Validation Set
0	R^2	0.686	0.667	0.659	0.641	0.635	0.626
	RMSE	2.430	2.401	2.524	2.406	2.841	2.685
	MRE/%	5.088	5.003	5.212	5.262	5.892	5.955
2	R^2	0.626	0.613	0.581	0.575	0.564	0.556
	RMSE	2.525	2.547	2.717	2.819	2.803	2.828
	MRE/%	5.470	5.274	5.719	6.089	5.887	6.273
4	R^2	0.691	0.690	0.671	0.659	0.654	0.562
	RMSE	2.101	2.028	2.327	2.381	2.532	2.583
	MRE/%	4.656	4.226	4.799	4.672	5.871	5.814
8	R^2	0.804	0.798	0.773	0.779	0.738	0.736
	RMSE	1.685	1.620	1.829	1.865	2.392	2.222
	MRE/%	3.402	3.540	3.959	3.969	5.039	5.038
16	R^2	0.661	0.656	0.630	0.621	0.619	0.610
	RMSE	2.729	2.836	2.789	2.837	3.257	3.363
	MRE/%	5.394	5.295	5.620	5.872	5.979	6.146
32	R^2	0.789	0.788	0.765	0.758	0.731	0.733
	RMSE	1.775	1.788	1.848	1.866	2.441	2.771
	MRE/%	3.659	3.670	3.689	4.169	5.321	5.276
64	R^2	0.858	0.859	0.803	0.809	0.777	0.779
	RMSE	1.281	1.366	1.563	1.663	1.808	1.978
	MRE/%	2.621	2.920	3.145	3.534	3.665	4.038
128	R^2	0.665	0.652	0.621	0.619	0.613	0.609
	RMSE	2.641	2.626	2.909	2.982	3.562	3.393
	MRE/%	5.923	5.873	6.155	6.212	6.853	7.271

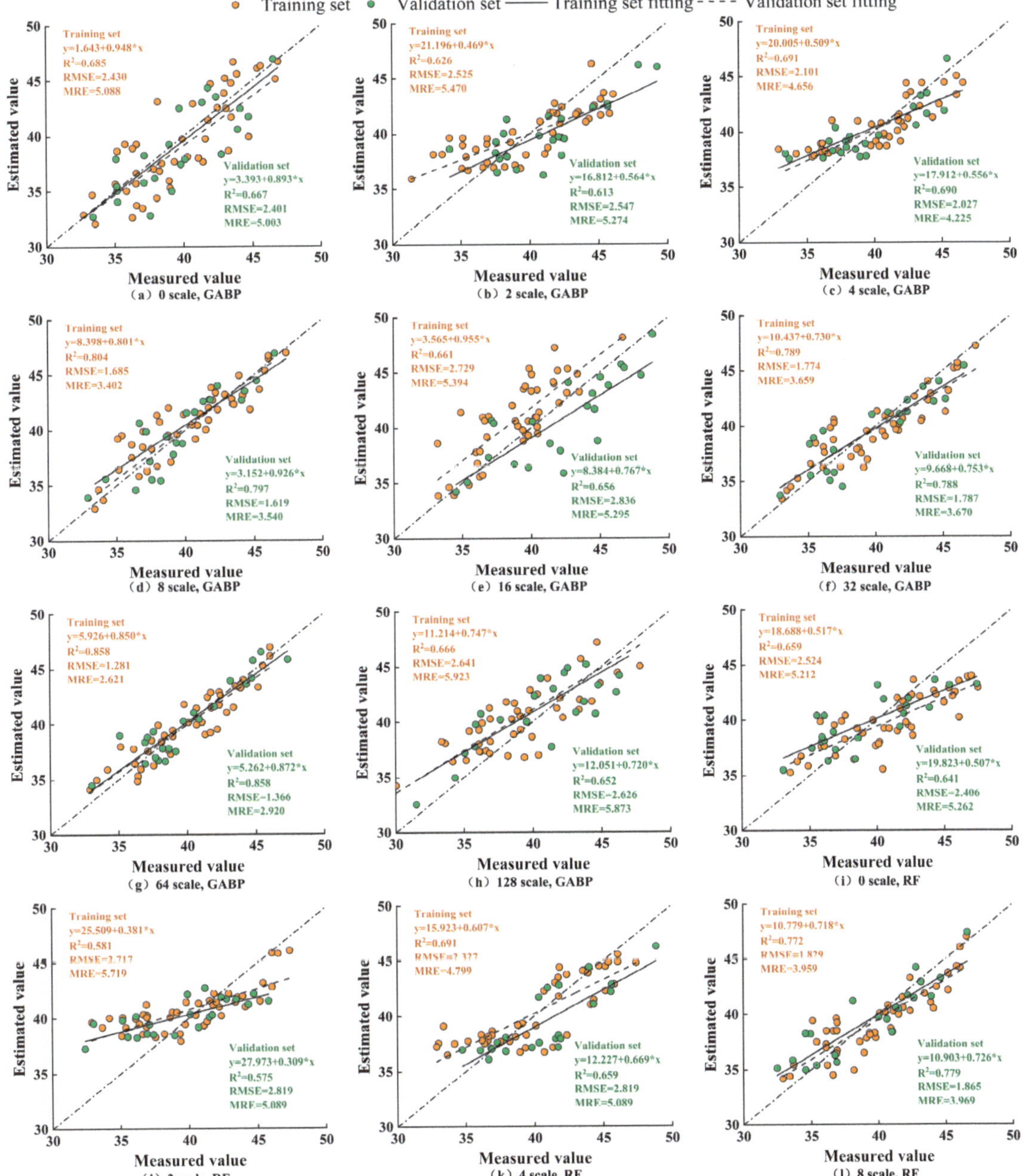

Figure 5. Modeling scatter plot. ((**a**–**h**) represents the GA-BP full-scale model, while (**i**–**l**) correspond to the original scale, 2^1-scale, 2^2-scale, and 2^3-scale RF models).

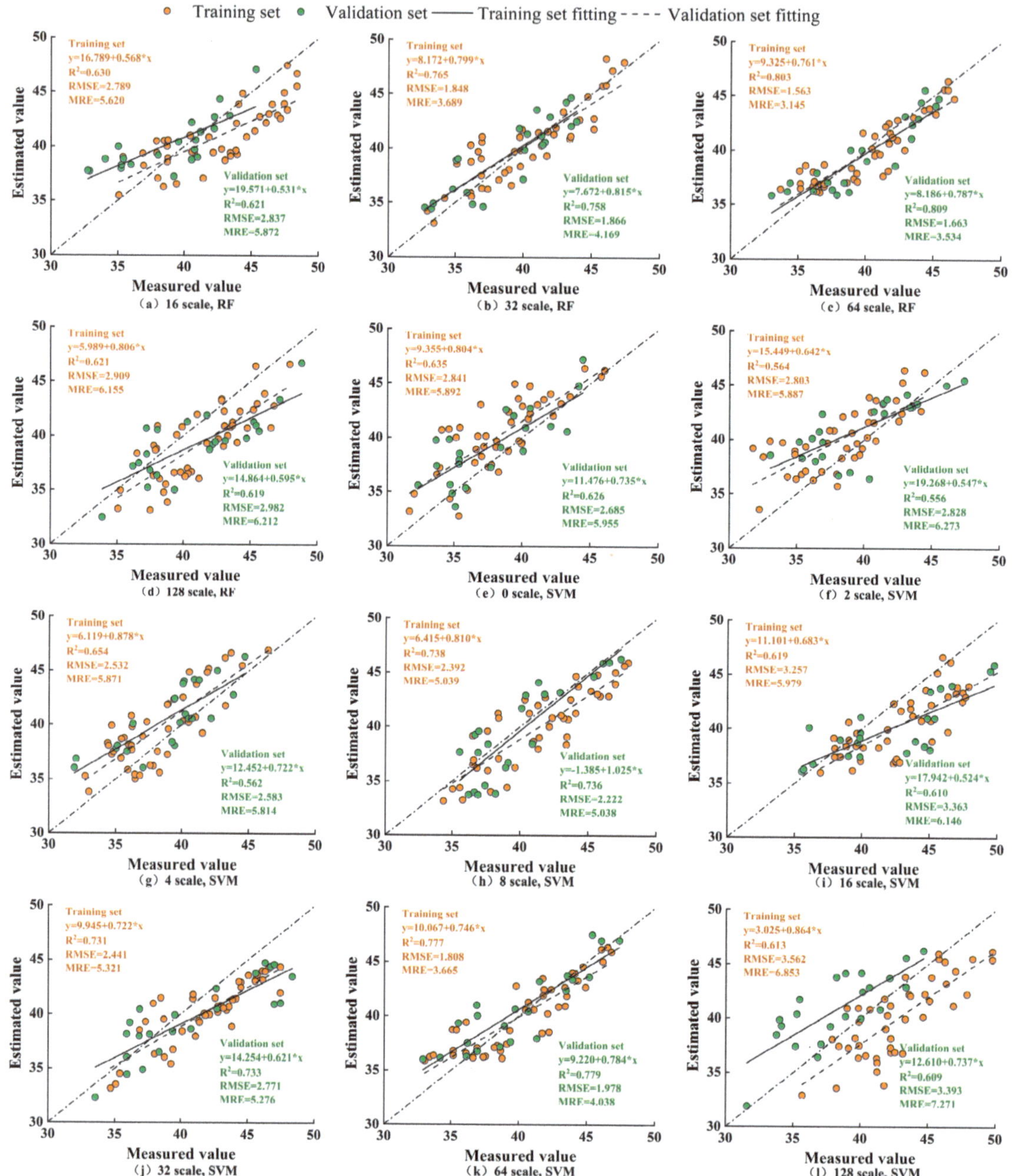

Figure 6. Modeling scatter plot. ((**a**–**d**) correspond to the 2^4-scale, 2^5-scale, 2^6-scale, and 2^7-scale RF models, while (**e**–**l**) represent the full-scale SVM model).

4. Discussion

Spectroscopic technology has shown great potential in monitoring crop growth and physiological indicators through Ahmad et al.'s research [38]. Among them, hyperspectral imaging, characterized by rich spectral information and strong band continuity, is commonly used for monitoring the crop chlorophyll content [39]. However, studies on the construction of chlorophyll content inversion models using hyperspectral reflectance have found that direct application of raw hyperspectral reflectance for chlorophyll content inversion modeling in crops can be affected by the soil background interference, atmospheric conditions, and lighting conditions; additionally, hyperspectral data contain a large number of bands, requiring extensive work and a high cost for direct application [40]. This is consistent with the research of Shu et al. [41], where the direct estimation of the chlorophyll content through hyperspectral remote sensing often exhibits poor model accuracy. For this reason, the introduction of continuous wavelet transform processing can decompose the hyperspectral data signals into frequency components of different scales, remove the noise in the data more easily, and at the same time identify the most informative features in the hyperspectral data, which can help simplify the dataset and improve the training efficiency and prediction accuracy of the model. In this study, the raw hyperspectral data were 2^1, 2^2, $2^3 \ldots \ldots 2^{10}$, ten scales of continuous wavelet transform, and we constructed an estimation model of chlorophyll content of winter wheat under the $2^1 \sim 2^7$ transform scales. With the improvement in the transform scales, the spectral index and chlorophyll content, as well as the model's accuracy, showed an increasing and then decreasing trend, and the optimal spectral index and chlorophyll content correlation under the 2^6-scale transform and the accuracy of the estimation model constructed under the 2^6-scale transform were higher than those in the other scales. This is consistent with the study by Li et al. [42], which found that the correlation coefficient was the highest at a scale of 6 using wavelet transform. This may be due to the fact that the continuous wavelet transform at small scales amplified subtle changes and noise, leading to loss or confusion of information; while at high scales, it may cause excessive smoothing, leading to loss of details. This situation may prevent the inversion model from accurately capturing the changes in chlorophyll content, thus reducing the accuracy of the model.

Wang et al. [43]'s research shows that the optimal spectral index constructed within the full spectral range contained a richer set of effective information related to the winter wheat chlorophyll content. The selected optimal spectral index combinations at various scales were used as the input variables, and three machine learning methods, GA-BP, SVM and RF, were employed to construct models for estimating the winter wheat chlorophyll content [44]. Among the three machine learning methods, the model based on the GA-BP method exhibited the highest accuracy for estimating the winter wheat chlorophyll content. Shi et al. [45] showed the highest accuracy of inversion of chlorophyll content in winter wheat using the GA-BP model, with an R^2 of 0.952, which was consistent with the results in this study. This indicates that the GA-BP method has a stronger capability to extract chlorophyll content information from spectral reflectance data. This is attributed to the integration of the genetic algorithm and back propagation neural network advantages in the GA-BP algorithm, which help in selecting the most representative input features, enhancing the model's generalization ability and prediction accuracy, and enabling the learning and capture of complex nonlinear relationships in the data. However, due to the combination of two algorithms, GA-BP often requires more time to complete training and may be constrained by the local optima, leading to suboptimal results. The prediction accuracy of the RF method is slightly lower than that of the GA-BP method, which may be due to the increased difference in the number of high and low content samples in the sample, which may form the model's preference for a certain category and affect the overall accuracy. The SVM method had the lowest prediction accuracy, which can be attributed to the high computational complexity of SVM algorithms when dealing with high-dimensional feature spaces or large-scale datasets. This leads to long training times and high memory consumption, and also increases the risk of overfitting the model.

Nowadays, modeling of the crop chlorophyll using hyperspectral data has achieved good results in practical applications of the model. R^2 can serve as a comprehensive evaluation of the model's predictive accuracy, providing a basis for accurate monitoring of the crop conditions. Meanwhile, RMSE and MRE reflect the stability of the model, offering relatively concentrated prediction results and providing a scientific basis for precision agriculture development [46]. But, currently, there are still many technical limitations, such as insufficient depth mining of spectral data information. In addition, because the performance of machine learning models is affected by the number of samples, parameter tuning, data quality, and other factors, there may be problems such as overfitting or underfitting [47]. To address the above problems, future research can try to improve the monitoring and estimation of crop chlorophyll content by evaluating the deep learning model for hyperspectral data analysis and designing the network structure [48], loss function [49] and training algorithm that are suitable for the characteristics of the field [50], as well as continuously optimizing and improving the machine learning model. These methods can more effectively utilize the information of hyperspectral data to enhance the understanding of the relationship between crop growth physiological state and chlorophyll content, and provide a reference for the use of multisource remote sensing such as hyperspectral to predict the chlorophyll content of winter wheat at the pulling stage with further exploration of hyperspectral reflectance continuous wavelet transform.

5. Conclusions

After performing continuous wavelet transform on the hyperspectral reflectance data, the correlation between the selected spectral indices and winter wheat chlorophyll content showed a significant improvement. When the modeling method was the same, but the input combinations differed, the predictive model accuracies ranked as follows: $2^6, 2^3, 2^5, 2^2, 2^0, 2^4$, and 2^7. On the other hand, when the input combinations were the same but the modeling methods varied, the predictive model accuracies ranked as follows: GA-BP, RF, and SVM. By considering various evaluation metrics, it was identified that the scale variation of 2^6 and the GA-BP method were the optimal transformation scale and modeling approach in this study. The predictive models constructed based on these parameters exhibited R^2 values of 0.858 and 0.859, RMSE values of 1.281 and 1.366, and MRE values of 2.621% and 2.920% for the training and validation sets, respectively.

Author Contributions: Data curation: X.L., Z.T. and Z.L.; Investigation: Z.T., X.H. and H.S.; Methodology: T.S., W.Y., S.C. and G.C.; Project administration: X.L.; Resources: Y.X. and F.Z.; Software: Z.T. and X.L.; Supervision: F.Z., Z.L. and Y.X.; Visualization: Z.L. and Y.X.; Writing—original draft: X.L.; Writing—review and editing: X.L. and Y.X. All authors have read and agreed to the published version of the manuscript.

Funding: This study was supported by the National Natural Science Foundation of China (No. 52179045).

Data Availability Statement: Data are contained within the article.

Conflicts of Interest: The authors declare no conflicts of interest.

References

1. Feng, Y.; Chen, B.; Liu, W.; Xue, X.; Liu, T.; Zhu, L.; Xing, H. Winter Wheat Mapping in Shandong Province of China with Multi-Temporal Sentinel-2Images. *Appl. Sci.* **2024**, *14*, 3940. [CrossRef]
2. Wang, F.; He, Z.; Sayre, K.; Li, S.; Si, J.; Feng, B.; Kong, L. Wheat cropping systems and technologies in China. *Field Crop. Red.* **2009**, *111*, 181–188. [CrossRef]
3. Coventry, D.R.; Gupta, R.K.; Yadav, A.; Poswal, R.S.; Chhokar, R.S.; Sharma, R.K.; Yadav, V.K.; Gill, S.C.; Kumar, A.; Mehta, A.; et al. Wheat quality and productivity as affected by varieties and sowing time in Haryana, India. *Field Crop. Res.* **2011**, *123*, 214–225. [CrossRef]
4. Tang, Z.; Lu, J.; Xiang, Y.; Shi, H.; Sun, T.; Zhang, W.; Wang, H.; Zhang, X.; Li, Z.; Zhang, F. Farmland mulching and optimized irrigation increase water productivity and seed yield by regulating functional parameters of soybean (*Glycine max* L.) leaves. *Agric. Water Manag.* **2024**, *298*, 108875. [CrossRef]
5. de Souza, R.; Peña-Fleitas, M.T.; Thompson, R.B.; Gallardo, M.; Grasso, R.; Padilla, F.M. The Use of Chlorophyll Meters to Assess Crop N Status and Derivation of Sufficiency Values for Sweet Pepper. *Sensors* **2019**, *19*, 2949. [CrossRef]
6. Xavier, W.D.; Castoldi, G.; Cavalcante, T.J.; Rodrigues, C.R.; Trindade, P.R.; Luiz, I.A.; Damin, V. Portable Chlorophyll Meter for Indirect Evaluation of Photosynthetic Pigments and Nitrogen Content in Sweet Sorghum. *Sugar Tech.* **2021**, *23*, 560–570. [CrossRef]

7. Tang, Z.; Wang, X.; Xiang, Y.; Liang, J.; Guo, J.; Li, W.; Lu, J.; Du, R.; Li, Z.; Zhang, F. Application of hyperspectral technology for leaf function monitoring and nitrogen nutrient diagnosis in soybean (*Glycine max* L.) production systems on the Loess Plateau of China. *Eur. J. Agron.* **2024**, *154*, 127098. [CrossRef]

8. Zhang, J.; Han, W.; Huang, L.; Zhang, Z.; Ma, Y.; Hu, Y. Leaf chlorophyll content estimation of winter wheat based on visible and near-infrared sensors. *Sensors.* **2016**, *16*, 437. [CrossRef]

9. Steele, M.R.; Gitelson, A.A.; Rundquist, D.C. A comparison of two techniques for nondestructive measurement of chlorophyll content in grapevine leaves. *Agron. J.* **2008**, *100*, 779–782. [CrossRef]

10. Fountas, S.; Wulfsohn, D.; Blackmore, B.; Jacobsen, H.; Pedersen, S. A model of decision-making and information flows for information-intensive agriculture. *Agr Syst.* **2006**, *87*, 192–210. [CrossRef]

11. Tang, Z.; Zhang, W.; Xiang, Y.; Liu, X.; Wang, X.; Shi, H.; Li, Z.; Zhang, F. Monitoring of Soil Moisture Content of Winter Oilseed Rape (*Brassica napus* L.) Based on Hyperspectral and Machine Learning Models. *J. Soil Sci. Plant Nutr.* **2024**, *24*, 1250–1260. [CrossRef]

12. Tang, Z.; Xiang, Y.; Zhang, W.; Wang, X.; Zhang, F.; Chen, J. Research on potato (*Solanum tuberosum* L.) nitrogen nutrition diagnosis based on hyperspectral data. *Agron. J.* **2024**, *116*, 531–541. [CrossRef]

13. Sampson, P.H.; Zarco-Tejada, P.J.; Mohammed, G.H.; Miller, J.R.; Noland, T.L. Hyperspectral remote sensing of forest condition: Estimating chlorophyll content in tolerant hardwoods. *Forest Sci.* **2003**, *49*, 381–391. [CrossRef]

14. Gitelson, A.A.; Gritz, Y.; Merzlyak, M.N. Relationships between leaf chlorophyll content and spectral reflectance and algorithms for non-destructive chlorophyll assessment in higher plant leaves. *J. Plant Physiol.* **2003**, *160*, 271–282. [CrossRef] [PubMed]

15. Yoder, B.J.; Pettigrew-Crosby, R.E. Predicting nitrogen and chlorophyll content and concentrations from reflectance spectra (400–2500 nm) at leaf and canopy scales. *Remote Sens. Environ.* **1995**, *53*, 199–211. [CrossRef]

16. Saberioon, M.; Amin, M.; Anuar, A.; Gholizadeh, A.; Wayayok, A.; Khairunniza-Bejo, S. Assessment of rice leaf chlorophyll content using visible bands at different growth stages at both the leaf and canopy scale. *Int. J. Appl. Earth Obs.* **2014**, *32*, 35–45. [CrossRef]

17. Yang, F.; Fan, Y.M.; Li, J.L.; Qian, Y.R.; Wang, Y.; Zhang, J. Hyperspectral data estimated rice and wheat leaf area index and chlorophyll density. *Trans. Chin. Soc. Agric. Eng.* **2010**, *26*, 237–243, (In Chinese with English Abstract). [CrossRef]

18. Briggs, F.H.; Bell, J.F.; Kesteven, M.J. Removing radio interference from contaminated astronomical spectra using an independent reference signal and closure relations. *J. Korean Astron. Soc.* **2000**, *120*, 3351. [CrossRef]

19. Nie, P.; Guo, Y.; Lou, B.; Yang, C.; Cao, L.; Pan, W. Tool wear monitoring based on scSE-ResNet-50-TSCNN model integrating machine vision and force signals. *Meas. Sci. Technol.* **2024**, *35*, 086117. [CrossRef]

20. Liu, D.D.; Nie, R.J.; Xu, X.B. Nitrogen estimation model for summer maize based on continuous wavelet transform and RF algorithm. *Barley Cereal Sci.* **2019**, *36*, 42–46, (In Chinese with English Abstract). [CrossRef]

21. Cheng, T.; Rivard, B.; Sánchez-Azofeifa, A. Spectroscopic determination of leaf water content using continuous wavelet analysis. *Remote Sens. Environ.* **2010**, *115*, 659–670. [CrossRef]

22. Bauer, M.E. Spectral inputs to crop identification and condition assessment. *Proc. IEEE* **1985**, *73*, 1071–1085. [CrossRef]

23. Xu, M.; Liu, R.; Chen, J.M.; Liu, Y.; Shang, R.; Ju, W.; Wu, C.; Huang, W. Retrieving leaf chlorophyll content using a matrix-based vegetation index combination approach. *Remote Sens. Environ.* **2019**, *224*, 60–73. [CrossRef]

24. Tang, Z.; Guo, J.; Xiang, Y.; Lu, X.; Wang, Q.; Wang, H.; Cheng, M.; Wang, H.; Wang, X.; An, J.; et al. Estimation of leaf area index and above-ground biomass of winter wheat based on optimal spectral index. *Agronomy* **2022**, *12*, 1729. [CrossRef]

25. Higdon, D. A process-convolution approach to modelling temperatures in the North Atlantic Ocean. *Envioron. Ecol. Stat.* **1998**, *5*, 173–190. [CrossRef]

26. Lawton, W. Applications of complex valued wavelet transforms to subband decomposition. *IEEE Trans. Signal Process.* **1993**, *41*, 3566–3568. [CrossRef]

27. Heil, C.E.; Walnut, D.F. Continuous and discrete wavelet transforms. *Siam Rev.* **1989**, *31*, 628–666. [CrossRef]

28. Rivera, J.P.; Verrelst, J.; Delegido, J.; Veroustraete, F.; Moreno, J. On the Semi-Automatic Retrieval of Biophysical Parameters Based on Spectral Index Optimization. *Remote Sens.* **2014**, *6*, 4927–4951. [CrossRef]

29. Tran, T.V.; Reef, R.; Zhu, X. A review of spectral indices for mangrove remote sensing. *Remote Sens.* **2022**, *14*, 4868. [CrossRef]

30. Liu, C.; Hu, Z.; Islam, A.T.; Kong, R.; Yu, L.; Wang, Y.; Chen, S.; Zhang, X. Hyperspectral characteristics and inversion model estimation of winter wheat under different elevated CO_2 concentrations. *Int J. Remote Sens.* **2021**, *42*, 1035–1053. [CrossRef]

31. Xu, G.; Liu, M.; Jiang, Z.; Söffker, D.; Shen, W. Bearing Fault Diagnosis Method Based on Deep Convolutional Neural Network and Random Forest Ensemble Learning. *Sensors* **2019**, *19*, 1088. [CrossRef]

32. Shi, K.; Qiao, Y.; Zhao, W.; Wang, Q.; Liu, M.; Lu, Z. An improved random forest model of short-term wind-power forecasting to enhance accuracy, efficiency, and robustness. *Wind Energy* **2018**, *21*, 1383–1394. [CrossRef]

33. Marjanović, M.; Kovačević, M.; Bajat, B.; Voženílek, V. Landslide susceptibility assessment using SVM machine learning algorithm. *Eng. Geol.* **2011**, *123*, 225–234. [CrossRef]

34. Zeng, W.; Jia, J.; Zheng, Z.; Xie, C.; Guo, L. A comparison study: Support vector machines for binary classification in machine learning. In Proceedings of the 2011 4th International Conference on Biomedical Engineering and Informatics (BMEI), Shanghai, China, 15–17 October 2011; IEEE: Piscataway, NJ, USA, 2011; Volume 3, pp. 1621–1625. [CrossRef]

35. Unger, R.; Moult, J. Genetic algorithms for protein folding simulations. *J. Mol. Biol.* **1993**, *231*, 75–81. [CrossRef] [PubMed]

36. Chicco, D.; Warrens, M.J.; Jurman, G. The coefficient of determination R-squared is more informative than SMAPE, MAE, MAPE, MSE and RMSE in regression analysis evaluation. *Peerj Comput. Sci.* **2021**, *7*, e623. [CrossRef] [PubMed]

37. Meng, X.L.; Rosenthal, R.; Rubin, D.B. Comparing correlated correlation coefficients. *Phychol Bull.* **1992**, *111*, 172. [CrossRef]

38. Ahmad, U.; Nasirahmadi, A.; Hensel, O.; Marino, S. Technology and Data Fusion Methods to Enhance Site-Specific Crop Monitoring. *Agronomy* **2022**, *12*, 555. [CrossRef]

39. Haboudane, D.; Tremblay, N.; Miller, J.R.; Vigneault, P. Remote estimation of crop chlorophyll content using spectral indices derived from hyperspectral data. *IEEE Trans. Geosci. Remote.* **2008**, *46*, 423–437. [CrossRef]

40. Berger, K.; Machwitz, M.; Kycko, M.; Kefauver, S.C.; Van Wittenberghe, S.; Gerhards, M.; Verrelst, J.; Atzberger, C.; van der Tol, C.; Damm, A.; et al. Multi-sensor spectral synergies for crop stress detection and monitoring in the optical domain: A review. *Remote Sens. Environ.* **2022**, *280*, 113198. [CrossRef]

41. Shu, M.; Zuo, J.; Shen, M.; Yin, P.; Wang, M.; Yang, X.; Tang, J.; Li, B.; Ma, Y. Improving the estimation accuracy of SPAD values for maize leaves by removing UAV hyperspectral image backgrounds. *Int. J. Remote Sens.* **2021**, *42*, 5862–5881. [CrossRef]

42. Li, C.H.; Shi, J.J.; Ma, C.Y.; Cui, Y.Q. Estimation of winter wheat chlorophyll content based on wavelet transform and fractional differential. *Trans. Chin. Soc. Agric. Mach.* **2021**, *52*, 172–182, (In Chinese with English Abstract). [CrossRef]

43. Wang, T.; Gao, M.; Cao, C.; You, J.; Zhang, X.; Shen, L. Winter wheat chlorophyll content retrieval based on machine learning using in situ hyperspectral data. *Comput. Electron. Agric.* **2022**, *193*, 106728. [CrossRef]

44. Chen, X.; Li, F.; Shi, B.; Fan, K.; Li, Z.; Chang, Q. Estimation of winter wheat canopy chlorophyll content based on canopy spectral transformation and machine learning method. *Agronomy* **2023**, *13*, 783. [CrossRef]

45. Shi, M.; Jing, X.; Shi, X.L. Inversion of winter wheat chlorophyll content based on hyperspectral and GA-BP neural network model. *Jiangsu J. Agric. Sci.* **2022**, *50*, 56–62. (In Chinese with English Abstract) [CrossRef]

46. Fu, Z.; Jiang, J.; Gao, Y.; Krienke, B.; Wang, M.; Zhong, K.; Cao, Q.; Tian, Y.; Zhu, Y.; Cao, W.; et al. Wheat Growth Monitoring and Yield Estimation based on Multi-Rotor Unmanned Aerial Vehicle. *Remote Sens.* **2020**, *12*, 508. [CrossRef]

47. Sehra, S.; Flores, D.; Montañez, G.D. Undecidability of underfitting in learning algorithms. In Proceedings of the 2021 2nd International Conference on Computing and Data Science (CDS), Stanford, CA, USA, 28–29 January 2021; IEEE: Piscataway, NJ, USA, 2021; pp. 591–594. [CrossRef]

48. Lan, B.; Zhou, X.; Yang, N.; Sun, S. Spectral radius is a better metric than weighted NODF to detect network nestedness: Linking species coexistence to network structure using a plant–larval sawfly bipartite. *Food Webs* **2023**, *36*, e00303. [CrossRef]

49. Zhu, Z.; Song, J.; He, S.; Liu, J.J.R.; Lam, H.-K. Event-Triggered Disturbance Rejection Control for Brain-Actuated Mobile Robot: An SSA-Optimized Sliding Mode Approach. *IEEE-AMSE Trans. Mechatron.* **2024**, 1–12. [CrossRef]

50. Reda, R.; Saffaj, T.; Ilham, B.; Saidi, O.; Issam, K.; Brahim, L.; El Hadrami, E.M. A comparative study between a new method and other machine learning algorithms for soil organic carbon and total nitrogen prediction using near infrared spectroscopy. *Chemom. Intell. Lab. Syst.* **2019**, *195*, 103873. [CrossRef]

 *agronomy*

Article

Efficient Damage Assessment of Rice Bacterial Leaf Blight Disease in Agricultural Insurance Using UAV Data

Chiharu Hongo [1,*], Shun Isono [2], Gunardi Sigit [3] and Eisaku Tamura [1]

[1] Center for Environmental Remote Sensing, Chiba University, Chiba 277-2835, Japan
[2] NHK (Japan Broadcasting Corporation), Osaka 540-8501, Japan
[3] Provincial Office of Food Crops and Horticulture of West Java Province, Bandung 40133, Indonesia; gunsigit@yahoo.com
* Correspondence: hongo@faculty.chiba-u.jp

Abstract: In Indonesia, where the agricultural insurance system has been in full operation since 2016, a new damage assessment estimation formula for rice diseases was created through integrating the current damage assessment method and unmanned aerial vehicle (UAV) multispectral remote sensing data to improve the efficiency and precision of damage assessment work performed for the payments of insurance claims. The new method can quickly and efficiently output objective assessment results. In this study, UAV images and bacterial leaf blight (BLB) rice damage assessment data were acquired during the rainy and dry seasons of 2021 and 2022 in West Java, Indonesia, where serious BLB damage occurs every year. The six-level BLB score (0, 1, 3, 5, 7, and 9) and damage intensity calculated from the score were used as the BLB damage assessment data. The relationship between normalized UAV data, normalized difference vegetation index (NDVI), and BLB score showed significant correlations at the 1% level. The analysis of damage intensities and UAV data for paddy plots in all cropping seasons showed high correlation coefficients with the normalized red band, normalized near-infrared band, and NDVI, similar to the results of the BLB score analysis. However, for paddy plots with damage intensities of 70% or higher, the biased numbering of the BLB score data may have affected the evaluation results. Therefore, we conducted an analysis using an average of 1090 survey points for each BLB score and confirmed a strong relationship, with correlation coefficients exceeding 0.9 for the normalized red band, normalized near-infrared band, and NDVI. Through comparing the time required by the current assessment method with that required by the assessment method integrating UAV data, it was demonstrated that the evaluation time was reduced by more than 60% on average. We are able to propose a new assessment method for the Indonesian government to achieve complete objective enumeration.

Keywords: food security; remote sensing; agricultural insurance; pest and diseases

Citation: Hongo, C.; Isono, S.; Sigit, G.; Tamura, E. Efficient Damage Assessment of Rice Bacterial Leaf Blight Disease in Agricultural Insurance Using UAV Data. *Agronomy* **2024**, *14*, 1328. https://doi.org/10.3390/agronomy14061328

Academic Editors: Jinling Zhao and Chuanjian Wang

Received: 31 March 2024
Revised: 6 June 2024
Accepted: 14 June 2024
Published: 19 June 2024

1. Introduction

Climate change is expected to expose humankind to various risks in the future. The Sixth Assessment Report released by the Intergovernmental Panel on Climate Change (IPCC) in August 2021 lists food security as a major threat [1]. Diseases, insects, droughts, and floods caused by extreme weather and other factors can damage crops and require considerable labor and costs to recover farmland and plant the next crop [2–4]. Farmers with insecure economic foundations are forced to leave their farms, which further threatens food security.

Serious damage is expected to occur primarily in developing countries, where a variety of policies have been proposed. In Indonesia, where rice is the primary crop [5], a farmer protection and empowerment law was enacted in 2013 [6]. As a result, an agricultural insurance system was launched in 2016, under which the government paid compensation for damage to rice paddies caused by pests, diseases, drought, and floods [7–9]. The current damage assessment method in Indonesia involves selecting only three paddy plots from the terminal irrigation area, which includes approximately 50–300 plots, and having an assessor called a pest observer visually assess the damage to 10 rice plants on the diagonal of one paddy plot. The average damage to the entire damaged area is calculated based on the evaluation of 30 rice plants. In Indonesia, an objective evaluation method acceptable to insurance subscribers is required because the results of the visual damage evaluation method may differ between assessors [10,11]. Recently, there has been a growing expectation to build more robust and efficient methods for detecting paddy rice diseases using machine and deep learning, as automated approaches to detect leaf diseases can help farmers detect diseases with or without human intervention [12–15].

Promptness is especially important when operating insurance programs in Southeast Asia. A prolonged evaluation time can lead to further losses, such as missing the next rice-planting season. Indonesia has wet and dry seasons, and two or three rice crops per year are commonly grown in two or three cropping seasons. Damaged paddy fields must be maintained in their current state until damage assessors complete their evaluation; however, the limited number of assessors limits their ability to quickly produce damage assessment results. Therefore, we have improved the current evaluation method to construct a new damage assessment method that can efficiently and quickly evaluate damage and output objective evaluation results.

Bacterial leaf blight (BLB) is a serious disease that occurs annually in West Java, Indonesia, where this study was conducted. Although resistant rice varieties have been cultivated and chemicals are sprayed to ensure stable rice production, the occurrence of BLB damage has not ended and a policy to protect farmers through agricultural insurance has been adopted. The key to agricultural insurance is objective, prompt, and inexpensive damage assessment, and remote sensing technology is expected to be utilized for its advantages, such as wide-area information, immediacy, and objectivity [16]. Research has already been conducted in Indonesia using satellite data to estimate paddy rice production, the transplanting dates of rice crops, and to assess drought damage [17–20], and it has been reported that satellite remote sensing data can be applied even in Indonesia, where the area of a single paddy field is small. Meanwhile, remote sensing data has been acquired using satellites, aircraft, helicopters, and unmanned aerial vehicles, although satellite remote sensing data has limitations in understanding in-field variability, such as in precision agriculture [21]. Among these platforms, multirotor drones are the most promising for smart farming [22], and there is a need to develop a damage assessment method using unmanned aerial vehicle (UAV) remote sensing for assessment of disease damage.

Previous studies on crop diseases using remote sensing data have reported that the sensitivity of near-infrared and short-wavelength infrared reflectance to the degree of BLB infection is high [23] and that the difference in reflectance between healthy and BLB-infected rice plants is pronounced in the 770–860 nm and 920–1050 nm wavelength ranges [24]. UAV imagery analysis in rice paddies in Bali, Indonesia, confirmed that vegetation indices, normalized difference vegetation index (NDVI), enhanced vegetation index (EVI), and normalized difference red edge (NDRE) had a strong linear correlation with BLB damage intensity [25]. Studies using aircraft and UAV observation data have evaluated the severity of BLB, rice blast, and rice spot disease [26–28]. Studies have also reported examining crop growth monitoring using an RGB camera on a UAV, from a low-cost perspective [29,30]. Studies using satellite observation data to detect crop diseases have included the detection of rice blast, rice sesame leaf blight, and yellow rust [31,32]. Studies using multispectral data from LISS-IV satellite observations to detect rice stress caused by BLB at the regional level [33] and a study evaluating the correlation between BLB damage severity and spectral indices from Sentinel-2 data have also been reported [34–36]. These reports show that reflectance data in the visible and near-infrared regions and vegetation indices calculated from reflectance are useful for understanding BLB and other diseases.

Therefore, this study reports a precise assessment of BLB damage using UAV multispectral remote sensing data in West Java, Indonesia, where serious BLB damage occurs annually, and examines the feasibility of reducing the damage assessment time via integrating current assessment methods and UAV data to create a new damage assessment estimation formula.

2. Materials and Methods

2.1. Study Area

The target area for this study was the Cihea Irrigation District (6°50′ S, 107°16′ E) in the northeastern part of Cihea, Cianjur Province, West Java, Republic of Indonesia (Figure 1a). This area is located just below the equator and has a tropical climate throughout the year, with a dry season from April to October and a rainy season from November to March. In this study, both dry and rainy season cropping from 2021 to 2022 were considered.

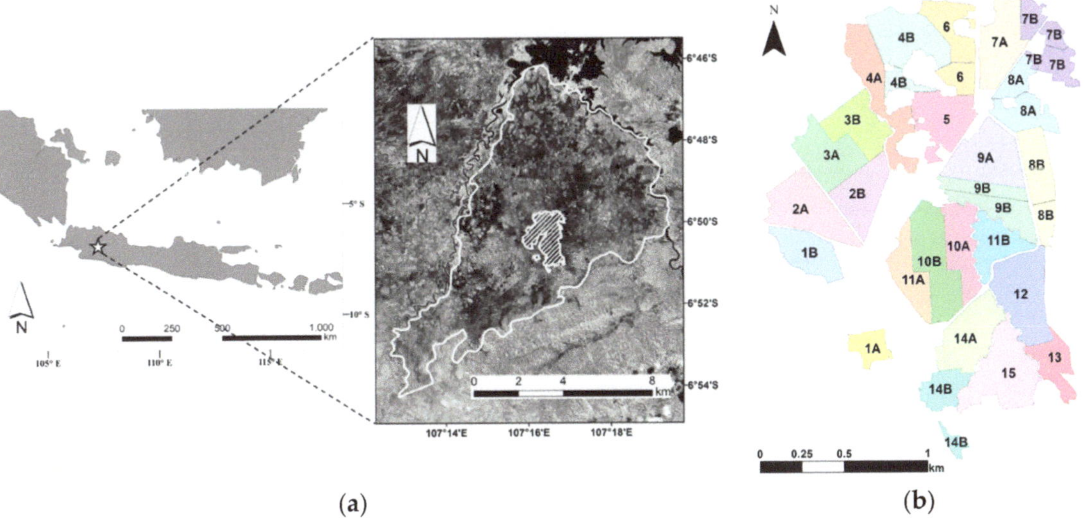

Figure 1. Study area; (**a**) Cihea Irrigation District, (**b**) 250-hectare test area with irrigation block name.

The target area was a large irrigated district of approximately 8000 ha. Severe damage occurred annually over a wide area because the BLB bacteria that developed at one location were carried downstream by irrigation water and spread over the entire area. In addition, the shape and size of the fields were not uniform, and each irrigated area was characterized by a mixture of fields at different growth stages owing to different planting times. In this study, the 250 ha irrigated area shown in Figure 1b was set as a test site, and several irrigated areas where BLB had occurred were selected for investigation. The rice variety grown in both years was Inpari 32, and the soil type in the area was identified as Inceptisol based on our previous soil auger and soil cross-sectional surveys [17].

2.2. Characteristics of the Target Disease; Bacterial Leaf Blight Disease of Rice

BLB is a disease caused by Xanthomonas oryzae pv. Studies on bacteriophages have shown that bacteria are transported through the flow of irrigation water. The disease has been confirmed to occur in rice-growing regions worldwide, including tropical and temperate Asia, West Africa, and Central and South America [37,38]. Breeding disease-resistant varieties is an effective and economical method for controlling BLB [39]. Resistant varieties have been introduced to the study area in recent years; however, disease outbreaks remain uncontrolled.

The disease symptoms of BLB include yellowing at the leaf margins, yellow irregular spots, or blotches that gradually expand and merge into wavy lesions that turn yellowish white or white and then turn grayish white and die at the tip of the leaf [40]. The border between the dead and healthy areas, that is, near the leaf margins and lesions, is marked by the appearance of small yellow granular mucilage masses that overflow from the pores and harden, distinguishing it from natural mortality. Figure 2 shows the symptoms of BLB with different degrees of damage observed in the field.

Figure 2. BLB outbreak plots (**left**: score 3, **right**: score 7).

BLB increases the risk of yield loss early in the infected growth stage. Therefore, proper cultivation and management are important. Major measures include the cultivation of resistant varieties, removal of weeds, stumps, and seedlings that harbor the fungus, and spraying with pesticides.

2.3. Current Damage Assessment Method in Agricultural Insurance in Indonesia

Damage assessments in agricultural insurance are performed by loss assessors known as pest observers, who are local state government employees. Visual assessment is conducted based on the method indicated in the guidelines for damage assessment procedures prepared by the Ministry of Agriculture of Indonesia. In recent years, the aging and under-staffing of pest observers have become major problems, and the area covered by a single pest observer can be as large as 5000–9000 ha.

An outline of the current BLB damage assessment methodology is shown in Figure 3. Three rice paddy plots were placed proportionally on the diagonal within the minimum level irrigation area, and these were set as the plots to be evaluated. Then, in the three selected plots, 10 plants in each plot diagonal were visually evaluated (a total of 30 plants). The degree of BLB damage was evaluated on a six-point scale (0, 1, 3, 5, 7, and 9) for each plant according to the area of diseased leaves. A higher score number indicated more severe BLB damage. Then, the BLB damage rate was calculated using the following formula, using the BLB scores of the 30 evaluated plants:

$$\text{BLB damage intensity}(\%) = \frac{n_1 + n_2 + \cdots + n_{30}}{9 \times 30} \times 100 \tag{1}$$

where n indicates the BLB damage score of each plant using the six grades.

The BLB damage intensity calculated from Equation (1) was the damage intensity for the entire area, reflecting the BLB scores of 30 plants in total for the three selected plots in the area.

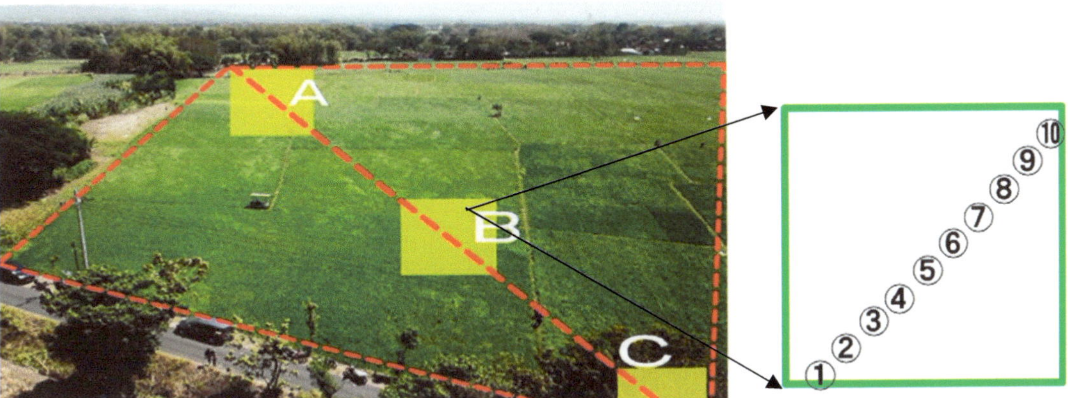

Figure 3. Outline of the current BLB damage assessment; Three paddy plots, A, B, and C, were proportionally located on the diagonal within the minimum irrigation zone and used as evaluation plots.

2.4. Field Survey Data

BLB damage scores were assessed in 109 plots: 9 plots during the wet season from 31 December 2020, to 5 March 2021; 30 plots during the dry season from 19 May to 12 August 2021; 18 plots during the transition from the dry to wet season from 12 October to 3 December 2021; 23 plots during the wet season from 1 March to 6 April 2022; and 29 plots during the dry season from July 7 to 11 September 2022. At approximately 10 d intervals during each growing season, pest observers evaluated the damage scores according to the current method, and aerial data were acquired using a UAV.

2.5. UAV Image Data

Aerial images were acquired synchronously with field survey data using a Bluegrass Fields (Parrot Inc., New York, NY, USA) device equipped with a Sequoia multispectral camera with an observation wavelength range from visible to near-infrared. A sunshine sensor module was attached to the top of the camera to calibrate the image according to the intensity of sunlight.

In addition to preliminary observations under various conditions, we used a radiative transfer model to estimate the effect of the solar radiation environment on the reflectance measurements acquired via the UAV camera. While the effect of solar altitude was not observed under cloudy conditions, where scattered light predominated, it was found that under clear skies, red reflectance increased when the solar altitude was high, and near-infrared reflectance increased when the solar altitude was low [41]. Therefore, it was found that taking images at times when the solar altitude was between 45 °and 65 °reduced the influence of the solar radiation environment; therefore, aerial images were obtained at times when the sun was at this altitude. The ground altitude and overlap ratio were determined through considering the time when UAV flights were possible, the area to be analyzed, and the resolution. The overlap–sidelap ratio was 80–90% and the altitude was 50–60 m, with a speed of 5 m/s and a ground sample distance (GDS) of 2.6–3.0 cm/pixel.

The acquisition dates of field data and UAV data used to create the estimation equation for BLB scores and the digitization footprint size of captured images are shown in Table 1.

Table 1. The data acquisition date and the digitization footprint size of captured images.

Field and UAV Data Acquisition Date	BLB Assessment Plot and Point	Digitization Footprint (Mb/ha)
25 February 2021	6, 60	2336.4
5 March 2021	3, 30	1845.6
19 July 2021	3, 30	3549.8
21 July 2021	3, 30	3106.2
22 July 2021	3, 30	3403.3
28 July 2021	1, 10	2570.0
29 July 2021	3, 30	3314.8
30 July 2021	3, 30	4015.8
7 August 2021	3, 30	3691.8
10 August 2021	3, 30	2835.5
11 August 2021	5, 50	3862.0
12 August 2021	3, 30	2857.5
22 November 2021	3, 30	3138.0
23 November 2021	6, 60	3860.1
30 November 2021	3, 30	3537.8
1 December 2021	3, 30	3299.5
2 December 2021	1, 10	3407.4
3 December 2021	2, 20	3280.1
5 March 2022	3, 30	3197.5
25 March 2022	3, 30	3356.9
28 March 2022	3, 30	2500.6
4 April 2022	6, 60	3160.7
5 April 2022	3, 30	3182.2
6 April 2022	5, 50	3393.6
16 August 2022	6, 60	3044.3
17 August 2022	3, 30	3418.4
29 August 2022	6, 60	3276.5
30 August 2022	6, 60	3447.0
31 August 2022	3, 30	3230.4
10 September 2022	2, 20	3886.7
11 September 2022	3, 30	3610.8

2.6. Creation of Orthomosaic Image

Metashape Professional version 1.8 (Agisoft) with integrated SfM-MVS (Structure from Motion and Multi-View Stereo) was used to create orthomosaic images. The software analysis procedure consisted of capturing the UAV images after shooting, reflectivity calibration, alignment adjustment, adjustment and optimization of camera parameters, high-density cloud construction, DSM creation, and orthomosaic image generation.

In addition, a geometric correction process using ground control points was applied between images to ensure that orthomosaic images from different time periods were not misaligned.

2.7. UAV Image Normalization Process

Because UAV data acquired in different years and seasons were used in this analysis, it was necessary to pay attention to the camera type, solar radiation conditions at the time of acquisition, and the altitude at which the data were collected. Normalized reflectance for ground-based and satellite data acquired with spectroradiometers can be used to reduce the differences in observation conditions, terrain effects, and atmospheric effects. The normalized reflectance was calculated taking advantage of the fact that the shape of the emission spectrum is similar to that of the ground-based spectra, while the emitted amount tends to vary depending on the observation conditions and due to using the additive average of the reflectance over all wavelength bands used. In this study, this method was applied to normalize UAV images to produce normalized reflectance images. The normalization procedure consisted of first calculating the additive averages of the four bands of reflectance from Equation (2) and then normalizing the reflectance of each band according to the additive averages. The additive mean and normalized reflectance of each band are expressed as follows:

$$r_0 = \frac{\text{Green} + \text{Red} + \text{Red edge} + \text{NIR}}{4} \tag{2}$$

$$\text{NGreen} = \frac{\text{Green}}{r_0} \tag{3}$$

$$\text{NRed} = \frac{\text{Red}}{r_0} \tag{4}$$

$$\text{NRed edge} = \frac{\text{Red edge}}{r_0} \tag{5}$$

$$\text{NNIR} = \frac{\text{NIR}}{r_0} \tag{6}$$

Comparing the correlation coefficients of the BLB scores with the UAV sample data before and after the normalization process, it was confirmed that the correlation coefficients were higher in all bands of reflectance; therefore, we decided to use the UAV images after normalization in this analysis.

2.8. Method for Identifying Survey Points and Extracting Reflectance Value on Images

To extract the reflectance from the orthomosaic images of the sites where the pest observers conducted the BLB damage assessment, the assessment sites were identified from the created orthomosaic images.

The resolution of the UAV image was 2.6–3.0 cm per pixel, which is high resolution. When the positional information acquired with a handheld GPS in the field was superimposed on this high-resolution image, it was difficult to capture the survey points accurately because of GPS positioning errors. Therefore, a red funnel, as shown in the aerial image (Figure 4), was placed at the survey site prior to the aerial photography, as shown in Figure 4. The diameter of the funnel was approximately 20 cm, the buffer radius was set to 50 cm, and the inner radius was set to 15 cm. The reflectance of the image corresponding to the donut-shaped portion with the inner diameter removed was extracted for analysis.

 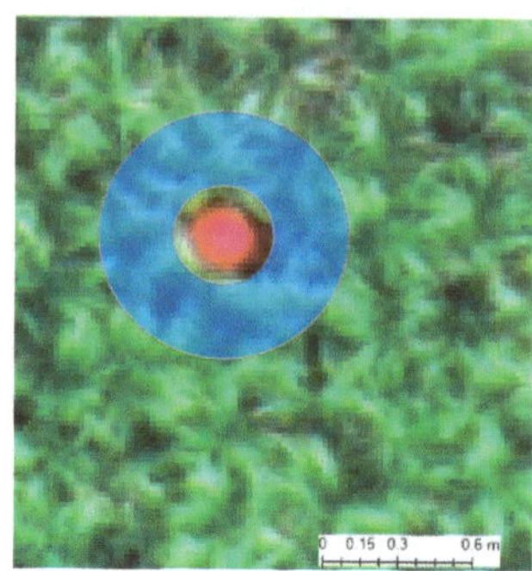 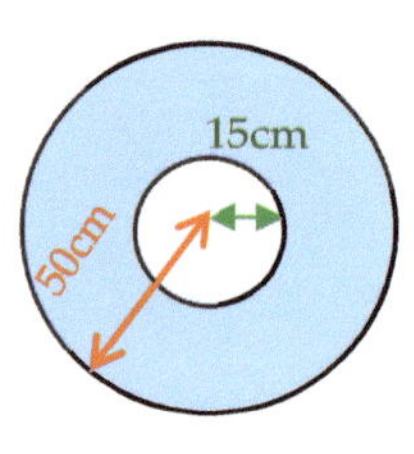

Figure 4. (**Left**) Red funnel installed at the survey site; (**Right**) red funnel on UAV image, and reflectance extraction area.

2.9. Creation of BLB Damage Assessment Estimation Equation

Multiple regression analysis was conducted using the normalized UAV green band, red band, red edge band, near-infrared band, and NDVI, GNDVI, and RGI indices as explanatory variables to develop an estimation equation for the degree of BLB damage from the averaged value calculated from 1090 survey points, followed by 5-fold cross-validation to evaluate the accuracy. Each index was calculated using the following equations:

$$NDVI = (NIR - R)/(NIR + R)$$

$$GNDVI = (NIR - G)/(NIR + G)$$

$$RGI = G \times Red\ Edge$$

As described in Section 2.3, the current BLB damage assessment method, as indicated by the Indonesian Ministry of Agriculture, outputs the damage intensity for a single plot through substituting a six-level BLB score (0, 1, 3, 5, 7, and 9) for each plant unit into the formula to calculate the damage intensity. Therefore, this analysis examined the relationship between the BLB score and UAV data and between the BLB damage intensity and UAV data.

Furthermore, the effectiveness of integrating UAV data into the current BLB evaluation method was examined through comparing the time required to output the evaluation results in order to understand the efficiency of the BLB damage evaluation process.

3. Results and Discussion

3.1. Relationship between Normalized Reflectance for Each Band and Indices and BLB

We analyzed the relationship between the normalized reflectance extracted from the orthomosaic images and calculated indices and the BLB score. The BLB score was assessed based on the number of surveyed plants, with 10 plants evaluated per plot.

Figure 5 and Table 2 show the relationship between the reflectance data obtained during the dry season and the BLB score. Figure 6 and Table 3 show the relationship between the reflectance data obtained during the rainy season and the BLB score. In the dry-season data, the correlation coefficients of Nred, NNIR, and NDVI were approximately 0.5, indicating a stronger correlation than those of the other normalized reflectance data and indices. In the rainy season data, there was no significant difference in the correlation between the single-year data and the 2-year data; however, when the 2-year rainy season data were combined, it became clear whether each normalized reflectance and index had a positive or negative correlation with the BLB score (Figure 6). The relationship between the UAV data and BLB scores for the dry and rainy seasons for all cropping seasons in 2021 and 2022 is shown in Figure 7 and Table 4. Regarding BLB disease incidence in the rainy season crops, no paddy rice with a high degree of damage, such as a score of 7 or 9, was identified, and the BLB scores of all evaluated rice were 5 or less. This is because there is more rainfall during the rainy season, which tends to wash away the rice BLB bacteria attached to the rice bodies, leaving fewer residual bacteria on the rice plants and in the field [42,43]. BLB enters rice through the water pore apertures in rice leaves and wind-driven wounds on the plant surface. Research using bacteriophages has shown that BLB can spread to other areas through irrigation [44].

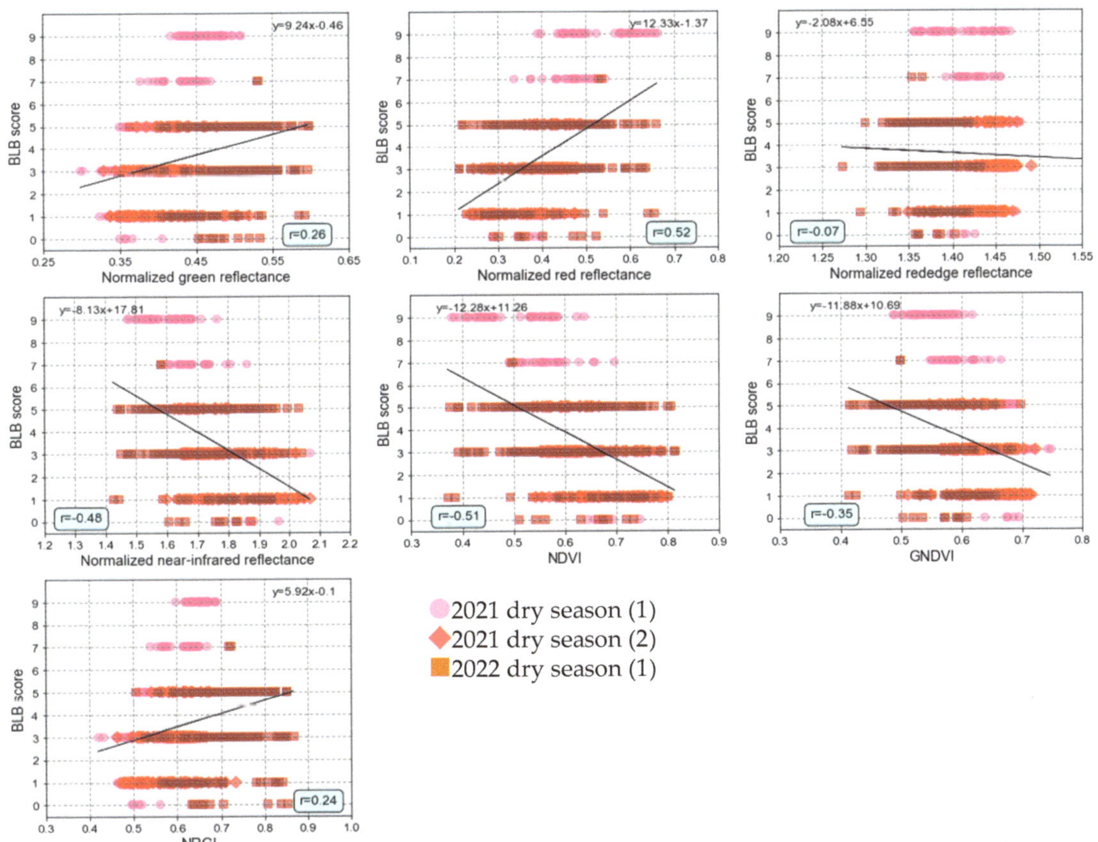

Figure 5. Relationship between UAV data and BLB scores acquired during the 2021 and 2022 dry seasons.

Table 2. Correlation coefficient between UAV data and BLB scores acquired during the 2021 and 2022 dry seasons (** significant at the 1% level).

	Ngreen	Nred	Nred Edge	NNIR	NDVI	GNDVI	NRGI
Correlation coefficient	0.261 **	0.516 **	−0.068	−0.478 **	−0.509 **	−0.354 **	0.235 **

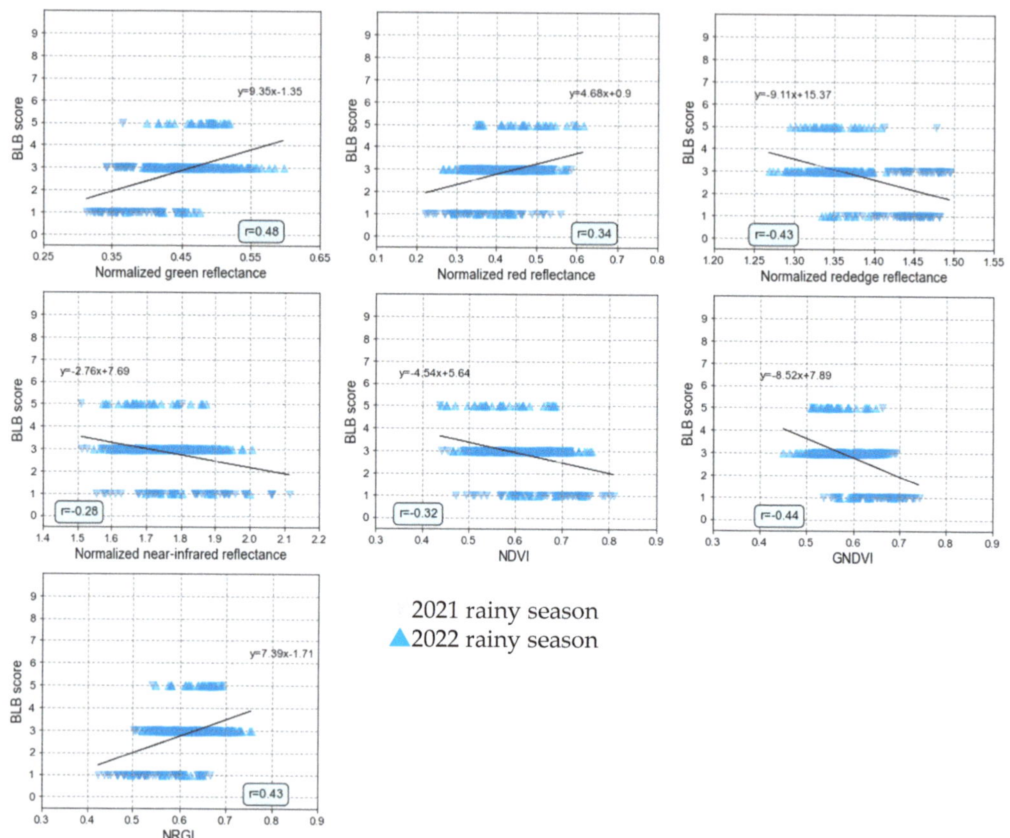

Figure 6. Relationship between UAV data and BLB scores acquired during the 2021 and 2022 rainy seasons.

Table 3. Correlation coefficient between UAV data and BLB scores acquired during the 2021 and 2022 rainy season (** significant at the 1% level).

	Ngreen	Nred	Nred Edge	NNIR	NDVI	GNDVI	NRGI
Correlation coefficient	0.481 **	0.337 **	−0.426	−0.283 **	−0.321 **	−0.443 **	0.435 **

In addition, the correlation coefficients were slightly lower than those obtained from single-year or seasonal data analyses. This was presumably due to the weighting of scores 3 and 5, because the total amount of data increased owing to the inclusion of data from all cropping seasons together, which corresponded to the increase in data with scores 3 and 5 for wet season crops. Furthermore, we observed a range of normalized reflectance and index values for the same BLB scores. One possible reason for this may be that the current assessment was based on the pest observers' visual evaluation of the BLB scores. It was necessary to be skilled in clearly separating the damage of scores 3 and 5, and it was inferred that judgment differed significantly among the evaluators. This suggests that it is necessary to incorporate UAV data into the current method and develop a damage assessment formula to obtain objective assessment results.

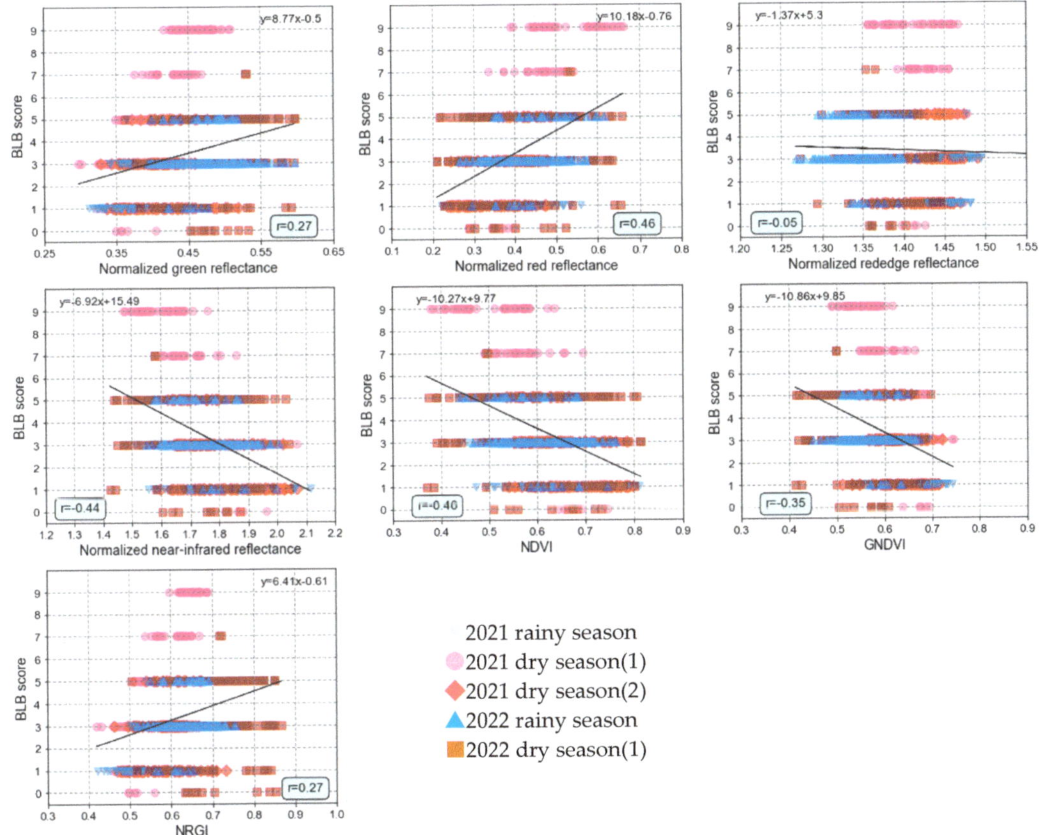

Figure 7. Relationship between UAV data and BLB scores acquired during the 2021 and 2022 dry and rainy seasons.

Table 4. Correlation coefficient between UAV data and BLB score acquired during the 2021 and 2022 dry and rainy seasons (** significant at 1% the level).

	Ngreen	Nred	Nred Edge	NNIR	NDVI	GNDVI	NRGI
Correlation coefficient	0.271 **	0.458 **	−0.049	−0.440 **	−0.456 **	−0.349 **	0.265 **

Next, a correlation analysis was performed between the UAV data and the field damage intensity determined from the BLB scores for all cropping seasons (Figure 8, Table 5). The correlation coefficients for Nred, NNIR, and NDVI exceeded 0.5, which was higher than those of the other normalized reflectance and indices. Similar to the BLB score analysis results, for plots with high correlation coefficients with Nred, NNIR, and NDVI and with more than 70% damage, the amount of data entered with scores 7 and 9 was lower than for other scores, indicating that they may have been difficult to estimate with high accuracy.

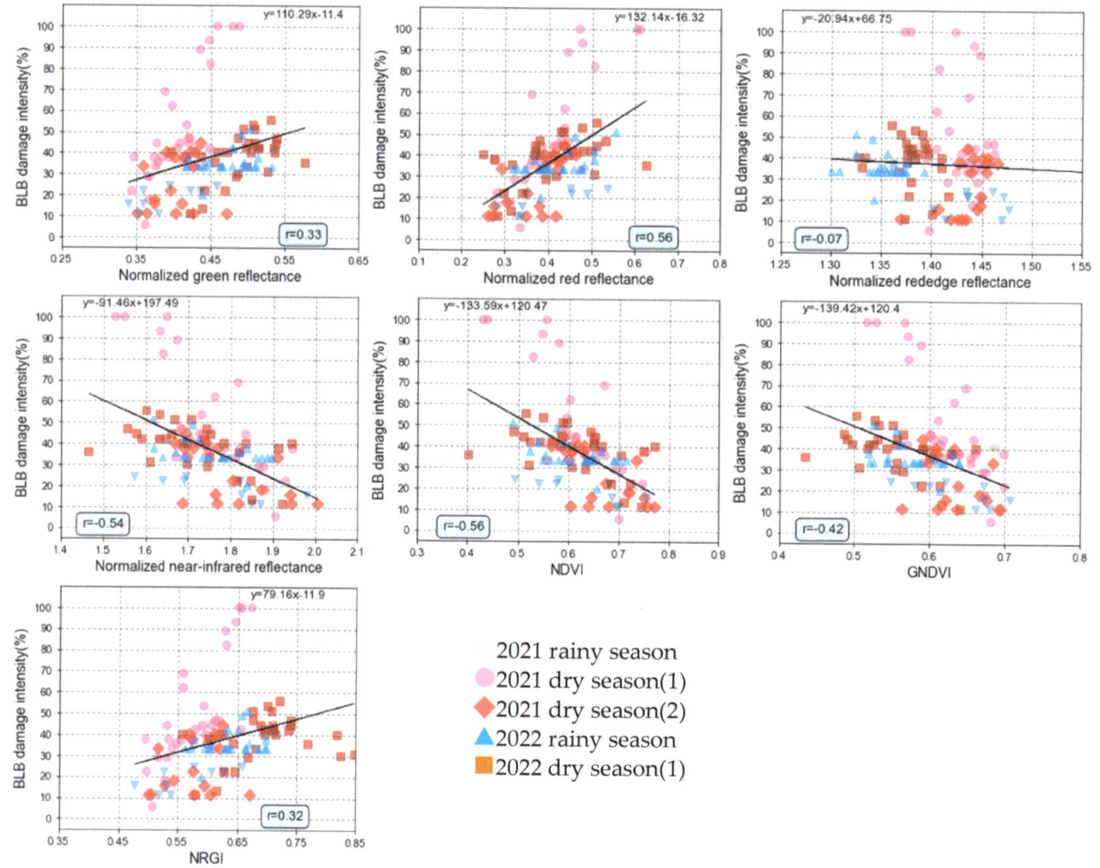

Figure 8. Relationship between UAV data and BLB damage intensity acquired during the 2021 and 2022 dry and rainy seasons.

Table 5. Correlation coefficient between UAV data and BLB damage intensity acquired during the 2021 and 2022 dry and rainy seasons (** significant at the 1% level).

	Ngreen	Nred	Nred Edge	NNIR	NDVI	GNDVI	NRGI
Correlation coefficient	0.329 **	0.558 **	−0.074	−0.537 **	−0.556 **	−0.425 **	0.316 **

As described above, the correlation analysis between the BLB scores and BLB damage intensities revealed normalized reflectance values and indices that may be effective indicators for BLB damage assessment. However, it was suggested that the strength of the correlations differed among the data groups used and that bias in the number of data obtained for each BLB score may have affected the evaluation results. Therefore, we calculated the mean values of the normalized reflectance and index for each BLB score for the 1090 survey points and performed a correlation analysis using the six mean scores (Figure 9, Table 6). For all crop season data, a strong relationship was confirmed for Nred, NNIR, and NDVI, as well as for the relationships among pre-average scores, damage intensities, and UAV data, with correlation coefficients exceeding 0.9. Based on the results of the analysis using the average score, Nred, NNIR, and NDVI had a strong correlation with the BLB score, increasing their possibility of being effective indicators for BLB damage assessment.

In addition, from the viewpoint that the current method is conducted according to visual assessment by the assessors, Nred in the visible range is shown to be an effective index for the estimation formula of the BLB damage assessment.

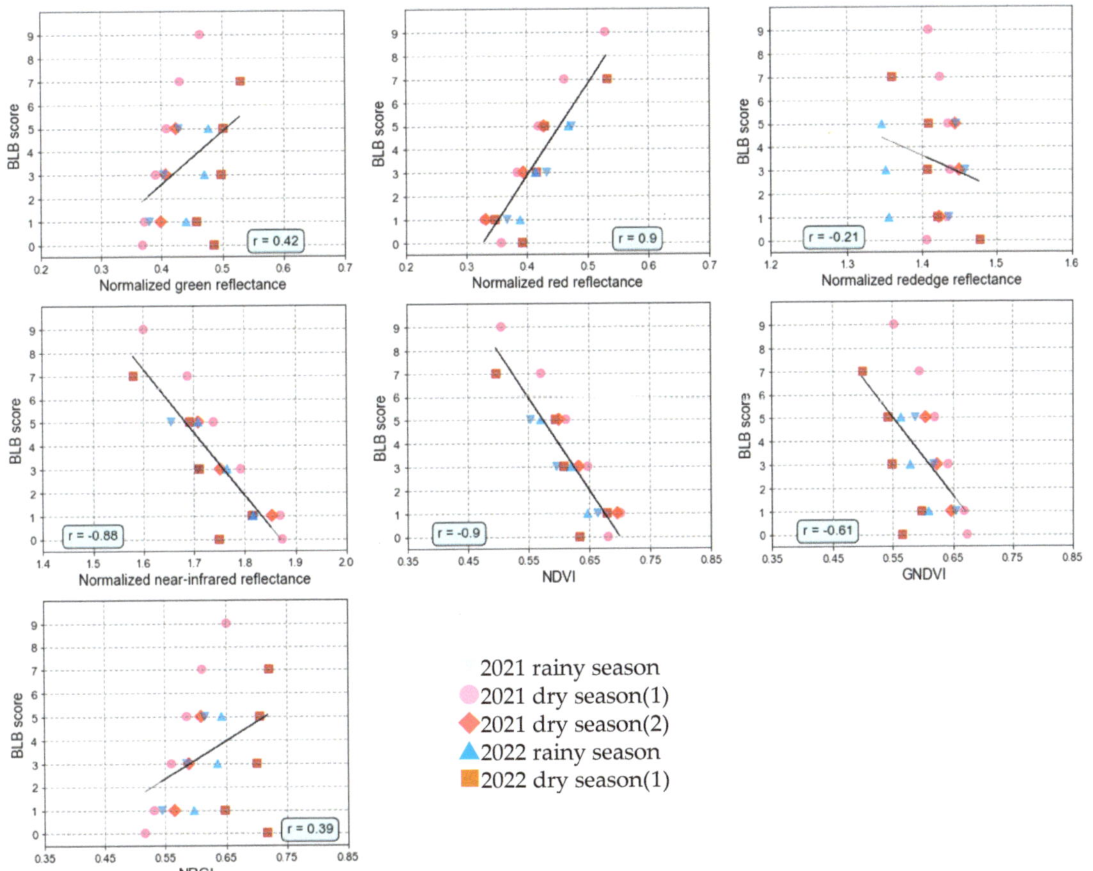

Figure 9. Relationship between UAV data and BLB mean score acquired during the 2021 and 2022 dry and rainy seasons.

Table 6. Correlation coefficient between UAV data and BLB mean score acquired during the 2021 and 2022 dry and rainy seasons (** significant at the 1% level).

	Ngreen	Nred	Nred Edge	NNIR	NDVI	GNDVI	NRGI
Correlation coefficient	0.418 **	0.895 **	−0.211	−0.876 **	−0.897 **	−0.605 **	0.389 **

Nred is the absorption band of chlorophyll; the higher the damage, the lower the chlorophyll content and the higher the reflectance, which may confirm a positive correlation [45–47]. In the case of BLB disease, the leaves fade, with yellowing at the leaf margins and a grayish white color in the case of advanced infection. A negative correlation between SPAD values obtained using a chlorophyll meter and BLB damage intensity [34], and the ability to classify paddy rice with high BLB severity based on the normalized green–red difference index obtained from the RGB base [48], have also been reported in research covering the same area as this study. Research using hyperspectral data has reported that BLB-infected rice plants show peculiar changes between 757 nm and 1039 nm, corresponding to the near-infrared region, and that reflectance decreases with increasing severity [49] the same trend was confirmed in this study. The NDVI is the vegetation index calculated from red and near-infrared, and the GNDVI is the vegetation index calculated from green and near-infrared. In this study, we found that Ngreen increased, Nred increased, and NNIR decreased with increasing BLB damage. These results confirmed a negative correlation, because the numerators of the NDVI and GNDVI were smaller when the damage to the BLB was larger. When rice is infected with BLB, the chlorophyll in the leaves decomposes, which leads to a loss of vitality in the rice plant, resulting in changes in the green color of the leaves. This could explain the positive correlation between the Nred band and degree of BLB infection. It is well known that the reflectance of the NIR band and vegetation index are related to the biomass volume and amount of crop production [50].

3.2. BLB Damage Assessment Estimation Equations

As described in Section 3.1, the correlation coefficients between UAV multispectral data categorized according to cropping season and average normalized reflectance in relation to BLB score were the highest for normalized red reflectance and NDVI. This corresponds to the selection of the red band, which is the absorption band of chlorophyll, because the current method allows the assessor to visually assess the degree of BLB damage based on the leaf color. It can also be reasoned that the NDVI, which is related to biomass, was selected because, as the degree of BLB damage progresses, the leaves wither and the crop body becomes smaller than the healthy plant. Therefore, to improve the efficiency of the current method using UAV data, we propose a formula to estimate the BLB score from the normalized red band as an evaluation method, reading the evaluator's decision as a visual one; that is, we used information in the visible range for damage assessment. The equation presented below has a coefficient of determination of 0.92 and RMSE of 1.46 ± 0.36:

$$\text{BLB score} = 50.841 \times \text{NRed(all season)} - 17.475 \tag{7}$$

For validation of the above equation, we used UAV data acquired in 2023, which were not used to create the equation. BLB scores were estimated from normalized red band values calculated from the 2023 UAV data. The estimated BLB score values and the BLB scores assessed by the pest observers were used to calculate the mean absolute percentage error (MAPE) [51], expressed with the following equation:

$$\text{MAPE} = \frac{1}{n} \sum \left| \frac{Gt - P}{Gt} \right| \tag{8}$$

MAPE is given by the average of the absolute value of the ratio of the difference between the ground truth data (Gt) and the estimation (P) to the ground truth, where n represents the number of the data. The MAPE of the BLB score estimated from the data of 70 survey points obtained in February and July 2023 was 9.1%, confirming that the Formula (2) for BLB damage assessment presented in this study was sufficient to be applicable in other years.

A visualization map of the damage assessment using the BLB score estimation equation is shown in Figure 10.

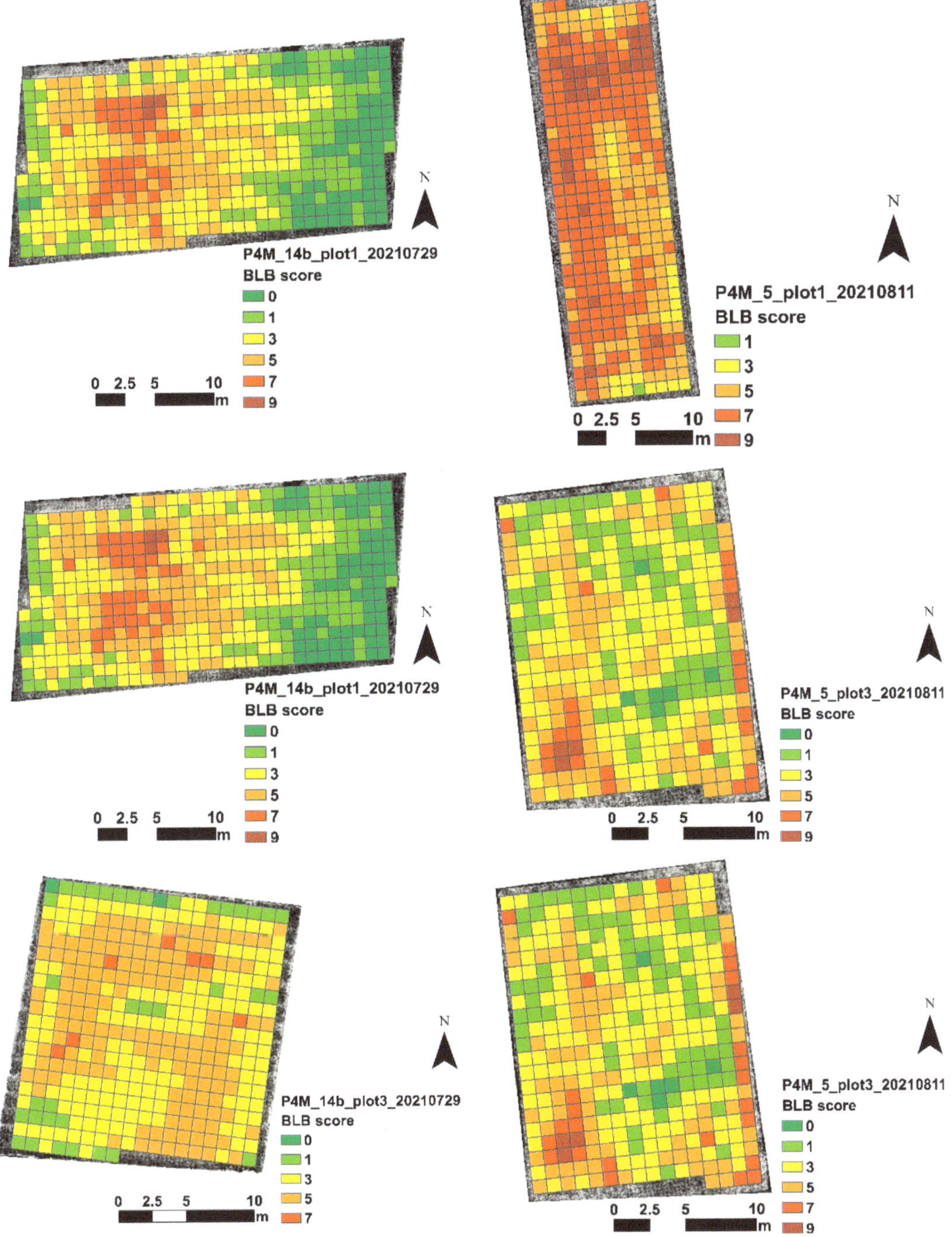

Figure 10. BLB damage assessment score map for dry season in 2021: (**Left**) Block 14b, Plots 1, 2, and 3 from top to bottom; (**Right**) Block 5, Plots 1, 2, and 3 from top to bottom.

3.3. Improvement of Efficiency and Objectivity of BLB Assessment

The effectiveness of using UAV data was examined through applying the constructed formula to estimate the BLB damage during the harvest season, using the red band as an indicator for each irrigated block.

The total time required for the BLB evaluation using UAV data was defined as the sum of the times required for the following processes: acquisition of field survey data and UAV data, uploading and downloading of acquired data, confirmation of acquired data content, preprocessing of UAV data, application of the estimation formula, and creation of evaluation maps. In addition, the assessors conducted a maximum of 12 field-based damage assessment surveys per day using the current method. However, because the number of plots that can be surveyed per day decreases depending on weather conditions, it is expected to take more time in rainy weather.

The time required for both methods and the efficiency gains are listed in Table 7. For example, Irrigation Block 1b contained 60 paddy plots. Using the current method, the evaluator would take 5 days to evaluate all the plots. However, if UAV imagery were integrated into the current method, the evaluation would be completed in 2 days and 3 h, with 57.5% time saving. The greater the number of plots, the greater the percentage of efficiency gain, with a maximum time–cost reduction of 76.4%.

Table 7. Comparison of time required to assess all fields using the current method and the integrated method with UAV data.

Irrigation Block Name	No. of Plots	Area (ha)	Time Required to Assess All Plots Using the Current Method	Time Required to Assess All Plots Using the Integrated Method with UAV Data	Percentage of Time Saved (%)
Block 1b	60	3.1	5 days	2 days and 3 h	57.5
Block 2a	43	2.2	4 days	2 days and 3 h	46.9
Block 3a	62	2.9	6 days	2 days and 3 h	64.6
Block 5	38	2.6	4 days	2 days and 2.5 h	47.4
Blcok 7a	54	2.7	5 days	2 days and 2.5 h	57.5
Block 9b	56	3.4	5 days	2 days and 3 h	57.5
Block 11a	67	2.6	6 days	2 days and 3 h	64.6
Block 11b	104	2.8	9 days	2 days and 3 h	76.4
Block 12a	68	2.9	6 days	2 days and 3 h	64.6
Block 14a	84	2.9	7 days	2 days and 3 h	69.6

An example of an evaluation map visualizing the damage to all paddy fields in an irrigated area obtained via inputting the BLB evaluation estimation equation into a normalized red-band image is shown in Figures 11–13. The numbers at the tops of the bars indicate the number of paddy plots with each damage intensity, and the numbers next to the stars indicate the damage intensities calculated from the three plots using the proposed method. The damage intensities for irrigation Blocks 1b, 9b, and 11b using the proposed method were 33.3%, 33.7%, and 40.7%, respectively. However, the damage intensities estimated from the UAV data differed from plot to plot, and incorporating the UAV data into the current method enabled an objective and complete enumeration.

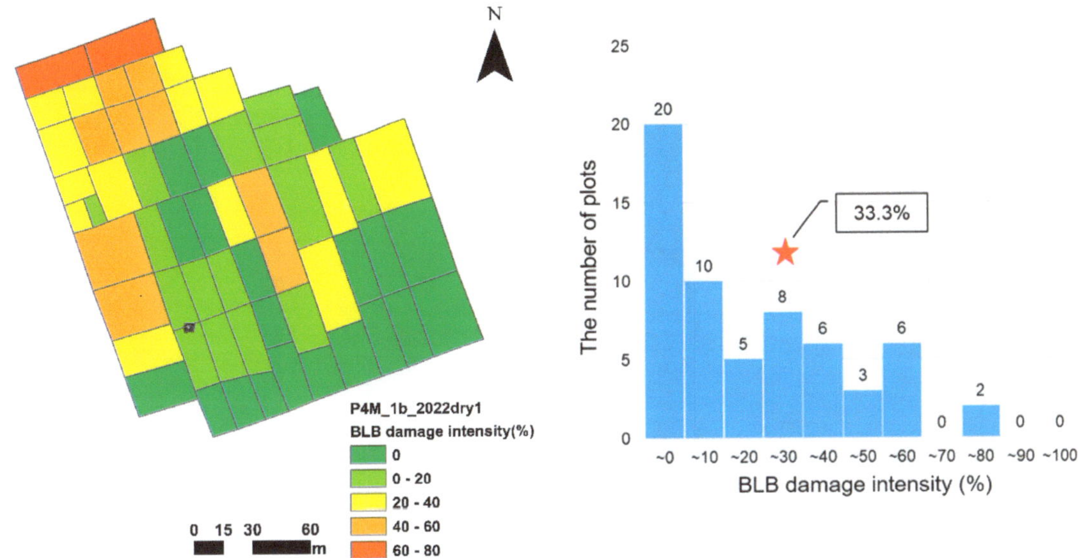

Figure 11. BLB damage assessment intensity map for the dry season in 2022: (**Left**) visualization map of Block 1b; (**Right**) number of paddy plots per damage intensity. (★) BLB damage intensity calculated from three plots using the current damage assessment method described in Section 2.3.

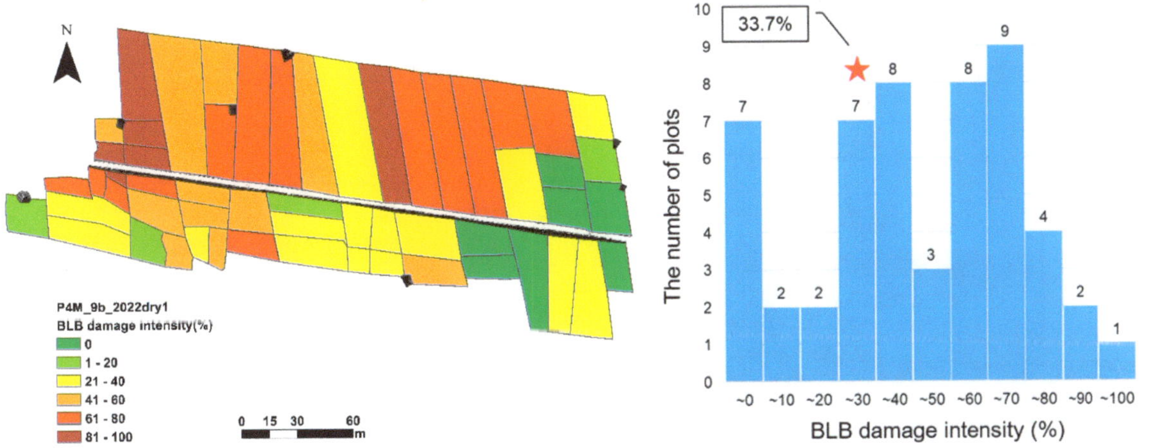

Figure 12. BLB damage assessment intensity map for the dry season in 2022: (**Left**) visualization map of Block 11b; (**Right**) number of paddy plots per damage intensity. (★) BLB damage intensity calculated from three plots using the current damage assessment method described in Section 2.3.

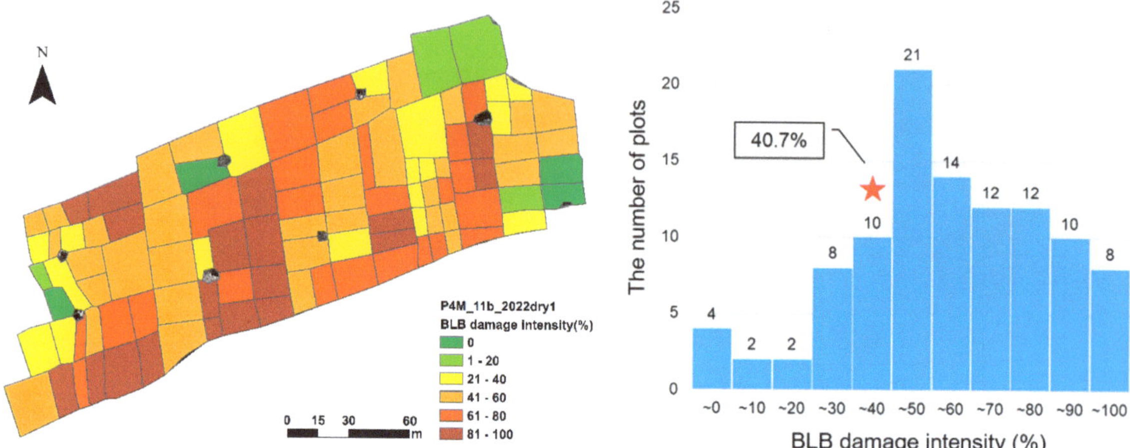

Figure 13. BLB damage assessment intensity map for the dry season in 2022: (**Left**) visualization map of Block 9b; (**Right**) number of paddy plots per damage intensity. (★) BLB damage intensity calculated from three plots using the current damage assessment method described in Section 2.3.

4. Conclusions

Agricultural insurance, which is expected to be a climate change adaptation measure, has been operating as an important social infrastructure for food security. However, there were many challenges in Indonesia soon after its operation. The biggest challenges were speeding up loss assessment, which is the core of agricultural insurance, and the objectivity of the assessment results. In this study, a rapid and objective damage assessment method was developed and implemented in West Java for BLB of rice, a disease covered by insurance.

Based on the analysis of the relationship between the BLB scores, damage intensities, and UAV data acquired during the dry and rainy seasons of 2021 and 2022, a BLB damage assessment estimation equation using normalized red bands was developed. Validation of the equation was performed using the MAPE. The MAPE of the BLB score estimated from the data of 70 survey points obtained in February and July 2023 was 9.1%, confirming that the formula (2) for BLB damage assessment presented in this study was suitable for application to other years' data. We integrated the process of UAV aerial photography and the estimation formula based on UAV data into the current damage assessment steps proposed by the Ministry of Agriculture of Indonesia and implemented in society. The evaluation time was reduced by more than 60% on average, and technical guideline is issued for a new damage evaluation process for agricultural insurance that reduces labor and ensures objectivity.

The national government is in charge of maintaining and improving the agricultural insurance system, while the state government is in charge of practices related to losses. It is necessary to obtain recognition and approval from the Crop Protection Bureau of the Ministry of Agriculture for the evaluation methodology. Therefore, we approached the central ministry regarding the actual operation of the new evaluation method and sent a letter from the Director of the Crop Protection Department of the Ministry of Agriculture supporting the use of the method established in the project to the Director of the West Java Provincial Department of Agricultural Policy, which was highly appreciated by the government. Currently, we aim to accumulate damage assessment data and improve the assessment method through incorporating machine learning and other methods.

Author Contributions: Conceptualization, methodology, and writing—original draft preparation, C.H. and S.I.; investigation, resources, and data curation, G.S., C.H. and S.I.; review and editing, E.T.; funding acquisition, C.H. and E.T.; supervision and project administration, C.H. All authors have read and agreed to the published version of the manuscript.

Funding: This research was funded by Japan Science and Technology Agency (JST) and Japan International Cooperation Agency (JICA), grant Number JPMJSA1604, and JSPS KAKENHI, grant number JP22KK0089.

Data Availability Statement: All data supporting the findings of this study are included in the text.

Acknowledgments: The authors gratefully acknowledge Danan Hidayat, Ir Kustiaman, Yogi Tojiri Septiadi, Hasam Supriatna, Gagan Gandana Wibawa, Adam Daniel, Denny Magria, and Taufik Rahmat from the Provincial Office of Food Crops and Horticulture of West Java Province, Indonesia for contribution to this research and field investigation at the study site.

Conflicts of Interest: Author Shun Isono was employed by the NHK (Japan Broadcasting Corporation). All authors declare that the research was conducted in the absence of any commercial or financial relationships that could be construed as a potential conflict of interest.

References

1. IPCC-Sixth Assessment Report. Available online. https://www.ipcc.ch/assessment-report/ar6/ (accessed on 23 February 2024).
2. Gurdeep, S.M.; Manpreet, K.; Prashant, K. Impact of Climate Change on Agriculture and Its Mitigation Strategies. *Sustainability* **2021**, *13*, 1318. [CrossRef]
3. Sameh, K.A.; Miriam, M.; Antonio, J.; Mariá, A.; Jonathan, D.P.; Laurence, J.; Zhenhua, Z.; Paulo, P.; Luuk, F.; Martine, P.; et al. Climate change impacts on agricultural suitability and yield reduction in a Mediterranean region. *Geoderma* **2020**, *374*, 114453. [CrossRef]
4. Muhammad, H.; Ashfaq, A.; Ahsan, R.; Muhammad, U.H.; Hesham, F.A.; Yahya, M.A.; Atif, A.B.; Khalid, R.H.; Saeed, A.; Wajid, N.; et al. Impact of climate change on agricultural production; Issues, challenges, and opportunities in Asia. *Plant Sci.* **2022**, *13*, 925548. [CrossRef] [PubMed]
5. FAOSAT, Food and Agriculture Organization of the United Nations. 2018. Available online: https://www.fao.org/faostat/en/#data/QC (accessed on 23 February 2024).
6. Muhammad, Y.Y.; Rahmat, F.; Teuku, S.B.; Hafizh, M.; Juli, F. Design of Islamic Agricultural Insurance Model: Evidence from Indonesia. *Int. J. Sustain. Dev. Plan.* **2022**, *17*, 2375–2384. [CrossRef]
7. Sahat, M.P. Developing rice farm insurance in Indonesia. *Agric. Agric. Sci. Procedia* **2010**, *1*, 33–41. [CrossRef]
8. Nyoman, Y.; Luh, P.K.P.; Made, B. Effectiveness of Agricultural Insurance Program as A Sustainable Agricultural Development Effort. *Sustain. Environ. Agric. Sci.* **2022**, *6*, 134–143. [CrossRef]
9. Abdul, H.; Rusli, R.; Dan, U.N. The Relationship between the Knowledge Level of Farmers and the Effectiveness of the Rice-Farming Business Insurance Program (AUTP) in Pinrang Regency, South Sulawesi, Indonesia. *Int. J. Soc. Sci. Educ. Res. Stud.* **2022**, *2*, 298–307.
10. Sahat, M.P. *Implementation of Indemnity-Based Rice Crop Insurance in Indonesia*; Food and Fertilizer Technology Center for the Asian and Pacific Region: Taipei, Taiwan, 2016; Available online: https://ap.fftc.org.tw/article/1079 (accessed on 10 March 2024).
11. Adhitya, M.; Sahara, A.D. Analysis of Implementation of Rice Farming Insurance: Case Study In Indonesia. *Dev. Ctry. Stud.* **2016**, *6*, 13–118.
12. Chinna, G.S.; Hari, K.K.; Valli, K.V.; Alakananda, M.; Preethi, A. Deep learning for rice leaf disease detection: A systematic literature review.on emerging trends, methodologies and techniques. *Inf. Process. Agric.* 2024, *in press*. [CrossRef]
13. Lin, S.; Yao, Y.; Li, J.; Li, X.; Ma, J.; Weng, H.; Cheng, Z.; Ye, D. Application of UAV-Based Imaging and Deep Learning in Assessment of Rice Blast Resistance. *Rice Sci.* **2023**, *30*, 652–660. [CrossRef]
14. Sourav, K.B.; Krishna, P.K.; Rajermani, T. A Machine Intelligent Framework for Detection of Rice Leaf Diseases in Field Using IoT Based Unmanned Aerial Vehicle System. *Sparkling Light Trans. Artif. Intell. Quantum Comput. (STAIQC)* **2022**, *2*, 42–51.
15. Shaodan, L.; Jiayi, L.; Deyao, H.; Zuxin, C.; Lirong, X.; Dapeng, Y.; Haiyong, W. Early Detection of Rice Blast Using a Semi-Supervised Contrastive Unpaired Translation Iterative Network Based on UAV Images. *Plants* **2023**, *12*, 3675. [CrossRef] [PubMed]
16. Hongo, C.; Tsuzawa, T.; Tokui, K.; Tamura, E. Development of Damage Assessment Method of Rice Crop for Agricultural Insurance Using Satellite Data. *J. Agric. Sci.* **2015**, *7*, 59–71. [CrossRef]
17. Hongo, C.; Gunardi, S.; Shikata, R.; Niwa, K.; Tamura, E. The Use of Remotely Sensed Data for Estimating of Rice Yield Considering Soil Characteristics. *J. Agric. Sci.* **2014**, *6*, 172–184. [CrossRef]
18. Sofue, Y.; Hongo, C.; Manago, N.; Gunardi, S.; Homma, K.; Baba, B. Estimation of Normal Rice Yield Considering Heading Stage Based on Observation Data and Satellite Imagery. In Proceedings of the 2021 IEEE International Geoscience and Remote Sensing Symposium IGARSS, Brussels, Belgium, 11–16 July 2021; pp. 6439–6442. [CrossRef]

19. Manago, N.; Hongo, C.; Sofue, Y.; Gunardi, S.; Budi, U. Transplanting date estimation using Sentinel-1 satellite data for paddy rice damage assessment in Indonesia. *Agriculture* **2020**, *10*, 625. [CrossRef]
20. Iwahashi, Y.; Gunardi, S.; Budi, U.; Iskandar, L.; Ahmad, J.; Bambang, H.T.; IMade, A.S.W.; Maki, M.; Hongo, C.; Homma, K. Drought Damage Assessment for Crop Insurance Based on Vegetation Index by Unmanned Aerial Vehicle Multispectral Images of Paddy Fields in Indonesia. *Agriculture* **2023**, *13*, 113. [CrossRef]
21. Inoue, T. Satellite- and drone-based remote sensing of crops and soils for smart farming—A review. *Soil Sci. Plant Nutr.* **2020**, *66*, 798–810. [CrossRef]
22. Inoue, Y. Remote Sensing of Plant and Soil Information by High-resolution Op-tical Satellite Sensors and Its Applications to Smart Agriculture. *J. Remote Sens. Soc. Jpn.* **2017**, *37*, 213–223.
23. Jean, R.F.M.; Yamashita, M.; Yoshimura, M.; Enrico, C.P. Leaf Spectral Analysis for Detection and Differentiation of Three Major Rice Diseases in the Philippines. *Remote Sens.* **2023**, *15*, 3058. [CrossRef]
24. Singh, B.; Singh, M.; Singh, G.; Suri, K.; Pannu, P.P.S.; Bal, S.K. Hyper-Spectral Data for The Detection of Rice Bacterial Leaf Blight (BLB) Disease. *Proc. AIPA* **2012**, *2012*, 177–182.
25. Wijaya, I.M.A.S.; Chandra, I.G.B.E.; Hongo, C. Assessment of bacterial leaf blight (BLB) diseases by unmanned aerial vehicle (UAV)-based vegetation index of paddy fields. In Proceedings of the 10th Asian-Australasian Conference on Precision Agriculture (ACPA10), Universiti Putra Malaysia, Seri Kembangan, Malaysia, 24–26 October 2023.
26. Yuti, G.; Hongo, C.; Saito, D.; Caasi, O.; Susilawati, P.N.; Shishido, M.; Sudiarta, I.P.; Wijaya, I.M.A.S.; Homma, K. Evaluating Multispectral Imaging for Assessing Bacterial Leaf Blight Damage in Indonesian Agricultural Insurance. In *E3S Web of Conferences*; EDP Sciences: Les Ulis, France, 2021; Volume 232.
27. Kobayashi, T.; Sasahara, M.; Kanda, E.; Ishiguro, K.; Hase, S.; Torigoe, Y. Assessment of Rice Panicle Blast Disease Using Airborne Hyperspectral Imagery. *Open Agric. J.* **2016**, *10*, 28–34. [CrossRef]
28. Zhao, D.; Cao, Y.; Li, J.; Cao, Q.; Li, J.; Guo, F.; Feng, S.; Xu, T. Early Detection of Rice Leaf Blast Disease Using Unmanned Aerial Vehicle Remote Sensing: A Novel Approach Integrating a New Spectral Vegetation Index and Machine Learning. *Agronomy* **2024**, *14*, 602. [CrossRef]
29. Yamaguchi, T.; Tanaka, Y.; Imachi, Y.; Yamashita, M.; Katsura, K. Feasibility of Combining Deep Learning and RGB Images Obtained by Unmanned Aerial Vehicle for Leaf Area Index Estimation in Rice. *Remote Sens.* **2021**, *13*, 84. [CrossRef]
30. Ning, L.; Jie, Z.; Zixu, H.; Dong, L.; Qiang, C.; Xia, Y.; Yongchao, T.; Yan, Z.; Weixing, C.; Tao, C. Improved estimation of aboveground biomass in wheat from RGB imagery and point cloud data acquired with a low-cost unmanned aerial vehicle system. *Plant Methods* **2019**, *15*, 17. [CrossRef] [PubMed]
31. Shi, Y.; Huang, W.; Ye, H.; Ruan, C.; Xing, N.; Geng, Y.; Dong, Y.; Peng, D. Partial Least Square Discriminant Analysis Based on Normalized Two-Stage Vegetation Indices for Mapping Damage from Rice Diseases Using PlanetScope Datasets. *Sensors* **2018**, *18*, 1901. [CrossRef]
32. Zheng, Q.; Huang, W.; Cui, X.; Shi, Y.; Liu, L. New Spectral Index for Detecting Wheat Yellow Rust Using Sentinel-2 Multispectral Imagery. *Sensors* **2018**, *18*, 868. [CrossRef] [PubMed]
33. Prabir, K.D.; Laxman, B.; Kameswara, S.V.C.R.; Seshasai, M.V.R.; Dadhwal, V.K. Monitoring of bacterial leaf blight in rice using ground-based hyperspectral and LISS IV satellite data in Kurnool, Andhra Pradesh, India. *Int. J. Pest Manag.* **2015**, *61*, 359–368. [CrossRef]
34. Caasi, O.; Hongo, C.; Suryaningsih, A.; Wiyono, S.; Homma, K.; Shishido, M. Relationships between bacterial leaf blight and other diseases based on field assessment in Indonesia. *Trop. Agric. Dev.* **2019**, *63*, 113–121. [CrossRef]
35. Caasi, O.; Hongo, C.; Wiyono, S.; Giamerti, Y.; Saito, D.; Homma, K.; Shishido, M. The potential of using sentinel-2 satellite imagery in assessing bacterial leaf blight on rice in West Java, Indonesia. *J. Int. Soc. Southeast Asia Agric. Sci.* **2020**, *26*, 1–16.
36. Hongo, C.; Takahashi, Y.; Gunardi, S.; Budi, U.; Tamura, E. Advanced Damage Assessment Method for Bacterial Leaf Blight Disease in Rice by Integrating Remote Sensing Data for Agricultural Insurance. *J. Agric. Sci.* **2022**, *14*, 1–18. [CrossRef]
37. Syed, A.H.N.; Rashida, P.; Ummad, D.U.; Owais, M.; Ateequr, R.; Sajid, W.; Taha, M. Determination of antibacterial activity of various broad spectrum antibiotics against Xanthomonas oryzae pv. Oryzae, a cause of bacterial leaf blight of rice. *Int. J. Microbiol. Mycol.* **2014**, *2*, 12–19.
38. Bai, X.; Zhou, Y.; Feng, X.; Tao, M.; Zhang, J.; Deng, S.; Lou, B.; Yang, G.; Wu, Q.; Yu, L.; et al. Evaluation of rice bacterial blight severity from lab to field with hyperspectral imaging technique. *Plant Sci.* **2022**, *13*, 1037774. [CrossRef]
39. Chukwu, S.C.; Rafii, M.Y.; Ramlee, S.I.; Ismail, S.I.; Hasan, M.M.; Oladosu, Y.A.; Magaji, U.G.; Akos, I.; Olalekan, K.K. Bacterial leaf blight resistance in rice: A review of conventional breeding to molecular approach. *Mol. Biol. Rep.* **2019**, *46*, 1519–1532. [CrossRef]
40. Mohammad, M.F.A.; Han, Y.L. Advanced diagnostic approaches developed for the global menace of rice diseases. *Can. J. Plant Pathol.* **2022**, *44*, 627–651. [CrossRef]
41. Hashimoto, N.; Saito, Y.; Maki, M.; Homma, K. Simulation of reflectance and vegetation indices for unmanned aerial vehicle (UAV) monitoring of paddy fields. *Remote Sens.* **2019**, *11*, 2119. [CrossRef]
42. Rice Knowledge Bank, IRRI. Available online: http://www.knowledgebank.irri.org/decision-tools/rice-doctor/rice-doctor-fact-sheets/item/bacterial-blight (accessed on 17 May 2024).
43. Tabasia, A.; Vishal, G.; Aarushi, S.; Sheikh, S.K. Effect of Weather Parameters on the Severity of Bacterial Leaf Blight of Rice. *Biol. Forum Int. J.* **2022**, *14*, 123–133.

44. Krishnan, N.; Gandhi, K.; Mohammed, F.P.; Muthuraj, R.; Kuppusamy, P.; Thiruvengadam, R. Management of Bacterial Leaf Blight Disease in Rice with Endophytic Bacteria. *World Appl. Sci. J.* **2013**, *28*, 2229–2241.

45. Qiao, L.; Tang, W.; Gao, D.; Zhao, R.; An, L.; Li, M.; Sun, H.; Song, D. UAV-based chlorophyll content estimation by evaluating vegetation index responses under different crop coverages. *Comput. Electron. Agric.* **2022**, *196*, 106775. [CrossRef]

46. Gu, Q.; Huang, F.; Lou, W.; Zhu, Y.; Hu, H.; Zhao, Y.; Zhou, H.; Zhang, X. Unmanned aerial vehicle-based assessment of rice leaf chlorophyll content dynamics across genotypes. *Comput. Electron. Agric.* **2024**, *221*, 108939. [CrossRef]

47. Nikolas, P.; Dionissios, K.; Rigas, G. Spatial Analysis of Agronomic Data and UAV Imagery for Rice Yield Estimation. *Agriculture* **2021**, *11*, 809. [CrossRef]

48. Nor, H.A.; Rohayu, H.N.; Tajul, R.R.; Siti, A.A.; Noorfatekah, T.; Zulkiflee, A.L.; Norhashila, H.; Khairulazhar, Z. Detection of Bacterial Leaf Blight Disease Using RGB-Based Vegetation Indices and Fuzzy Logic. In Proceedings of the 2023 19th IEEE International Colloquium on Signal Processing & Its Applications, Kedah, Malaysia, 3–4 March 2023. [CrossRef]

49. Chwen, M.Y. Assessment of the severity of bacterial leaf blight in rice using canopy hyperspectral reflectance. *Precis. Agric.* **2010**, *11*, 61–81. [CrossRef]

50. Taifeng, D.; Jiangui, L.; Budong, Q.; Liming, H.; Jane, L.; Rong, W.; Qi, J.; Catherine, C.; Heather, M.; Jarrett, P.; et al. Estimating crop biomass using leaf area index derived from Landsat 8 and Sentinel-2 data. *J. Photogramm. Remote Sens.* **2020**, *168*, 236–250. [CrossRef]

51. Brenon, D.S.B.; Gabriel, A.S.F.; Lucas, C.; Yiannis, A.V.V.; Luana, M.S. UAV-based coffee yield prediction utilizing feature selection and deep learning. *Smart Agric. Technol.* **2021**, *1*, 100010. [CrossRef]

MDPI AG
Grosspeteranlage 5
4052 Basel
Switzerland
Tel.: +41 61 683 77 34

Agronomy Editorial Office
E-mail: agronomy@mdpi.com
www.mdpi.com/journal/agronomy